T0143085

Communications in Computer and Information Science 1629

More information about this series at https://link.springer.com/bookseries/7899

Yang Wang · Guobin Zhu · Qilong Han ·
Liehui Zhang · Xianhua Song ·
Zeguang Lu (Eds.)

Data Science

8th International Conference
of Pioneering Computer Scientists, Engineers
and Educators, ICPCSEE 2022
Chengdu, China, August 19–22, 2022
Proceedings, Part II

 Springer

Editors
Yang Wang
Southwest Petroleum University
Chengdu, China

Qilong Han
Harbin Engineering University
Harbin, China

Xianhua Song
Harbin University of Science and Technology
Harbin, China

Guobin Zhu
University of Electronic Science
and Technology of China
Chengdu, China

Liehui Zhang
Southwest Petroleum University
Chengdu, China

Zeguang Lu
National Academy of Guo Ding Institute
of Data Sciences
Beijing, China

ISSN 1865-0929   ISSN 1865-0937 (electronic)
Communications in Computer and Information Science
ISBN 978-981-19-5208-1   ISBN 978-981-19-5209-8 (eBook)
https://doi.org/10.1007/978-981-19-5209-8

This Springer imprint is published by the registered company Springer Nature Singapore Pte Ltd.
The registered company address is: 152 Beach Road, #21-01/04 Gateway East, Singapore 189721, Singapore

Preface

As the chairs of the 8th International Conference of Pioneer Computer Scientists, Engineers and Educators 2022 (ICPCSEE 2022, originally ICYCSEE), it is our great pleasure to welcome you to the conference proceedings. ICPCSEE 2022 was held in Chengdu, China, during August 19–22, 2022, and hosted by the Southwest Petroleum University, the University of Electronic Science and Technology of China, and the National Academy of Guo Ding Institute of Data Sciences, China. The goal of this conference series is to provide a forum for computer scientists, engineers, and educators.

This year's conference attracted 263 paper submissions. After the hard work of the Program Committee, 63 papers were accepted to appear in the conference proceedings, with an acceptance rate of 24.7%. The major topic of this conference is data science. The accepted papers cover a wide range of areas related to basic theory and techniques for data science including mathematical issues in data science, computational theory for data science, big data management and applications, data quality and data preparation, evaluation and measurement in data science, data visualization, big data mining and knowledge management, infrastructure for data science, machine learning for data science, data security and privacy, applications of data science, case studies of data science, multimedia data management and analysis, data-driven scientific research, data-driven bioinformatics, data-driven healthcare, data-driven management, data-driven e-government, data-driven smart city/planet, data marketing and economics, social media and recommendation systems, data-driven security, data-driven business model innovation, and social and/or organizational impacts of data science.

We would like to thank all the Program Committee members, a total of 261 people from 142 different institutes or companies, for their hard work in completing the review tasks. Their collective efforts made it possible to attain quality reviews for all the submissions within a few weeks. Their diverse expertise in each research area helped us to create an exciting program for the conference. Their comments and advice helped the authors to improve the quality of their papers and gain deeper insights.

We thank the team at Springer, whose professional assistance was invaluable in the production of the proceedings. A big thanks also to the authors and participants for their tremendous support in making the conference a success.

Besides the technical program, this year ICPCSEE offered different experiences to the participants. We hope you enjoyed the conference.

June 2022

Liehui Zhang
Hongzhi Wang
Yang Wang
Guobin Zhu
Qilong Han

Organization

The 8th International Conference of Pioneering Computer Scientists, Engineers and Educators (http://2022.icpcsee.org) was held in Chengdu, China, during August 19–22, 2022, and hosted by the Southwest Petroleum University, the University of Electronic Science and Technology of China and the National Academy of Guo Ding Institute of Data Sciences, China.

General Chairs

Liehui Zhang	Southwest Petroleum University, China
Hongzhi Wang	Harbin Institute of Technology, China

Program Chairs

Yang Wang	Southwest Petroleum University, China
Guobin Zhu	University of Electronic Science and Technology of China, China
Qilong Han	Harbin Engineering University, China

Program Co-chairs

Xiaohua Xu	University of Science and Technology of China, China
Zhao Kang	University of Electronic Science and Technology of China, China
Yingjie Zhou	Sichuan University, China
Jinshan Tang	George Mason University, USA
Jingfeng Jiang	Michigan Technological University, USA
Xiaohu Yang	Hebei University, China

Organization Chairs

Jie Gong	Southwest Petroleum University, China
Yishu Zhang	Southwest Petroleum University, China
Jiong Mu	Sichuan Agricultural University, China
Xianhua Niu	Xihua University, China
Wei Pan	China West Normal University, China

Organization Co-chairs

Bo Peng	Southwest Petroleum University, China
Jian Zhang	Southwest Petroleum University, China
Fei Teng	Southwest Jiaotong University, China
Xin Yang	Southwestern University of Finance and Economics, China
Yongqing Zhang	Chengdu University of Information Technology, China
Hongyu Han	Sichuan Normal University, China
Chuanlin Liu	Southwest Petroleum University, China

Publication Chairs

Xianhua Song	Harbin University of Science and Technology, China
Zeguang Lu	National Academy of Guo Ding Institute of Data Sciences, China

Publication Co-chairs

Xiaoou Ding	Harbin Institute of Technology, China
Dan Lu	Harbin Engineering University, China

Forum Chairs

Lei Chen	Chengdu Supercomputing Center, China
Xiaoliang Chen	Xihua University, China
Hai Li	iQIYI Inc., China
Pinle Qin	North University of China, China

Oral Session Chairs

Fan Min	Southwest Petroleum University, China
Ping Li	Southwest Petroleum University, China
Xin Wang	Southwest Petroleum University, China

Registration and Financial Chair

Zhongchan Sun	National Academy of Guo Ding Institute of Data Sciences, China

Academic Committee Chair

Hongzhi Wang   Harbin Institute of Technology, China

Academic Committee Vice President

Qilong Han   Harbin Engineering University, China

Academic Committee Secretary General

Zeguang Lu   National Academy of Guo Ding Institute of Data
                                Sciences, China

Academic Committee Under Secretary General

Xiaoou Ding   Harbin Institute of Technology, China

Academic Committee Secretaries

Dan Lu   Harbin Engineering University, China
Zhongchan Sun   National Academy of Guo Ding Institute of Data
                                Sciences, China

Academic Committee Executive Members

Cham Tat Huei   UCSI University, Malaysia
Xiaoju Dong   Shanghai Jiao Tong University, China
Lan Huang   Jilin University, China
Ying Jiang   Kunming University of Science and Technology,
                                China
Weipeng Jiang   Northeast Forestry University, China
Min Li   Central South University, China
Junyu Lin   Institute of Information Engineering, CAS, China
Xia Liu   Hainan Province Computer Federation, China
Rui Mao   Shenzhen University, China
Qiguang Miao   Xidian University, China
Haiwei Pan   Harbin Engineering University, China
Pinle Qin   North University of China, China
Xianhua Song   Harbin University of Science and Technology,
                                China
Guanglu Sun   Harbin University of Science and Technology,
                                China
Jin Tang   Anhui University, China
Ning Wang   Xiamen Huaxia University, China

Xin Wang	Tianjin University, China
Yan Wang	Zhengzhou University of Technology, China
Yang Wang	Southwest Petroleum University, China
Shengke Wang	Ocean University of China, China
Yun Wu	Guizhou University, China
Liang Xiao	Nanjing University of Science and Technology, China
Junchang Xin	Northeastern University, China
Zichen Xu	Nanchang University, China
Xiaohui Yang	Hebei University, China
Chen Ye	Hangzhou Dianzi University, China
Canlong Zhang	Guangxi Normal University, China
Zhichang Zhang	Northwest Normal University, China
Yuanyuan Zhu	Wuhan University, China

Steering Committee

Jiajun Bu	Zhejiang University, China
Wanxiang Che	Harbin Institute of Technology, China
Jian Chen	ParaTera, China
Wenguang Chen	Tsinghua University, China
Xuebin Chen	North China University of Science and Technology, China
Xiaoju Dong	Shanghai Jiao Tong University, China
Qilong Han	Harbin Engineering University, China
Yiliang Han	Engineering University of CAPF, China
Yinhe Han	Institute of Computing Technology, Chinese Academy of Sciences, China
Hai Jin	Huazhong University of Science and Technology, China
Weipeng Jing	Northeast Forestry University, China
Wei Li	Central Queensland University, Australia
Min Li	Central South University, China
Junyu Lin	Institute of Information Engineering, Chinese Academy of Sciences, China
Yunhao Liu	Michigan State University, USA
Zeguang Lu	National Academy of Guo Ding Institute of Data Sciences, China
Rui Mao	Shenzhen University, China
Qiguang Miao	Xidian University, China
Haiwei Pan	Harbin Engineering University, China
Pinle Qin	North University of China, China

Zheng Shan	The PLA Information Engineering University, China
Guanglu Sun	Harbin University of Science and Technology, China
Jie Tang	Tsinghua University, China
Feng Tian	Institute of Software, Chinese Academy of Sciences, China
Tao Wang	Peking University, China
Hongzhi Wang	Harbin Institute of Technology, China
Xiaohui Wei	Jilin University, China
Lifang Wen	Beijing Huazhang Graphics & Information Co., Ltd, China
Liang Xiao	Nanjing University of Science and Technology, China
Yu Yao	Northeastern University, China
Xiaoru Yuan	Peking University, China
Yingtao Zhang	Harbin Institute of Technology, China
Yunquan Zhang	Institute of Computing Technology, Chinese Academy of Sciences, China
Baokang Zhao	National University of Defense Technology, China
Min Zhu	Sichuan University, China
Liehuang Zhu	Beijing Institute of Technology, China

Program Committee

Witold Abramowicz	Poznan University of Economics, Poland
Chunyu Ai	University of South Carolina Upstate, USA
Jiyao An	Hunan University, China
Ran Bi	Dalian University of Technology, China
Zhipeng Cai	Georgia State University, USA
Yi Cai	South China University of Technology, China
Zhao Cao	Beijing Institute of Technology, China
Wanxiang Che	Harbin Institute of Technology, China
Wei Chen	Beijing Jiaotong University, China
Hao Chen	Hunan University, China
Xuebin Chen	North China University of Science and Technology, China
Chunyi Chen	Changchun University of Science and Technology, China
Yueguo Chen	Renmin University, China
Siyao Cheng	Harbin Institute of Technology, China
Byron Choi	Hong Kong Baptist University, China

Vincenzo Deufemia	University of Salerno, Italy
Gong Dianxuan	North China University of Science and Technology, China
Xiaofeng Ding	Huazhong University of Science and Technology, China
Jianrui Ding	Harbin Institute of Technology, China
Hongbin Dong	Harbin Engineering University, China
Lei Duan	Sichuan University, China
Xiping Duan	Harbin Normal University, China
Xiaolin Fang	Southeast University, China
Ming Fang	Changchun University of Science and Technology, China
Jianlin Feng	Sun Yat-sen University, China
Jing Gao	Dalian University of Technology, China
Yu Gu	Northeastern University, China
Qi Han	Harbin Institute of Technology, China
Meng Han	Georgia State University, USA
Qinglai He	Arizona State University, USA
Wei Hu	Nanjing University, China
Lan Huang	Jilin University, China
Hao Huang	Wuhan University, China
Feng Jiang	Harbin Institute of Technology, China
Bin Jiang	Hunan University, China
Cheqing Jin	East China Normal University, China
Hanjiang Lai	Sun Yat-sen University, China
Shiyong Lan	Sichuan University, China
Hui Li	Xidian University, China
Zhixu Li	Soochow University, China
Mingzhao Li	Royal Melbourne Institute of Technology, Australia
Peng Li	Shaanxi Normal University, China
Jianjun Li	Huazhong University of Science and Technology, China
Xiaofeng Li	Sichuan University, China
Zheng Li	Sichuan University, China
Min Li	Central South University, China
Zhixun Li	Nanchang University, China
Hua Li	Changchun University of Science and Technology, China
Rong-Hua Li	Shenzhen University, China
Cuiping Li	Renmin University of China, China
Qiong Li	Harbin Institute of Technology, China

Yanli Liu	Sichuan University, China
Hailong Liu	Northwestern Polytechnical University, China
Guanfeng Liu	Macquarie University, Australia
Yan Liu	Harbin Institute of Technology, China
Zeguang Lu	National Academy of Guo Ding Institute of Data Sciences, China
Binbin Lu	Sichuan University, China
Junling Lu	Shaanxi Normal University, China
Jizhou Luo	Harbin Institute of Technology, China
Li Mohan	Jinan University, China
Tiezheng Nie	Northeastern University, China
Haiwei Pan	Harbin Engineering University, China
Jialiang Peng	Norwegian University of Science and Technology, Norway
Fei Peng	Hunan University, China
Yuwei Peng	Wuhan University, China
Shaojie Qiao	Southwest Jiaotong University, China
Li Qingliang	Changchun University of Science and Technology, China
Zhe Quan	Hunan University, China
Yingxia Shao	Peking University, China
Wei Song	North China University of Technology, China
Yanan Sun	Oklahoma State University, USA
Minghui Sun	Jilin University, China
Guanghua Tan	Hunan University, China
Yongxin Tong	Beihang University, China
Xifeng Tong	Northeast Petroleum University, China
Vicenc Torra	University of Skövde, Sweden
Leong Hou	University of Macau, China
Hongzhi Wang	Harbin Institute of Technology, China
Yingjie Wang	Yantai University, China
Dong Wang	Hunan University, China
Yongheng Wang	Hunan University, China
Chunnan Wang	Harbin Institute of Technology, China
Jinbao Wang	Harbin Institute of Technology, China
Xin Wang	Tianjin University, China
Peng Wang	Fudan University, China
Chaokun Wang	Tsinghua University, China
Xiaoling Wang	East China Normal University, China
Jiapeng Wang	Harbin Huade University, China
Huayu Wu	Institute for Infocomm Research, Singapore

Yan Wu	Changchun University of Science and Technology, China
Sheng Xiao	Hunan University, China
Ying Xu	Hunan University, China
Jing Xu	Changchun University of Science and Technology, China
Jianqiu Xu	Nanjing University of Aeronautics and Astronautics, China
Yaohong Xue	Changchun University of Science and Technology, China
Li Xuwei	Sichuan University, China
Mingyuan Yan	University of North Georgia, USA
Yajun Yang	Tianjin University, China
Gaobo Yang	Hunan University, China
Lei Yang	Heilongjiang University, China
Ning Yang	Sichuan University, China
Xiaochun Yang	Northeastern University, China
Bin Yao	Shanghai Jiao Tong University, China
Yuxin Ye	Jilin University, China
Xiufen Ye	Harbin Engineering University, China
Minghao Yin	Northeast Normal University, China
Dan Yin	Harbin Engineering University, China
Zhou Yong	China University of Mining and Technology, China
Lei Yu	Georgia Institute of Technology, USA
Ye Yuan	Northeastern University, China
Kun Yue	Yunnan University, China
Xiaowang Zhang	Tianjin University, China
Lichen Zhang	Shaanxi Normal University, China
Yingtao Zhang	Harbin Institute of Technology, China
Yu Zhang	Harbin Institute of Technology, China
Wenjie Zhang	University of New South Wales, Australia
Dongxiang Zhang	University of Electronic Science and Technology of China, China
Xiao Zhang	Renmin University of China, China
Kejia Zhang	Harbin Engineering University, China
Yonggang Zhang	Jilin University, China
Huijie Zhang	Northeast Normal University, China
Boyu Zhang	Utah State University, USA
Jian Zhao	Changchun University, China
Qijun Zhao	Sichuan University, China
Bihai Zhao	Changsha University, China

Contents – Part II

Infrastructure for Data Science

Education Track

Regulatory Technology in Finance

Contents – Part I

Machine Learning for Data Science

Multimedia Data Management and Analysis

Big Data Management and Applications

Research on the Realization Path and Application of a Data Governance System Based on Data Architecture

Fang Miao[1(✉)], Wenhui Yang[2], Yan Xie[3], and Wenjie Fan[1]

[1] Big Data Research Institute, Chengdu University, Chengdu 610103, China
miaofang@126.com
[2] Cyber Security College, Chengdu University of Technology, Chengdu 610059, China
[3] Sichuan Provincial Big Data Center, Chengdu 610000, China

Abstract. The construction and development of the digital economy, digital society and digital government are facing some common basic problems. Among them, the construction of the data governance system and the improvement of data governance capacity are short boards and weak links, which have seriously restricted the construction and development of the digital economy, digital society and digital government. At present, the broad concept of data governance goes beyond the scope of traditional data governance, which "involves at least four aspects: the establishment of data asset status, management system and mechanism, sharing and openness, security and privacy protection". Traditional information technologies and methods are powerless to comprehensively solve these problems, so it is urgent to improve understanding and find another way to reconstruct the information technology architecture to provide a scientific and reasonable technical system for effectively solving the problems of data governance. This paper redefined the information technology architecture and proposed the data architecture as the connection link and application support system between the traditional hardware architecture and software architecture. The data registration system is the core composition of the data architecture, and the public key encryption and authentication system is the key component of the data architecture. This data governance system based on the data architecture supports complex, comprehensive, collaborative and cross-domain business application scenarios. It provides scientific and feasible basic support for the construction and development of the digital economy, digital society and digital government.

Keywords: Data sharing and transaction · Data governance system · Data architecture · Data registration · Public key encryption define data ownership

1 Introduction

The digital economy, digital society, digital government construction and digital transformation are all related to data governance. In the past, the concept of data governance was limited to within a system or within an enterprise, but in the near future, we will be facing data governance at different levels of countries, industries and organizations. Mei,

H. pointed out that "big data is becoming a trend to promote economic development, improve social governance and improve government services and regulatory capacity". However, "one of the biggest weaknesses restricting the development of big data is that the data governance system is far from being formed. For example, the establishment of the status of data assets has not reached consensus, and the data ownership confirmation, circulation and control of data face multiple challenges". "The construction of big data governance system involves three levels of countries, industries and organizations, including at least four aspects: the establishment of data asset status, management system and mechanism, sharing and openness, security and privacy protection. Support needs to be provided from the perspectives of systems and regulations, standards and norms, application practice and support technology"; "the construction of data governance system should be carried out first" [1–3]. The content of data governance can be simply summarized into data: "rights confirmation, management, sharing and security". This paper focuses on the technical realization of data governance and its supporting role in systems, regulations, standards and application practice.

2   Research Status and Development of Data Governance

Liu, G. et al. combed and summarized the concepts, elements, models and frameworks of data governance at home and abroad, compared and discussed the data governance frameworks of DGI, DAMA and CALib, and pointed out that on the basis of learning from foreign research results, domestic research gradually shifted from theoretical exploration to empirical research [4]. Dai, H. et al. studied the standard of big data governance system [5].

Al-Ruither, M. et al. proposed that the only way to solve the problem of persistent data complexity is to implement effective data governance from the perspective of literature research in academic and industrial circles. At present, data governance is still in the research stage, and research in this field needs to be promoted [6]. Stockdale, S. analyzed data governance and clarified the concept and function of data governance [7]. Khatri, V. and Brown, C. V. provide an overall framework for practitioners to develop effective data governance methods, strategies and designs [8].

Zhang, N. and Yuan, Q. believe that at present, big data governance research mainly focuses on framework model design, value discussion and application in different fields, with less empirical research and a lack of optimization of data governance framework model design. Future research should focus on the framework system, policy standards, maturity model, data quality and governance of massive heterogeneous data [9]. Zheng, D. et al. proposed elaborating the connotation of the concept of big data governance from the four dimensions of purpose, power level, governance scope and solving practical problems and constructed the reference framework of big data governance from the three dimensions of big data life cycle, stakeholders and circulation mode [10], which has great reference value.

Yin, J. et al. proposed a panoramic framework for big data governance, including full-dimensional collection, high-quality cleaning and multichannel feedback; multi-modal fusion, analytic empowerment and group empowerment; open sharing, interactive collaboration, transparency and credibility [11].

Gan, S. et al. put forward a more comprehensive planning and function realization for the big data governance system at the enterprise level, but there are still some deficiencies in the role of metadata and data security protection [12].

Referring to the cultivation of the data element market by data governance, Mei, H. proposed that asset status establishment, data right confirmation, data sharing and circulation, data security and privacy protection are the challenges faced and put forward four aspects of work: straightening out the boundary of rights and responsibilities, promoting market allocation, promoting application implementation and improving technical ability [13].

In recent years, because government governance is inseparable from the wide application of data, people at home and abroad have begun to pay attention to the problem of government data governance. Huang, H. studied the data governance of the US government from the two dimensions of governance structure and policy system, including data openness, information disclosure (freedom), personal privacy protection, e-government, information security and information resource management [14].

At the level of government data governance in China, both the Office of the Central Cyberspace Affairs Commission and the National Information Center pay special attention to the important role of data governance in the economy, society and government. For example, it is proposed that "data governance is the primary link to release data value"; "Accurate, timely, complete and consistent data resources are an important prerequisite for the value release of data elements" [15]. It is pointed out that "technology is the core and key of data governance"; and "Explore the construction of data governance framework and fundamentally solve the problems of 'data separation', 'data island' and 'data fragmentation' caused by the lack of data governance in China" [16].

An, X. et al. put forward the government big data governance framework at three levels: macroscopic (institutional arrangement), intermediate perspective (mechanism rules) and microcosmic (implementation tools). It has certain significance and value through the practice verification of Guizhou Province [17]. Liang, Z. proposed the big data governance path in national governance from the aspects of data mining, data integration, data analysis, data sharing and data push [18].

Yang, M. and Du, X. discussed the concept, theory, mechanism and framework of government big data governance from the two aspects of the management and utilization of big data. In particular, they proposed further exploring the issues of government data ownership, government data asset management, commercial utilization of government data and government data pricing mechanisms [19].

Fan, L. et al. proposed that data fusion, data disclosure prevention and privacy protection, and data pricing are the basic and core technologies for government big data governance [20].

From the perspective of data security, Li, Z. et al. proposed a government big data management reference model based on privacy protection, covering seven dimensions: concept and strategy, data governance, data quality, data application, data life cycle, data security, platform and architecture [21]. Ma. C. et al. discussed the fusion of fragmented data in big data and the data protection mechanism based on the Bell LaPadula model from the perspective of data mode conversion and security protection [22].

From the aspect of technical implementation, Du, X. et al. proposed that data integration is one of the key technologies of big data governance, especially mentioning that the data integration system architecture based on intermediate mode is very similar to the data architecture proposed in this paper [23].

In summary, the understanding, theory, development and practice of data governance at home and abroad in the past decade have expanded from past organizations to the government and national levels, from past data governance to the all-round transformation of policy systems, governance structures, framework models and technical realization, and from governance considering only data sources, quality and management to the establishment of asset status. All round changes in data management systems and mechanisms, open data sharing, data security and personal privacy protection. China has a profound and forward-looking understanding of the concept of broad data governance and the role of data governance in government governance.

The problem of data "right confirmation, management, sharing and security" that needs to be solved by the data governance system is the bottleneck and weak link that restricts the development of the digital economy, digital society and digital government. It is difficult for traditional information technology to form a package solution, which must be solved through data architecture.

The implementation path of the data governance system and supporting digital system (sublimation from information system to data system) needs to break through the traditional implementation mode of local, internal, limited and a large number of manual management means.

This paper redefines the information technology architecture and proposes that a new data architecture must be built between the hardware architecture and the software architecture. The data registry is the core component for perceiving, managing and sharing global data, while data encryption by public keys and ownership authorization are the key components for confirming data rights and protecting sensitive data. With this data architecture as the basic support, it can improve the governance capability of the data life cycle and can effectively solve the basic problems of data rights confirmation, management, sharing and security faced by generalized data governance. Furthermore, it can support complex, comprehensive, collaborative and cross-domain business application scenarios and provide scientific and feasible basic common support for the construction and development of the digital economy, digital society and digital government. It is an effective implementation path of the data governance system.

3 Functional Composition of Data Architecture

The data architecture constructs a data governance system with the functions of data "rights confirmation, management, sharing and security" to solve the common basic problems faced in collaborative management and services across levels, regions, systems, departments and businesses in the process of digital economy, digital society, digital government and digital transformation; for example, it is difficult to gather data, share data, confirm data rights, trade data, secure data, etc.

Data governance or big data governance is faced with massive, heterogeneous, multisource and real-time changing data, which come from different business application systems and serve their own and different business application systems. Therefore, the technical implementation of a data governance system should build a basic, simple and universal infrastructure for data "rights confirmation, management, sharing and security". We call it "data architecture" (DA) to meet the common technical requirements of data governance.

Obviously, to achieve data "sharing", first, data must be "managed". Data "rights confirmation" and data "security" are related to the interests associated with. Therefore, we can consider "management" and "sharing" together and "rights confirmation" and "security" together. Therefore, the data governance technology system based on data architecture includes the following functions.

3.1 The Method of Integrating Heterogeneous Data with Internal and External Centralized Data Management

Data governance faces all kinds of data from all aspects. There must be a unified data aggregation and management method to support the needs of various business applications and cross-domain data sharing. The function includes unified aggregation and management of internal and external multisource data; unified aggregation and management of multitype heterogeneous data; the logical centralization and physical centralization of data are combined; the provision of data access services for business applications; the data directory system for business application; etc.

3.2 Data Security Classification, Data Right Confirmation and Authorization Methods

Data governance should enable data circulation on the premise of ensuring data security. If data circulation is used by others, we must confirm the right. The function includes the following: The basic concept of data ownership; Data right confirmation and data security; Data security classification; Data security application; Data authorization application method; Data process watermark; Data transactions; Data traceability; Discovery of illegal use of data; etc.

3.3 Basic Composition and Form of the Data Architecture

The data governance system provides common and basic data governance capabilities for various application scenarios. It needs to be supported by the underlying data architecture that does not depend on specific software and hardware, which requires research on the data architecture. The function includes: Data architecture composition; Data architecture function; Scientific data structure; Completeness of data architecture; etc.

3.4 Data Governance System Supported by Data Architecture

The data governance system is supported by the data architecture. The function includes defining the functions of the data governance system, construction of the data governance system, self-consistency of the data governance system, application conditions of the data governance system, etc.

4   Understanding the Data Architecture Supporting the Data Governance System

4.1   The Impact of Data on Human Civilization – Recognizing the Importance of Data

Figure 1 shows that the impact of data on human civilization is epoch-making and far-reaching. Mankind is in the stage of transition from material civilization to data civilization. When solving new problems, using traditional ideas and methods will certainly encounter obstacles and confusion. It is this subversive and fundamental impact that when we study data problems, we should focus on the future and think and explore according to the changes brought by data civilization and the laws that should be followed.

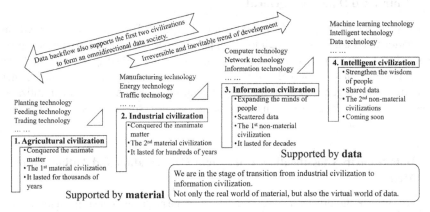

Fig. 1. The essence of the four civilizations is material civilization and data civilization [24, 25]

Figure 2 further points out that the future society is a society in which the real world composed of materials and the virtual world composed of data are integrated. To make the virtual society of data orderly, there are still too many things to do in developing data productivity and production relations suitable for data, including building a data governance system. These are unprecedented challenges and opportunities and put forward higher requirements for us.

4.2   Comparison Between Data and Material – Similarity and Particularity of Data and Material

Materials and data have complete correspondence and differences. Table 1 comprehensively compares materials and data, especially defining the characteristics of data sharing and safety, and needs to follow its laws to carry out research.

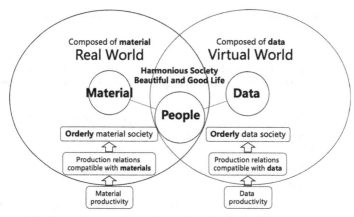

Fig. 2. An orderly, harmonious and beautiful society for mankind in the future – the integration of the material world and the data world

Table 1. Comparison of material and data [25].

Category	Material	Data
Civilization support	Agriculture, Industrial	**Information, Intelligent**, and Agriculture, Industrial
Carrying the world	The **real world**	The **virtual world**, Cyber space, and Intertwined with real world
Ownership attributes	Yes	**Yes**, but you have it does not mean you own it.
Value attributes	Yes	**Yes**, data itself is valuable, and also its derived information and knowledge are valuable.
Asset attributes	Yes	**Yes**, virtual digital property protection and inherit are very urgent
Law	Property Law	**Not yet**. Data Asset Law should be needed and urgent
Copy	Not easy	**Easy**, almost zero cost, but not all data can be copied (classified)
Transmission	High cost	**Easy**, almost zero cost, but some data needs restrict to transmission
Sharing nature	Bad, time sharing	**Good**, time sharing and simultaneous, the key nature of data
Fulfill security	Perimeter protection, system protection	**Perimeter protection**, system protection (sharing constraints; Loopholes, Trojan horses, back doors, stealing)
Individual protection	Put into the safe	**Data encryption** (Management and use constraints)
Key keeping	Owner / credible bank	**Owner?**/ Administrator (Believable? Encryption is still unsafe)

4.3 Comparison of Data and Information - the Difference Between Data and Information Determines the Difference in Research Data Governance Methods

We say that data are the carrier of information and that information is valuable. However, usually, from the perspective of computers and technology, there seems to be little difference between information and data. Table 2 compares information and data from multiple perspectives, from which we can see that there are still great differences between information and data. This requires us to treat data more carefully and treat the research of the data governance system more comprehensive.

Table 2. Comparison of characteristics of data with information

	Information	Data
Concept	Extracting from data is a kind of valuable data for human beings	Carrying all information and knowledge, it is the content of multimedia
Subject	People	People and computer
Quantity	Small, extracted from data	Large, can produce different information
Application	Vary from person to person, utilitarian	Universality
Range	Only for the benefit of relevant personnel, small range	Useful to more people, wide range
Business	Only single business is supported	Support multiple businesses
Value	Dominance, high value and high density	Recessive, high value, low density
Technology	Acquisition, transmission, storage, processing, calculation, analysis and display	Acquisition, transmission, storage, processing, calculation, analysis and display, more resources and capabilities are needed
System	Small, relatively simple	Large and complex
Relationship	Data generates information, which comes from data	Information and data both are data
Visual angle	Business, people, personality, subjective	Computer, machine, universal, objective
Business relation	Business is closely related, cross business is not applicable, and the effectiveness is poor	Loose business association, cross business availability and long-term
Storage	Small scale, random storage, disposable after use/disposable	Large scale, continuous storage, long-term preservation
Resources	Decentralized, easy to manage and limited to use	Centralized, difficult to manage and widely used
Standard	Data standards are business related	It is impossible to achieve a unified data standard for all businesses, but a unified data management standard can be achieved
Catalog	Business perspective, serving the business	You can also provide data directories of different businesses from a business perspective

(*continued*)

Table 2. (*continued*)

	Information	Data
Register	It can register the data of the information system to facilitate data management	From the data perspective, the information obtained by the data system is convenient for data management, unified and standardized
Ownership	Clear, no need to specify	Unclear, it must be pointed out that the basis of data flow
Share	Small scale, information exchange	Large scale, data circulation, data transaction
Security	Information technology security, system security, network security	Data security, system security, network security

4.4 The Relationship Between Data and Applications – the Systematicness, Integrity, Relevance and Essence of Data

Data support applications. Applications generate data. Data and applications should be an ecological relationship. Figure 3 shows the ecological relationship between data and application and puts forward higher requirements for data governance. On the premise of data security, it can realize that the "fertile" data "soil" grows the "lush" application "forest".

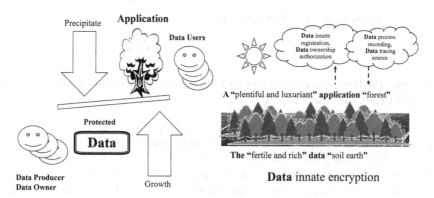

Fig. 3. Data application ecosystem [25]

5 Implementation Path of the Data Governance System Supported by the Data Architecture

5.1 Distinguish the Information System and Data System, and Develop the Data System by Using the Data-Oriented Software Engineering Method

The change in the software development method is to recognize the difference between the information system and the data system. Adopt the new data-oriented software engineering development method according to the law that the information system should follow, take the data as the core, and develop the data-driven data system. Figure 4 compares the two systems. It can be seen that the information system meets different business needs, similar to "planting flowers in a pot", and its limitations and "isolated island" are obvious. The data system can meet various business needs at the same time and build a data ecology. In theory, there is no "data island" in the data system. The data required by various business systems can be found in the "fertile" data "soil".

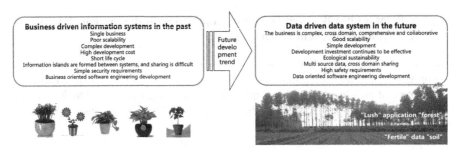

Fig. 4. Comparison between the information system and data system

5.2 Unify the Database of the Existing Information System, Build a Data System and Crack the "Data Island"

It seems that the scale and cost of building a data system are too large, but the data system can be "transformed" from the existing information system. The "trick" is to unify the database of the existing information system, as shown in Fig. 5. Here, by transforming the information system of each department and registering the data of each department through the unified data registration standard, we can achieve the function of a unified "flower pot" so that all the uniformly registered department data can be associated and accessed through their respective data registration center (DRC) to build an integrated data system, break the "information island" and "data island" and realize data sharing by technical means.

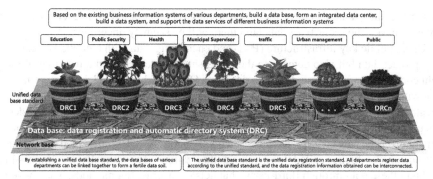

Fig. 5. Method of constructing data system – unifying the database standard of each information system.

5.3 The Separation of Data Management and Use is Adopted to Simplify the Complexity of Data and Business

As shown in Fig. 6, the function of data architecture is to realize the separation of "management" and "use" of data. It can manage complex data sources and data categories and support complex and collaborative application requirements.

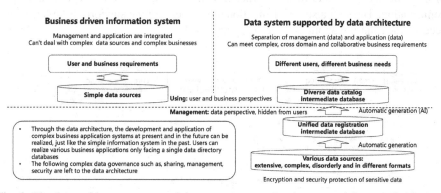

Fig. 6. The data architecture supported data management: data is separated from applications to form the data system.

5.4 The Public Key Infrastructure of the Domestic Commercial Key Algorithm is Used to Realize Data Right Confirmation, Data Ownership Authorization and Data Protection

The "Cryptography Law of the People's Republic of China" came into effect on January 1, 2020. It is to encourage and promote the research, development and application of commercial cryptography in the economic and social fields. To protect data security, the direct idea is to encrypt data. Although the security of data is greatly improved after data encryption, the management of the key becomes very important. Whoever master the key will master the security of data, and there will be security uncertainty.

Using public key infrastructure (PKI) and encrypting data with a public key can not only protect data but also determine the ownership of data. Here, the meaning of data right confirmation is "whose data who decides". On the "whose data", by encrypting the data with the "whose" public key, the data belong to the "who". Only by using his/her private key can he/she decrypt the data and use the data; on the issue of "who decides", it is up to "who" to decide whether to give the data to others. If it is decided to give it to someone, the data are encrypted with "someone's" public key. The encrypted data belong to "someone", which is called "ownership authorization". Theoretically, it is impossible to obtain these data through third-party channels or by cracking methods. The above is an important function of the data architecture.

5.5 Data Architecture Supports the Construction of a Data Governance System and Solves Data Governance Problems in a Package

Figure 7 is the schematic diagram of the "one body and two wings" data architecture. "One body" refers to the integration of data and people; the left wing of the "two wings" refers to the innate registration of data, or data native registration. Through a series of functional modules, it can support open data sharing applications. The right wing of the "two wings" refers to data ownership encryption, or data encryption and right confirmation. Through a series of functional modules and the support of the left wing, security data sharing applications can be supported.

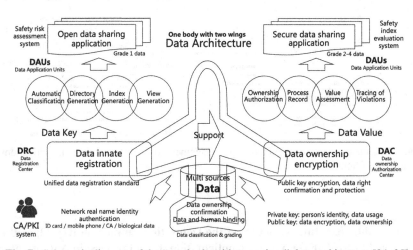

Fig. 7. Schematic diagram of the "one body with two wings" data architecture [24, 25].

Based on the data architecture, a data governance system with the functions of "rights confirmation, management, sharing and security" of data can be built, as shown in Fig. 8. The data governance system is composed of data resources, classification and grading, data rights confirmation, data registration, data directories, data access and so on.

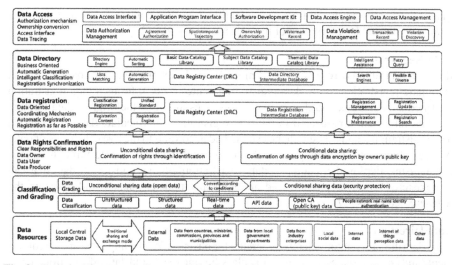

Fig. 8. Technical framework diagram of the data governance system based on the data architecture.

Data Resources

It is mainly composed of data stored in the local center and external data. External data include data from national ministries and commissions, provinces and cities, as well as data from local government departments, industry enterprises, social data, Internet of Things perception data, other data, etc. These external data can be physically centralized in the local data center through traditional sharing and exchange methods. The data aggregation method in this data governance system is to use the data registration center in the system to realize logical centralized aggregation for all data stored in the local center and accessible external data (see the description of "data registration").

Data Classification

The aggregated data sources are classified into five categories: unstructured data, structured data, real-time data, API data and open CA (public key) data.

Data Grading

Classify the grading data according to the security level, which is divided into unconditional shared data (open data) and conditional shared data (security protection), which can be converted to each other according to conditions. For example, some sensitive data involving personal privacy can become open data after desensitization; after some insensitive individual data are gathered together, they may become sensitive data and need security protection.

Data Rights Confirmation

To clarify the responsibilities and rights of the data owner or user, it is necessary to clarify the data owner, authorized data user, authorized data producer and authorized data agent. The data agent must be authorized by the data owner to exercise part or all of the rights of the data owner on behalf of the data owner as needed. For unconditional shared data, the right is confirmed by identification. For conditional shared data, the right is confirmed by encryption of the owner's public key [26].

Data Registration

It is a data-oriented core function and mechanism. It obtains data information through automatic registration in the data registration center (DRC), forms a data registration intermediate database, and then manages, accesses and uses the original data through the data information registered in the data registration intermediate database. Unified data registration standards should be established according to the five categories of data. All accessible data, including internal and external data, coordinate through various mechanisms, deploy the automatic data registration engine, register all data that can be registered as far as possible and automatically update, manage and maintain the registered contents [27].

Data Directory

It is a data directory automatically generated for business. According to business requirements and characteristics, it intelligently classifies the registered data information, forms an intermediate database of data directories, and synchronizes with the registration information of data. It mainly includes the functions of the directory engine, search engine, fuzzy query, intelligent assistance, automatic sorting, automatic generation and list matching. The data directory can follow the relevant standards of the data directory and the responsibility list mechanism for automatic matching and automatic association. At the same time, based on the data registration center (DRC), the basic data catalogs, the subject data catalogs and the thematic data catalogs are formed to facilitate the subsequent business calling the original data.

Data Access

Including data authorization management, data violation management and data access methods. Data authorization management also includes agreement authorization (legal authorization in the traditional sense), ownership authorization (i.e., transferring the ownership of data, realizing data ownership transfer or ownership authorization by encrypting data with the user's public key), space-time trajectory (space-time imprint of data authorization), and watermark recording (for important data, the watermark of adding space-time imprint to the original data is used to record the authorization information). Data violation management mainly includes data transaction records, violation discovery, tracking and traceability. Data access methods include the data access interface (DAI), application program interface (API), software development kit (SDK), data access engine and data access management. Among them, the data access software development kit (SDK) includes automatic generation and management tools of CA/PKI key pairs, public key encryption tools, private key decryption tools, data registration tools, data directory tools, data authorization tools, data access tools, and so on.

5.6 Application Example of Data Governance System Based on Data Architecture in County New Smart City

Figure 9 shows a schematic diagram of the new smart city technology architecture supported by the data governance system. As seen from the figure, the data governance system designed in this paper is above the data resources of the digital base and below the support platform. It plays a connecting role between data and application and can effectively solve the problems of data right confirmation, data management, data sharing, data security and so on.

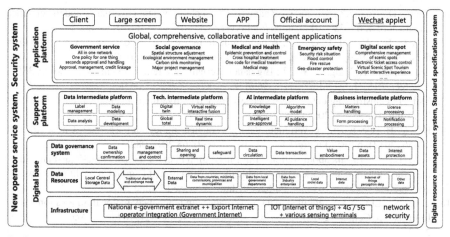

Fig. 9. Schematic diagram of the new smart city technology architecture supported by the data governance system.

6   Conclusion

By using the data architecture of one body and two wings to build the technical system of data governance, we can effectively solve the problems of data ownership, management, sharing and security. It is an effective path to realize the technical system of generalized data governance.

References

1. Mei, H.: Big data development and digital economy. China's Ind. Informatiz. **5**, 60–66 (2021)
2. Mei, H.: Big data and digital economy. Qiushi. http://www.qstheory.cn/dukan/qs/2022-01/16/c_1128261786.htm. Accessed 7 June 2022
3. Mei, H., Yang, J.: Promote the construction of big data governance system at different levels and dimensions. China Educ. Netw. **7**, 38–39 (2018)
4. Liu, G., Qian, J., Lu, Z.: Research progress of data governance at home and abroad: connotation, elements, model and framework. Libr. Inf. Work **21**, 137–144 (2017)
5. Dai, H., Zhang, Q., Yin, Z.: Study on big data governance standard system. Big Data **5**(3), 47–54 (2019). https://doi.org/10.11959/j.issn.2096-0271.2019023
6. Al-Ruithe, M., Benkhelifa, E., Hameed, K.: A systematic literature review of data governance and cloud data governance. Pers. Ubiquit. Comput. **23**(5–6), 839–859 (2018). https://doi.org/10.1007/s00779-017-1104-3

7. Stockdale, S.: Deconstructing data governance (2015). https://digitalrepository.unm.edu/hslic-posters-presentations/11/. Accessed 7 June 2022

8. Khatri, V., Brown, C.V.: Designing data governance. Commun. ACM **53**(1), 148–152 (2010). https://doi.org/10.1145/1629175.1629210

9. Zhang, N., Yuan, Q.: Review of data governance research. Intell. Mag. **36**(5), 129–134, 163 (2017)

10. Zheng, D., Huang, L., Zhang, C., Zhang S.: Concept of big data governance and its reference architecture **4**, 65–72 (2017). https://doi.org/10.13581/j.cnki.rdm.2017.04.005

11. Yin, J., Zhu, H., Yu, J., Qiu, S.: A panoramic framework of big data governance. Big Data **6**(2), 19–26 (2020). https://doi.org/10.11959/j.issn.2096-0271.2020011

12. Gan, S., Che, P., Yang, T., Wu, J.: Big data governance system. Comput. Appl. Softw. **35**(6), 1–8, 69 (2018). https://doi.org/10.3969/j.issn.1000-386x.2018.06.001

13. Mei, H.: Constructing data governance system and cultivating data element market ecology. Sci. Chin. **16**, 36–37 (2021)

14. Huang, H.: US federal government data governance: policy and structure. China Adm. **8**, 47–56 (2017). https://doi.org/10.3782/j.issn.1006-0863.2017.08.07

15. Song, L.: Experts talk about National Informatization Plan for the 14th Five-Year-Plan: Stimulate the value of data elements and enable the construction of Digital China. http://www.cac.gov.cn/2022-01/21/c_1644368244622007.htm. Accessed 7 June 2022

16. Tang, S., Liu, Y.: Strengthen data governance and enhance national innovation. State Information Center (2018). http://www.sic.gov.cn/News/612/9707.htm. Accessed 7 June 2022

17. An, X., Guo, M., Hong, X., Wei, W.: Framework of government big data governance system and effective way of implementation. Big Data **5**(3), 3–12 (2019). https://doi.org/10.11959/j.issn.2096-0271.2019019

18. Liang, Z.: Big data governance: the proper meaning of the modernization of national governance capacity. J. Jishou Univ. (Soc. Sci. Ed.) **36**(2), 34–41 (2015). https://doi.org/10.13438/j.cnki.jdxb.2015.02.00

19. Yang, M., Du, X.: Big data governance in governments: a new form of the government administration. Big Data **6**(2), 3–18 (2020). https://doi.org/10.11959/j.issn.2096-0271.2020010

20. Fan, L., Hong, X., Huang, Z., Hua, G., Li, G.: Challenge and countermeasure of governing government big data. Big Data **2**(3), 27–38 (2016). https://doi.org/10.11959/j.issn.2096-0271.2016028

21. Li, Z., Hong, Y.: Study on big data management for government based on privacy protection. Big Data **6**(2), 69–82 (2020). https://doi.org/10.11959/j.issn.2096-0271.2020015

22. Ma, Z., et al.: Research on data schema and security in data governance. Big Data **2**(3), 83–95 (2016). https://doi.org/10.11959/j.issn.2096-0271.2016033

23. Du, X., Cheng, Y., Fan, J., Lu, W.: Data wrangling: a key technique of data governance. Big Data **5**(3), 13–22 (2019). https://doi.org/10.11959/j.issn.2096-0271.2019020

24. Miao, F., Yang, W., Xie, Y., Fan, W.: Consideration and research on data architecture for the future cyber society. In: IEEE SmartWorld, Ubiquitous Intelligence & Computing, Advanced & Trusted Computing, Scalable Computing & Communications, Cloud & Big Data Computing, Internet of People and Smart City Innovation (SmartWorld/SCALCOM/UIC/ATC/CBDCom/IOP/SCI), Leicester, United Kingdom, pp. 1671–1676 (2019). https://doi.org/10.1109/SmartWorld-UIC-ATC-SCALCOM-IOP-SCI.2019.00298

25. Miao, F., Yang, W., Xie, Y., Fan, W.: Data architecture for big data service operations management (the new vision of data architecture for the future human society). In: Emrouznejad, A., Charles, V. (eds.) Big Data and Blockchain for Service Operations Management. Studies

in Big Data, vol. 98, pp. 95–137. Springer, Cham (2022). https://doi.org/10.1007/978-3-030-87304-2_4

26. Miao, F., et al.: Digital copyright works management system based on DOSA. In: CSAE 2018, Proceedings of the 2nd International Conference on Computer Science and Application Engineering, Article no. 179, Hohhot, China. Association for Computing Machinery, ACM (2018). ISBN 978-1-4503-6512-3/18/10. https://doi.org/10.1145/3207677.3278047

27. Panpeng, V., Fang, M., Phaphuangwittayakul, A., Rattanadamrongaksorn, T.: Preliminary study and implementation of Chiang Mai tourism platform based on DOSA. In: Yang, X.-S., Sherratt, S., Dey, N., Joshi, A. (eds.) Proceedings of Fifth International Congress on Information and Communication Technology. AISC, vol. 1184, pp. 511–521. Springer, Singapore (2021). https://doi.org/10.1007/978-981-15-5859-7_51

Data Quality Identification Model for Power Big Data

Haijie Zheng$^{(\boxtimes)}$, Bing Tian, Xiaobao Liu, Wenbin Zhang, Shenqi Liu, and Cong Wang

State Grid Shandong Electric Power Company Information and Communication Company, Jinan 250000, Shandong, China
hjiezheng@126.com

Abstract. Data quality identification is an important task in power big data. Abnormal data exist and hamper the effective utilization of power big data. Moreover, the lack of labeled data makes the detection of abnormal data more challenging. Then, a data quality identification model for power big data is proposed. It can detect abnormal data from massive power big data. In this model, power data are grouped and then mapped into different feature spaces based on data augmentation technology. Tri-training is applied to detect abnormal data from different power data from different feature spaces. Experiments and simulations are performed to demonstrate the effectiveness of the proposed model.

Keywords: Data quality · Tri-training · Data augmentation

1 Introduction

With the orderly development of power IoT (Internet of Things) construction, power companies have collected massive amounts of heterogeneous business data. These data are of great value. Smart business analysis applications are constructed to take full advantage of these valuable data. Then, the power company carries out periodic data governance work [6,17], and spends considerable time and manpower on data quality identification and data governance to provide high-quality data.

Data quality is the core issue for efficient data analysis. Dirty data identification is the primary and normal work in data quality identification. Experts with experience could be used for data quality identification. However, this is not cost-efficient. With the gradual expansion of the scale of data in power companies, manual identification approaches have become infeasible. Data mining, deep learning and other technologies can be used for abnormal data identification, such as decision trees [4] and neural networks [12]. However, the labels are

Supported by the Science and Technology Project of State Grid Shandong Electric Power Company: "Research on the Key Technology of Heterogeneous Graph Anomaly Pattern Recognition Governance Based on Attention Mechanism" (Grant No. 2020A-135).

absent or few in many scenarios. Many supervised machine learning approaches construct the data identification model with the labeled examples. It is a challenge to construct the data identification model in data sets with very few labels.

To address the above issues, this paper proposes a data quality identification model for power big data that can detect abnormal data from massive power big data. First, the massive power data are divided into different groups according to the behavior pattern of electricity consumption. Then, the power data are mapped into different feature spaces based on data augmentation technology. In different feature spaces, the power data could have similar characteristics. Tri-training is applied to the power data in different feature spaces to detect abnormal power data. The main contribution of this paper is organized as follows:

- To detect abnormal power data, a data quality identification model is proposed. In this model, power data are grouped and then mapped into different feature spaces based on data augmentation technology. Tri-training is applied to detect abnormal data from different power data from different feature spaces.
- Experiments and simulations are performed to demonstrate the effectiveness of the proposed model.

The organization of this paper is as follows: Sect. 2 gives the background and makes a survey of the related works. The problem definition is given in Sect. 3. The data quality identification model is proposed in Sect. 4. Experiments and simulations are given in Sect. 5. Finally, Sect. 6 presents the conclusion and provides the future research work.

2 Background and Related Works

This section provides the background of the related technologies. The related works are also surveyed.

2.1 Related Works

Data quality is an important topic in many areas. The quality of data affects the performance of subsequent data processing [1,20] and is a hot topic. It attracted researchers from scholarly and industrial areas. In the long run, the quality should be managed continuously [19]. Data quality model in big data [2,15] is presented from a systematic view. Tools are also made to perform quality control and data preprocessing, such as SeQual [8]. Data quality plays an important role in big data [21].

In data quality identification areas, anomaly detection is a commonly used approach to detect abnormal data from massive data. Some researchers review the related works about anomaly detection from different viewpoints, such as the trustworthy viewpoint [24] and image viewpoint [14]. Anomaly detection can be used in many areas, such as medical areas [10] and smart grid areas [26].

Anomaly detection could be combined with other technologies to enhance the performance. A knowledge graph can represent the rich connections between entities. It can be used to enhance the performance of anomaly detection [16,27].

2.2   Related Technologies

In many scenarios, labeled examples are rare. Lack of labeled examples is a common situation. For this situation, many approaches have been proposed, such as tri-training [29]. In this approach, the training could be mapped into three labeled training sets through bootstrapping sampling. The classifier is generated and trained for every training set. Then, these classifiers are refined using these unlabeled examples in the tri-training process. Tri-training has been used in many areas, such as heartbeat classification [13], label noise detection [30], recommendation [23] and other areas [18].

Data augmentation is a technology that can generate new examples from the original examples. For example, in image processing areas, new images can be generated from the original images through flip, rotation, scale, crop, translation, Gaussian noise, and conditional GANs. Data augmentation can be used in many areas, such as NLP (natural language processing) [9], time series data [22], text data [3], and graph machine learning [7,28].

3   Problem Definition

In smart grid companies, the power data from different end users can be collected and analyzed. Data quality could be measured, and abnormal data could be detected. However, due to the expensive cost of labeling and other technologies, there are plenty of unlabeled examples.

For a specific end user x_m of power consumption, the power consumption data in a time period $[t_i, t_j]$ could be modeled as $L_{x_m}^{t_i \sim t_j} = \{l_{x_m}^{t_n} | t_i \leqslant t_n \leqslant t_j\}$, where $l_{x_m}^{t_n}$ denotes the power load that is consumed by end user x_m at time point t_n.

The power data of end user x_m can be given a label, as denoted by $L_{x_m,tag}^{t_i \sim t_j} = \{L_{x_m}^{t_i \sim t_j}, tag\}$. The tag could be true, false or null. True means $L_{x_m}^{t_i \sim t_j}$ is normal data. False means $L_{x_m}^{t_i \sim t_j}$ is abnormal data. Null means the type of $L_{x_m}^{t_i \sim t_j}$ is not known thus far. The tag could be determined by experts. However, this cost much. In most cases, the tag is null in most situations. The number of power data with true or false $tags$ is small. For a specific end user x_m, the power data collected could be discrete in the time dimension, such as $L_{x_m,tag}^{t_i \sim t_j}$ and $L_{x_m,tag}^{t_m \sim t_n}$. $[t_i, t_j]$ and $[t_m, t_n]$ are disjoint time intervals. The power data from all end users and different time intervals could form a set $L_{original}$.

Then, the problem can be modeled as follows. Given the original power data set $L_{original}$, the tag is null in many power data. The goal is to set the null tag of $L_{x_m,tag}^{t_i \sim t_j}$ to true or false.

4   Proposed Approach

This section gives the proposed approach. First, the data quality identification architecture is given. Then, the process of data quality identification is presented.

4.1 Data Quality Identification Architecture

The proposed data quality identification architecture is shown in Fig. 1. Given the original power data, preprocessing and grouping are the first stage of the proposed architecture. The original data could be proposed and divided into different groups. The power data in the same group have similar characteristics. After grouping, power data in the same group are augmented through different ways into different feature spaces. Finally, the tri-training-based approach is applied to detect abnormal data from the power data set with few *tag* labels.

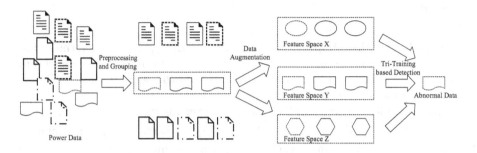

Fig. 1. Data quality identification architecture.

4.2 Data Preprocessing and Grouping

Given the original power data from $L_{original}$, there are many steps here. First, data standardization is the first step. Then, the power data can be transformed into new power data. After that, the power data could be divided into different groups.

Data Preprocessing. The power data should be standardized through the min-max approach. The interval is set to $[0, 1]$. Given all the power data from $L_{original}$, the maximum value of power data is denoted by L_{max}, and the minimum value of power data is denoted by L_{min}. The power data can be standardized through Eq. 1. After standardization, obtain the standard power data set $L_{normalized}$ from $L_{original}$.

$$l_{x_m}^{t_n} = \frac{l_{x_m}^{t_n} - L_{min}}{L_{max} - L_{min}} \tag{1}$$

Then, the power data from $L_{normalized}$ could be transformed into a larger data set $L_{transformed}$, which could be used in the following processing. Given any data $L_{x_m,tag}^{t_i \sim t_j}$ from $L_{normalized}$, randomly assign a time point t_n in time interval $[t_i, t_j]$, where $t_i \leqslant t_n \leqslant t_j$. Then, the data $L_{x_m,tag}^{t_i \sim t_j}$ could be fragmented into two data $L_{x_m,tag}^{t_i \sim t_n}$ and $L_{x_m,tag}^{t_n \sim t_j}$. The *tag* in $L_{x_m,tag}^{t_i \sim t_n}$ and $L_{x_m,tag}^{t_n \sim t_j}$ remains the same as *tag* in $L_{x_m,tag}^{t_i \sim t_j}$. Given a fragmentation threshold $\tau_{partition}$, where $0 \leqslant \tau_{partition} \leqslant 1$. At least $\tau_{partition}$ percent of data in $L_{normalized}$ are chosen for fragmentation to get $L_{transformed}$.

Grouping. Power data in $L_{transformed}$ could be divided into different groups using time-series clustering approaches. There are many studies on the clustering of time series data [11,25]. Then, the data in $L_{transformed}$ could be clustered and formed a new grouped set $L_{grouped}$.

4.3   Data Augmentation

Power data within the same group of $L_{grouped}$ are augmented through feature extraction and wavelet decomposition. Then, two new feature spaces are constructed.

Feature Extraction Approach. For every data $L_{x_m,tag}^{t_i \sim t_j}$ in $L_{grouped}$, the statistics information can be computed to form the new data in the statistics feature space. $L_{x_m,tag}^{t_i \sim t_j}$ can be transformed into $S_{x_m,tag}^{t_i \sim t_j}$ in the statistics feature space, which can form a set $S_{statistics}$.

The statistics feature could include: the average value of $L_{x_m,tag}^{t_i \sim t_j}$ in time period $[t_i, t_j]$, denoted by $l_{x_m,avg}^{t_i \sim t_j}$, as shown in Eq. 2, where $N(t_i, t_j)$ is the number of data points between time period $[t_i, t_j]$; the minimum value of $L_{x_m,tag}^{t_i \sim t_j}$ in time period $[t_i, t_j]$, denoted by $l_{x_m,min}^{t_i \sim t_j}$, as shown in Eq. 3; the maximum value of $L_{x_m,tag}^{t_i \sim t_j}$ in time period $[t_i, t_j]$, denoted by $l_{x_m,max}^{t_i \sim t_j}$, as shown in Eq. 4; the median value of $L_{x_m,tag}^{t_i \sim t_j}$ in time period $[t_i, t_j]$, denoted by $l_{x_m,median}^{t_i \sim t_j}$, as shown in Eq. 5; the mode value of $L_{x_m,tag}^{t_i \sim t_j}$ in time period $[t_i, t_j]$, denoted by $l_{x_m,mode}^{t_i \sim t_j}$, as shown in Eq. 6; the standard deviation value of $L_{x_m,tag}^{t_i \sim t_j}$ in time period $[t_i, t_j]$, denoted by $l_{x_m,standardDeviation}^{t_i \sim t_j}$, as shown in Eq. 7.

$$l_{x_m,avg}^{t_i \sim t_j} = \frac{1}{N(t_i, t_j)} \sum_{t_n=t_i}^{t_j} l_{x_m}^{t_n} \tag{2}$$

$$l_{x_m,min}^{t_i \sim t_j} = min\{l_{x_m}^{t_n} | t_i \leqslant t_n \leqslant t_j\} \tag{3}$$

$$l_{x_m,max}^{t_i \sim t_j} = max\{l_{x_m}^{t_n} | t_i \leqslant t_n \leqslant t_j\} \tag{4}$$

$$l_{x_m,median}^{t_i \sim t_j} = median\{l_{x_m}^{t_n} | t_i \leqslant t_n \leqslant t_j\} \tag{5}$$

$$l_{x_m,mode}^{t_i \sim t_j} = mode\{l_{x_m}^{t_n} | t_i \leqslant t_n \leqslant t_j\} \tag{6}$$

$$l_{x_m,standDev}^{t_i \sim t_j} = standardDeviation\{l_{x_m}^{t_n} | t_i \leqslant t_n \leqslant t_j\} \tag{7}$$

All these statistics values could be formed as a new data $s_{x_m}^{t_i \sim t_j}$ in statistical feature space, as shown in Eq. 8.

$$s_{x_m}^{t_i \sim t_j} = \{l_{x_m,avg}^{t_i \sim t_j}, l_{x_m,min}^{t_i \sim t_j}, l_{x_m,max}^{t_i \sim t_j}, l_{x_m,median}^{t_i \sim t_j}, l_{x_m,mode}^{t_i \sim t_j}, l_{x_m,standDev}^{t_i \sim t_j}\} \qquad (8)$$

with the *tag* in the original data $L_{x_m,tag}^{t_i \sim t_j}$, obtain the mapping data in statistical feature space $S_{x_m,tag}^{t_i \sim t_j}$, where $S_{x_m,tag}^{t_i \sim t_j} = \{s_{x_m}^{t_i \sim t_j}, tag\}$. All the $S_{x_m,tag}^{t_i \sim t_j}$ formed the data set $S_{statistics}$ in statistical feature space.

Wavelet Decomposition Approach. Time series data can be transformed into the frequency domain. Wavelet decomposition is applied to transform the data $L_{x_m,tag}^{t_i \sim t_j}$ in the original feature space into a new data $F_{x_m,tag}^{t_i \sim t_j}$ in frequency domain feature space. All $F_{x_m,tag}^{t_i \sim t_j}$ form the set $F_{frequency}$ in the frequency domain feature space.

Based on the wavelet decomposition, the data $L_{x_m,tag}^{t_i \sim t_j}$ in the original feature space can be mapped into the frequency set, shown as follows: $f_{x_m}^{t_i \sim t_j} = \{CA3_{x_m}^{t_i \sim t_j}, CD1_{x_m}^{t_i \sim t_j}, CD2_{x_m}^{t_i \sim t_j}, CD3_{x_m}^{t_i \sim t_j}\}$. $CA3_{x_m}^{t_i \sim t_j}$, $CD1_{x_m}^{t_i \sim t_j}$, $CD2_{x_m}^{t_i \sim t_j}$, $CD3_{x_m}^{t_i \sim t_j}$ is the corresponding frequency component.

With the *tag* in the original data $L_{x_m,tag}^{t_i \sim t_j}$, obtain the mapping data in frequency feature space $F_{x_m,tag}^{t_i \sim t_j}$, where $F_{x_m,tag}^{t_i \sim t_j} = \{f_{x_m}^{t_i \sim t_j}, tag\}$. All the $F_{x_m,tag}^{t_i \sim t_j}$ formed the data set $F_{frequency}$ in frequency feature space.

4.4   Tri-Training Based Detection

Given the three different feature spaces of the same data, the tri-training- based approach could be used for anomaly detection. The corresponding feature spaces are the original grouped set $L_{grouped}$ in the original feature space, $S_{statistics}$ in the statistical feature space and $F_{frequency}$ in the frequency feature space. The data in corresponding feature spaces are $L_{x_m,tag}^{t_i \sim t_j}$, $S_{x_m,tag}^{t_i \sim t_j}$ and $F_{x_m,tag}^{t_i \sim t_j}$.

The tri-training-based detection approach is as follows:

Initialize the corresponding classifiers in each feature space. In the original feature space $L_{grouped}$, initialize a classifier based on long-short term memory (LSTM for short), denoted by $C_{grouped}$. Initialize two different classifiers based on neural network for feature space $S_{x_m,tag}^{t_i \sim t_j}$ and $F_{x_m,tag}^{t_i \sim t_j}$, denoted by $C_{statistics}$ and $C_{frequency}$.

Training the three classifiers using labeled data in the corresponding feature space. For simplicity, denote the three different classifiers as $Classifier_a$, $Classifier_b$ and $Classifier_c$, the data from the corresponding feature spaces $Data_a$, $Data_b$ and $Data_c$ are denoted by $data_a$, $data_b$ and $data_c$. Sample the data with true or false *tag* from $Data_a$, $Data_b$ and $Data_c$ and train the three corresponding classifiers with these labeled data. Then, the trained classifiers are obtained.

Tri-training-based detection. For any data $data_a$ with null *tag* in $Data_a$, use the trained $Classifier_b$ and $Classifier_c$ to predict the *tag*. Select the $data_b$ and $data_c$ with the same predicted *tag* label and put the *tag* label on the corresponding $data_a$, $data_b$ and $data_c$. Then, the labeled data in $Data_a$, $Data_b$ and $Data_c$ are increased in this way.

The above steps are repeated until the stop criterion is reached. The newly labeled data are used to retrain the three classifiers $Classifier_a$, $Classifier_b$ and $Classifier_c$. The re-trained classifiers are used to label new samples in a tri-training way. Given the maximum number of iterations $Num_{maxIteration}$. After reaching $Num_{maxIteration}$, the training process is stopped. Then, the $data_a$ with a false tag is the predicted abnormal data.

5   Experiments

5.1   Training Data and Baselines

The proposed approach is evaluated in the production data from a specific power company. The power company collects a large amount of power consumption data from different industries. End users of different types have different power consumption patterns. To measure the performance of the proposed approach, these collected power data are labeled. In our experiments, some labels are manually made to null.

Tri-training [29] and co-training [5] are classical approaches. These approaches are used as baselines with our proposed approach. In our proposed approach, an LSTM-based classifier and neural network-based classifiers are used.

5.2   Analysis

The metric used in our experiments is the classification error rates used in Ref. [29]. The elementary experiment results are shown in Table 1. With the help of clustering and data augmentation, our proposed approach could detect abnormal data in a more effective way.

Table 1. The classification error rates

Approaches	Classification error rates
Co-training	0.1785
Tri-training	0.1342
Proposed approach	0.1057

6   Conclusion

To improving the quality of power big data, this paper proposes a data quality identification model that can distinguish abnormal data from massive power data with small labels. Considering the difference in abnormal data in different areas, the power data are first divided into different groups. The abnormal data in different groups may have different patterns. Then, data augmentation technologies are applied to map the power data into different feature spaces, while the mapped data retain the original characteristics. Then, tri-training technology is applied to the power data of different feature spaces to detect abnormal data.

However, there are some issues to be addressed in the future research. First, the proposed model uses tri-training, which is a classic approach. There are many improvements of this proposed model. Second, the rich knowledge and pattern contained in the power big data are not used. A knowledge graph could be applied to capture and represent the rich knowledge in power big data. A power knowledge graph could be used to identify the quality of power data in a more reasonable way. Third, the quality standards and criteria of power data evolve with the development of the power industry, society and others. Then, the data quality identification model should be adaptive to this evolving and dynamic situation. These are future research issues.

References

1. Azeroual, O., Jha, M.: Without data quality, there is no data migration. Big Data Cogn. Comput. **5**(2), 24 (2021)
2. Batini, C., Rula, A.: From data quality to big data quality: a data integration scenario. In: Greco, S., Lenzerini, M., Masciari, E., Tagarelli, A. (eds.) Proceedings of the 29th Italian Symposium on Advanced Database Systems, SEBD 2021, Pizzo Calabro (VV), Italy, 5–9 September 2021. CEUR Workshop Proceedings, vol. 2994, pp. 36–47. CEUR-WS.org (2021)
3. Bayer, M., Kaufhold, M., Reuter, C.: A survey on data augmentation for text classification. CoRR abs/2107.03158 (2021)
4. Biswal, B.N., Behera, H.S., Bisoi, R., Dash, P.K.: Classification of power quality data using decision tree and chemotactic differential evolution based fuzzy clustering. Swarm Evol. Comput. **4**, 12–24 (2012)
5. Blum, A., Mitchell, T.M.: Combining labeled and unlabeled data with co-training. In: Bartlett, P.L., Mansour, Y. (eds.) Proceedings of the Eleventh Annual Conference on Computational Learning Theory, COLT 1998, Madison, Wisconsin, USA, 24–26 July 1998, pp. 92–100. ACM (1998)
6. Chemnitz, N.Ø., Bonnet, P., Büttrich, S., Shklovski, I., Watts, L.: Unionized data governance in virtual power plants: poster. In: de Meer, H., Meo, M. (eds.) e-Energy 2021: The Twelfth ACM International Conference on Future Energy Systems, Virtual Event, Torino, Italy, 28 June - 2 July 2021, pp. 282–283. ACM (2021)
7. Ding, K., Xu, Z., Tong, H., Liu, H.: Data augmentation for deep graph learning: a survey. CoRR abs/2202.08235 (2022)
8. Expósito, R.R., Galego-Torreiro, R., González-Domínguez, J.: SeQual: big data tool to perform quality control and data preprocessing of large NGS datasets. IEEE Access **8**, 146075–146084 (2020)
9. Feng, S.Y., Gangal, V., et al.: A survey of data augmentation approaches for NLP. In: Zong, C., Xia, F., Li, W., Navigli, R. (eds.) Findings of the Association for Computational Linguistics: ACL/IJCNLP 2021, Online Event, 1–6 August 2021. Findings of ACL, vol. ACL/IJCNLP 2021, pp. 968–988. Association for Computational Linguistics (2021)

10. Fernando, T., Gammulle, H., Denman, S., Sridharan, S., Fookes, C.: Deep learning for medical anomaly detection - a survey. ACM Comput. Surv. **54**(7), 141:1–141:37 (2022)
11. Hallac, D., Vare, S., Boyd, S.P., Leskovec, J.: Toeplitz inverse covariance-based clustering of multivariate time series data. In: Proceedings of the 23rd ACM SIGKDD International Conference on Knowledge Discovery and Data Mining, Halifax, NS, Canada, 13–17 August 2017, pp. 215–223. ACM (2017)
12. Lee, G., Lee, S.J., Lee, C.: A convolutional neural network model for abnormality diagnosis in a nuclear power plant. Appl. Soft Comput. **99**, 106874 (2021)
13. Li, J., Wang, G., Chen, M., Ding, Z., Yang, H.: Mixup asymmetric tri-training for heartbeat classification under domain shift. IEEE Signal Process. Lett. **28**, 718–722 (2021)
14. Mohammadi, B., Fathy, M., Sabokrou, M.: Image/video deep anomaly detection: a survey. CoRR abs/2103.01739 (2021)
15. Montero, O., Crespo, Y., Piatini, M.: Big data quality models: a systematic mapping study. In: Paiva, A.C.R., Cavalli, A.R., Ventura Martins, P., Pérez-Castillo, R. (eds.) QUATIC 2021. CCIS, vol. 1439, pp. 416–430. Springer, Cham (2021). https://doi.org/10.1007/978-3-030-85347-1_30
16. Nesen, A., Bhargava, B.K.: Knowledge graphs for semantic-aware anomaly detection in video. In: 3rd IEEE International Conference on Artificial Intelligence and Knowledge Engineering, AIKE 2020, Laguna Hills, CA, USA, 9–13 December 2020, pp. 65–70. IEEE (2020)
17. Qiao, L., Zhou, Q., Song, C., Wu, H., Liu, B., Yu, S.: Design of overall framework of self-service big data governance for power grid. In: Zhai, X.B., Chen, B., Zhu, K. (eds.) MLICOM 2019. LNICST, vol. 294, pp. 222–234. Springer, Cham (2019). https://doi.org/10.1007/978-3-030-32388-2_19
18. Saito, K., Ushiku, Y., Harada, T.: Asymmetric tri-training for unsupervised domain adaptation. In: Precup, D., Teh, Y.W. (eds.) Proceedings of the 34th International Conference on Machine Learning, ICML 2017, Sydney, NSW, Australia, 6–11 August 2017. Proceedings of Machine Learning Research, vol. 70, pp. 2988–2997. PMLR (2017)
19. Taleb, I., Serhani, M.A., Bouhaddioui, C., Dssouli, R.: Big data quality framework: a holistic approach to continuous quality management. J. Big Data **8**(1), 1–41 (2021). https://doi.org/10.1186/s40537-021-00468-0
20. Talha, M., Kalam, A.A.E.: Big data between quality and security: dynamic access control for collaborative platforms. J. Univers. Comput. Sci. **27**(12), 1300–1324 (2021)
21. Talha, M., Elmarzouqi, N., Kalam, A.A.E.: Quality and security in big data: challenges as opportunities to build a powerful wrap-up solution. J. Ubiquit. Syst. Perv. Netw. **12**(1), 9–15 (2020)
22. Wen, Q., et al.: Time series data augmentation for deep learning: a survey. In: Zhou, Z. (ed.) Proceedings of the Thirtieth International Joint Conference on Artificial Intelligence, IJCAI 2021, Virtual Event / Montreal, Canada, 19–27 August 2021, pp. 4653–4660. ijcai.org (2021)
23. Yu, J., Yin, H., Gao, M., Xia, X., Zhang, X., Hung, N.Q.V.: Socially-aware self-supervised tri-training for recommendation. In: Zhu, F., Ooi, B.C., Miao, C. (eds.) KDD 2021: The 27th ACM SIGKDD Conference on Knowledge Discovery and Data Mining, Virtual Event, Singapore, 14–18 August 2021, pp. 2084–2092. ACM (2021)
24. Yuan, S., Wu, X.: Trustworthy anomaly detection: a survey. CoRR abs/2202.07787 (2022)

25. Zakaria, J., Mueen, A., Keogh, E.J.: Clustering time series using unsupervised-shapelets. In: Zaki, M.J., Siebes, A., Yu, J.X., Goethals, B., Webb, G.I., Wu, X. (eds.) 12th IEEE International Conference on Data Mining, ICDM 2012, Brussels, Belgium, 10–13 December 2012, pp. 785–794. IEEE Computer Society (2012)
26. Zhang, J.E., Wu, D., Boulet, B.: Time series anomaly detection for smart grids: a survey. CoRR abs/2107.08835 (2021)
27. Zhao, B., Shi, Y., Zhang, K., Yan, Z.: Health insurance anomaly detection based on dynamic heterogeneous information network. In: Yoo, I., Bi, J., Hu, X. (eds.) 2019 IEEE International Conference on Bioinformatics and Biomedicine, BIBM 2019, San Diego, CA, USA, 18–21 November 2019, pp. 1118–1122. IEEE (2019)
28. Zhao, T., Liu, G., Günnemann, S., Jiang, M.: Graph data augmentation for graph machine learning: a survey. CoRR abs/2202.08871 (2022)
29. Zhou, Z., Li, M.: Tri-training: exploiting unlabeled data using three classifiers. IEEE Trans. Knowl. Data Eng. 17(11), 1529–1541 (2005)
30. Zhu, H., Liu, J., Wan, M.: Label noise detection based on tri-training. In: Sun, X., Pan, Z., Bertino, E. (eds.) ICCCS 2018. LNCS, vol. 11063, pp. 613–622. Springer, Cham (2018). https://doi.org/10.1007/978-3-030-00006-6_56

Data Security and Privacy

Effective and Lightweight Defenses Against Website Fingerprinting on Encrypted Traffic

Chengpu Jiang, Zhenbo Gao, and Meng Shen[✉]

Beijing Institute of Technology, Beijing, China
{jiangchengpu,gaozhenbo07,shenmeng}@bit.edu.cn

Abstract. Recently, website fingerprinting (WF) attacks that eavesdrop on the web browsing activity of users by analyzing the observed traffic can endanger the data security of users even if the users have deployed encrypted proxies such as Tor. Several WF defenses have been raised to counter passive WF attacks. However, the existing defense methods have several significant drawbacks in terms of effectiveness and overhead, which means that these defenses rarely apply in the real world. The performance of the existing methods greatly depends on the number of dummy packets added, which increases overheads and hampers the user experience of web browsing activity.

Inspired by the feature extraction of current WF attacks with deep learning networks, in this paper, we propose TED, a lightweight WF defense method that effectively decreases the accuracy of current WF attacks. We apply the idea of adversary examples, aiming to effectively disturb the accuracy of WF attacks with deep learning networks and precisely insert a few dummy packets. The defense extracts the key features of similar websites through a feature extraction network with adapted Grad-CAM and applies the features to interfere with the WF attacks. The key features of traces are utilized to generate defense fractions that are inserted into the targeted trace to deceive WF classifiers. The experiments are carried out on public datasets from DF. Compared with several WF defenses, the experiments show that TED can efficiently reduce the effectiveness of WF attacks with minimal expenditure, reducing the accuracy by nearly 40% with less than 30% overhead.

Keywords: Encrypted traffic · Website fingerprinting · Privacy

1 Introduction

Tor, a free anonymous communication software based on onion routing [1,4], is leveraged to hide browsing traces and prevent users from being eavesdropped on. It encapsulates the transmitted contents into Tor packets with the same length to mask the obvious length features, which is commonly used in website fingerprinting (WF). Website fingerprinting analysis is a type of attack where

Y. Wang et al. (Eds.): ICPCSEE 2022, CCIS 1629, pp. 33–46, 2022.
https://doi.org/10.1007/978-981-19-5209-8_3

the adversary attempts to eavesdrop on the web browsing activity of the user by observing the sequence of generated packets. Existing WF attacks [2,6,12] usually deploy machine learning methods to extract features for multiclass classification tasks. Recently, WF attacks with deep learning networks have been used to automatically extract available information from local sequence features of traffic for classification tasks, which invalidates deceptive traffic patterns of traditional WF defense methods.

Several defenses [2,3,5,9,16,18] on Tor have been proposed to deal with WF attacks [6,7,10,12,14,17,19]. The main approaches of defenses are either to pad dummy packets to original traces or deliberately delay the sending of the packet to change the formats of traces. The former can increase the bandwidth overheads. A mass of dummy packets may cause network jams. The latter can increase the delay with packets, which inevitably impacts the user experience and is generally not recommended in practice [5].

However, two main challenges remain to be solved for WF defenses in the real world. First, unlike attacks with deep learning networks [13,15], which have no restriction on input feature vectors as long as the features can depict patterns of traces effectively, defenses should consider whether the features can be restored to the original sequence. At this point, it is necessary to simultaneously prove the effectiveness of the feature and propose a restoration scheme. Second, we need to raise a valid coroutine for users or Tor nodes to meet real-time requirements in the real world. Because duplex communication is asynchronous, it is difficult to generate bidirectional sequences cooperatively. Several WF defenses [18] make the network perform in half-duplex mode and delay normal Tor packets to construct the defended traces, which is difficult to deploy and increases the time of communication.

In this paper, we present a **T**argeted Lightw**E**ight **D**efense method named **TED** based on adversary examples with a key feature dictionary, and the key contributions of our work can be described as follows:

- We propose TED, a new and generic WF defense method based on a deep learning network, to extract the key features of websites, emphasizing precisely padding the least dummy packets to disturb WF attacks. Unlike existing defense, the novel idea of TED is to disguise the traffic of targeted websites as the traffic from similar websites by imitating key features. The method can learn the most distinguishing features between traces on a similar scale, generating accurate defensive traces for targeted traces. We also provide deployments of TED.
- We propose a trace-level dynamic defense instead of a website-level defense, which increases the ability against adversarial training and is different from existing defenses. The website-level defense believes that traces generated by multiple visits to the same website are consistent, only constructing a unique defensive trace for each website. However, we consider that the traces generated by visiting the same website may be different. The trace-level defense is proposed to generate a dynamic defensive trace for each original trace, even if these traces belong to the same website. Each time TED randomly picks up

the defensive key features from different traces of a similar website to protect the defended trace.

- We prove the effectiveness of TED with little overhead for existing WF attacks. We implement some advanced attacks, such as k-NN [2], CUMUL [12], and DF [17], to verify the effectiveness of our defense. For these attacks, TED can effectively reduce the accuracy by nearly 40% for each WF attack with a bandwidth overhead of 29.46%. Therefore, TED can achieve a great confusion effect.

The paper is organized as follows: we first introduce related works about WF in Sect. 2. Then, we propose the threat model for the WF defense scenario in Sect. 3. We give a detailed description of the overall architecture of TED in Sect. 4. Then, we evaluate our approach in Sect. 5. Finally, we conclude the paper in Sect. 6.

2 Background and Related Work

2.1 WF Attacks

WF attacks are proposed to identify the website that Tor users visit through encrypted traffic. We select three representative WF attacks as follows:

k-FP. Hayes et al. [6] proposed applying Hamming distance to achieve a WF attack. They extract timing information and statistical features from incoming and outgoing traffic sequences, respectively. They evaluate feature importance and regard the 30 most important features as website fingerprints.

CUMUL. Panchenko et al. [12] provide a WF attack by the support vector machine (SVM) classifier based on the feature of accumulated packet length. Considering the classification accuracy and computational efficiency, they utilize the first 100 accumulated packet lengths as well as basic features.

DF. Sirinam et al. [17] presented a novel website fingerprinting attack based on the deep learning technology Convolutional Neural Networks (CNN) with a sophisticated architecture design. The feature they applied to train CNN is the direction sequence of packets.

2.2 WF Defense

WF Defense can be considered a kind of information hiding technology that is applied to protect the privacy of the user. Several defenses reduce the bandwidth and time overhead while ensuring the defense effect to a certain extent. We pick four representative WF defenses.

Walkie-Talkie. Walkie-Talkie [18] is an efficient WF defense, which makes the browser communicate with the web server in the half-duplex mode instead of the usual full-duplex mode. The key idea of the defense is adding dummy packets and delays to create collisions to make two or more sites have the same features.

Walkie-Talkie requires a network with the half-duplex communication protocol. When a web resource is loaded, another request will be sent to continue loading only after the current resource has finished loading.

WTF-PAD. WTF-PAD was proposed by Juarez et al. [8] to defend against WF attacks in Tor. This method leverages adaptive padding and only adds padding when the utilization rate of channels is low.

Front. Gong et al. [5] presented a novel and zero-delay WF defense Front, which decorates the feature-rich trace front with dummy packets. This makes the trace different from each other trace to impede the attackers learning process.

Adversarial Examples. The tailored defense proposed by Nasr et al. [11] is a universe WF defense method for different attack models based on adversarial examples. It produces blind adversarial perturbations by training the generator for a specific WF attack model and inserting the perturbations lively and fixedly.

3   Threat Model

In our consideration, the adversary concerns the web browsing activity of the user through the encrypted proxy Tor. Although Tor can encapsulate the information in packets and standardize the sizes of packets, metadata information such as directed packet sequences and statistical features can still expose the information of websites that the user has visited. As Fig. 1 shows, the adversary is in a local position, and all their attacks are passive. Local adversaries capture traces on network nodes between the client and the entry node of the proxy, such as eavesdroppers at the same local network as clients, administrators, or even Internet Service Provider (ISP). They can identify traces from specific clients passively but cannot make any modifications to the trace, which will conceal their attacks from being detected.

A complete trace is denoted as a sequence of packets: $S = \langle p_1, p_2 \ldots, p_n \rangle$. All information of a packet can be expressed as $p_i = (t_i, \pm l_i)$. The information of the i-th packet p_i contains a timestamp t_i and a packet direction $\pm l_i$, where $+1$ or -1 means an incoming or outgoing Tor packet.

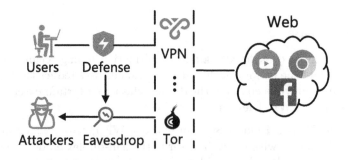

Fig. 1. The threat model of website fingerprinting.

We aim to deploy our defense on the Tor client proxy and Tor exit node to prevent eavesdropping from vulnerable network nodes. We suppose that the WF defense methods can process the trace in advance and that the adversary can collect the defended traces as training samples.

4   The Propose TED

In this section, we start with the parts and an overall workflow of TED. Then, we describe the details of parts of TED, such as scale clustering, interconversion, and key feature extraction. Finally, we introduce the detailed implementation of the proposed TED.

4.1   Overview of TED

As Fig. 2 depicts, before feature extraction, we first convert the original traces to cumulative traces in the interconversion process. In the interconversion process, we propose a transformation method that can interconvert the original traces and continuous cumulative packet sequence with the contextual informational feature. Considering that the scale of the website is a distinct feature for WF attacks, in the scale clustering process, we assemble the traces of similar websites into several clusters, aiming to decrease the unnecessary overhead of the defense. In the feature extraction process, we adapt Grad-CAM, an explainable deep learning technology, which locates key features of preprocessed sequences and generates defensive traces for each trace.

When an original trace comes, the TED will be categorized as a cluster by the label of the trace. Then, the defense will randomly pick out key features of similar websites from the cluster and make Tor proxy and exit node reach an agreement. Then, the Tor nodes independently send padding packets to collaboratively construct a defensive trace.

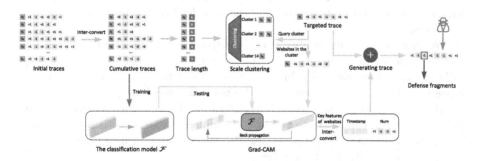

Fig. 2. Block diagram of the TED method.

4.2   Similar Scale Traces Clustering

Note that there is a large gap on trace scales of different kinds of websites that attackers can easily distinguish by counting the packets of traces. For example, the scale of traffic generated by Google is far less than YouTube but relatively similar to other search engines such as Bing. To this end, it is necessary to cluster websites before adding packets. Moreover, it can also decrease the cost of modifying the traces.

The aim of clustering websites is to improve the effectiveness of our defenses. If the length of the original trace is far less than the confused trace that offers key features, it is possible that before the key features are inserted into the original trace, the transmission of the original trace has been completed. Therefore, we first compute the average length of different website traces and assemble the traces into clusters. The scale clustering process is based on the following principles: traces with similar data scale should be grouped into the same cluster; each cluster contains two more traces.

4.3   Interconversion

As described above, before extracting key features by deep learning, it is necessary to preprocess the original traces in TED. In detail, we design a trace-to-feature process to construct input vectors for the FCN discriminator and a feature-to-trace process to restore a trace.

Trace-to-Feature. Considering that the convolution layer in deep learning classifiers can extract information in receptive fields, a single Tor packet consisting of a pair of timestamps and directions exposes no sequential information, which has little effect on classification tasks. Therefore, we define an element of a cumulative continuous packet sequence that contains 2 numbers as the minimum granularity of the packet sequence that carries on sequence information. The transformation from the original trace to the cumulative continuous packet sequence can reserve full sequence information, leading to the high accuracy of classifiers.

We take an example to illustrate the method of feature extraction, such as a trace with original trace without timestamp $\{+1, +1, -1, -1, -1..., -1, -1\}$. Adjacent packets in the same direction can be thought of as a continuous transmission of data. We can accumulate the original trace into $\{+2, -3..., -2\}$ by several linear transformations. A key feature of the transformed trace can be $\{+2, -3\}$, which can be restored into a complete segment with a timestamp. The numbers mean that two communicators first transfer 2 incoming Tor packets and then transfer 3 outgoing Tor packets.

Feature-to-Trace. Feature extraction can attain the key features of different websites, which can be used for deceiving the WF attacks classifiers. However, there is a gap to locate when the Tor proxy and exit node should add confused

---

Algorithm 1. Generate defense trace

Input: dictionary of key features' sequence $\mathcal{D}$, real-time traces $\mathcal{R}$.
Output: protected traces $\mathcal{P}_t$.
1: Match a confused trace $\mathcal{F}$ in the same cluster with $\mathcal{R}$ from $\mathcal{D}$ and obtain the confused segments $\mathcal{F}_i$.
2: **for** R_t in $\mathcal{R}$ **do**
3:     Sum up bidirectional cumulative packet total $\mathcal{T} = \{\mathcal{T}_{forward}, \mathcal{T}_{back}\}$.
4:     **if** IsMatch$(\mathcal{T}, \mathcal{F}_i)$ **then**
5:         $R_t \leftarrow f_i$.
6:     **end if**
7: **end for**
8: Restore generated traces p_i.
9: **return** The protected traces $p_i \in \mathcal{P}$.

---

segments. To restore the timestamp to insert padding packets, it is necessary to compute the statistical quantile of key features for targeted traces in advance. The process of restoring bidirectional traces can be described as algorithm 1.

Considering the fluctuation of the network, we replace the direct timestamp with the cumulative bidirectional packet amount to perform the cumulative packet quantile. For example, when a real-time trace comes, the defense will offer confused segments from a confused trace in the cluster of the original trace. The segments include the cumulative continuous packets and the timestamp measured by bidirectional packet counts $\{+x, -y\}$. As the Tor packets are sent, the TED will judge whether there are x incoming packets and y outgoing packets accumulated in the communication. If matched, the normal sequence of the original trace will be decorated with confused segments.

4.4 Key Feature Extraction

In this part, we mainly introduce the feature extraction module, which can screen out the distinctive key segments from whole processed traces. The Key Feature Extraction module is based on an explanation of the classification results of deep learning networks. We adapt the FCN network and Grad-CAM for one-dimensional sequences and extract the features from processed trace sequences. Without the fully connected layer, the FCN not only significantly simplifies the parameters in the network but also has no limit on the length of the input vectors, which corresponds with variable-length traces. In the FCN, there are five blocks for feature extraction. In the first four blocks, there are three convolution layers and one max pooling layer, while in the last block, there is one convolution layer and one max pooling layer. In the last block, there are two convolution layers that replace the fully connected layers to implement classification. The explainable work for deep learning is migrated by Grad-CAM. Note that prior work has applied Grad-CAM for explainable deep learning models in computer vision. However, we first modify the Grad-CAM for website fingerprinting defense. We resize the weighted activation values and the gradients into a one-dimensional vector, which can align with the input vector for the FCN network. Considering that the trace does not have information of the contour feature in the computer version, we remove the guided backpropagation of Grad-CAM, which used to be applied in the computer vision and simplify the structure of Grad-CAM. Let $A^k \in \mathcal{R}^{C*T}$ of channels C and classes T be the weighted activation values of a convolution layer. The neuron importance weights a_k^c of target c and channel k are represented as Eq. 1:

$$a_k^c = \overbrace{\frac{1}{Z} \sum_i}^{global\ pooling} \frac{\partial y^c}{\partial A_i^k} \tag{1}$$

The $L_{Grad-CAM}^c \in \mathcal{R}^{1*L}$ of input length L for any class c is represented as Eq. 2:

$$L_{Grad-CAM}^c = ReLU(\sum_k a_k^c A^k) \tag{2}$$

Therefore, we have a universal outline to extract key features for deep learning models of one-dimensional sequences.

Specifically, for websites that cannot match any cluster, we consider adding various interference features of websites. Thereafter, for each target trace, the client and server negotiate a pseudo trace based on a dictionary library. The key features of the original defended trace are masked by the confused sequences at the corresponding location of the confused trace, while the key features of the confused sequences are inserted.

5   Performance Evaluation

Here, we evaluate the accuracy of our defense model against known WF attacks with deep learning models and the bandwidth and time overhead of the deployment process.

We first describe preliminary work on the setup of the environment, known WF attacks, and defenses for comparison in Sect. 5.1. In Sect. 5.2, we expound on metrics to evaluate the practicality of WF defense methods. In Sect. 5.3, we describe the process and verify the necessity of clustering the websites. In Sect. 5.4, we compare defensive capabilities with WF defenses against the above WF attacks to show the advantages of TED.

5.1   Preliminary

Setup and Dataset. In the experiment, we use a machine with an Nvidia RTX 2070 GPU for matrix calculation of training FCN, a 32 GB RAM, and an AMD Ryzen7 7200X CPU processor for feature processing. We carry out simulation experiments for the Tor anonymous network. The Tor datasets come from DF [17], which include 95 websites and 1,000 traces per site.

To evaluate the WF defenses, it is necessary to deploy WF attack models first. We perform several representative WF attack models as benchmarks, such as DF, k-NN, and CUMUL. These methods are based on different eavesdropping principles that have been introduced in Sect. 2.1 and apply traditional machine learning methods or deep learning methods to feature extraction. Considering that TED is a reflection on adversary examples of deep learning models and DF can reach the best attack effects of WF defense methods among these attacks, we mainly conclude the experimental results of TED against DF.

For comparison with WF defenses, we perform state-of-the-art WF defenses such as WTF-PAD, Walkie-Talkie, Front, which can randomize or pad dummy packets to generate interferential sequences.

5.2   Performance Metrics

Intuitively, if WF defense methods pad more packets into a trace, it is more likely to induce WF attacks to make false classifications. However, more dummy packets means higher bandwidth utilization, which may cause network congestion, thus influencing the user experience. Balancing overheads and effectiveness is the main task for WF defenses to be discussed. Therefore, we take precision and overhead into consideration of performance metrics.

- Precision. We define precision as the misidentification rate for WF attack models influenced by WF defenses.
- Bandwidth overhead. We compute the overhead by counting the ratio of the average number of modified packets and the original trace sizes of the defense models.
- Time overhead. We compute the time overhead by counting the time of delaying normal Tor packets or appending dummy packets to original traces of defense models.

5.3   Parameter Tuning

Here, we mainly discuss parameter tuning of TED that influences the effect of TED, such as the number of websites in clusters and the number of key features for a trace. Before measuring the number of websites, we design a set of experiments to prove the validity of the clustering method. We also show the parameters of the FCN and Grad-CAM in this part.

5.4   Evaluation

In this part, we evaluate the efficiency of TED against several WF attacks. We mainly discuss the performance in terms of effectiveness and overhead. In addition, we analyze the time overhead of the whole deployment process of the TED (Table 1).

Table 1. Parameters configuration of the experiments.

Hyperparameters	Range	Final
Learning rate	[0.001,0.002]	0.001
Batch size	[16,128]	64
Training epoch	[20,60]	50
Input vector	[(1,500),(1,2000)]	(1,2000)
Block1[Conv1_1,Conv1_2]	–	[32,32]
Block2[Conv2_1,Conv2_2]	–	[64,64]
Block3[Conv3_1,Conv3_2]	–	[128,128]
Block4[Conv4_1,Conv4_2]	–	[256,256]
Block5[Conv5_1,Conv5_2]	–	[500,95]
Output of adapted grad-CAM	–	(1,2000)
Original trace length	–	[86,13680]
Processed trace length	–	[22,1915]
Original maximum timestamp	–	120

Table 2. Accuracy and overhead of WF attacks in scenarios without defense and against different strategies, where we present some representative results.

Defenses	Overhead		Accuracy of WF attacks on defended datasets		
	Bandwidth	Time	k-NN	CUMUL	DF
NoDef	–	–	93.64%	95.11%	97.44%
Walkie-Talkie	31.0%	34.0%	26.11%	24.48%	71.02%
WTF-PAD	108.0%	0%	40.94%	59.80%	90.85%
Front	64.0%	0%	4.37%	30.61%	76.85%
TED	**29.46%**	4.87%	30.78%	31.01%	**50.71%**

Precision. We start with the evaluation of the accuracy of different WF attacks and compare the TED with several state-of-the-art defense methods. Let $A_{\mathcal{F}}^d$ be the accuracy of the WF attacks $\mathcal{F}$ for dataset d. The effectiveness P of WF defenses D can be described by the following formula:

$$P = A_{\mathcal{F}}^d - A_{\mathcal{F}}^{D(d)},$$

As shown in Table 2, for undefended traces, all WF attacks have high accuracy rates, which shows the effectiveness of WF attacks. Relatively speaking, k-NN and CUMUL have lower accuracy rates, while DF is more effective. When state-of-the-art defense methods are deployed, the attacks have interfered to different degrees.

For WF attacks such as k-NN and CUMUL, the proposed TED can decrease the accuracy by over 40% for these attacks, which is close to these existing methods. However, compared with existing methods, TED can apparently reduce the attack efficiency of the DF method with deep learning, reducing the accuracy of DF from 97.44% to 50.71%. To this end, we can prove the validity of the model features by resisting deep learning WF attacks.

In conclusion, the experiments have proven the effectiveness of the targeted feature extraction. The first phenomenon can be explained by the fact that TED can insert a sufficient number of deceptive segments, which changes the basic format of all traces. Toward DF with deep learning, TED has a similar mechanism of extracting features, which means the segments can have great effects on confusion.

Bandwidth Overhead. The bandwidth overhead of different websites is measured by their scales of original traces. We define the bandwidth overhead B of WF defenses of Tor as the ratio of the lengths of inserting dummy packets $\mathcal{L}_c$ and the lengths of targeted traces $\mathcal{L}_t$, which can be described as:

$$B = \frac{1}{N} \sum_{i}^{N} \frac{\mathcal{L}_c}{\mathcal{L}_c + \mathcal{L}_t},$$

To avoid the random selection of the confused traces of TED and the deviations of key feature extraction of different websites, we compute the ratio by averaging all samples of the targeted websites in each cluster.

To compare different WF defenses on the same scale, we adjust the parameters of the compared defenses to reach relatively low bandwidth overhead. As shown in Table 2, our deployed TED model has the lowest overhead compared with current state-of-the-art defense methods, which proves that the targeted WF defense is more precise by simulating a particular distribution because TED tends to customize a set of defense schemes for each website of the clusters. In addition, Walkie-Talkie requires approximately 31% bandwidth overhead, which is superior to other defenses.

Time Overhead. In TED, we do not deliberately block the delivery of normal packets. Conversely, we hope to take advantage of normal packets to construct defensive traces, which means that the defense is nearly real-time. The timestamp to pad dummy packets is equivalent to the count of cumulative bidirectional packets. However, when the normal packets conflict with the confused packets, we tend to suspend the normal packets to construct a defensive trace. We define the time of single packet delivery t and the number of inserted packets n; then, in the worst condition, the time overhead will be n * t. In the deployment of experiments, we set the delay time to 0.01 ms, which represents the response time of sending a dummy packet. Table 2 shows that the time overhead of TED is less than 5%, which is close to zero-delay defenses, while Walkie-Talkie needs only approximately 34% time overhead. However, Walkie-Talkie needs the half-duplex mode, which means that the network can only transmit packets in one direction, thus increasing the time overhead in practice.

Finally, we discuss the possible time overhead of deployment in practice. In our design, the negotiation of the confused website in the same cluster of the targeted website is accompanied by the establishment of a Tor socket. In addition, the time overhead of searching the key features of the confused website from servers is acceptable; since the buffers can quickly and concurrently process requests by the structure of message queues, the time overhead is acceptable.

6   Conclusion

In this paper, based on the idea of adversary examples, we proposed a new targeted WF defense that takes advantage of a deep learning method to find the key feature of traces and apply the features to confuse WF attacks. In short, we first clustered the different websites by the average scale of traces, which aims to carry on defense on similar websites on the scale, improving the defense effectiveness and decreasing the overhead of inserting dummy packets. Then, we preprocessed the original traces to express key features and abstracted the features of traces by adapting Grad-CAM on the variant of FCN. We also discussed the practical deployment, including the negotiation of the Tor proxy node and Tor exit node and the data structure of processing requests. Then, we determined the optimal choice of parameters through hyperparameter tuning. We demonstrated the feasibility of the TED method through a set of experiments on effectiveness and overhead. The experimental results showed that TED had a smaller overhead, which could prove that TED with the help of deep learning methods is an accurate means of obfuscation.

Acknowledgement. This work is partially supported by the National Key R&D Program of China under Grant 2020YFB1006101, the Beijing Nova Program under Grant Z201100006820006, and the NSFC Project under Grant 61972039.

References

1. Tor metrics website, February 2021. https://metrics.torproject.org/
2. Cai, X., Nithyanand, R., Wang, T., Johnson, R., Goldberg, I.: A systematic approach to developing and evaluating website fingerprinting defenses. In: Ahn, G., Yung, M., Li, N. (eds.) Proceedings of the 2014 ACM SIGSAC Conference on Computer and Communications Security, Scottsdale, AZ, USA, 3–7 November 2014, pp. 227–238. ACM (2014). https://doi.org/10.1145/2660267.2660362
3. Cherubin, G.: Bayes, not naïve: security bounds on website fingerprinting defenses. Proc. Priv. Enhancing Technol. **2017**(4), 215–231 (2017). https://doi.org/10.1515/popets-2017-0046
4. Dingledine, R., Mathewson, N., Syverson, P.F.: Tor: the second-generation onion router. In: Blaze, M. (ed.) Proceedings of the 13th USENIX Security Symposium, San Diego, CA, USA, 9–13 August 2004, pp. 303–320. USENIX (2004)
5. Gong, J., Wang, T.: Zero-delay lightweight defenses against website fingerprinting. In: Capkun, S., Roesner, F. (eds.) 29th USENIX Security Symposium, USENIX Security 2020, 12–14 August 2020, pp. 717–734. USENIX Association (2020)

6. Hayes, J., Danezis, G.: k-fingerprinting: a robust scalable website fingerprinting technique. In: Holz, T., Savage, S. (eds.) 25th USENIX Security Symposium, USENIX Security 2016, Austin, TX, USA, 10–12 August 2016, pp. 1187–1203. USENIX Association (2016)
7. Herrmann, D., Wendolsky, R., Federrath, H.: Website fingerprinting: attacking popular privacy enhancing technologies with the multinomial Naïve-Bayes classifier. In: Sion, R., Song, D. (eds.) Proceedings of the first ACM Cloud Computing Security Workshop, CCSW 2009, Chicago, IL, USA, 13 November 2009, pp. 31–42. ACM (2009). https://doi.org/10.1145/1655008.1655013
8. Juarez, M., Imani, M., Perry, M., Diaz, C., Wright, M.: Toward an efficient website fingerprinting defense. In: Askoxylakis, I., Ioannidis, S., Katsikas, S., Meadows, C. (eds.) ESORICS 2016. LNCS, vol. 9878, pp. 27–46. Springer, Cham (2016). https://doi.org/10.1007/978-3-319-45744-4_2
9. Li, S., Guo, H., Hopper, N.: Measuring information leakage in website fingerprinting attacks and defenses. In: Lie, D., Mannan, M., Backes, M., Wang, X. (eds.) Proceedings of the 2018 ACM SIGSAC Conference on Computer and Communications Security, CCS 2018, Toronto, ON, Canada, 15–19 October 2018, pp. 1977–1992. ACM (2018). https://doi.org/10.1145/3243734.3243832
10. Lu, L., Chang, E.-C., Chan, M.C.: Website fingerprinting and identification using ordered feature sequences. In: Gritzalis, D., Preneel, B., Theoharidou, M. (eds.) ESORICS 2010. LNCS, vol. 6345, pp. 199–214. Springer, Heidelberg (2010). https://doi.org/10.1007/978-3-642-15497-3_13
11. Nasr, M., Bahramali, A., Houmansadr, A.: Defeating DNN-based traffic analysis systems in real-time with blind adversarial perturbations. In: Bailey, M., Greenstadt, R. (eds.) 30th USENIX Security Symposium, USENIX Security 2021, 11–13 August 2021, pp. 2705–2722. USENIX Association (2021)
12. Panchenko, A., et al.: Website fingerprinting at internet scale. In: 23rd Annual Network and Distributed System Security Symposium, NDSS 2016, San Diego, California, USA, 21–24 February 2016. The Internet Society (2016)
13. Shen, M., Gao, Z., Zhu, L., Xu, K.: Efficient fine-grained website fingerprinting via encrypted traffic analysis with deep learning. In: 29th IEEE/ACM International Symposium on Quality of Service, IWQOS 2021, Tokyo, Japan, 25–28 June 2021, pp. 1–10. IEEE (2021). https://doi.org/10.1109/IWQOS52092.2021.9521272
14. Shen, M., Liu, Y., Zhu, L., Du, X., Hu, J.: Fine-grained webpage fingerprinting using only packet length information of encrypted traffic. IEEE Trans. Inf. Forensics Secur. 16, 2046–2059 (2021). https://doi.org/10.1109/TIFS.2020.3046876
15. Shen, M., Zhang, J., Zhu, L., Xu, K., Du, X.: Accurate decentralized application identification via encrypted traffic analysis using graph neural networks. IEEE Trans. Inf. Forensics Secur. 16, 2367–2380 (2021). https://doi.org/10.1109/TIFS.2021.3050608
16. Shmatikov, V., Wang, M.-H.: Timing analysis in low-latency mix networks: attacks and defenses. In: Gollmann, D., Meier, J., Sabelfeld, A. (eds.) ESORICS 2006. LNCS, vol. 4189, pp. 18–33. Springer, Heidelberg (2006). https://doi.org/10.1007/11863908_2
17. Sirinam, P., Imani, M., Juárez, M., Wright, M.: Deep fingerprinting: undermining website fingerprinting defenses with deep learning. In: Lie, D., Mannan, M., Backes, M., Wang, X. (eds.) Proceedings of the 2018 ACM SIGSAC Conference on Computer and Communications Security, CCS 2018, Toronto, ON, Canada, 15–19 October 2018, pp. 1928–1943. ACM (2018). https://doi.org/10.1145/3243734.3243768

18. Wang, T., Goldberg, I.: Walkie-talkie: an efficient defense against passive website fingerprinting attacks. In: Kirda, E., Ristenpart, T. (eds.) 26th USENIX Security Symposium, USENIX Security 2017, Vancouver, BC, Canada, 16–18 August 2017, pp. 1375–1390. USENIX Association (2017)
19. Xu, Y., Wang, T., Li, Q., Gong, Q., Chen, Y., Jiang, Y.: A multi-tab website fingerprinting attack. In: Proceedings of the 34th Annual Computer Security Applications Conference, ACSAC 2018, San Juan, PR, USA, 03–07 December 2018, pp. 327–341. ACM (2018). https://doi.org/10.1145/3274694.3274697

Data Hiding in the Division Domain: Simultaneously Achieving Robustness to Scaling and Additive Attacks

Shu Yu, Junren Qin, Jiarui Deng, Shanxiang Lyu[✉], and Fagang Li[✉]

College of Cyber Security, Jinan University, Guangzhou 510632, China
lsx07@jnu.edu.cn, lifagang13@mails.ucas.ac.cn

Abstract. Data hiding plays an important role in privacy protection and authentication, but most data hiding methods fail to achieve satisfactory performance in resisting scaling attacks and additive attacks. To this end, this paper proposes a new quantization index modulation (QIM) variant based on division domains (D-QIM). It can not only resist the above two attacks well, but also adjust the performance trade-offs by controlling the parameters. Simulation results confirm the performance gain of D-QIM in terms of the bit error rate (BER).

Keywords: Data hiding · Robust watermarking · Quantization index modulation (QIM)

1   Introduction

Currently, the rapid development of networks has led to increasing requirements for their security, which has aroused high interest in secure communication. To guarantee secret communication between the sender and receiver, a solution for multimedia is data hiding, in which the secret messages are embedded into a multimedia signal with unnoticeable methods. The critical design purposes of data hiding schemes the imperceptibility and robustness, which are developed into two different fields: steganography and watermarking [5,6]. In steganography, the core is to make the embedded message imperceptible to avoid detection, and there are many attempts, such as [10,11,16]. Digital watermarking focuses on robustness against various types of interference, such as amplitude-scaling attacks, pepper attacks and additive white Gaussian noise (AWGN) attacks. Based on the recent watermarking schemes with different methods [7,13,17], improvements are mainly aimed at resistance for AWGN attacks and specific attacks on which the watermarking scheme has a weak performance.

Recently, the quantization index modulation (QIM) algorithm [2] has attracted the attention of researchers. In QIM-based schemes, a cover sample is addressed at the nearest location in a code book indexed by a message with a quantizer [12]. Compared with other data hiding algorithms, QIM has many advantages including higher capacity, better robustness against rotation

Y. Wang et al. (Eds.): ICPCSEE 2022, CCIS 1629, pp. 47–58, 2022.
https://doi.org/10.1007/978-981-19-5209-8_4

attacks, and better trade-off between capacity and distortion [7,9]. However, a serious drawback of basic QIM watermarking is its sensitivity to amplitude-scaling attacks, in which signal samples are multiplied by a gain factor ρ. If the factor ρ is constant for all samples, the attack is called a fixed-gain attack (FGA). In amplitude-scaling attacks, the detector does not possess the factor ρ, which causes a mismatch between the encoder and decoder of lattice quantization and dramatically affects the QIM-detector performance. To solve this issue, researchers have proposed many attempts. For example, Fabrício et al. [14] proposed Angle-QIM (A-QIM), where the QIM is not implemented in real numbers but in the angular domain generated by a hyperspherical coordinate system. Comesana et al. [4] processed QIM into a logarithmic domain to resist scaling attacks. Hwang et al. [8] proposed a blind digital audio watermarking algorithm that utilizes the QIM and the singular value decomposition (SVD) of stereo audio signals, which is robust against volumetric scaling attacks. Terchi et al. [15] proposed an efficient transform-based blind audio watermarking technique by introducing a parametric QIM.

Nevertheless, these attempts have reduced the sensitivity of QIM to amplitude-scaling attacks, and some new problems emerge from the above two algorithms. For the A-QIM algorithm in [14], since additive white Gaussian noise (AWGN) causes an irregular movement of watermarked data, hyperspherical coordinate transformation would make it more sensitive to this type of attack. Meanwhile, the operation to shift into the angular domain is too complex, which causes a higher computational burden in practice. For the method of the logarithmic domain in [4], huge damage occurs by shifting from the real number to the logarithm, which causes weak robustness performance.

To solve these problems while resisting amplitude-scaling attacks, in this paper, we propose a method based on a novel domain called the division domain. First, a review of QIM and some related work on resist scaling attacks is elaborated in Sect. 2. Then, to resist scaling attacks, we propose a new domain called the division domain for data hiding and a new QIM-based scheme called D-QIM in this novel domain in Sect. 3, which outperforms the resistance of AWGN and amplitude-scaling attacks. Some simulations to verify its effectiveness are introduced in Sect. 4.

The notations used in this paper are listed as follows:

- **1**: a vector with all elements equal to 1.
- $\oplus$: the addition in division domain.
- $\ominus$: subtraction in the division domain.
- $\otimes$: the multiplication in division domain.
- $\oslash$: the division in division domain.
- $\lfloor \cdot \rfloor$ denotes the floor function.
- $| \cdot |$ denotes getting the absolute value.
- Matrices and column vectors are denoted by uppercase and lowercase boldface letters.

2   Preliminaries

2.1   QIM

The process of QIM in [2] can be described as follows. Assume that the messages $m \in \{0,1\}$ are embedded into the host signal sample s by a quantizer with a step size Δ. Then the watermarked signal sample is given by

$$x = Q(s; m, \Delta) \tag{1}$$

where $Q(\cdot)$ represents a quantization function. As depicted in Fig. 1, the host signal sample s is moved to the nearest cross when $m = 1$ and to the circle when $m = 0$.

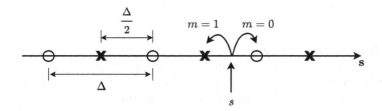

Fig. 1. Embed one bit into a sample with the original QIM.

For the extracting process, since the received signal y is transmitted through the AWGN channel, the noise n would be contained, i.e., $y = x + n$. Therefore, a minimum distance decoder to acquire the estimated value of m, $\hat{m}$, can be denoted as

$$\hat{m} = \arg\min_{m \in \{0,1\}} |y - Q(y; m, \Delta)|. \tag{2}$$

If $|n| < \Delta/4$, $\hat{m}$ is correct.

Figure 2 shows that the maximum error caused by embedding is $\Delta/2$. Furthermore, if the quantization errors are distributed over $[-\Delta/2, \Delta/2]$, the mean-squared distortion owing to embedding is $D = \Delta^2/12$. In addition, QIM can also be extended to N-dimensional integer sets by independently performing embedding and extraction in each dimension.

2.2   Related Work on Resisting Scaling Attacks

Due to the sensitivity of the QIM scheme, many attempts have been proposed to resist scaling attacks. Although the solutions of these schemes are diverse, all of them are designed with the same idea that the form of original host signals are transformed before QIM embedding to eliminate or mitigate the effects of scaling attacks by the mathematical properties of this domain. Since the transformation of the host signal does not alter the QIM algorithm, in general they have the same embedding and extracting process as QIM. In these attempts, the most representative are A-QIM [14] and L-QIM [4].

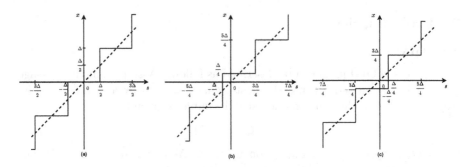

Fig. 2. A watermarked sample x given s and $m \in \{0, 1\}$, using the original QIM method. (a) Prototype symmetric function. (b) Embedding function for $m = 0$. (c) Embedding function for $m = 1$.

A-QIM. In A-QIM, the message would be embedded into the angle formed by the host-signal vector with the origin of the hyperspherical coordinate system [3] since the operation of angular domain shifting has natural resistance to scaling attacks. For the angular domain shifting, first, we consider a point $\mathbf{x}$ in the two dimensional Euclidean space. These two samples x_1 and x_2 can be viewed as a point on a 2-dimensional plane, whose Cartesian coordinates (x_1, x_2) can be converted to polar coordinate (r, θ). The angle θ and radius r are, respectively given as

$$\begin{cases} \theta = \tan^{-1}(\dfrac{x_2}{x_1}) \\ r = \sqrt{x_1^2 + x_2^2} \end{cases}. \tag{3}$$

After employing Eq. (1) on the angle θ and obtaining the watermarked θ_w, the watermarked data $\mathbf{x_w}$ are placed as

$$\mathbf{x}_w = \begin{bmatrix} r\cos\theta_w \\ r\sin\theta_w \end{bmatrix}. \tag{4}$$

Assume that the noisy observation of $\mathbf{x}_w$ is $\mathbf{y}$. The operation that $\hat{\theta} = \tan^{-1}(\frac{y_2}{y_1})$ offsets the affects caused by scaling attacks when $\mathbf{y} = \rho\mathbf{x}_w$. When $\mathbf{y} = \mathbf{x}_w + \mathbf{n}$, the difference $\mathcal{D}$ between the received signal and the watermarked signal in the angular domain can be calculated as

$$\mathcal{D} = \theta_{\text{Noisy}} - \theta_w = \tan^{-1}\left(\frac{rn_1\cos\theta_w - rn_2\sin\theta_w}{r^2(rn_2\cos\theta_w + rn_1\sin\theta_w)} \right). \tag{5}$$

L-QIM. Similarly, logarithmic quantization acquires the signal for data hiding by converting the host signal to a logarithmic domain with logarithmic function such as $f(x) = ln(|x|)$ before quantization. In [4], Comesana et al. addressed the logarithmic version of Dither Modulation (DM), whose embedding function is given by

$$\log\left(|y_i|\right) = Q_\Delta(\log\left(|x_i|\right) - \frac{b_i\Delta}{2} - d_i) + \frac{b_i\Delta}{2} + d_i; \qquad (6)$$

Then, the scaling resistance scheme becomes a differential watermarking method in the logarithmic domain, where the embedding procedure can be described as

$$\log\left(|y_i|\right) = Q_\Delta(\log\left(|x_i|\right) - \log\left(|y_{i-1}|\right) - \frac{b_i\Delta}{2} - d_i) + \log\left(|y_{i-1}|\right) + \frac{b_i\Delta}{2} + d_i; \qquad (7)$$

In both cases, $y_i = sign(x_i) \cdot e^{log(|y_i|)}$.

At the receiver's side, we have the received signal denoted as z_i. Similar to A-QIM, the operation that $\log\left(|z_i|\right) - \log\left(|z_{i-1}|\right)$ offsets the noise's affects when $z_i = \rho y_i$. When $z_i = y_i + n_i$, the difference $\mathcal{D}$ between the received signal and watermarked signal in the logarithmic domain can be calculated as

$$\mathcal{D} = \log\left(|z_i|\right) - \log\left(|z_{i-1}|\right) - \log\left(|y_i|\right) + \log\left(|y_{i-1}|\right) = \log\left(|\frac{y_{i-1}(y_i + n_i)}{y_i(y_{i-1} + n_{n-1})}|\right). \qquad (8)$$

3   Proposed Method

3.1   Division Domain for Data Hiding

Considering the structures of A-QIM and L-QIM, their resistance to scaling attacks is based on the same reason that the operation of domain conversion offsets the effects caused by the gain factor ρ in scaling attacks. However, regardless of the conversions to a hyperspherical coordinate system or logarithmic domain, the operation is too complex to implement in practice, which can be simplified by the division operation. Therefore, we define a novel domain called the division domain.

Let $\mathbf{x}$ and $\mathbf{y}$ be the inputs to the division-domain transform. There are two functions to derive the vector in the division domain. One is defined as

$$\mathbf{x} \otimes \mathbf{y} \triangleq \mathrm{diag}(\mathbf{x}) \cdot \mathrm{diag}(\mathbf{y}) \cdot \mathbf{1}, \qquad (9)$$

another is

$$\mathbf{x} \oslash \mathbf{y} \triangleq \mathrm{diag}(\mathbf{x}) \cdot \mathrm{diag}(\mathbf{y})^{-1} \cdot \mathbf{1}, \qquad (10)$$

where $\mathrm{diag}(\mathbf{x})$ denotes the diagonal matrix with the same elements of $\mathbf{x}$.

Then, two operations to realize the conversion from the space domain to the division domain, called division domain transformation and inverse division domain transformation, can be defined in the next case.

To shift the host vector into the division domain, a pair of host vectors $(\mathbf{x}, \mathbf{y})$ would be inputted to generate a pair of division vectors $(\mathbf{v_x}, \mathbf{v_y})$, which is represented as

$$\begin{cases} \mathbf{v_x} = \mathbf{x} \oslash \mathbf{y} \\ \mathbf{v_y} = \mathbf{y} \end{cases}. \qquad (11)$$

By this operation, the pair of host vectors $(\mathbf{x}, \mathbf{y})$ would be mapped to the division domain, and we can obtain the generated vector pair $(\mathbf{v_x}, \mathbf{v_y})$. Obviously, the scaling attacks would be eliminated since the implementation of the division operation. Therefore, it is possible to resist scaling attacks for watermarking schemes with division domains. Meanwhile, the effect of the noise would be greatly reduced by the division operation, which improves the robustness of watermarking schemes against AWGN attacks.

To restore the division vector pair $(\mathbf{v_x}, \mathbf{v_y})$ to the host vector pair $(\mathbf{x}, \mathbf{y})$, the inverse domain transformation is implemented and can be described as

$$\begin{cases} \mathbf{x} = \mathbf{v_x} \otimes \mathbf{v_y} \\ \mathbf{y} = \mathbf{v_y} \end{cases}. \tag{12}$$

Based on the definition of the division domain, to design a watermarking scheme with better performance on the robustness of additive and scaling attacks, a novel algorithm called D-QIM is proposed, which has inherent resistance to scaling attacks and strong resistance to additive attacks. Here, we introduce the embedding and extracting process of D-QIM.

3.2   D-QIM

For the embedding process, with inputting a pair of host vectors $(\mathbf{s}_a, \mathbf{s}_b)$ and message vectors $\mathbf{m}_i$, the embedding process of D-QIM is listed as follows:

1 Transform the host vector $(\mathbf{s}_a, \mathbf{s}_b)$ to $(\mathbf{s}_a, \alpha \mathbf{s}_b)$, where the parameter α is used to reduce the effect caused by data overflow.
2 Employ Eq. (11) to obtain the division vector pair $(\mathbf{d}_a, \mathbf{d}_b)$.
3 Regard $\mathbf{d}_a$ as the cover and implement QIM with the same embedding process, which gains a watermarked vector $\mathbf{d}_{w,a}$ in the division domain.
4 Employ Eq. (12) to restore the division vector pair $(\mathbf{d}_{w,a}, \mathbf{d}_b)$ into the watermarked vector pair.

For the extracting process, since the watermarked vector pair is received with the transmission through scaling and additive attack channel, we assume that the received vector pair $(\mathbf{y}_{w,a}, \mathbf{y}_b) = (\beta \mathbf{s}_{w,a} + n, \beta \mathbf{s}_b + n)$. The extraction process of D-QIM is listed as follows:

1 Transform the received vector $(\mathbf{y}_{w,a}, \mathbf{y}_b)$ to $(\mathbf{y}_{w,a}, \alpha \mathbf{y}_b)$
2 Employ Eq. (11) to obtain the division vector pair $(\mathbf{d}_{\mathbf{y},w,a}, \mathbf{d}_{\mathbf{y},b})$.
3 Implement the decoder in Eq. (2) to gain the message $\mathbf{m}'_i$. If the noise is small enough such that

$$\frac{\beta \mathbf{s}_{w,a} + n}{\beta \mathbf{s}_b + n} \rightarrow \frac{\beta \mathbf{s}_{w,a}}{\beta \mathbf{s}_b} = \frac{\mathbf{s}_{w,a}}{\mathbf{s}_b}, \tag{13}$$

the extracted message $\mathbf{m}'_i$ is correct.

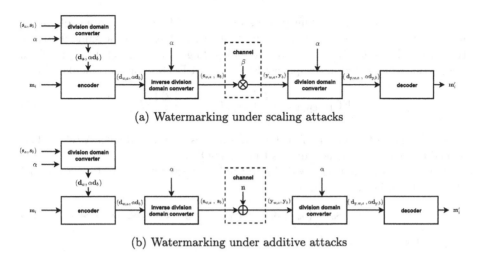

(a) Watermarking under scaling attacks

(b) Watermarking under additive attacks

Fig. 3. The proposed watermarking scheme under different attacks.

After introducing the whole scheme of D-QIM, Fig. 3 depicts the whole model of D-QIM with the transmission of scaling and additive attack channel, in which a message vector $\mathbf{m}_i$ is embedded into a pair of host signals $(\mathbf{s}_a, \mathbf{s}_b)$, transmitted through a scaling, additive attack channel and extracted for $\mathbf{m}'_i$. Notably, the division domain converter employs steps 1 and 2 in extraction process or embedding process of D-QIM. The inverse division domain converter employs step 4 in the embedding process of D-QIM. Furthermore, the encoder executes step 3 in the embedding process of D-QIM, while the decoder executes step 3 in extraction process of D-QIM. Note that the domain transformation of D-QIM is much simpler because it only needs to perform multiplication and division operations. However, A-QIM needs to perform rooting operations and trigonometric and inverse trigonometric function operations, and L-QIM needs to perform logarithmic operations and exponentiation operations. In addition, D-QIM, like the other two schemes, naturally has the ability to resist scaling attacks in structure, but it has better resistance to additive attacks.

3.3 Theoretical Analysis for D-QIM

To analyze D-QIM theoretically, we assume that the cover $\mathbf{s}$ follows an 0-mean Gaussian distribution, i.e., $\mathbf{s} \sim (\mathbf{0}, \sigma_{\mathbf{s}}^2)$, the probability density function of $\mathbf{s}$ is denoted as

$$f(\mathbf{s}) = \frac{1}{\sqrt{2\pi\sigma_{\mathbf{s}}^2}} \exp\left\{-\frac{\|\mathbf{s}\|^2}{2\sigma_{\mathbf{s}}^2}\right\}. \tag{14}$$

By considering the embedding process of D-QIM where two samples of watermarked signal contain one sample of message, $\mathbf{s}_a$ and $\mathbf{s}_b$ have the same distribution of $\mathbf{s}$ independently. Therefore the embedding ratio of D-QIM is 2:1, which means that the embedding capacity $\mathrm{EC_{D-QIM}}$ of D-QIM can be calculated as

$$\mathrm{EC_{D-QIM}} = \left\lfloor \frac{\mathcal{L}}{2} \right\rfloor \log_2 \left(2^N \right) = \left\lfloor \frac{\mathcal{L}}{2} \right\rfloor N, \tag{15}$$

where $\mathcal{L}$ denotes the length of host signal samples. Compared with A-QIM, their embedding capacity is the same as that of D-QIM.

Since the message is embedded into the host vector $\mathbf{s}_a$, the watermark signal of D-QIM can be calculated as

$$\mathbf{w} \triangleq \mathbf{s}_{w,a} - \mathbf{s}_a = (\mathbf{s}_a \oslash \mathbf{s}_b + \mathbf{w}_{QIM}) \otimes \mathbf{s}_b - \mathbf{s}_a = \mathbf{w}_{QIM} \otimes \mathbf{s}_b. \tag{16}$$

Its power is

$$
\begin{aligned}
\sigma_{\mathbf{w}}^2 &\triangleq E \left[\|\mathbf{w}\|^2 \right] = E \left[\|\mathbf{w}_{QIM} \otimes \mathbf{s}_b\|^2 \right] \\
&= \sum_{i=1}^{N} E(e_i^2 s_i^2) \\
&= \sum_{i=1}^{N} \left[\frac{1}{2} \sum_{m_i=0}^{1} \int_{-\infty}^{+\infty} (1 - m_i \Delta \alpha)^2 s_i^2 f_S(s_i) \, ds_i \right] \\
&= N \cdot \left[1 - 2\Delta \alpha + \Delta^2 \alpha^2 \right] \int_{-\infty}^{+\infty} s_i^2 f_S(s_i) \, ds_i \\
&= \left[1 - 2\Delta \alpha + \Delta^2 \alpha^2 \right] \sigma_{\mathbf{s}}^2
\end{aligned}
\tag{17}
$$

where $\sigma_{\mathbf{s}}^2$ denotes the power of $\mathbf{s}$, and subscript i represents the index of the dimensional component.

4   Simulations

To verify the effectiveness of D-QIM, we perform some numerical simulations in this section. First, we carry out a robustness comparison for scaling attacks and AWGN attacks by employing A-QIM and L-QIM as benchmarks. A uniformly distributed sequence of $m_i \in \{0, 1\}$ would be set as the embedded messages. The images downloaded from the Signal and Image Processing Institute (SIPI) database [1], depicted in Fig. 4, are chosen as the cover of all the experiments with the consideration of their different features on the arrangement of pixels. For all these experiments, we employ $\Delta = 2$ as the step size.

To test the robustness performance in scaling and additive attacks, the bit error rate (BER) metric is used in the experiment, which is defined as follows:

$$BER = \frac{\sum_{i=1}^{M} \sum_{j=1}^{N} (x_{ij} \text{ xor } x_{ij}')}{M * N}, \tag{18}$$

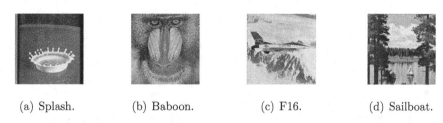

(a) Splash.   (b) Baboon.   (c) F16.   (d) Sailboat.

Fig. 4. Images for experiments

where M and N are the sizes of the host-signal, and x_{ij} and x'_{ij} represent the original and extracted message samples, respectively.

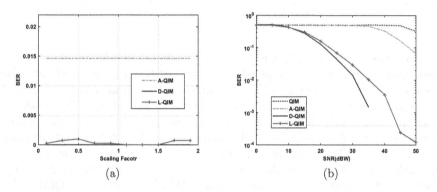

(a)   (b)

Fig. 5. BER comparison of different methods under different attacks. (a) Scaling attacks. (b) AWGN attacks.

The performance of the robustness against the amplitude scaling attacks is illustrated in Fig. 5(a) by plotting the average value of BER with A-QIM, L-QIM and D-QIM in the scaling attacks of different intensities. Similar to or even better than A-QIM and L-QIM, it shows that D-QIM has the ability to resist amplitude scaling attacks since the vector $\mathbf{v_x}$ in Eq. (11) did not change after all the amplitudes were scaled by the same scaling factor. For robustness against AWGN attacks, Fig. 5(b) plots the average BER of 4 images with QIM, A-QIM, L-QIM and D-QIM in AWGN attacks of different intensities. Note that the parameters α of D-QIM are both $\alpha = 0.75$ in Fig. 5. By observing two comparisons in this figure, a conclusion that D-QIM has a better robustness against AWGN attacks while resisting scaling attacks can be deduced.

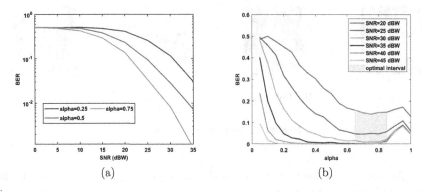

Fig. 6. (a) The comparison between the BERs of D-QIM using different α, which are 0.25, 0.5 and 0.75. (b) The BER of the D-QIM comparison using different α values under several AWGN attack conditions.

Meanwhile, to explore how the parameter α affects the robustness of D-QIM against AWGN attacks, we compared the average BER of 4 images with different α. The conclusion is depicted in Fig. 6(a), in which the parameters 0.25, 0.5 and 0.75 are implemented. This shows that robustness against AWGN attacks is positively correlated with the parameter α. Furthermore, to explore the positive correlation between the BER of D-QIM and α, we plot Fig. 6(b) with SNR from 20 dBW to 45 dBW, which reveals that as the noise gradually increases the applicable range of α gradually moves closer to $[0.65, 0.85]$. We could derive a better performance on robustness against AWGN attacks of D-QIM when $\alpha \in [0.65, 0.85]$.

To evaluate the performance of imperceptibility, all the images in Fig. 4 have been implemented with D-QIM for simulation, and the comparisons between the original and embedded images are shown in Fig. 7. The observation can be derived that D-QIM has a good performance on imperceptibility, which is almost the same as observed in Fig. 7(a) and (b).

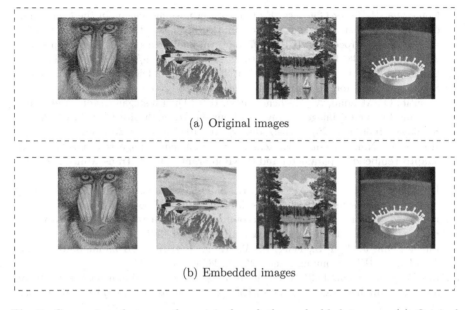

(a) Original images

(b) Embedded images

Fig. 7. Comparison between the original and the embedded images. (a) Original images. (b) Watermarked images embedded with 2048 bits of data with SWRs of 37.2079, 38.8954, 37.2444 and 39.1359.

5 Conclusions

In this paper, a new domain for data hiding is proposed. A watermark scheme D-QIM is derived, which resists scaling attacks and additive attacks. The robustness and distortion performance of D-QIM are theoretically analyzed and well verified in experiments. Experiments show that the performance of D-QIM is better than that of QIM, A-QIM and L-QIM in resisting scaling attacks and additive attacks.

Acknowledgment. This work was supported in part by the National Natural Science Foundation of China (61902149, 61932010 and 62032009) and the Natural Science Foundation of Guangdong Province (2020A1515010393).

References

1. Image processing institute (SIPI) database. https://sipi.usc.edu/database/. Accessed 30 Mar 2022
2. Chen, B., Wornell, G.W.: Quantization index modulation: a class of provably good methods for digital watermarking and information embedding. IEEE Trans. Inf. Theory **47**(4), 1423–1443 (2001)
3. Chen, B., Wornell, G.W.: Quantization index modulation methods for digital watermarking and information embedding of multimedia. J. VLSI Signal Process. **27**(1–2), 7–33 (2001)

4. Comesaña, P., Pérez-González, F.: On a watermarking scheme in the logarithmic domain and its perceptual advantages. In: Proceedings of the International Conference on Image Processing, ICIP 2007, San Antonio, Texas, USA, 16–19 September 2007, pp. 145–148. IEEE (2007). https://doi.org/10.1109/ICIP.2007.4379113
5. Cox, I., Miller, M., Bloom, J., Fridrich, J., Kalker, T.: Digital Watermarking and Steganography. Morgan Kaufmann, San Francisco (2007)
6. Evsutin, O., Melman, A., Meshcheryakov, R.V.: Digital steganography and watermarking for digital images: a review of current research directions. IEEE Access **8**, 166589–166611 (2020). https://doi.org/10.1109/ACCESS.2020.3022779
7. Huang, Y., Niu, B., Guan, H., Zhang, S.: Enhancing image watermarking with adaptive embedding parameter and PSNR guarantee. IEEE Trans. Multim. **21**(10), 2447–2460 (2019)
8. Hwang, M., Lee, J., Lee, M., Kang, H.: SVD-based adaptive QIM watermarking on stereo audio signals. IEEE Trans. Multim. **20**(1), 45–54 (2018)
9. Kricha, Z., Kricha, A., Sakly, A.: Accommodative extractor for QIM-based watermarking schemes. IET Image Process. **13**(1), 89–97 (2019)
10. Lin, J., Qin, J., Lyu, S., Feng, B., Wang, J.: Lattice-based minimum-distortion data hiding. IEEE Commun. Lett. **25**(9), 2839–2843 (2021)
11. Liu, W., et al.: Secure halftone image steganography with minimizing the distortion on pair swapping. Signal Process. **167** (2020). https://doi.org/10.1016/j.sigpro.2019.107287
12. Moulin, P., Koetter, R.: Data-hiding codes. Proc. IEEE **93**(12), 2083–2126 (2005)
13. Naderahmadian, Y., Beheshti, S.: Robustness of wavelet domain watermarking against scaling attack. In: 2015 IEEE 28th Canadian Conference on Electrical and Computer Engineering (CCECE), pp. 1218–1222 (2015). https://doi.org/10.1109/CCECE.2015.7129451
14. Ourique, F., Licks, V., Jordan, R., Pérez-González, F.: Angle QIM: a novel watermark embedding scheme robust against amplitude scaling distortions. In: 2005 IEEE International Conference on Acoustics, Speech, and Signal Processing, ICASSP 2005, Philadelphia, Pennsylvania, USA, 18–23 March 2005, pp. 797–800. IEEE (2005). https://doi.org/10.1109/ICASSP.2005.1415525
15. Terchi, Y., Bouguezel, S.: A blind audio watermarking technique based on a parametric quantization index modulation. Multimedia Tools Appl. **77**(19), 25681–25708 (2018). https://doi.org/10.1007/s11042-018-5813-z
16. Verma, V., Muttoo, S.K., Singh, V.B.: Enhanced payload and trade-off for image steganography via a novel pixel digits alteration. Multim. Tools Appl. **79**(11-12), 7471–7490 (2020). https://doi.org/10.1007/s11042-019-08283-9
17. Wang, X., Wang, P., Zhang, P., Xu, S., Yang, H.: A blind audio watermarking algorithm by logarithmic quantization index modulation. Multimedia Tools Appl. **71**(3), 1157–1177 (2012). https://doi.org/10.1007/s11042-012-1259-x

BMSC: A Novel Anonymous Trading Scheme Based on Zero-Knowledge Proof in Ethereum

Yang Li[1,2(✉)], Yinyun Zhang[1], Mengmeng Wang[1], Jianming Zhu[1,2], and Xiuli Wang[1,2]

[1] School of Information, Central University of Finance and Economics, Beijing 100081, China
liyang@cufe.edu.cn

[2] Engineering Research Center of State Financial Security, Ministry of Education, Central University of Finance and Economics, Beijing 102206, China

Abstract. Blockchains are widely used because of their openness, transparency, nontampering and decentralization. However, there is a high risk of information leakage when trading on blockchain, and the existing anonymous trading schemes still have some problems. To meet the high requirement of anonymity, the cost of proof submitted by the user is too large, which does not apply to blockchain storage. Meanwhile, transaction verification takes too long to ensure the legitimacy of the transaction. To solve these problems, this paper presents a novel anonymous trading scheme named Block Maze Smart Contract (BMSC) based on the zero-knowledge proof system zk-SNARKs to propose efficiency. This scheme can hide account balances, transaction amounts, and the transfer relationships between transaction parties while preventing overspending attacks and double-spending attacks. Compared with other anonymous schemes, this scheme has less cost of proof and takes less time for transaction verification while meeting the high requirements of anonymity and security.

Keywords: Zero-knowledge proof · Ethereum · Account-model · Anonymous trading

1 Introduction

Block chain technology, which integrates distributed data storage, P2P networks, consensus mechanisms and encryption algorithms, has a series of features, such as decentralization, nontampering and open transparency, and is widely used in many fields, such as the Internet of Things, finance, intellectual property protection and secret bidding. However, the formation of their decentralized consensus relies on public transaction records submitted by users, and the simple anonymous way of replacing accounts with addresses is unreliable, so users run the risk of disclosing information. Attackers can analyze transaction relationships between accounts by analyzing a large amount of transaction data. Once a user's identity is exposed, information about that user recorded on all transactions related to it will be disclosed. To protect the privacy of the blockchain, anonymous transaction schemes have become a research hotspot.

To date, many anonymous trading schemes based on the UTXO model have been proposed. to protect privacy on blockchains. However, the balance calculation of the traditional UTXO-model system is more complex, and the lack of Turing completeness is not as convenient as account-model management. Obviously, in today's increasingly complex trading model, Ethereum, which is based on an account model and can deploy smart contracts, is a better choice for users. However, the scheme design is difficult because the limitations of the account model are not easy to hide addresses naturally. Therefore, there are few transaction schemes based on account models that can meet the high requirements of anonymity and security while being practical and efficient. Therefore, it is necessary to design an anonymous transaction scheme that can be implemented in Ethereum with less cost of proof and less time for transaction verification.

2   Related Works

Some anonymous trading schemes based on the UTXO model have been proposed, such as Dash [1], Zcoin [2], Monero [3] and Zcash [4]. Dash uses the mixed currency mechanism to hide the transfer relationship, which is lightweight. Zcoin uses zk-SNARKs to cut off the connection between the generated currency and the redeemed currency to hide the relationship between the two sides of the transaction. Monero uses Bulletproof to hide the transaction amounts and uses ring signature technology to hide account addresses. Zcash uses zk-SNARKs to hide the account address information and transaction amounts. However, these schemes are based on the UTXO model, which has a more complex balance calculation and is not user-friendly. Therefore, the above scheme cannot meet the needs of today's increasingly complex transaction mode.

Zether [5] proposed by Biinz et al. uses homologous encryption to hide account balances and transaction amounts. It uses anonymous sets to hide the accounts of both parties. The scheme proposes a Σ-Bullets zero-knowledge proof system to verify the legality of transactions. The scheme divides time into epochs to prevent conflicts between transactions. Balance changes of accounts within each epoch are stored in the pending transfer table and will be transferred in the next epoch. To prevent double-cost attacks, each transaction has a sequence number based on the current epoch and the account private key. This means that even honest users within the same epoch can only complete one transaction at most, and each transaction must be completed within the specified time. In addition, the validation time of the zero-knowledge proof of the scheme is linearly related to the size of the anonymous set, which takes a long time to ensure sufficient anonymity.

ZETH [6] proposed by Rondelet et al. implements Zerocash on Ethereum. This scheme uses a smart contract to design and implement a transaction mechanism similar to the UTXO model in Ethereum. The core of the solution is to deploy a Mixer Intelligence Contract, which constructs and maintains the Merkel tree internally. It hides account addresses with one-time addresses and validates transactions with zk-SNARKs. The scheme uses secure multiparty computing to enhance the security of zk-SNARKs trusted settings. However, the disadvantage is that the construction is very complex and inefficient.

The principle of an efficient NIZK scheme DSC [7] is to use homomorphic encryption to hide account balance and transaction amount. Then, it uses an effective noninteractive zero-knowledge proof system to prove the validity of the transaction. The disadvantage of the scheme is that the cost of proof is high, and the anonymity does not hide the transfer relationship between the two parties.

BlockMaze [8] scheme proposed by Shuangzhang Guan et al. uses zk-SNARKs to verify the effectiveness of the transaction and uses the double balance model and two capital transfer operations to hide the account balance, transaction amount and the transfer relationship between the two parties. However, the scheme redefines the underlying data structure of blockchain, so the implementation of the scheme needs to modify the existing blockchain system.

Based on related works, this paper proposes a new scheme, the BMSC. Compared with Zether, BMSC has less cost of proof and takes less time for transaction verification. Compared with ZETH, BMSC is more efficient and practical. Compared with DSC, BMSC can hide the transfer relationships. Compared with BlockMaze, BMSC does not need to modify the underlying blockchain system.

3 Zreo-Knowledge Proof

Among the existing privacy protection technologies, homomorphic commitment, ring signature, coin mixer and zero knowledge proof are commonly used in the privacy protection of blockchain. Homomorphic encryption means that the result of encryption before operation is the same as that of encryption after operation. It can be used to hide account balances. However, the efficiency of fully homomorphic encryption is low at present, so it cannot be well used in blockchain. A zero-knowledge proof means that the certifier proves to the verifier that he truly knows the correct information without disclosing the specific information. Ring signatures are often used to hide the address of transaction accounts. It hides the user's account address in an anonymous set. The disadvantage is that it relies too much on an anonymous set and has poor scalability. The principle of a currency mixer is to put some unrelated transactions into one transaction, thus hiding the transfer relationship between the two parties. However, the mixed transaction time of the currency mixer is generally longer, and there is a certain security risk for the currency mixer as a third-party intermediary. The zero-knowledge proof must satisfy completeness, reliability and zero knowledge. Completeness refers to the fact that an honest witness and an honest verifier can be trusted to be true if the statement is true while following the agreement to perform the verification steps. Reliability means that if a statement of proof is false, no dishonest witness can convince an honest verifier of his statement. Zero knowledge refers to the fact that the verifier only knows if the statement of the verifier is true during the whole process of the execution of the certificate, and no other information can be obtained. Compared with other privacy protection technologies, its performance is more powerful.

The zero-knowledge succinct noninteractive arguments of knowledge are called zk-SNARKs, which are popular due to their fast verification speed. It is the most widely used zero-knowledge proof systems on the blockchain. The core idea of traditional zk-SNARKs is to transform the calculation problem into an arithmetic circuit, and then build a rank-1 constraint system, named R1CS, according to the arithmetic circuit. Then it is transformed into a QAP problem, that is, the vector equation is transformed into a polynomial equation by using polynomial interpolation, and the constraints between numbers are transformed into constraints between polynomials. zk-SNARKs transform the problem from checking whether two vectors are equal to checking whether two polynomials of the same scale are equal, which can amplify the error by using the leverage of polynomial. To verify the equality of two vectors, we need to check each item, while to verify the equality of two polynomials, we only need to take a random point in a sufficiently large range for verification. It can be considered that the probability of two different polynomials being equal at a certain point is zero, which is the fundamental principle of ZK snarks with high efficiency.

Traditional zk-SNARKs have a faster verification speed and smaller proof size, and their computational complexity is also not high. However, they require a one-time initialization preset by a trusted third party to generate a common reference string for a particular circuit. This indicates that the relational circuit constructed by traditional zk-SNARKs during initialization cannot be upgraded and needs to be reset once the relationship changes. Generally, traditional zk-SNARKs are nonuniversal settings, and the circuit cannot be changed. Sonic [9] and its improved versions Marlin [10], Plonk [11] and SuperSonic [12] are universal settings, and Bulletproof [13] is also a universal setting. This means that their reference strings can be updated and their circuits can be upgraded. Theoretically, they support infinite arbitrary circuits. zk-STARK [14], Fractal [15] and Halo [16] are transparency settings. Theoretically, zk-STARK and Fractal can protect against quantum adversaries. However, their disadvantage is that their proof is very large, so they cannot be applicable to block chains.

This paper evaluates four mainstream zero-knowledge proof systems, zk-SNARKs, BulletProof and zk-STARK, from three aspects: the need for trusted presets, the complexity of verification, the complexity of proof and the size of proof. We use the Groth16 [17] scheme with the fastest proof speed and the least proof quantity as the representative of traditional zk-SNARKs. The symbol G_B denotes an element in a group with a bilinear map (pairing), G_P one in a prime order group with known order. EXP refers to exponentiation of λ-bit numbers in these groups, and H is either the size of a hash output or the time it takes to compute a hash (Table 1).

Table 1. Comparison of zero-knowledge proofs systems

Scheme	Trusted preset	Proof	Verification	Size of proof
Groth16	Yes	$EXP \cdot O(n)$	$1\,Pairing$	$G_B \cdot O(1)$
Bulletproof	No	$EXP \cdot O(n)$	$EXP \cdot O(n)$	$2G_P \cdot log(n)$
zk-STARK	No	$H \cdot O(\lambda T)$	$H \cdot O(\lambda log^2(T))$	$H \cdot O(\lambda log^2(T))$

We compare the concrete proofs of different zero-knowledge proof systems for an NP relation with arithmetic complexity 2^{20}. The results are shown in the table below (Table 2).

Table 2. Comparison of concrete proofs

Scheme	Groth16	Bulletproof	zk-STARK
$n = 2^{20}$	192b	1.7 KB	600 KB

It can be seen from the above table that the size of the proof of Groth16 is fixed, and it is the smallest to date. Its verification speed is also the fastest, which is suitable for packaging on the blockchain. The cost of its proof calculation under the chain is also not high.

4 BMSC: Anonymous Transaction Scheme

The basic scheme of anonymous transactions is to deploy a smart contract named BMSC on Ethereum. A series of tables and parameters are stored in the contract to maintain the state. The contract internal account does not store the balance directly but stores a commitment related to the balance. The ethernet account can transfer a certain number of ETHs to an account in the contract, and the transfer between internal accounts of the smart contract is completed in two steps. When transferring money to the recipient, the sender's first step is to publish the encrypted transferred funds in a public list and update his encrypted account balance, using zero-knowledge proof to ensure its legitimacy. The second step is that the recipient confuses the corresponding transferred funds with other published transferred funds and updates the corresponding amount into its own encrypted account balance. This scheme can hide account balances, transaction amounts, and the transfer relationships between transaction parties.

4.1 Groth16

According to the previous introduction, the Groth16 scheme with the smallest proof size and the fastest verification speed is selected. The algorithm is divided into initialization $Setup_z$, key generation $KeyGen$, proof $GenProof$ and verification $VerProof$. The scheme is represented by ($Setup_z$, $KeyGen$, $GenProof$, $VerProof$).

(1) $Setup_z(1^\lambda) \rightarrow pp_z$

In the initialization phase, algorithm initialization generates a common parameter list ($p, e, G_1, G_2, G_T, g, h, Z_p$) for a given security parameter λ, where p is a prime, (G_1, G_2, G_T) is a cyclic group of order p, e is a bilinear pair $G_1 \times G_2 \rightarrow G_T$, g and h are generators of G_1 and G_2, and Z_p is a finite field. Each problem to be proved is described by a specific arithmetic circuit.

(2) $KeyGen(C) \rightarrow (pk_z, vk_z)$

Each circuit needs to generate the proof key and verification key according to the list of common parameters. In the process of transforming the R1CS problem into a QAP problem, polynomial interpolation is used to calculate $u_i(X)$, $v_i(X)$, $w_i(X)$, x^i and $t(X)$. In this way, the relationship given by the circuit is defined as ($p, e, G_1, G_2, G_T, g, h, l, \{u_i(X), v_i(X), w_i(X)\}_{i=0}^m, t(X)$), where l is the prefix length to be declared to public in the vector $\vec{a}$ composed of all variables. In the finite field Z_p, the public statement $\vec{a_1}$ is $(a_1, \ldots, a_l) \in Z_p^l$, and the private witness $\vec{a_2}$ is $(a_{l+1}, \ldots, a_m) \in Z_p^{m-l}$. When $a_0 = 1$, equation $\sum_{i=0}^m a_i u_i(X) \cdot \sum_{i=0}^m a_i v_i(X) = \sum_{i=0}^m a_i w_i(X) + h(X)t(X)$ is true. Randomly select $\alpha, \beta, \gamma, \delta, x \leftarrow Z_p^*$, generate σ and τ, which are points on $\mathbb{G}_1$ and $\mathbb{G}_2$. The specific process is as follows.

$$\tau = (\alpha, \beta, \gamma, \delta, x) \tag{1}$$

$$\sigma = ([\sigma_1]_1, [\sigma_2]_2) \tag{2}$$

$$\sigma_1 = (\alpha, \beta, \delta, \left\{ x^i \right\}_{i=0}^{n-1}, \{ \frac{\beta u_i(x) + \alpha v_i(x) + w_i(x)}{\gamma} \}_{i=0}^l,$$

$$\{ \frac{\beta u_i(x) + \alpha v_i(x) + w_i(x)}{\delta} \}_{i=l+1}^m, \{ \frac{x^i t(x)}{\delta} \}_{i=0}^{n-2}) \tag{3}$$

$$\sigma_2 = (\beta, \gamma, \delta, \left\{ x^i \right\}_{i=0}^{n-1}) \tag{4}$$

The key pair formed by the certification key and the verification key is (pk_z, vk_z). The key pair is calculated from formulas (1), (2), (3), and (4).

$$pk_z = ((\beta, \delta, \left\{ x^i \right\}_{i=0}^{n-1}, \{ \frac{\beta u_i(x) + \alpha v_i(x) + w_i(x)}{\delta} \}_{i=i+1}^m, \frac{x^i t(x)}{\delta})_1, (\left\{ x^i \right\}_{i=0}^{n-1})_2) \tag{5}$$

$$vk_z = ((\alpha, \{ \frac{\beta u_i(x) + \alpha v_i(x) + w_i(x)}{\gamma} \}_{i=0}^l)_1, (\beta, \gamma, \delta)_2) \tag{6}$$

According to formulas (5) and (6), the size of pk_z is related to the size of all variables m, while the size of vk_z is only related to the number of variables in the publicly declared state l.

(3) $GenProof(pk_z, \vec{a_1}, \vec{a_2}) \rightarrow \pi$

The proof algorithm generates a zero-knowledge proof π based on the pk_z, public statement $\vec{a_1}$, and the prover's private witness $\vec{a_2}$ to show that they satisfy the relationship R_C of circuit C construction. The proof process is to randomly select two parameters r and s, and calculate $\pi = \prod \sigma = ([A]_1, [B]_2, [C]_1)$.

$$A = \alpha + \sum_{i=0}^m a_i u_i(X) + r\delta \tag{7}$$

$$B = \beta + \sum_{i=0}^m a_i v_i(X) + s\delta \tag{8}$$

$$C = \frac{\sum_{i=l+1}^m a_i(\beta u_i(x) + \alpha v_i(x) + w_i(x)) + h(x)t(x)}{\delta} + As + rB - rs\delta \tag{9}$$

The π and $\overrightarrow{a_1}$ in the algorithm are publicly visible to anyone. According to formulas (7), (8), and (9), if the algorithm generates a successful proof, it outputs π, and if it fails to generate a proof, it returns $\perp$. The calculation process of the proof is generally performed off-chain. After the calculation is completed, the generated proof is submitted to the miner for verification.

(4) $VerProof(vk_z, \overrightarrow{a_1}, \pi) \rightarrow \{1, 0\}$

This is the verification algorithm. Anyone can verify its validity with the verification key vk_z based on the submitted zero-knowledge proof π and the public state statement $\overrightarrow{a_1}$. If the verification is successful, 1 will be output, otherwise 0 will be output.

$$[A]_1 \cdot [B]_2 = [\alpha]_1 \cdot [\beta]_2 + \sum_{i=0}^{i} a_i \left[\frac{\beta u_i(x) + \alpha v_i(x) + w_i(x)}{\gamma}\right]_1 \cdot [\gamma]_2 + [C]_1 \cdot [\delta]_2$$

$$(10)$$

The verification process is to verify whether Eq. (10) is true. The verification algorithm takes a very short time, and the result given by the Groth16 scheme is roughly between 1–10 ms.

4.2 Scheme Construction

A smart contract BMSC is deployed on Ethereum to maintain a state inside the contract and store some necessary data. When a new account is created in BMSC, a pair of keys (PK, SK) are generated for it. The account is identified by the public key and has nothing to do with the Ethereum key. There are two kinds of commitment in the BMSC: balance commitment, which is used to hide account balance, and transfer commitment, which is used to hide transaction amount. There are two steps for fund transfer in the BMSC. The first step is for the sender to generate a transfer transaction and write the transfer commitment and related data into the list. The second step is for the receiver to generate a deposit transaction and transfer the amount in the transfer commitment to his own account. In addition, the BMSC will also use the lock operation to associate the account in the contract with the address on Ethereum to realize interoperability with users in Ethereum and other smart contracts.

Asymmetric encryption, digital signatures and hash functions are used in the scheme. Asymmetric encryption schemes use public keys to encrypt messages and private keys to decrypt messages. The digital signature authorizes a message by signing it, with the sender signing it with a private key and the receiver verifying it with a public key. A hash function is a one-way function that can compress arbitrary length data into a fixed length. The commitment function with hidden and bound features, PRF pseudorandom function and Merkle tree construction can be instantiated by a hash algorithm.

The detailed construction of the BMSC includes the global setting algorithm, user algorithm and public methods. Circuits can be used to describe the relationships that need to be proved in zero-knowledge proofs. The circuit can be composed of basic plus, minus, and less than gate circuits, as well as Hash, Merkel tree and other structured circuit components. The two commitment functions *COMM* and *COMM$_V$*, pseudorandom function *PRF*, anti-collision hash function *CRH* and Merkel tree used in this scheme can all be instantiated by the hash algorithm. These circuit components can be found and used in the Libsnark library of C++ [5].

4.2.1 Setup

The initial setting algorithm generates a series of public parameters according to the security parameter λ, including five key pairs with zero-knowledge proof. Set an empty account table *acc* to map the public key to the balance commitment. Set a lock table *Lock* to represent the mapping of the public key to the Ethernet address. Set a *CMSet* to represent the mapping from the transfer commitment cmtv to the structure (pk_A, aux_A, $auth_{enc}$) which represents the sender's public key, the ciphertext of the shared parameter, and the hash value of the ciphertext. Set a *SNSet* to represent the mapping of sequence number to $\{1, \perp\}$, where 1 means used and $\perp$ means not used. Finally, set a *btotal* to represent the total amount of eth stored in the smart contract, and use *MAX* to represent the maximum value of ETH stored in the smart contract (Fig. 1).

Setup
INPUT: Security parameter λ (in unary)
1. $pp_{zk} \leftarrow Setup_{zk}(1^\lambda)$
2. $pp_\varepsilon \leftarrow Setup_\varepsilon(1^\lambda)$
3. $pp_s \leftarrow Setup_\varepsilon(1^\lambda)$
4. For each i $\in$ {CreateAccount,Fund, Burn, Send, Deposit}
   (1)Construct a circuit C_i
   (2)Compute (pk_{Zi}, vk_{Zi})=KeyGen(C_i)
5. Initialize
   • an empty account table, acc
   • an empty lock table, Lock
   • an empty transfer commitment table, CMSet
   • an empty serial number table, SNSet
   • total balance btotal=0
6. Set pp=(pp_{zk}, pp_ε, pp_s, $\cup pk_{Zi}$, $\cup vk_{Zi}$)
7. Deploy smart contract BMSC with pp, acc, lock, CMSet, SNSet, btotal, MAX.

Fig. 1. Setup

4.2.2 User Algorithm

User algorithms are algorithms used by users to interact with smart contracts. All algorithms take security parameters as input, and the result of the algorithm is to output the original transaction. BMSC is driven by transactions, which are signed by the Ethereum account that sent them. Users need to use the public data in the contract state and their own private data to generate zero knowledge proof and submit it to the contract. $cmt = COMM(pk, b, sn, r)$ is the balance commitment, which is used to bind the balance status of the account. pk is the public key. b is the balance of the commitment. sn is the serial number associated with the commitment. $sn = PRF(sk, r)$, and

PRF is a pseudorandom function. r is a random number. sk is the private key of the account. $cmtv = COMM_V(pk_A, pk_B, v, sn_v, r_v, sn_A)$ is a transfer commitment, which is used when Alice transfers money to Bob. pk_A and pk_B are the public keys of the two persons. v is the amount of funds transferred. sn_v is the serial number related to the transfer commitment with random number r_v. sn_A is the serial number in the account balance commitment before Alice transfers.

(1) *CreateLockTx* $\rightarrow tx_{lock}$

This algorithm allows smart contract users to lock their accounts to an Ethernet address. Only users who have the account's private key can complete this operation. The algorithm uses the private key to sign the address with the current balance commitment to ensure that the transaction cannot be replayed. Input the public parameter list *pp*. Compute σ_{lock} from ethereum address *addr*, balance commitment *cmt* and x by using the digital signature function. Output the public keys *pk*, *addr* and σ_{lock} as tx_{lock} (Fig. 2).

CreateLockTx
- INPUTS: pp
  - state st of BMSC
  - public key pk
  - ethereum address addr
  - private key sk
- OUTPUT: tx_{lock}=(pk, addr, σ_{lock})
1. Let cmt=acc[pk]
2. Compute σ_{lock}=Sign(x,(addr, cmt))

Fig. 2. User algorithm

(2) *CreateUnlockTx* $\rightarrow tx_{unlock}$

This algorithm is used to unlock. Input *pp*. Output *pk* as tx_{unlock} (Fig. 3).

CreateUnlockTx
- INPUTS: pp
  - Public key pk
- OUTPUT: tx_{unlock}=(pk)

Fig. 3. Unlock algorithm

(3) *CreateAccount* $\rightarrow tx_{account}$

This algorithm provides users with a pair of keys that identify themselves in the smart contract BMSC. It will generate an initial balance commitment with a balance of 0 and a zero-knowledge proof. Input *pp*. Output *pk*, *cmt*, *b* and π_0 as $tx_{account}$. The specific calculation process is shown below.

The public evidence is $\overrightarrow{x_0} = (cmt, pk, b)$, and the private evidence is $\overrightarrow{w_0} = (sk, sn, r)$. The zero-knowledge proof shows that the following statements are true (Fig. 4):

① $sn = PRF(sk, r)$

② $cmt = COMM(pk, b, sn, r)$

CreateAccountTX
- INPUTS: pp
- OUTPUT: $tx_{account}$=(pk, cmt, b, π_0)
1. Compute (sk,pk)=KeyGen(pp$_\varepsilon$)
2. Generate a new random number r
3. Compute sn=PRF(sk, r)
4. Set b=0
5. Compute cmt=COMM(pk, b, sn, r)
6. Set $\overrightarrow{x_0}$=(cmt, pk, b) and $\overrightarrow{w_0}$=(sk, sn, r)
7. Compute π_0=GenProof(pk$_{Z0}$, $\overrightarrow{x_0}$, $\overrightarrow{w_0}$)

Fig. 4. Create account algorithm

(4)   *CreateFundTx* → tx_{fund}

This algorithm is used when a certain Ethereum account provides a certain amount of funds to a certain BMSC account identified by a public key. It generates a new balance commitment for the account after being funded, and generates a zero-knowledge proof. Input *pp*. Output *pk*, cmt^*, *sn*, *v* and π_1 as tx_{fund}. The specific calculation process is shown below.

The public evidence is $\overrightarrow{x_1} = (pk, cmt, cmt^*, v, sn)$, and the private evidence is $\overrightarrow{w_1} = (sk, b, r, sn^*, r^*)$. The zero-knowledge proof shows that the following statements are true (Fig. 5):

① $sn = PRF(sk, r)$

② $cmt = COMM(pk, b, sn, r)$

③ $sn^* = PRF(sk, r^*)$

④ $cmt^* = COMM(pk, b + v, sn^*, r^*)$

CreateFundTx
- INPUTS: pp
  - −state st of BMSC
  - −public key pk
  - −private key sk, b, sn and r of cmt
  - −amount v
- OUTPUT: tx_{fund}=(pk,cmt^*, sn, v, π_1)
1. Let cmt=acc[pk]
2. Generate a new random number r^*
3. Compute sn^*=PRF(sk, r^*)
4. Compute cmt^*=COMM(pk, b+v, sn^*, r^*)
6. Set $\overrightarrow{x_1}$=(pk, cmt,cmt^*, v, sn) and $\overrightarrow{w_1}$=(sk, b , r,sn^*, r^*)
7. Compute π_1=GenProof(pk$_{Z1}$, $\overrightarrow{x_1}$, $\overrightarrow{w_1}$)

Fig. 5. Fund algorithm

(5) *CreateBurnTx* → tx_{burn}

This algorithm is used when a certain BMSC account wants to withdraw a sum of funds and send it to a certain Ethereum account. It calculates a new balance commitment for the account after withdrawing the amount, and generates a zero-knowledge proof. Input *pp*. Output *pk*, cmt^*, *sn*, *v* and π_2 as tx_{burn}. The specific calculation process is shown below.

The public evidence is $\overrightarrow{x_2} = (pk, cmt, cmt^*, v, sn)$, and the private evidence is $\overrightarrow{w_2} = (sk, b, r, sn^*, r^*)$. The zero-knowledge proof shows that the following statements are true (Fig. 6):

① $sn = PRF(sk, r)$
② $cmt = COMM(pk, b, sn, r)$
③ $sn^* = PRF(sk, r^*)$
④ $cmt^* = COMM(pk, b + v, sn^*, r^*)$
⑤ $b \geq v$

CreateBurnTx
• INPUTS: pp
  −state st of BMSC
  −public key pk
  −private key sk, b, sn and r of cmt
  −amount v
• OUTPUT: tx_{burn}=(pk,cmt^*, sn, v, π_2)
1.Let cmt=acc[pk]
2.Generate a new random number r^*
3.Compute sn^*=PRF(sk, r^*)
4.Compute cmt^*=COMM(pk, b-v, sn^*, r^*)
6.Set $\overrightarrow{x_2}$=(pk, cmt,cmt^*, v, sn) and $\overrightarrow{w_2}$=(sk, b , r,sn^*, r^*)
7.Compute π_2=GenProof(pk_{z2}, $\overrightarrow{x_2}$, $\overrightarrow{w_2}$)

Fig. 6. Burn algorithm

(6) *CreateSendTx* → tx_{send}

This algorithm is used when Alice transfers money to Bob. Alice generates transfer fund commitment *cmtv*, calculates new balance commitment after transfer, and generates zero knowledge proof. After the zero-knowledge proof submitted by Alice is verified by the BMSC, *cmtv* will be published on the public list of the contract. Others can only know that the sender of *cmtv* is Alice, but they cannot know the information of the receiver. Alice will inform Bob offline of *cmtv* so that Bob can use *cmtv* for subsequent operations. Input *pp*. Output pk_A, cmt_A^*, sn_A, *cmtv*, aux_A, $auth_{enc}$ and π_3 as tx_{send}. The specific calculation process is shown below.

The public evidence is $\overrightarrow{x_3} = (pk_A, cmt_A, cmt_A^*, cmtv, sn_A, aux_A, auth_{enc})$, and the private evidence is $\overrightarrow{w_3} = (pk_B, sk_A, b_A, r_A, sn_A^*, r_A^*, v, sn_v, r_v)$. The zero-knowledge proof shows that the following statements are true:

① $sn_A = PRF(sk_A, r_A)$
② $cmt_A = COMM(pk_A, b_A, sn_A, r_A)$
③ $sn_A^* = PRF(sk_A, r_A^*)$
④ $cmt_A^* = COMM(sk_A, b_A - v, sn_A^*, r_A^*)$
⑤ $b_A \geq v$
⑥ $r_v = CRH(pk_A|r_A^*)$
⑦ $sn_v^* = PRF(sk_v, r_v)$
⑧ $cmtv = COMM_V(pk_A, pk_B, v, sn_v, r_v, sn_A)$
⑨ $auth_{enc} = PRF(sk_A, h_{enc})$

CreateSendTx
- INPUTS: pp
  - state st of BMSC
  - public key pk_A, pk_B
  - private key sk_A, b_A, sn_A and r_A of cmt_A
  - amount v
- OUTPUT: tx_{send}=(pk_A,cmt_A^*, sn_A, cmtv, aux_A, $auth_{enc}$, π_3)
1. Let cmt_A=acc[pk_A]
2. Generate a new random number r_A^*
3. Compute sn_A^*=PRF(sk_A, r_A^*)
4. Compute cmt_A^*=COMM(sk_A, b_A-v, sn_A^*, r_A^*)
5. Compute r_v=CRH($pk_A|r_A^*$)
6. Compute sn_v^*=PRF(sk_v,r_v)
7. Require: $sn_v \notin$ SNSet
8. Compute cmtv=COMM$_V$(pk_A,pk_B, v, sn_v, r_v, sn_A)
9. Compute aux_A=Enc$_{pkB}$({pk_A, v, sn_v, r_v, sn_A})
10. Compute h_{enc}=CRH(aux_A)
11. Compute $auth_{enc}$=PRF(sk_A,h_{enc})
12. Set $\vec{x_3}$=(pk_A, cmt_A, cmt_A^*, cmtv,sn_A,aux_A, $auth_{enc}$)
13. Set $\vec{w_3}$=(pk_B,sk_A, b_A, r_A, sn_A^*, r_A^*, v, sn_v, r_v)
14. Compute π_3=GenProof(pk_{Z3}, $\vec{x_3}$, $\vec{w_3}$)

Fig. 7. Send algorithm

(7) *CreateDepositTx* $\rightarrow tx_{deposit}$

This algorithm is used when Bob needs to withdraw the funds from *cmtv* to his own account. Bob mixes *cmtv* into a *cmtv* sequence set constructed with other transfer commitments, and this sequence set constructs a Merkel tree. Then, the value of the root node of the tree and the path from *cmtv* to the root are calculated. The BMSC contract does not operate on the target *cmtv*, so no one knows which *cmtv* value Bob will transfer to his own account. The algorithm will also calculate Bob's new balance commitment after successful withdrawal, and it will generate zero knowledge proof together with Merkel tree root and other data to be submitted to the BMSC contract for verification. Input *pp*. Output pk_B, *mtseq*, rt_{cmtv}, sn_v, cmt_B^*, sn_B and π_4 as $tx_{deposit}$. The specific calculation process is shown below (Fig. 7).

The public evidence is $\overrightarrow{x_4} = (pk_B, cmt_B, cmt_B{}^*, sn_B, rt_{cmtv}, sn_v)$, and the private evidence is $\overrightarrow{w_3} = (pk_A, sn_A, cmtv, v, r_v, path, sk_B, b_B, r_B, sn_B{}^*, r_B{}^*)$. The zero-knowledge proof shows that the following statements are true (Fig. 8):

① $sn_B = PRF(sk_B, r_B)$
② $cmt_B = COMM(pk_B, b_B, sn_B, r_B)$
③ $sn_B{}^* = PRF(sk_B, r_B{}^*)$
④ $cmt_B{}^* = COMM(sk_B, b_B + v, sn_B{}^*, r_B{}^*)$
⑤ $cmtv = COMM_V(pk_A, pk_B, v, sn_v, r_v, sn_A)$
⑥ the path from $cmtv$ to rt_{cmtv} is recorded on MT.

CreateDepositTx
• INPUTS: pp
  −state st of BMSC
  −public key pk_B
  −private key sk_B, b_B, sn_B and r_B of cmt_B
  −shuffled set mtseq=$\{cmtv_1,...,cmtv_i,...,cmtv_n\}$ from existed cmtv,
  which includes cmtv
• OUTPUT: $tx_{deposit}$=$(pk_B, mtseq, rt_{cmtv}, sn_v, cmt_B{}^*, sn_B, \pi_4)$
1.Let cmt_B=acc[pk_B]
2.Let $(pk_A, aux_A, auth_{enc})$= acc[cmtv]
3.Compute (v, sn_v, r_v, sn_A)= $Dec_{skB}(aux_A)$
4.Construct a Merkle tree MT over mtseq
5.Compute path=Path(cmtv) and rt_{cmtv} over MT
6.Generate a new random number $r_B{}^*$
7.Compute $sn_B{}^*$=PRF($sk_B, r_B{}^*$)
8.Compute $cmt_B{}^*$=COMM(sk_B, b_B+v, , $r_B{}^*$)
9.Set $\overrightarrow{x_4}$=($pk_B, cmt_B, cmt_B{}^*, sn_B, rt_{cmtv}, sn_v$)
10.Set $\overrightarrow{w_3}$=($pk_A, sn_A, cmtv, v, r_v, path, sk_B, b_B, r_B, sn_B{}^*, r_B{}^*$)
11.Compute π_4=GenProof($pk_{Z4}, \overrightarrow{x_4}, \overrightarrow{w_4}$)

Fig. 8. Deposit algorithm

4.2.3 Public Methods

BMSC has seven common methods: *Lock*, *Unlock*, *CreateAccount*, *Fund*, *Burn*, *Send* and *Deposit*. In addition, the contract also defines an internal assistant function checklock, which is used to check whether the Ethereum address that initiated the transaction is legal to the account in the contract. Each public method needs to call this function before it is used. When the verification fails, the state of the smart contract will be rolled back to the state before it is modified. The public method verifies whether the conditions are met according to the data submitted by the user. After the verification is passed, the status of the smart contract will be updated, such as updating the account balance status, publishing the existing serial number, and adding transfer commitment.

(1) *Checklock*

This is the internal assistant function of the smart contract. Before calling other public methods, you need to call this function to check the lock status of the account. This will not be repeated in the following methods (Fig. 9).

```
CheckLock
• INPUTS:
  −public key pk
  −ethereum address addr
• OUTPUT:
  1 if account pk can be operated by addr;
  0 otherwise
1.If lock[y] = addr or lock[y]=⊥:
  −Output 1
  else
  −Output 0
```

Fig. 9. Check lock method

(2) *Lock*

Check whether the signature is valid to verify whether the user's lock request is appropriate. If the verification is successful, the contract will change the address of the account public key in the lock table (Fig. 10).

```
Lock
• INPUTS: tx_{lock}=(pk, addr, σ_{lock})
1.Require:
  −CheckLock(pk, msg.sender)=1
  − Verify_{sig}(pk, msg.sender)=1
2.Set lock[pk]=addr
```

Fig. 10. Lock method

(3) *Unlock*

Check that the Ethereum address of the sending message is the same as the address corresponding to the account public key in the lock table, and then set the public key mapping in the lock table to ⊥ to release the lock (Fig. 11).

```
Unlock
• INPUTS: tx_{unlock}=(pk)
1.Require
  −CheckLock(pk,msg.sender)=1
2.Set lock[pk]=⊥
```

Fig. 11. Unlock method

(4) *CreateAccount*

This initial balance is required to be set to 0. If the verification is successful, the mapping of the new account public key to the initial balance commitment is added to the account table (Fig. 12).

CreateAccount
- INPUTS: $tx_{account}$=(pk, cmt, b, π_0)
1. Set $\vec{x_0}$=(cmt, pk, b)
2. Require:
   - CheckLock(pk, msg.sender)=1
   - b=0
   - VerProof(vk_{Z0}, $\vec{x_0}$, π_0)=1
3. Set acc[pk]=cmt

Fig. 12. Create account method

(5) *Fund*

The funding amount is required to be greater than 0. It is required that the sequence number associated with the old balance commitment is not used (that is, the mapping value of the sequence number in the table is returned as $\perp$). The total amount of ETH stored in the contract must be less than MAX. Verify zero knowledge proof. After all conditions are met, the smart contract updates the balance commitment of the account and the total amount of ETH stored. In addition, it adds the old serial number to the serial number table (Fig. 13).

Fund
- INPUTS: tx_{fund}=(pk,cmt^*, sn, v, π_1)
1. Let cmt=acc[pk]
2. Set $\vec{x_1}$=(pk, cmt,cmt^*, v, sn)
3. Require:
   - CheckLock(pk, msg.sender)=1
   - v>0
   - b_{total}+v≤MAX
   - SNSet[sn]=$\perp$
   - VerProof(vk_{Z1}, $\vec{x_1}$, π_1)=1
4. Set acc[pk]= cmt^*
5. Set SNSet [sn]=1
6. Set b_{total}=b_{total}+v

Fig. 13. Fund method

(6) *Burn*

The redemption amount must be greater than 0 and the serial number associated with the old balance commitment has not been used. Verify zero knowledge proof. After all conditions are met, the smart contract updates the balance commitment of the account and the total amount of ETH stored. In addition, it adds the old serial number to the serial number table (Fig. 14).

Burn
- INPUTS: $tx_{burn}=(pk, cmt^*, sn, v, \pi_2)$
1. Let cmt=acc[pk]
2. Set $\overrightarrow{x_2}=(pk, cmt, cmt^*, v, sn)$
3. Require:
    −CheckLock(pk, msg.sender)=1
    −v>0
    −SNSet[sn]=⊥
    −VerProof($vk_{Z2}, \overrightarrow{x_2}, \pi_2$)=1
4. Set acc[pk]= cmt^*
5. Set SNSet[sn]=1
6. Set $b_{total}=b_{total}-v$

Fig. 14. Burn method

(7) *Send*

Alice's old balance commitment serial number must have not been used. Calculate the hash of the shared encryption parameters to ensure that the shared encryption parameters have not been tampered with. After the verification is passed, Alice's account balance commitment is updated, and the serial number of Alice's old balance commitment is added to the serial number table. Add a new mapping to the transfer commitment table, which is the transfer commitment to the structure composed of encrypted shared parameters, ciphertext hash and Alice's public key (Fig. 15).

Send
- INPUTS:$tx_{send}=(pk_A, cmt_A^*, sn_A, cmtv, aux_A, auth_{enc}, \pi_3)$
1. Let $cmt_A=acc[pk_A]$
2. Compute $h_{enc}=CRH(aux_A)$
3. Set $\overrightarrow{x_3}=(pk_A, cmt_A, cmt_A^*, cmtv, sn_A, h_{enc}, auth_{enc})$
4. Require:
    −CheckLock(pk, msg.sender)=1
    −SNSet(sn_A)=⊥
    −VerProof($vk_{Z3}, \overrightarrow{x_3}, \pi_3$)=1
5. Set acc[pk_A]= cmt_A^*
6. Set SNSet[sn_A]=1
7. Set CMSet[cmtv]=($pk_A, aux_A, auth_{enc}$)

Fig. 15. Send method

(8) *Deposit*

Check that the serial number of Bob's current balance commitment has not been used. Check that the transfer commitment serial number to be published has not been used, that is, the unknown target *cmtv* has not been spent. Check that the root node is indeed the root node of the set of transfer commitment sequences, that is, the unknown transfer commitment does exist. Verify zero knowledge proof. Then Bob's balance status is updated. Publish the serial number of Bob's old balance commitment and the serial number of the unknown transfer commitment (Fig. 16).

Deposit
- INPUTS: $tx_{deposit}$=(pk_B, mtseq,rt_{cmtv}, sn_v, $cmt_B{}^*$,sn_B, π_4)
1. Let cmt_B=acc[pk_B]
2. Set $\vec{x_3}$=(pk_B,rt_{cmtv}, sn_v, cmt_B, $cmt_B{}^*$,sn_B)
3. Require:
   $-$CheckLock(pk, msg.sender)=1
   $-$SNSet(sn_B)=$\perp$
   $-$SNSet(sn_v)=$\perp$
   $-$for every cmtv in mtseq CMSet[cmtv]$\neq \perp$
   $-rt_{cmtv}$ is the root over mtseq
   $-$VerProof(vk_{Z4}, $\vec{x_4}$, π_4)=1
4. Set acc[pk_B]= $cmt_B{}^*$
5. Set SNSet[sn_B]=1
6. Set SNSet[sn_v]=1

Fig. 16. Deposit method

5 Analysis of Anonymity and Security

5.1 Analysis of Anonymity

5.2 Hide Account Balances and Transaction Amounts

The BMSC does not store the balance information of the account directly, but binds the balance with other account information to generate the balance commitment as the balance status of the account. Balance commitment $cmt = COMM\,(pk, b, sn, r)$, pk is the public key, b is the balance, and sn is the serial number of balance commitment, which is generated by random number r and account private key. According to the hidden and binding characteristics of the commitment scheme and the zero knowledge characteristic of the zero-knowledge proof, the probability of the adversary successfully resolving the balance value b according to the published cmt or submitted zero- knowledge proof π can be ignored.

Similar to hiding the account amount, the transaction amount is hidden in the transfer commitment $cmtv = COMM_V(pk_A, pk_B, v, sn_v, r_v, sn_A)$ together with other relevant information. The transaction amount v cannot be parsed by others from the published transfer commitment itself and the zero-knowledge proof submitted by the transfer deposit transaction. The encrypted sharing parameter $aux_A = Enc_{pkB}(\{pk_A v, sn_v, r_v, sn_A\})$ corresponding to the transaction commitment can only be decrypted by receiver Bob using private key sk_B, so the adversary cannot know the transaction amount v in it.

5.3 Hide the Transfer Relationships

A complete fund transfer process from Alice to Bob includes two steps, which cut off the association between the sender and the receiver, thus hiding the transfer relationship between the two parties.

The first step is that Alice generates and publishes $cmtv$ and other transaction related information, in which Bob's public key pk_B is hidden. Depending on the hidden feature of the commitment scheme and the zero knowledge feature of the zero-knowledge proof, the adversary cannot resolve pk_B, so they do not know who the receiver of the transaction is.

The second step is Bob's deposit process. After Alice generates the transfer commitment, Bob can complete the withdrawal at any time. Bob mixes the target *cmtv* into a specified size transfer commitment sequence set, generates a Merkel tree MT, and calculates the root node value rt_{cmtv} and the *path* from the target *cmtv* to the root node. *cmtv* and *path* are used as private evidence to generate a zero- knowledge proof together with other data. Relying on the characteristic of zero- knowledge proof, the adversary can only know that Bob has taken out a sum of money from the existing transfer commitment and published its serial number but cannot know which transfer commitment it is. Therefore the adversary cannot know the sender of the promise, Alice.

5.4 Analysis of Security

5.5 Overspending Attack

The attack means that the attacker does not execute the user algorithm according to the regulations, so that the transferred funds exceed the balance held by the account. In the BMSC scheme, each transaction needs to submit zero knowledge proof to prove its effectiveness, stating that the funds transferred or redeemed by itself do not exceed its existing balance. Then, according to the KOE security assumption in the zero- knowledge proof scheme Groth16, the probability of the attacker forging the zero- knowledge proof to complete the overdraft consumption attack can be ignored.

5.6 Double-Spending Attack

The attack means that the attacker spends a sum of money twice. In the BMSC scheme, both balance commitment and fund transfer commitment are associated with a specific serial number. Maintain a sequence table *SNSet* in the smart contract. Once a transaction is verified, both the old balance commitment serial number *sn* and the used transfer commitment serial number sn_v will be added to the sequence table *SNSet*, indicating that the serial number has been used. To prevent double spending attacks, each transaction verification process in a smart contract needs to prove that the serial number has not been used. Therefore, a sum of money can only be spent once.

6   Conclusion

To protect the privacy of users on blockchain, this paper proposes a novel anonymous transaction scheme in Ethereum named BMSC based on zk-SNARKs. The scheme hides account balances, transaction amounts and transfer relations in terms of anonymity. In terms of security, it meets the requirements of preventing overspending attacks and preventing double-spending attacks. Compared with other anonymous schemes based on the account model, this scheme has less cost of proof and takes less time for transaction verification while meeting the high requirements of anonymity and security.

Acknowledgment. This work is supported by the Emerging Interdisciplinary Project of CUFE, the National Natural Science Foundation of China (No. 61906220) and Ministry of Education of Humanities and Social Science project (No. 19YJCZH178).

References

1. Duffield, E., Diaz, D.: Dash: A Payments-Focused CryptoCurrency [EB/OL] (2015). https://github.com/dashpay/dash/wiki/Whitepaper
2. Miers, I., Garman, C., Green, M., et al.: Zerocoin: anonymous distributed E-cash from bitcoin. In: 2013 IEEE Symposium on Security and Privacy, pp. 397–411. IEEE (2013)
3. Nicolas Saberhagen, N.: Cryptonote v2.0 [EB/OL] (2013). https://cryptonote.org/whitepaper.pdf
4. Ben-Sasson, E., Chiesa, A., Garman, C., et al.: Zerocash: decentralized anonymous payments from bitcoin. In: 2014 IEEE Symposium on Security and Privacy, pp. 459–474. IEEE (2014)
5. Biinz, B., Agrawal, S., Zamani, M., et al.: Zether: toward privacy in a smart contract world. IACR Cryptology ePrint Archive (2019)
6. Rondelet, A., Zajac, M.: ZETH: on integrating zerocash on ethereum. arXiv preprint arXiv: 1904.00905 (2019)
7. Ma, S., Deng, Y., He, D., et al.: An efficient NIZK scheme for privacy-preserving transactions over account-model blockchain. IEEE Trans. Dependable Secure Comput. **18**, 641–651 (2020)
8. Guan, Z.: Research on privacy protection of account model blockchain system based on zero knowledge proof. Shandong University (2020)
9. Mailer, M., Bowe, S., Kohlweiss, M., et al.: Sonic: zero-knowledge SNARKs from linear-size universal and updatable structured reference strings. In: Proceedings of the 2019 ACM SIGSAC Conference on Computer and Communications Security, pp. 2111–2128 (2019)
10. Chiesa, A., Hu, Y., Mailer, M., et al.: Marlin: preprocessing zkSNARKs with universal and updatable SRS. IACR Cryptology ePrint Archive (2019)
11. Gabizon, A., Williamson, Z.J., Ciobotam, O.: PLONK: permutations over lagrange-bases for oecumenical non interactive arguments of knowledge. IACR Cryptology ePrint Archive (2019)
12. Bunz, B., Fisch, B., Szepieniec, A.: Transparent snarks from dark compilers. IACR Cryptology ePrint Archive (2019)
13. Biinz, B., Bootle, J., Boneh, D., et al.: Bulletproofs: short proofs for confidential transactions and more. In: 2018 IEEE Symposium on Security and Privacy, pp. 315–334. IEEE (2018)
14. Ben-Sasson, E., Bentov, I., Horesh, Y., et al.: Scalable, transparent, and postquantum secure computational integrity. IACR Cryptology ePrint Archive (2018)
15. Chiesa, A., Ojha, D., Spooner, N.: Fractal: postquantum and transparent recursive proofs from holography. IACR Cryptology ePrint Archive (2019)
16. Bowe, S., Grigg, J., Hopwood, D.: Halo: recursive proof composition without a trusted setup. IACR Cryptology ePrint Archive (2019)
17. Groth, J.: On the size of pairing-based non-interactive arguments. In: Fischlin, M., Coron, J.S. (eds.) EUROCRYPT 2016. LNCS, vol. 9666, pp. 305–326. Springer, Heidelberg (2016). https://doi.org/10.1007/978-3-662-49896-5_11

Research on the Design and Education of Serious Network Security Games

Keyong Wang[1], Pei Wang[1], Yuanheng Zhang[2], and Dequan Yang[3(✉)]

[1] School of Continuing Education, Beijing Institute of Technology, Beijing 100081, China
[2] The Affiliated High School of Peking University, Beijing 100080, China
[3] Network Information Technology Center, Beijing Institute of Technology, Beijing 100081, China
yangdequan@bit.edu.cn

Abstract. Through the correct teaching and game design, not only can innovation be carried out, but the highly effective network security teaching and the training solution countermeasure can also develop the student's correct social behavior and value idea. Furthermore, through rule design and role play, students can improve the overall level of network security awareness and enhance their sense of cooperation during their studies. This paper defines the concept of a serious game and its significance to the network security curriculum of schools and designs the corresponding card game according to the feature of network security. The purpose of this paper is to strengthen overall network security awareness, cultivate professional skills, and enhance network security.

Keywords: Serious game · Network security · Teaching technique

1 Introduction

Due to the complexity of the network security situation and the rapidly changing trend, serious games, as an educational means, can supplement teaching by creating virtual reality problem situations, providing corresponding training and simulation, and exercising the practical ability of participants. Therefore, by developing a teaching model based on serious games [1], the participants' game-based useful attack and defense strategy can improve the understanding of the attack and defense of network security application scenarios.

The settings of different games can enable the participants to adapt and modify accordingly. Compared with the development of system games, physical card games are more suitable for students to use and are more practical. It is essential for career-oriented teaching to let participants experience how to think like an attacker or react to an attack and related strategies.

Supported by Hainan Provincial National Science Foundation of China, 621MS0789.

2   Related Work

In the 1970s, serious games were first proposed [2]. As a new method of education, it is skill teaching through the game of curriculum content and the form of games. Compared with traditional education, serious games have stronger acceptability and generality [3]. Moreover, they provide an exciting and pleasant educational environment, which can be supplemented by classroom education or vocational training. Therefore, this concept has attracted the attention of academia and gradually developed into an important research direction in the field of education [4].

In 2009, the first serious game Innovation Summit was held in Beijing [5]. With the introduction of the concept of serious games into China, it has gradually developed into vocational education and training. To date, the domestic research on serious games mostly focuses on empirical research, and the theoretical research is relatively few, but it has initially formed a certain scale and has broad development prospects [6]. The main place of vocational education in China is still the school classroom. Indoctrination teaching makes students' awareness of autonomous learning weak, and learning has no motivation. In contrast, game teaching can effectively improve the efficiency and quality of the whole education by building students' independent knowledge. Serious games [7] show a powerful educational function and can also play a significant role in cultivating internal educational activities such as the spirit of division of labor and cooperation among students [8].

Practice and operation ability are critical in the career-oriented network information security course [9]. Learners need to master solid practical operation ability to improve their personal and professional quality and industry competitiveness. Furthermore, in the application of serious games, students, as participants of the game, learn how to use vulnerabilities to attack in a dynamic environment and think about the relevant countermeasures of network security defense on the spot. Therefore, it is beneficial for participants to quickly learn and master the concept of network security and improve their attack and defense practical operation ability.

Therefore, serious games are beneficial for participants to quickly learn, master relevant concepts, and improve their practical ability. Suppose we can integrate serious games into the classroom as teaching devices and learning methods through the simulation of story situations, the application, and understanding of game art and game process design content [10]. In that case, we can fully attract students' interest and let them invest more time and energy in the learning process. Realize the practical unity of the game and the learning process.

3   Feasibility Analysis

3.1   Educational Dilemma

- Lack of educational resources.

The course on network security needs practice. Only by letting students operate it by themselves can they truly understand and apply theoretical knowledge. For students, boring textbook knowledge is not enough to attract them to learn effectively. Only through practical teaching can we digest theory in practice and truly master the content of network security courses. The game's characteristic is unique entertainment, and the challenge may add some pleasure to the boring study content. The game-based teaching method is in line with the learning characteristics of students in vocational schools. It is conducive to students obtaining a more pleasant and relaxed learning experience. Therefore, the development of game teaching has received increasing attention, and the demand for game teaching resources has become increasingly greater.

- The teaching model is single.

The traditional teaching mode takes teachers as the main body and students passively accept knowledge. This mode ensures the dominant position of teachers in classroom teaching, is conducive to their organization, management and control of classroom teaching, and targets learning. However, due to the lack of interaction between students and teachers, it is easy for students to lose interest in learning. Traditional "cramming" teaching [9] focuses on theoretical learning, lacks practical teaching means and is not combined with real life, which greatly reduces teaching quality and is not conducive to the cultivation of students' creativity and imagination [11].

- Lack of a perfect and effective evaluation system.

At present, the teaching investigation of the introductory course of network security focuses on the theory, and the examination plan only uses the examination paper to decide the student's achievement. On the other hand, the talent training goal of vocational schools mainly focuses on cultivating working ability. Therefore, a serious game is a practical and effective learning method for vocational school students. Serious games mainly focus on skill training. It requires students to start from a relatively simple learning task and promote the acquisition of curriculum theoretical knowledge by completing the task. In terms of teaching evaluation, it evaluates whether it is necessary to master the classroom content according to the game task. It can cultivate their ability to explore independently and help form a sense of teamwork in learning from others.

3.2 Education Status

- Student Character.

Vocational school students treat learning, often bored and excluded, compared with ordinary school students, belonging to a relatively particular group of students. Traditional teaching methods can not only attract them, play can not only bring interest but also build confidence [12], achieve spiritual and spiritual satisfaction, and choose game-based teaching, using modern information technology and game resources to carry out various teaching activities, integrating game elements, reorganizing the teaching process, constructing interaction before, during and after class, and sharing experiences between teachers and students. Designing a game-based teaching environment in a specific space-time context is helpful for students to carry out situation cognition, concept understanding and knowledge construction, give full play to students' initiative and enthusiasm, and let students feel that the learning process is no longer a burden but a kind of innovation, wisdom and development. The heart will gradually produce the desire to learn so that students in learning can regain confidence [13].

- Subject Characteristics.

With the rapid development of computer network technology, network attack and prevention technology have changed rapidly. The course of network security must keep up with the actual situation of technology development and update in time and effectively. Network security protection is a protection measure for the whole computer system. Ensuring computer information security is the core work of computer network security protection, usually monitoring and comparing data. It can protect the use and transmission of data on the technical level by encrypting the data to protect the relevant data and ensure the security and reliability of the system. The network security curriculum cannot leave the practice. It only lets the student operate. They can understand the application theory knowledge truly. In comparison, vocational education pays more attention to promoting professional ability. At the same time, the practice teaching method is more suitable for the cultivation and transportation of enterprise talent to better integrate into the work environment, from the practice of Digestion Textbook Knowledge, a fundamental understanding of network security courses, a better grasp of technology.

4 Theoretical Basis

4.1 Constructivist Theory

Constructivist theory, commonly known as structuralism, is a branch of cognitive psychology. Constructivism theory holds that the formation and change of schema is the essence of people's mental world, and the three processes of assimilation, adaptation, and balance affect people's cognitive development [14].

The core of constructivist cognition is that learners acquire knowledge using meaning construction under the specific social and cultural background and with the help of others. In other words, significant changes have taken place in the teaching of the constructivist learning environment. Therefore, instructional design is one of the most critical aspects of instructional design to consider the analysis of instructional objectives and the environment conducive to constructing students' meaning.

Constructivism is a learner-centered teaching method that requires teachers' guidance in teaching. While emphasizing learners as the main body, we cannot ignore the guiding role of teachers. In a complete learning situation, teachers do not simply imitate knowledge but promote the construction of curriculum content. Students are the main body of learning. They need not only external stimulation but also learning to construct the situation actively.

The game provides a more experimental deconstruction method, allowing students to practice in the virtual environment, reducing resource consumption, and effectively improving practical ability. Based on this new concept, many researchers have put forward serious games.

4.2 Situational Cognition

The theory of situated cognition is a kind of teaching theory. By setting up situations consistent with the teaching objectives, we can deal with different complex problems with the learned knowledge. According to the idea of situated cognition, the essence of learning refers to the process of learners' interaction with others and the external environment when they participate in practical activities.

According to the theory of situated cognition, the relationship between context, teaching, activities, and knowledge is close. In social life, learning is a process in which people interact with others and gain experience and skills. Therefore, if people want to acquire knowledge, they must constantly practice and acquire knowledge in practice. In other words, learning is a participatory process, a "Legitimate marginal participation" [15] in certain situations. "Legitimate" means that learners have the real identity of learning in the activity and can use the shared resources in action. "Marginalized" means that learners can acquire knowledge and skills by participating in specific situations. According to the theory of situated cognition, the learning environment of learners also influences the degree of absorption of knowledge. Therefore, constructing a real, complex, and meaningful learning environment is the premise of situation cognition and learning. For example, in teaching activities, teachers or experts can use case studies to teach knowledge so that students can better understand and use it.

5 Teaching Design

In the process of network security, teachers first use teaching methods to teach students basic theoretical knowledge and then flexibly embed serious games into classroom teaching through interactive teaching as the game leader. Teachers judge and interpret the process of the game and make the practice content gamified to increase the interest in practical skills training. At the end of the game, students take the lead in flipping the classroom, discussing the process of the game and the effect of skill training, and evaluating the teaching. Finally, the teacher guides students to understand and improve their network security awareness.

5.1 Design Principle

According to the characteristics of vocational school students and professional disciplines, the design principles of teaching games are as follows:

(1) Acceptability

Vocational school students are younger, regardless of their learning style and character development, and have a certain degree of plasticity. In addition, vocational school students in the learning of weak cultural foundation, learning patience and lack of learning ability and other characteristics are in the rebellious period they are challenging to change entirely in a short period. To help them build up their self-confidence and rebuild their learning passion, we must effectively solve the students' difficulties in the learning process. Combine games with relevant courses to form a game group. It can effectively stimulate students' learning enthusiasm and help them eliminate boredom. It can easily and happily understand concepts and knowledge construction in the learning experience and improve learning efficiency.

(2) Interestness

When the traditional teaching method cannot meet the current teaching demand, the classroom teaching and the game-based organic compound teaching model add many fun to the Boring Textbook classroom study. Different teaching scenes are set according to the course training's working environment and course content. The game teaching method simulates the real business situation, combined with the actual cases, to complete practical teaching. Experience the content of the work and have a deeper understanding of textbook knowledge. Therefore, separating the practice content from the teaching content and carding the game-based business cases greatly enhance the interest in the practice teaching and arouse the students' interest in the practice teaching process.

(3) Teaching applicability

The game-based teaching method based on confrontation makes the knowledge structure of the textbook more intuitive and makes the complicated contents in the text easier to understand. It also stimulates the motivation of students to master knowledge, makes students easy to accept and enjoy it, makes students learn efficiently and happily, and achieves the teaching goal and teaching effect.

(4) Interactivity

The teaching content is more intuitive and exciting in experiencing the game through programmed learning in the situational teaching mode. For teachers, it simultaneously strengthens communication and cooperation with students and improves enthusiasm for teaching. Moreover, the cooperative games between the students can enable them to see different viewpoints, promote learning and develop a sense of cooperation and dialectical thinking. The existence of other points of view and the diversity of problem-solving methods have a deeper understanding.

5.2 Design Principle

The game design aims to create a learning environment to help players improve their network security awareness and understand the possible countermeasures to prevent or mitigate attacks [16].

- Master, the attacker can take advantage of the vulnerability and breadth of attack methods.
- To understand the diversity of network security prevention and detection methods and grasp the various ways to reduce or mitigate network attacks.
- Practice how to attack and exploit vulnerabilities and defend against them.

- Have the ability of network security demand analysis and network security implementation.
- More in-depth understanding of network security risks, risk management thinking, and understanding.
- Raising awareness of network security.

6 Game Design

Unlike traditional installation teaching, let the students study the contents of time simulation exercises in the learning Knowledge Acquisition Initiative. Through the teacher in the game to guide students, students can understand the network security attack and defense work principle to achieve a deeper understanding of memory [17]. In the game design, students play different roles through the game's storyline, in which to practice attack scenarios and defense strategies, to find solutions to "How to attack" and "How to prevent or mitigate attacks," Improve the player's relative abilities. To achieve this goal, we design the game theme, set the corresponding game story, and use cards to simulate the natural working environment. Players participate in the game and consciously use book knowledge for analysis to improve their understanding of network security technology and related operating principles [18]·

6.1 Game Theme

The theme of the game includes the description of various network security events. The scenarios of network security events are adjustable according to the teaching requirements and are suitable for various working situations. They can be combined with practical events and integrated into ideological and political education, providing students with a positive learning environment and building knowledge through personal experiences, such as the following:

- Scenario 1:

The content of Network Fraud Identification includes the network security foundation, network threat behavior, security protection skills of common network passwords, etc.

- Scenario 2:

Network attack identification includes network security factors, computer system vulnerabilities, and insecure factors, including the classification and form of hacker attacks and network virus classification and virus detection technology.

- Scenario 3:

Network security construction and management, including firewall classification, critical technology, network access control strategy, intrusion detection system, VPN technology, and other applications [19].

6.2 Game Components

There are six card types: identity card, attack card, defense card, information card, and question bank card.

(1) Identity Cards

The game has three characters: attacker, defender, and game master. Hackers extract attack cards as attackers, and security guards extract defense cards as defenders.

(2) Attack Cards

The game's attack cards include the following four types of security events [20]. See Table 1 below for details:

- Pest events. It refers to the untested code program being inserted into the information system, causing harm to the data and application programs in the system.
- Network attack events. It refers to using the defects of information systems to attack the system through network technology.
- Information destruction. Information destruction relates to tampering, forgery, and theft of data in the information system through various technical means.
- Information content security. It refers to destroying the information system due to its defects or the artificial use of nontechnical means.

Table 1. Distribution of attack cards.

Harmful program	Attackers exploit known vulnerabilities to attack devices to create botnets
	The attackers used a large botnet for a denial-of-service attack
	An attacker sends an e-mail that contains malicious links and files
	An attacker decoys a user to connect to a malicious wireless access point
	An attacker lures a user to a malicious website
Cyberattack	IP spoofing causes the server to connect to illegal user accounts, affecting legitimate user connections
	The attackers used radio interference to interfere with local wi-fi
	An attacker tries out all the possible combinations to crack sensitive information such as a user's account name, Password, etc.
	The attacker manipulates and generates different session ids to gain privileged access

(*continued*)

Table 1. (*continued*)

Information destruction	An attacker modifies sensitive information in a database
	The attacker stores the wrong data in an unencrypted removable storage medium
	An attacker breaks the authentication message exchange to gain access to the system
	Attackers falsify real-time security logs to hide activity
	An attacker falsifies a document by using the digital signature of a legitimate user
Information content security	The attacker gains physical access to the user's computer
	The attacker breached the network routing configuration, making it inaccessible to the webserver
	An attacker reads the security logs to gain access to sensitive data
	The attacker uses poor access configuration to attack and read sensitive data
	An attacker can access a public-facing database
	The attacker uses the stolen encryption key to access all the user's sensitive data

(3) Defense Cards

See Table 2 below for details:

Table 2. Defensive cards.

Face of cards	Function
Suspicious links/attachments	It's not receiving. It's not opening
Firewalls	To filter potentially destructive code and prevent unauthorized access to the internal private network
Patch management	Install and update system patches to combat low-level network statistics
Antivirus software protection	Install and update software promptly
Safe Culture	Train the staff's awareness of network security, enhance the ability to identify and deal with network fraud
Secure configuration	Ensure the system configuration can meet the needs of the work and limit users to modify the network configuration
Audit system	Record the contents and activities related to network security for review

(*continued*)

Table 2. (*continued*)

Face of cards	Function
Network monitoring	The server monitors and analyzes network activity and services in the form of round-the-clock alerts
Security Policy	Establish network rules and guidelines to protect the storage of data and information
Physical security	Security Control and access to the data center are subject to strict personnel restrictions
Back up your data regularly	Copy or archive files and data to restore the original folder in the event of data loss
Network restriction protection	Limits the transfer of information from the network to the internal network of the user's organization
To encrypt the transmission	Therefore, even if the attacker intercepts it, it is not easy to crack it
High-security Server	Carry on the flow protection to the server of the machine room, resist the network attack
Shut it down in time	Shut down devices or networks when they are not in use

(4) Information Cards

Information description of network security events. See Table 3 for details:

Table 3. Information cards.

Security breach	Information description
Unknown Code or Third-Party Code	Use Unverified Code created by a third party
Physical security is weak	There's no access control in the data center
Weak security configuration	UNENCRYPTED transfer file allows users to modify network configuration
Unsecured wi-fi	A Wi-Fi network, which can be accessed randomly without any protection
Lack of consciousness	The staff was unable to identify malicious e-mails and links
Network monitoring flaws	There was no round-the-clock or regular network monitoring and review
Deficiencies in the formulation of security policy	There are no specific and clear rules governing the use of networks and data

(*continued*)

Table 3. (*continued*)

Security breach	Information description
UNSAFE USB drive	There are no clear rules for the safe use of USB drives
Access control defect	Users who do not have strict access control
Additional information	Additional information provided by the game master, including assistance in handling related security incidents

(5) Question Bank Card

Each question bank card contains a question about network security, and the content is replaceable.

6.3 Rules of the Game

The game is composed of odd players, one of whom acts as "God" (usually played by the teacher), grasps the rules and progress of the game, is responsible for publishing the correct answers of the question bank cards, and records the scores of the two teams. The rest drew their identities and formed two camps for attack and defense. The attacker simulates the hacker's motivation and behavior, detects the weakness of the enterprise network, makes use of and expands in depth, obtains the business data within the scope of authorization, server control organization, business control organization and business control organization, and monitors the attack, responds and handles it through device monitoring and log flow analysis [13].

The game adopts the integral system. At the end of the round, the winner is judged by the score. During the game, students can obtain the information card by answering security questions or consuming a point. To a certain extent, the information card can help the attacker or defender make the correct card selection. At the beginning of the first round, God draws a question bank card for both offensive and defensive sides to answer, and the one who answers correctly obtains the information card. In the process of defense, the defense party can play three cards at the same time, and its defense strength decreases gradually with the value of points. The points are 3, 2 and 1, of which the card with a score of 3 is the main card, which is considered to be the most suitable defense method. After playing cards, the game master will judge the appropriateness of the defense method. The score of the main card method is 3 points, the score of the second defense method is 2 points, and the score of the third defense method is 1 point. If the method is invalid, no score will be given. The score obtained by the attacker is opposite to that of the defender. The scores in the four cases are 0, 1, 2 and 3.

The game is dominated by teachers, providing corresponding game help, promoting players' knowledge construction, allowing players to actively reflect on their attack and defense strategies, promoting discussions between players, and providing strategies for the correctness and effectiveness of real-time feedback. The attack and defense board covers a wide range of attack and defense measures, and at the beginning of each round, it has to answer security questions for educational purposes.

6.4 Hands-On Game

The game is generally divided into three stages. The first stage is for players to draw identity cards and answer security questions under the leadership of the game master. The second link is the attack and defense link, which is also the most important part of the game. The third stage is for the game master to judge and calculate scores.

The situation simulation is as follows:

Answer stage. At the beginning of the game, players draw their own identity cards from the disrupted identity card group. The number of identity cards can be increased or reduced according to the teaching needs. They are divided into two teams, the attacker and the defender, with 3–5 people in each team. The game master extracts the cards in the question bank, and both parties rush to answer. The other party can obtain an information card or a point.

Attack and defense stage. The game master can help the attacker to develop the attack scene. After discussion, the attacker player selects a card from the attack suit. The face of this card is a "send email containing malicious links and files". The attacker collects the relevant information of the target through various channels and uses this information to send an e-mail to gain trust. By pretending to be an IT service and using urgency, the attacker tries to ask the e-mail receiver to click the link provided in the e-mail for network maintenance. Since the target can access sensitive information, the click of a malicious link may enable the attacker to obtain a large amount of sensitive information or paralyze the target system. After the attacker plays cards, the defender must choose countermeasures for the phishing attack proposed by the attacker (the game master will only give limited time to decide the defense), select 3 cards from all defense cards, and describe and evaluate their effectiveness. After discussion, the defender agreed to choose "security configuration" as the main card, that is, the main defense method. Under the condition of ensuring that the system configuration basically meets the work needs, the defender can protect the network security by limiting users to modify the network configuration. Security awareness training and regular backup of data are the second and third defense methods, that is, training employees to identify malicious e-mails faster, improve network security awareness and regularly copy or archive files and data.

Scoring stage. After the description of the defender, the game master decided that the second defense method is the most effective. Therefore, the defender obtains 2 points, and the attacker obtains 1 point.

At the end of the round, the cards were drawn again and entered into the answer stage.

Students play different roles in the game to understand how other attacks and defenses play a role in network security management. By exchanging parts, all students can have the opportunity to play different roles and form a general intuitive impression. Traditional teaching is a more vivid and attractive image. Let students feel the joy of learning inspires them to continue to learn this course with enthusiasm.

7 Conclusion

This paper proposes a card game scheme for students. It complements the learning of students of relevant majors, mainly focusing on improving network security awareness and supplemented by skill goals. The game design adds question bank cards, which increase the game's challenge, improve students' enthusiasm for learning, and make the game more educational. The disadvantage is that the game's audience is relatively limited, and the environment significantly affects the game's progress. The classroom effect needs to be grasped by teachers in the whole process, which has high requirements for teachers. With the rapid development of network security technology, it is essential to improve network security awareness. Students learn relevant technologies and enhance their understanding of protection through learning in class. The network security course uses the game teaching mode, which can effectively avoid the problems existing in traditional classroom teaching, improve students' interest in learning, realize the combination of work and rest, and consolidate the purpose of learning. The reform of the game teaching method has changed the traditional view of education and learning of the network engineering specialty and put forward new problems and challenges for teachers and students.

References

1. Wim, W.: Why and how serious games can become far more effective: accommodating productive learning experiences, learner motivation and the monitoring of learning gains. Educ. Technol. Soc. **22**(1), 59–69 2019
2. Lezama, O.B.P., Manotas, E.N., Mercado-Caruzo, N.: Analysis of design patterns for educational application development: serious games. Procedia Comput. Sci. 175, 641–646 (2020)
3. De Troyer, O., Van Broeckhoven, F., Vlieghe, J.: Linking serious game narratives with pedagogical theories and pedagogical design strategies. J. Comput. High. Educ. **29**(3), 549–573 (2017). https://doi.org/10.1007/s12528-017-9142-4
4. Checa, D., Miguel-Alonso, I., Bustillo, A.: Immersive virtual-reality computer-assembly serious game to enhance autonomous learning. Virtual Real. (2021). https://doi.org/10.1007/s10 055-021-00607-1. (prepublish)
5. Yang, Y.: Analysis on the dilemma development strategies of higher vocational education in China. In: Analysis on the Dilemma Development Strategies of Higher Vocational Education in China, Chengdu, Sichuan, China (2021)
6. Zhou, Z.: The development of serious games in China. Intelligence (03), 247 (2019). (in Chinese)
7. De Gloria, A., Bellotti, F., Berta, R.: Serious Games for education and training. Int. J. Serious Games **1**(1) (2014)
8. Laamarti, F., Eid, M., Saddik, A.E.: An overview of serious games. Inventi Impact Comput. Games Technol. **2015** (2015)
9. Deng, D.: The research and practice of serious games embedded in classroom teaching in colleges and universities in the era of 'Internet+'. Educ. Teach. Forum (22), 242–243 (2020). (in Chinese)
10. Zhang, Q., Du, L.: The application and challenges of serious games in education. Educ. Technol. Equip. China (24), 111–112+117 (2018). (in Chinese)

11. Luo, Z., et al.: Improved flipped classroom network security technology game teaching research. Exp. Technol. Manag. **34**(09), 164–168+172 (2017). (in Chinese)
12. Liu, L.: Application of project games in secondary vocational classrooms. J. Chengde Pet. Coll. **20**(06), 81–84 (2018). (in Chinese)
13. Raquel, M., et al.: Design of a serious games to improve resilience skills in youngsters. Entertain. Comput. **40** (2022)
14. Sitthiphong, S., et al.: Expectations on online orthopedic course using constructivism theory: a cross-sectional study among medical students. Ann. Med. Surg. **67**, 102493 (2021)
15. Chen, X.: From the 'legitimate marginal participation' to see the learning difficulties of beginners. Glob. Educ. Outlook **42**(12), 3–10 (2013). (in Chinese)
16. Hart, S., et al.: Riskio: a serious game for cyber security awareness and education. Comput. Secur. **95**, 101827 (2020). (prepublish)
17. Luh, R., et al.: PenQuest: a gamified attacker/defender meta model for cyber security assessment and education. J. Comput. Virol. Hacking Tech. **16**(4), 19–61 (2020)
18. Valentino, S., et al.: Card game analysis for fast multi-criteria decision making. RAIRO Oper. Res. **55**(3), 1213–1229 (2021)
19. Yong, T., et al.: Research on network security risk analysis method based on full traffic. J. Phys. Conf. Ser. **1792**(1), 012038 (2021)
20. Shanjie, C.: Network security protection technology under the background of computing big data. J. Phys. Conf. Ser. **1982**(1), 012207 (2021)

KPH: A Novel Blockchain Privacy Preserving Scheme Based on Paillier and FO Commitment

Yang Li[1,2], Mengmeng Wang[1,2(✉)], Jianming Zhu[1,2], and Xiuli Wang[1,2]

[1] School of Information, Central University of Finance and Economics, Beijing 100081, China
Doublemeng9535@163.com
[2] Engineering Research Center of State Financial Security, Ministry of Education, Central University of Finance and Economics, Beijing 102206, China

Abstract. Blockchain is a shared database with excellent characteristics, such as high decentralization and traceability. However, data leakage is still a major problem for blockchain transactions. To address this issue, this work introduces KPH (Paillier Homomorphic Encryption with Variable k), a privacy protection strategy that updates the transaction amount using the enhanced Paillier semihomomorphic encryption algorithm and verifies the transaction using the FO commitment. Unlike the typical Paillier algorithm, the KPH scheme's Paillier algorithm includes a variable k and combines the L function and the Chinese remainder theorem to reduce the time complexity of the algorithm from $O(|n|^{2+e})$ to $O(logn)$, making the decryption process more efficient.

Keywords: Blockchain · Paillier homomorphic encryption · Chinese remainder theorem · FO commitment

1 Introduction

As an important infrastructure for building the Internet of Value in the future, blockchain deeply integrates cutting-edge technologies such as point-to-point communication, distributed architecture, consensus mechanisms and cryptography and is becoming the forefront of technological innovation [1]. Blockchain is tamper-resistant and transparent.

Integrated and distributed security features have become a new hotspot in global technological and economic development in recent years. However, with the widespread popularity of applications driven by blockchain, the technical challenges from performance, cost, expansion and other aspects of service quality and privacy security are becoming increasingly severe. The existing blockchain privacy protection schemes based on traditional homomorphic encryption obviously have poor performance in terms of service quality and privacy protection, such as a low verification rate, poor scalability, high energy consumption, and high risk of data leakage, which makes it difficult to maintain transparency and privacy. There is a contradiction between providing satisfactory services to users and maintaining fairness, reliability, sustainability, privacy and computing efficiency in the process of performing tasks, thus hindering the widespread application of blockchain technology. There are two types of privacy vulnerabilities in

Y. Wang et al. (Eds.): ICPCSEE 2022, CCIS 1629, pp. 92–104, 2022.
https://doi.org/10.1007/978-981-19-5209-8_7

the blockchain: identity privacy leakage and transaction information leaking. To address these two issues, this work employs the modified Paillier method to homomorphically encrypt transaction amount information and conduct interval verification using the FO commitment in the zero-knowledge proof.

Homomorphic encryption is a method that allows performing calculations on ciphertext without any plaintext information. Although fully homomorphic encryption technology has the excellent properties of both additive homomorphism and multiplicative homomorphism, it has high model complexity and thus few practical applications. Compared with fully homomorphic encryption technology, semihomomorphic encryption technology has higher algorithm efficiency. Transaction verification in the blockchain only requires the property of additive homomorphism. Combined with the results of the paper raised by Zongyu et al. [2], this paper compares the existing semihomomorphic encryption algorithms from four aspects: encryption system, security assumption, homomorphic properties, and ciphertext expansion rate, as shown in Table 1.

Table 1. Comparison of semihomomorphic encryption algorithms

Semihomomorphic encryption algorithm	Encryption system	Homomorphic properties	Security assumption	Message expansion	Decryption complexity
RSA	Deterministic	Multi	IF	1	$O(log(n)^3)$
ElGamal	Probabilistic	Multi	DL, SS	2	$O(n)^3$
GM	Probabilistic	Addition	QR, SS	n	$O(ln^3)$
OU	Probabilistic	Addition	IF, P-S, SS	3	$O(log^3 n)$
Paillier	Probabilistic	Addition	QR, DCR, SS	2	$O(|n|^{2+e})$
BGN	Probabilistic	Addition	QR, SD, SS	n/r	$O(\sqrt{n})$

It is evident that the RSA, ElGAmal and Paillier algorithms have lower ciphertext expansion rates. The RSA and ElGamal algorithms are multiplicative homomorphic encryption algorithms, and the Paillier algorithm is an additive homomorphic encryption algorithm, which can be better applied to verify the transaction amount in blockchain. Therefore, this paper chooses Paillier as the encryption algorithm of the blockchain privacy protection scheme. Moreover, studies have been conducted to enhance algorithm efficiency by speeding up the decryption phase. This paper proposes an improved Paillier algorithm. This scheme introduces a variable k to simplify the modular operation. The L function and the Chinese remainder theorem are combined in this variable, which effectively minimizes the algorithm's time complexity.

The contributions of this paper are manifold:

- First, this paper expounds on the privacy leakage problem of blockchain.
- It summarizes the existing and typical blockchain privacy protection schemes, summarizes the characteristics of several typical semihomomorphic encryption algorithms,

and expounds the reasons for choosing the Paillier encryption algorithm for scheme design.

- In addition, based on papers [15–17, 21], this paper proposes a KPH scheme with higher decryption efficiency. This scheme applies the improved Paillier encryption algorithm, which makes the decryption efficiency of the scheme higher.
- Finally, we compare the proposed scheme with the papers [15–17, 21]. Compared with these schemes, our scheme is more efficient and has lower time complexity

2 Related Works

Blockchain technology relies on the openness and transparency of the global ledger to ensure that nodes on the chain can verify the legitimacy of transactions at any time and jointly maintain the global ledger, which leads to the problem of privacy leakage. To protect transaction privacy on blockchain, extensive research has been carried out in academia. In 2013, Maxwell [3] proposed a currency mixing scheme, which aims to hide the input and output addresses by mixing multiple transaction information to keep the safety of user identity, but the transaction amount is still open and transparent, and the currency mixing operation brings additional costs. In contrast, the scheme that uses encryption algorithms to encrypt and protect transaction information is more secure and reliable. This technology can be divided into four categories: ring signature, secret address, zero-knowledge proof and homomorphic encryption.

Rivest [4] and others first proposed ring signature technology. The sender signs a transaction using its own private key and public keys of the ring members, making it impossible for the verifier to figure out who the sender is and thereby concealing the transaction's identify information. However, ring signature technology does not hide the transaction amount, and there is a large amount of redundant data in the transaction process, which wastes storage space. Additionally, since there are no administrators on the ring, the technology cannot police transactions.

Todd [5] first proposed the secret address scheme. Whenever the sender wants to initiate a transaction, it must use the recipient's private key to encrypt a one-time address, and the recipient monitors the network at any time and uses the private key to check the information related to itself. Since the addresses used in each transaction are different, the technique ensures that both parties to the transaction's identification information are protected.

Goldwasser [6] and others proposed the prototype of zero-knowledge proof, which proves the correctness of the assertion without providing valid information to the verifier. Zero-knowledge proofs are divided into two categories: interactive and noninteractive. The former lacks efficiency, while the latter is more widely used on the blockchain. Zero-Cash [7] introduces the noninteractive zero-knowledge proof algorithm zk-SNARKs to protect the transaction amount and transaction address of both parties, achieving an excellent privacy protection effect. However, the elliptic encryption algorithm adopted in this scheme costs overlong time and reduces the transaction efficiency.

The homomorphic encryption technique can execute mathematical operations on ciphertext directly and obtain the same result as plaintext operations. This feature makes homomorphic encryption widely used in the area of cloud service data privacy protection. Anjan et al. [8] used Paillier and elgamal encryption algorithms for double-layer encryption for privacy protection. Tsai et al. [9] proposed a separable and reversible data hiding in an encrypted image scheme based on histogram translation and Paillier cryptosystem. Fang et al. [10] explored an average consensus algorithm based on Paillier encryption. Jing et al. [11] applied a homomorphic encryption scheme to design a novel federated learning protection scheme Gradually, domestic and foreign scholars have also begun to explore the application of on-state encryption technology in blockchain privacy protection. Bernal et al. [12] concluded that the current blockchain blockchain is subject to different scalability, security and potential privacy issues, such as transaction linkability, encryption key management, on-chain data privacy and other issues. Aiming at the problem of privacy leakage in smart grid data aggregation and incentives, Zhu et al. [13] proposed a privacy protection scheme based on the Paillier algorithm. This scheme creates bidirectional anonymity of smart grid client transaction information and client information by introducing blockchain and exchanging signatures. Ghadamyari et al. [14] proposed using a combination of a Paillier encryption algorithm and blockchain to solve the problem of medical privacy leakage in the process of health data analysis. Jianzhen et al. [15], Yan et al. [16] and others used the Paillier homomorphic encryption technique to secure transaction data on blockchain and achieved remarkable results. Diao et al. [17] and others proposed group signature and homomorphic encryption technology on the alliance chain, which not only hides the transaction amount but also protects the personal information of both parties to the transaction. Liu et al. [18] proposed that the Chinese remainder theorem may be used to speed up the Paillier homomorphic encryption technique, thereby improving the system performance of the algorithm. On their basis, this paper proposes a more efficient KPH scheme, which applies the improved Paillier scheme of the Chinese remainder theorem, adds a variable k to update the transaction amount, and uses the FO commitment to verify transaction validity. Compared with the above schemes, the scheme in this paper reduces the time complexity of decryption on the basis of security, making the entire transaction process more efficient.

3 Preliminaries

This paper proposes a more efficient KPH scheme that applies the improved Paillier scheme of the Chinese remainder theorem and adds a variable k to update the transaction amount and the FO commitment to verify transaction validity.

3.1 Paillier Cryptosystem

Pascal Paillier [19] proposed an additive homomorphic cryptosystem in 1999, which is a probabilistic symmetric algorithm for public cryptography. The Paillier cryptosystem has three stages, which are as follows:

- Key generation

Randomly choose two large prime numbers p and q, and calculate $n = pq$ and $\lambda = lcm(p - 1, q - 1)$ such that $gcd(pq, (p - 1)(q - 1)) = 1$.

Choose a random number $g \in Z_{n^2}^*$, and confirm that n divides the order of g by looking for the following modular multiplication inverse: $\mu = \left(L(g^\lambda \bmod n^2)\right)^{-1} \bmod n$ where $L(x) = (x-1)/n$.

(n, g) performs the function of a public key.

(λ, μ) performs the function of a private key.

Setting $g = n + 1$, $\lambda = \phi(n)$, and $\mu = \phi(n)^{-1} \bmod n$ where $\phi(n) = (p-1)(q-1)$, given that p, q are of the same length, is a simple version of the previous key generation processes.

- Encryption

Let m be the encrypted message, and where $0 \le m < n$.

Choose a random integer r where $0 < r < n$.

Calculate ciphertext as $c = g^m \cdot r^n \bmod n^2$.

- Decryption

Suppose c is the ciphertext to decrypt, where $c \in Z_{n^2}^*$.

Calculate the plaintext message as $m = L(c^\lambda \bmod n^2) \cdot \mu \bmod n$.

Ouput m.

3.2 Chinese Remainder Theorem

Under the condition that the divisors are pairwise coprime, the Chinese remainder theorem [20] asserts that if one knows the remainders of the division of an integer n by multiple integers, one can uniquely calculate the remainder of the division of n by the product of these integers.

Suppose $n_1, n_2, \cdots, n_k$ is a set of pairwise relatively prime natural numbers, and let $b_1, b_2, \cdots, b_k \in Z$. Put $N = n_1 n_2 \cdots n_k$, the product of the moduli. Then there is an individual $x \bmod N$ such that $x \equiv b_i \bmod n_i$ for all $1 \le i \le k$ as follows:

$$x = b_1 m_1 m_1^{-1} + b_2 m_2 m_2^{-1} + \cdots + b_k m_k m_k^{-1} (\bmod N) \tag{1}$$

where $m_i = N/n_i$ and $m_i m_i^{-1} = 1 \bmod n_i$ for all $1 \le i \le k$.

3.3 CRT-Based Paillier Cryptosystem

The Chinese remainder theorem has been widely applied in various encryption methods to facilitate either the decryption or encryption procedure. Here, we focus on the decryption phase.

Recalling the L function in Sect. 3.1 Paillier Cryptosystem, we define $L_p(x) = (x-1)/p$ and $L_q(x) = (x-1)/q$. Since $\mu = \left(L(g^\lambda \bmod n^2)\right)^{-1} \bmod n$, let $\mu = h_p + h_q$, where

$$h_p = \left(L_p(g^{p-1} \bmod p^2)\right)^{-1} \bmod p \tag{2}$$

$$h_q = \left(L_q \left(g^{q-1} \bmod q^2 \right) \right)^{-1} \bmod q \tag{3}$$

which are precomputed.

Since $m = L\left(c^{\lambda} \bmod n^2\right) \cdot \mu \bmod n$, we have

$$m_p = L\left(c^{p-1} \bmod p^2\right) \cdot h_p \bmod p \tag{4}$$

$$m_q = L\left(c^{q-1} \bmod q^2\right) \cdot h_q \bmod q \tag{5}$$

Applying the Chinese remainder theorem, we can solve the above equation set and obtain the decryption output

$$m = CRT\left(m_p, m_q\right) \bmod pq \tag{6}$$

3.4 An Optimized Paillier Cryptosystem

Ogunseyi et al. [21] proposed an enhanced decryption technique based on the Paillier and CRT-based systems that leverages the variable k and the L function to decrease decryption overload.

- Parameter Setting

$g = n+1$, $\lambda = \phi(n)$, $\mu = \phi(n)^{-1} \bmod n$, in which $\phi(n) = (p-1)(q-1)$ and the sizes of p and q are the same.

- Key Generation

Select huge stochastic prime integers p and q in which $p, q \in 2^{n-1}$.
Let $n = pq$, $\phi(n) = (p-1)(q-1)$ and generate $g = n+1$, define $L(y) = (y-1)/n$.
(n, g) performs as a public key.
(p, q, n) performs as a private key.

- Encryption

Given public key (n, g) and a message $m \in \{0, 1\}^{|n|-1}$.
Select a stochastic integer r, where $0 < r < n$ and set the $gcd(r, n) = 1$.
Calculate ciphertext as $c = g^m \cdot r^n \bmod n^2$.

- Decryption

Given a ciphertext c and the private key (p, q, n).
Set variable $k_y = L(y) \cdot \phi(n)^{-1} \bmod n^2 = (y-1)/n \cdot \phi(n)^{-1} \bmod n^2$.
Define $L_p(x) = (x-1)/p$ and $L_q(x) = (x-1)/q$.
Compute $k_p = L_p\left((p-1)^{-1}\right) \bmod p^2$ and $k_q = L_q\left((q-1)^{-1}\right) \bmod q^2$.
Compute $m_p = L_p\left(c^{p-1} \bmod p^2\right) \cdot k_p \bmod p$ and $m_q = L_q\left(c^{q-1} \bmod q^2\right) \cdot k_q \bmod q$.
Output $m = CRT\left(m_p, m_q\right) \bmod pq$.

3.5 Fujisaki-Okamoto Commitment

FO Commitment is insecure when applied to blockchain transactions with no strings attached. Using the homomorphism of FO commitment, it is easy to verify the balance property of transactions; that is, the verifier only needs to compare the commitment of the input amount with the commitment of the output to determine the correctness of the transaction. This scheme is simple and verifies transactions but is not secure. Because committed additions and subtractions operate modulo N in a cyclic group, there is a possibility of overflow or negative amounts. Usually, the solution is to attach a range proof condition to prove that the transaction amount is a positive integer or within a certain integer range, e.g., within $[0, 2^{10}]$.

The followings refer to this paper raised by Wu et al. [22]. Suppose $t, l, s1, s2$ are public security parameters and n is a large composite number where the prover and verifier do not know its factorization. The two commitment winds of x are (7), where r_i meets condition (8), which makes (9).

The prover is attempting to persuade the verifier that the commitment value x in a commitment is satisfied x $\in [a, b]$. The prover calculates (10) (11) (12) (13) and sends (u, E_2, E_3, F) for the verifier. Then, the verifier calculates (1) when $i = 1$ and (13). Finally, the prover and verifier calculate $PK_i (i = 1, 2, 3)$ separately, and the verifier checks to prove

$$E_i(x_i, r_i) = g_i^x h_i^{r_i} \bmod n \tag{7}$$

$$r_i \in [-2^{s_1}n + 1, 2^{s_1}n - 1] \tag{8}$$

$$r_3 - r\alpha^2 + r_1\alpha + r_2 \in [-2^s n + 1, \cdots, 2^s n - 1] \tag{9}$$

$$E_1 = g^{m-a}h^r \bmod n \tag{10}$$

$$E_i = E_{i-1}^\alpha h^{r_{i-1}} \bmod n \tag{11}$$

$$F = g^\omega h^{r_3} \bmod n \tag{12}$$

$$U = \frac{g^u}{E_3} = g^\omega h^{-r\alpha^2 - r_1\alpha - r_2} \bmod n \tag{13}$$

In a transaction, we need to verify whether the transaction amount is greater than 0 and whether the transaction initiator Alice's account balance is greater than the transaction amount. Using a zero-knowledge proof, when verifying that the transaction amount m is greater than 0, $y = x - a$ is transformed into $y = m - 0$.

- The prover sets $u = \alpha^2 y + \omega > 2^{t+l+s+T}$, where $\alpha \neq 0$ and randomly chooses $\omega \in (0, 2^{s+T}))$, $r_1, r_2, r_3 \in [-2^s n + 1, \cdots, 2^s n - 1]$ which makes $r_3 - r\alpha^2 + r_1\alpha + r_2 \in [-2^s n + 1, \cdots, 2^s n - 1]$. The prover calculates $E_1 = g^{m-a}h^r \bmod n$, $E_2 = E_1^\alpha h^{r_1} \bmod n$, $E_3 = E_2^\alpha h^{r_2} \bmod n$, $F = g^\omega h^{r_3} \bmod n$, $U = \frac{g^u}{E_3} = g^\omega h^{-r\alpha^2 - r_1\alpha - r_2} \bmod n$, and sends (u, E_2, E_3, F) to the verifier.

- The verifier calculates $E_1 = \frac{E(m,r)}{g^a} = g^y h^r \bmod n$, $U = \frac{g^u}{E_3} = g^\omega h^{-r\alpha^2 - r_1\alpha - r_2} \bmod n$
- The prover and verifier calculate separately

$$PK_1\{\alpha, r_1, r_2 : E_2 = E_1^\alpha h^{r_1} \bmod n \wedge E_3 = E_2^\alpha h^{r_2} \bmod n\} \qquad (14)$$

$$PK_2\{\omega, r^{-r\alpha^2 - r_1\alpha - r_2} : F = g^\omega h^{r_3} \bmod n \wedge U = g^\omega h^{r^{-r\alpha^2 - r_1\alpha - r_2}} \bmod n\} \qquad (15)$$

$$PK_3\{\omega, r_3 : F = g^\omega h^{r_3} \bmod n \wedge \omega \in \left[-2^{t+l+s+T}, 2^{t+l+s+T}\right]\} \qquad (16)$$

- The verifier checks the correctness of $PK_i(i = 1, 2, 3)$ and $u > 2^{t+l+s+T}$ to verify $x > a$, namely, $m > 0$.

3.6   Blockchain Data Sharing Model

The blockchain infrastructure [23] is shown in Fig. 1. There are two forms of bookkeeping in blockchain: the UXTO model and the ACCOUNT model. The former only has the concept of the transfer amount but not the balance, while in the latter scenario, as long as the transfer amount is less than the current balance, the transaction can be carried out, so this article chooses the accounting model based on the alliance chain.

Applica- tion layer	Programmable currency		Progammable finance		Programmable society	
	Script code			Smart contract		
Consensus layer	Consensus mechanism		Issuing mechanism		Incentive mechanism	
	PBFT	PoW		PoS	DPoS	
Network layer	Transmission mechanism			Verification mechanism		
	P2P network					
Data layer	Data block			Chain structure		
	Hash function	Timestamp		Merkle tree	Asymmetric encryption	

Fig. 1. Blockchain infrastructure

4   Privacy Protection Model for Blockchain Data Sharing

4.1   Hidden Amount

(1)   RSA public key encryption

In the process of blockchain transactions, it is necessary to hide the balance of each account on the public ledger and the transfer amount in a transaction. For the former, the improved Paillier homomorphic encryption algorithm is used for protection; for the latter, the improved Paillier homomorphic encryption algorithm and the RSA public key algorithm are used for encryption to verify transactions and update the local balance. The schematic diagram is shown in Fig. 2.

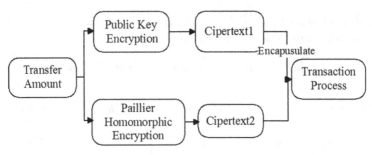

Fig. 2. Data encryption schematic

The principle of RSA public key decryption and encryption is as follows:

- The prover selects both p and q as large prime numbers, then calculates $\varphi(n) = (p-1)(q-1)$, and arbitrarily picks a small integer e that satisfies $gcd(e, \varphi(n)) = 1$ and $1 < e < \varphi(n)$. Then, find the integer $d < \varphi(n)$ so that $de = k\varphi(n) + 1$
- The prover takes (e, n). as the public key pk and makes it public and (d,n) as the private key sk.
- The verifier computes the plaintext m $c = m^e \bmod n$ and delivers the ciphertext to the prover.
- After the prover receives the ciphertext c, he calculates $m = c^d \bmod n = (m^e \bmod n)^d \bmod n = m^{ed} \bmod n = m \bmod n$.

During the transaction, the sender utilizes the receiver's public key pk to encrypt the amount and add it to the transaction information. After the transaction is completed, the receiver utilizes its own private key sk to obtain the plaintext of the transfer amount and updates the local balance accordingly. This process is shown in Fig. 3.

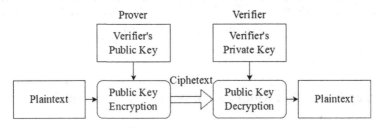

Fig. 3. Verification process

(2) Optimized Paillier Homomorphic Encryption

The optimized Paillier homomorphic encryption mainly includes three stages: key generation, plaintext encryption and ciphertext decryption.

The public key $pk_d = (n, g)$ and private key $sk_d = (p, q, n)$ are obtained in the key generation stage. When encrypting, for plaintext $m \in \{0, 1\}^{|n|-1}$, use public key encryption to obtain ciphertext $c = g^m r^n \bmod n^2$. Since the r's value is random, the same plaintext m can be encrypted to obtain different ciphertexts c, and the same m can be restored after decryption, thus ensuring the semantic security of the encryption system.

When decrypting, the following values are precomputed:

$$k_p = L_p\left((p-1)^{-1}\right) \bmod p^2 \tag{17}$$

$$k_q = L_q\left((q-1)^{-1}\right) \bmod q^2 \tag{18}$$

We can solve the following equation set by the Chinese remainder theorem

$$m_p = L_p\left(c^{p-1} \bmod p^2\right) \cdot k_p \bmod p \tag{19}$$

$$m_q = L_q\left(c^{q-1} \bmod q^2\right) \cdot k_q \bmod q \tag{20}$$

Finally, the decrypted plaintext is $m = CRT\left(m_p, m_q\right) \bmod pq$.

4.2 Transaction Verification

In a transaction, we need to verify whether the transaction amount is greater than 0 and whether the transaction initiator Alice's account balance is greater than the transaction amount. Using a zero-knowledge proof, when verifying that the transaction amount m is greater than 0, $y = x - a$ is transformed into $y = m - 0$.

- The prover sets $u = \alpha^2 y + \omega > 2^{t+l+s+T}$, where $\alpha \neq 0$ and randomly chooses $\omega \in (0, 2^{s+T}))$, $r_1, r_2, r_3 \in [-2^s n + 1, \cdots, 2^s n - 1]$ which makes $r_3 - r\alpha^2 + r_1\alpha + r_2 \in [-2^s n + 1, \cdots, 2^s n - 1]$. The prover calculates $E_1 = g^{m-a}h^r \bmod n$, $E_2 = E_1^\alpha h^{r_1} \bmod n$, $E_3 = E_2^\alpha h^{r_2} \bmod n$, $F = g^\omega h^{r_3} \bmod n$, $U = \frac{g^u}{E_3} = g^\omega h^{-r\alpha^2 - r_1\alpha - r_2} \bmod n$ and sends (u, E_2, E_3, F) to the verifier.
- The verifier calculates $E_1 = \frac{E(m,r)}{g^a} = g^y h^r \bmod n$, $U = \frac{g^u}{E_3} = g^\omega h^{-r\alpha^2 - r_1\alpha - r_2} \bmod n$
- The prover and verifier calculate separately

$$PK_1\left\{\alpha, r_1, r_2 : E_2 = E_1^\alpha h^{r_1} \bmod n \wedge E_3 = E_2^\alpha h^{r_2} \bmod n\right\},$$
$$PK_2\left\{\omega, r^{-r\alpha^2 - r_1\alpha - r_2} : F = g^\omega h^{r_3} \bmod n \wedge U = g^\omega h^{-r\alpha^2 - r_1\alpha - r_2} \bmod n\right\},$$
$$PK_3\left\{\omega, r_3 : F = g^\omega h^{r_3} \bmod n \wedge \omega \in \left[-2^{t+l+s+T}, 2^{t+l+s+T}\right]\right\}$$

- The verifier checks the correctness of $PK_i (i = 1, 2, 3)$ and $u > 2^{t+l+s+T}$ to verify $x > a$, namely, $m > 0$.

If the above verification is successful, it means that the transaction amount is greater than 0. In the same way, let $y = b - x = b - m$, and repeat the above process to verify whether the transaction amount is more than the account balance.

If both zero-knowledge proofs pass the verification, the correctness of the transaction amount needs to be verified through the additive homomorphism of the improved version of Paillier.

Let the transaction amount be *price*, the prover's current account balance is *before_A*, the account balance after the transaction is completed is *after_A*, the verifier's current account balance is *before_B*, and the account balance after the transaction is completed is *after_B*.

Then the following two equations need to be verified:

$$price + after\_A = before\_A \tag{21}$$

$$price + before\_B = after\_B \tag{22}$$

According to additive homomorphism, the following two equations need to be verified:

$$c_{pk_A}(price) \times c_{pk_A}(after\_A) = c_{pk_A}(before\_A) \tag{23}$$

$$c_{pk_B}(price) \times c_{pk_B}(before\_B) = c_{pk_B}(after\_B) \tag{24}$$

Among them, represents the ciphertext obtained by homomorphically encrypting $c_{pk_A}(*)$ using the prover's public key.

Thus far, we have completed the verification of the legitimacy of the transaction. In the process, only the ciphertext of the transaction amount and account balance is used, which greatly protects the user's transaction privacy.

4.3 Update Account Balance

As mentioned above, on the one hand, the account balance is encrypted by the improved Paillier algorithm and stored on the global ledger, and on the other hand, it is stored locally in plaintext.

After completing a transaction, the prover's account balance in the global ledger will be updated to $c_{pk_A}(after\_A)$. The local account balance is updated to *after_A*. The verifier utilizes the RSA private key to decrypt the data encrypted by the public key and obtains the trading money *price*, updates the local balance to *pirce + before_B*, and updates his account balance on the global ledger to $c_{pk_B}(price) \times c_{pk_B}(before\_B)$.

4.4 Performance Analysis

To safeguard the privacy of the blockchain, Jianzhen et al. [15], Yan et al. [16] and Diao et al. [17] employed the Paillier homomorphic encryption algorithm and used a zero-knowledge proof to validate the blockchain transactions. On this basis, the KPH system, which employs optimized Paillier homomorphic encryption, is proposed in this study. The algorithm introduces a variable k, adds it to the L function, and then combines them with the Chinese remainder hypothesis, which ultimately decreases the decryption efficiency of the algorithm, making the algorithm's decryption time complexity from $O(|n|^{2+e})$ to $O(log\ n)$, and uses the FO commitment for transaction verification. To ensure the security of blockchain transactions, the optimization performance of the KPH scheme is shown in Table 2.

Table 2. Performance analysis results

Schemes	Semihomomorphic encryption algorithm	Computational complexity
Jianzhen et al. [15] Yan et al. [16] Diao et al. [17]	PHES	$O(|n|^{2+e})$
Ogunseyi et al. [21]	CRT-PHES	$O(|n|^2|\alpha|)$
KPH (Scheme in this paper)	K-CRT-PHES	$O(logn)$

5 Conclusion

This paper combines the additive homomorphism of the improved Paillier homomorphic encryption algorithm with the consortium chain of the account model and proposes a scheme that fully guarantees transaction information. To effectively protect the balance information of nodes, this paper uses homomorphic encryption to encrypt and update the public ledger data on the chain so that the balance information of all nodes is invisible to the outside world in the form of ciphertext. The transaction amount is passed to update the personal balance information maintained by the node in clear text. When verifying the legitimacy of the transaction, first, the noninteractive zero-knowledge proof application is applied to determine the legitimacy of the amount and the ability of the payer to pay, and then the homomorphism of the improved Paillier homomorphic encryption algorithm is used to verify the correctness of the balance between the two parties after the transaction is completed. Only the ciphertext information of the balance is used in this process, which fully protects the privacy of the transaction amount. The time complexity of the encryption algorithm suggested in this study is lower than that of the classical Paillier homomorphic encryption algorithm and CRT-PH, and the decryption efficiency is higher.

Acknowledgment. This research is funded by the Emerging Interdisciplinary Project of CUFE, the National Natural Science Foundation of China (No. 61906220) and Ministry of Education of Humanities and Social Science project (No. 19YJCZH178).

References

1. Naiquan, L.: Overview of privacy data security based on blockchain. Netw. Secur. Technol. Appl. **01**, 19–21 (2022)
2. Zongyu, L., Xiaolin, G., Yingjie, G., Xuesong, L., Xuejun, Z.: Homomorphic encryption technology and its application in cloud computing privacy protection. J. Softw. **29**(07), 1830–1851 (2018)
3. Maxwell, G.: CoinJoin: bitcoin privacy for the real world. Post on Bitcoin Forum (2013)
4. Rivest, R.L., Shamir, A., Adleman, L.M.: A method for Obtaining Digital Signatures and Public Key Cryptosystems, pp. 217–239. Routledge (2019)

5. Todd, P.: Stealth addresses. Post on Bitcoin development mailing list (2014). https://www. mail-archive.com/bitcoindevelopment@lists.sourceforge.net/msg03613.html
6. Goldwasser, S., Micali, S., Rackoff, C.: The knowledge complexity of interactive proof systems. SIAM J. Comput. **18**(1), 186–208 (1989)
7. Sasson, E.B., et al.: Zerocash: decentralized anonymous payments from bitcoin. In: 2014 IEEE Symposium on Security and Privacy, pp. 459–474. IEEE (2014)
8. Koundinya, A.K., Gautham, S.K.: Two-layer encryption based on paillier and elgamal cryptosystem for privacy violation. Int. J. Wirel. Microw. Technol. (IJWMT), **11**(3), 9–15 (2021)
9. Tsai, C.S., Zhang, Y.S., Weng, C.Y.: Separable reversible data hiding in encrypted images based on Paillier cryptosystem. Multimed Tools Appl. **81**, 18807–18827 (2022)
10. Fang, W., Zamani, M., Chen, Z.: Secure and privacy preserving consensus for second-order systems based on paillier encryption. Syst. Control Lett. **148**, 104869 (2021)
11. Ma, J., Naas, S., Sigg, S., Lyu, X.: Privacy-preserving federated learning based on multi-key homomorphic encryption. Int. J. Intell. Syst. (2022)
12. Bernal Bernabe, J., Canovas, J.L., Hernandez-Ramos, J.L., Torres Moreno, R. Skarmeta, A.: Privacy-preserving solutions for blockchain review and challenges. IEEE Access **7**, 164908– 164940 (2019)
13. Zhu, S., Wang, H.: Smart grid data aggregation and incentive scheme based on Paillier algorithm. Comput. Eng. **11**, 166–174 (2021)
14. Ghadamyari, M., Samet, S.: Privacy-preserving statistical analysis of health data using paillier homomorphic encryption and permissioned blockchain. In: 2019 IEEE International Conference on Big Data (Big Data), pp. 5474–5479. IEEE (2019)
15. Jianzhen, L.: Applied research on privacy protection of blockchain transactions based on paillier homomorphic encryption, Master's thesis, Southeast University. (2019)
16. Yan, X., Wu, Q., Sun, Y.: A homomorphic encryption and privacy protection method based on blockchain and edge computing. Wirel. Commun. Mob. Comput. (2020)
17. Diao, Y., Ye, A., Zhang, J., Deng, H., Zhang, Q., Cheng, B.: A dual privacy protection method based on group signature and homomorphic encryption for alliance blockchain. J. Comput. Res. Dev. **01**, 172–181 (2022)
18. Yao, L., Shuai, X.: Accelerate the paillier cryptosystem in CryptDB by Chinese remainder theorem. In: 2018 20th International Conference on Advanced Communication Technology (ICACT), pp. 74–77. IEEE (2018)
19. Paillier, P.: Public-key cryptosystems based on composite degree residuosity classes. In: Stern, J. (ed.) EUROCRYPT 1999. LNCS, vol. 1592, pp. 223–238. Springer, Heidelberg (1999). https://doi.org/10.1007/3-540-48910-X_16
20. The wikipedia website. https://en.wikipedia.org/wiki/Chineseremaindertheorem
21. Ogunseyi, T.B., Bo, T.: Fast decryption algorithm for paillier homomorphic cryptosystem. In: 2020 IEEE International Conference on Power, Intelligent Computing and Systems (ICPICS), pp. 803–806. IEEE (2020)
22. Wu, Q., Zhang, J., Wang, Y.: Simple proof that a commitment value is in a specific interval. Electron. J. **07**, 1071–1073 (2004)
23. Zhang, R., Xue, R., Liu, L.: Security and privacy on blockchain. ACM Comput. Surv. (CSUR) **52**(3), 1–34 (2019)

Blockchain Access Control Scheme Based on Multi-authority Attribute-Based Encryption

Yang Li[1,2(✉)], Baoyue Qi[3], Mengmeng Wang[1], Jianming Zhu[1,2], and Xiuli Wang[1,2]

[1] School of Information, Central University of Finance and Economics, Beijing 100081, China
liyang@cufe.edu.cn
[2] Engineering Research Center of State Financial Security, Ministry of Education, Central University of Finance and Economics, Beijing 102206, China
[3] Computer Science Department, University College London, Gower Street, London WC16BT, UK

Abstract. Blockchain has been widely used in many fields because it can solve the problem of information asymmetry and enable users who do not trust each other to collaborate without the participation of third-party intermediaries. Existing blockchain access control schemes usually use attribute-based encryption, but most of them adopt traditional single-attribute authority for attribute authorization, which has the problem that the authority is overburdened and must be fully credible. This paper proposes a blockchain access control scheme based on multi-authority attribute-based encryption by improving the existing blockchain privacy protection method. Autonomous identity management is performed through the blockchain to complete the initialization of user identity and the issuance of attribute certificates. Attribute authorities are selected using the reputation proof consensus mechanism. The distributed key generation protocol is used to generate keys, and the linear secret sharing scheme is improved. The hierarchical relationship of the access structure is used to encrypt and access control the private data that need to be uploaded to the blockchain. According to the comparison with other blockchain access control schemes, the scheme proposed in this paper has been improved in terms of security and efficiency.

Keywords: Access control · Blockchain · Attribute-based encryption · Multi-authority

1 Introduction

The 21st century is an information age where technologies such as mobile networks, e-commerce, and social platforms are developing vigorously. The expansion of Internet applications has enabled us to generate a large amount of data every day. While enjoying the convenience it brings, we are collecting and analyzing personal privacy data all the time to promote industry innovation and stimulate economic growth. As time goes by, the privacy of personal data has attracted increasing attention from the public. Since third-party organizations collect a large amount of personal sensitive information, and individuals can hardly control the storage of their data and how it is used, protecting users' personal data has become a research hotspot today.

Users, on the one hand, need to use third-party applications to obtain convenient services; on the other hand, they need to protect data privacy. Since the data on the blockchain are difficult to change and can be traced, it has become an important technology to solve the above problems. The blockchain enables nodes that do not trust each other to carry out peer-to-peer trustworthy value transfers without relying on trusted third-party institutions. Therefore, in recent years, it has been widely used in the fields of finance, supply chain, healthcare, and e-discovery. For example, in the financial field, storing data on the blockchain can improve data security, and it can also help regulatory agencies collect data from the blockchain for supervision. However, at the same time, these data are very sensitive and important. The data owner does not want to disclose all the data to everyone because once the data are leaked, it will cause considerable losses. Since all data in the blockchain are disclosed to all nodes, the risk of privacy leakage is also significantly increased. Therefore, it is necessary to adopt appropriate encryption technology to achieve flexible and fine-grained access control to better apply to the blockchain environment and protect the security of information.

2 Related Works

Due to the distributed, decentralized, and tamper-proof characteristics of blockchain, it has been widely used in environments other than digital currencies in recent years [1], such as existence verification, smart contracts, identity certification, market forecasting, the e-commerce industry, social platforms and data storage. Because the data in the blockchain are public, it is necessary to use certain access control techniques to achieve secure access to the access target.

In the traditional asymmetric encryption mechanism, the encryption and decryption of information are peer-to-peer, which can guarantee the confidentiality of data, but the encryption operation has a large amount of calculation, and it is difficult to implement fine-grained access control [2]. To adapt to the distributed application environment, the one-to-many communication mode is adopted to reduce the huge cost of encrypting data for each user. The traditional encryption mechanism based on public key infrastructure (PKI) can protect the confidentiality of data, but the resource provider must obtain the user's real public key certificate. In addition, although broadcast encryption [3] partially solves the efficiency problem, it compromises user privacy when obtaining a list of user identities. To solve the above problems, Shamir [4] et al. proposed the identity-based encryption (IBE) mechanism based on the bilinear pairing technique, which directly uses the user's identity as the public key so that the resource provider does not need to query the user's public key certificate online. However, high data encryption costs and user privacy leakage problems remain. Therefore, the above methods cannot be well applied to data sharing in the blockchain environment.

Sahai and Waters proposed a mechanism of attribute-based encryption (ABE) [5] based on IBE technology, which could further solve the problems existing in the IBE mechanism. Compared with other encryption methods, the major difference between attribute-based encryption (ABE) is that it uses attributes as public keys. There is no need to know the identity information of the data visitor during each encryption process, and as long as the attributes of the accessing visitor comply with the access policy described by the data owner, the data can be decrypted. The ABE mechanism can be flexibly used to represent the corresponding access control policies and greatly reduce the overhead of access control, which is more suitable for fine-grained access control on blockchain [6]. For example, a financial company encrypts its own data information and sends it to the blockchain. The data on the chain are encrypted, so other people can only see the encrypted garbled code, which has no use value. Correspondingly, financial companies can control access to data by themselves and only need to issue access permissions to data partners and regulatory agencies.

There are many studies on ABE-based blockchain privacy protection mechanisms in existing solutions, but most of them use traditional single-attribute authorization agencies for attribute authorization. For example, Rahulamathavan [7] et al. combined blockchain with ABE to solve the data protection problem in blockchain and realize end-to-end privacy protection in IoT systems. However, the single authorization authority mechanism has certain problems, as it may not be able to satisfy all demands at the same time when the user demand is large, and a single point of failure will directly lead to the collapse of the system. In addition, a single authority is responsible for the management and release of attribute keys, which will make it too powerful and pose a certain security risk. Adopting such a centralized organization contradicts the distributed and decentralized characteristics of blockchain. To reduce the security risks caused by the excessive authority of a single-attribute authority, it is more appropriate to use the multi-authority attribute-based encryption (MA-ABE) scheme proposed by Chase [8] for blockchain data sharing and access control. However, if the work of multiple authorization centers is independent, the system will be threatened by collision attacks. To solve this problem, one option is to choose to set up a central authority (CA) such as Chase to introduce a global identity locator (GID) that manages all identities and attributes and performs authentication, but the CA must be prevented from becoming the target of malicious attacks. Another option is the scheme without CA proposed by Lewko and Waters [9], and each attribute authority works together, which may lead to high communication cost and low system scalability.

There are many extension studies of MA-ABE scheme applications in different scenarios. Li and Huang propose an MA-ABE-based accountability encryption algorithm [18] that tracks users who maliciously compromise their keys, solving the problem that attributes cannot be associated with a specific individual for decryption keys. Li [19] et al.'s research is geared toward mobile cloud computing, introducing cloud-based semi-trusted authorities (STAs) between mobile users and attribute issuing authorities for a large amount of communication and computation. Rouselakis et al. proposed an efficient large-universe MA-ABE encryption system [20]. In this scheme, any string can be used as an attribute of the system and does not have to be enumerated at setup time, overcoming the "use each attribute only once" limitation. Zhong [21] et al. improved the MA-ABE scheme to hide policies in cloud storage to protect data privacy and access policy privacy, while this scheme can efficiently revoke users.

To solve the problem that the single attribute authorization center in the existing scheme is overburdened and needs to be fully trusted, this paper will adopt a blockchain access control scheme based on MA-ABE and replace the CA with blockchain for identity management. The calculation of the key is handed over to a number of the most reputable blockchain nodes as the attribute authorities complete. Each attribute authority is only responsible for the distribution of part of the private key, which reduces the burden while improving the confidentiality of user data. In addition, this paper also makes use of certain hierarchical relationships between attributes. The data owner formulates an access control strategy and performs an encryption operation on the file. Visitors can access some or all of the files according to the attributes they meet. In this way, the encryption and decryption cost is reduced, and it is more suitable for certain application scenarios, such as the enterprise environment that requires supervision and auditing.

3   Preliminaries

3.1   Bilinear Mapping

Define a cyclic group G_1 whose order is a prime number p, its generator is g, and there is a bilinear mapping $e: G_1 \times G_1 \rightarrow G_2$, which satisfies the following properties.

Bilinear: for any $x, y \in G_1$ and $m, n \in Z_p$, we have $e(x^m, y^n) = e(x, y)^{mn}$.

Non-degeneracy: there exists $x, y \in G_1$ such that $e(x, y) \neq 1$.

Computability: $e(x, y)$ can be efficiently computed for any $x, y \in G_1$.

3.2   Access Control Structure

Let $\{P_1, P_2, \ldots, P_n\}$ be the set of all participants, $P = 2^{\{P_1, P_2, \ldots, P_n\}}$. Define an access structure X as a non-empty subset of the set of participants, i.e. $X \subseteq P \backslash \{\varnothing\}$. When the access structure X is monotone, then for any Y and Z, if $Y \in X$ and $Y \subseteq Z$, then $Z \in X$.

An access control tree is commonly used to describe access control policies. A leaf node is used to denote an attribute node, and a non-leaf node is used to denote a threshold node. For a (t, n) threshold node, its policy is that it has n children nodes, and this threshold policy is satisfied when the attributes of $t(1 \leq t \leq n)$ of these children nodes are satisfied. For example, the AND threshold node is (n, n), and the OR threshold node is $(n, 1)$. The access control tree shown in Fig. 1 represents the policy of owning attribute A and owning at least two of the attributes $\{B, C, D, E\}$.

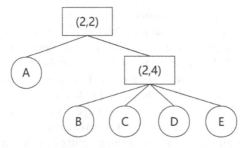

Fig. 1. Access control tree

The access control tree can be converted into an access control matrix by the marking method [10], where the number of matrix rows is equal to the number of leaf nodes and the number of matrix columns is equal to the number of threshold nodes in the access control tree.

3.3 Linear Secret Sharing Scheme

Suppose there exists an $l \times n$ secret sharing generation matrix M, as shown in Formula (1). The function $\rho(i)$ denotes the attribute mapping corresponding to the i-th row of the matrix M. Let $s \in Z_p$ be the secret to be shared, and the vertical vector $v = (s, r_2, \ldots, r_n)$, where $r_2, \ldots, r_n \in Z_p$ are random numbers. The matrix M is multiplied by the vector v to obtain the secret s shared to the l participants with secret shares $\lambda_2, \ldots, \lambda_n$.

$$M = \begin{bmatrix} m_{11} & m_{12} & \cdots & m_{1n} \\ m_{21} & m_{22} & \cdots & m_{2n} \\ \vdots & \vdots & \ddots & \vdots \\ m_{l_1} & m_{l_2} & \cdots & m_{l_n} \end{bmatrix}, v = \begin{bmatrix} s \\ r_2 \\ \vdots \\ r_n \end{bmatrix}, Mv = \begin{bmatrix} \lambda_1 \\ \lambda_2 \\ \vdots \\ \lambda_n \end{bmatrix} \tag{1}$$

Linear reconfigurability: let the access structure of a linear secret sharing scheme be T with a set of attributes $S \in T$ and a set of attributes $I = \{i : \rho(i) \in S\} \subset \{1, 2, \ldots, l\}$ of participants. Then, it is possible to find a set of constants $\omega_i \in Z_p (i \in I)$ within polynomial time, and if λ_i is the effective secret share of secret s, then the $\sum \omega_i \lambda_i = s(i \in I)$ equation holds.

4   Blockchain Access Control Scheme Based on MA-ABE

4.1   Scheme Overview

This paper completes the design of a blockchain access control scheme based on multi-authority attribute-based encryption by improving the existing privacy protection scheme. Autonomous identity management is performed via blockchain to complete the initialization of user identity and the issuance of attribute certificates. Use the proof of reputation (PoR) consensus mechanism to select the attribute authorities. The distributed key generation (DKG) protocol is used to generate keys, and the private data to be recorded by the blockchain are encrypted and access controlled by an improved hierarchical linear secret sharing scheme (LSSS).

4.1.1   Scheme Entity

(1)   *Data owner (DO)*

The data owner (DO) refers to the user who owns and shares the data file, encrypts the data file he wishes to upload using symmetric encryption algorithms, and encrypts the symmetric key using attribute-based encryption algorithms. After that, the generated data ciphertext and key ciphertext are uploaded to the cloud server, and the data upload record is uploaded to the blockchain.

(2) *Data visitor (DV)*

The data visitor (DV) refers to a user who wishes to access the data file and decrypt it based on the private key of the attributes he possesses.

(3) *Attribute authority (AA)*

The attribute authority (AA) is not fully trusted. All AAs generate their master keys and public keys through mutual negotiation, and they are responsible for the generation of the private key components.

(4) *Cloud server (CS)*

The cloud server (CS) is not fully trusted and is responsible for storing various ciphertexts and providing download services to DV.

4.1.2 Scheme Process

(1) *Identity registration management*

All entities send registration requests to the blockchain platform. User identity initialization, attribute certificate issuance and revocation are performed through the autonomy identity management scheme.

(2) *Attribute authorities selection*

Using the PoR consensus mechanism, each node issues verifiable reputation certificates to other nodes and selects a certain number of nodes with the highest reputation as the attribute authorities.

(3) *Password system initialization*

The DKG protocol is used to generate a series of keys by entering security parameters, including system public parameters, system master key, system master public key, the public key of each AA, the private key of each AA, the public key for attributes managed by each AA, the global public key of each user, and the global private key of each user.

(4) *Attribute key generation*

The user sends an attribute key generation request to all AAs. AAs validate the user's attribute certificate and generate the user's attribute private key components to send to the user.

(5) *Data encryption and uploading*

The DO integrates multiple access control trees into a single access control tree according to the hierarchical relationship of data files and divides the data into corresponding parts according to the hierarchy to form a data file set. The data files of each part are encrypted separately using a symmetric encryption algorithm, and each encryption key is encrypted by a hierarchical LSSS improvement scheme. The data ciphertext set is uploaded to the CS along with the key ciphertext set. The DO receives the data storage path and uploads the data records to the blockchain to ensure the integrity of the data files.

(6) *Data Access*

When a DV wants to access the data files shared by other users, he first sends his attribute private key to the CS. The CS verifies the DV's attribute certificate and access structure, generates the corresponding decryption token and sends it. When the attributes of the DV satisfy part of the access structure, the symmetric key decryption token combined with this part can be obtained, and when the attributes satisfy the whole access structure, all the symmetric key decryption tokens can be obtained. The DV uses the decryption token set and its global private key to obtain the encryption key set and uses this key set to decrypt the data ciphertext set with the symmetric encryption algorithm to obtain some or all of the data plaintexts.

4.1.3 Scheme Structure

The structure of the blockchain access control scheme based on MA-ABE is shown in Fig. 2.

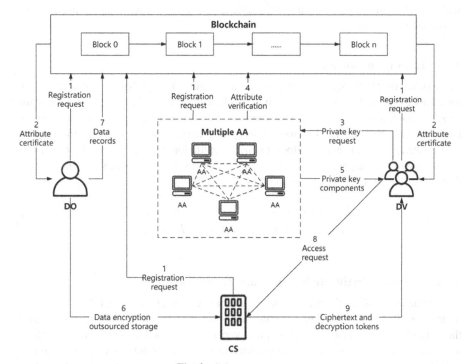

Fig. 2. Scheme structure

4.2 Autonomous Identity Management

In the early Internet environment, the initiation and authentication of user identity were undertaken by central government agencies. As technology further evolved, central authorities (CAs) began to certify the commercial Internet. To decentralize to some extent, some organizations chose to adopt a hierarchical structure. While this has been managed by a secondary organization, the centrality of the root organization remains unchanged, and our identity is still centrally controlled by the website. We would prefer to have autonomous control over our identity.

In the blockchain environment, distributed ledger technology offers the possibility to achieve this goal. Unlike the traditional MA-ABE scheme, users on the blockchain manage their own identities as well as attributes individually, and identity management no longer relies on a CA but on user-centered access control. At present, there are many solutions for blockchain autonomous identity management, and they are being continuously improved.

In the autonomous identity management model of this paper, the identity certificate includes global identity identifiers and corresponding attributes. Each user in the blockchain can act as the owner and publisher of the identity certificate, and each AA and CS act as the verifier of the identity certificate. The model mainly contains three steps: identity initialization, attribute certificate issuance and revocation, and attribute certificate validation.

4.2.1   Identity Initialization

The blockchain platform uses a default approach to generate a global identity locator GID for all nodes, as well as the corresponding blockchain public and private keys PK_{GID} and SK_{GID}, which are used for authentication and communication during transactions. At the same time, the identity document $File_{GID}$ is generated, which contains information such as PK_{GID}, cloud server endpoint CS_{GID}, reputation certificate link $RepL_{GID}$, etc., and is signed by the private key SK_{GID} and published to the blockchain. The identity document $File_{GID}$ is defined as shown in Formula (2)

$$File_{GID} = \{GID, PK_{GID}, CS_{GID}, RepL_{GID},$$
$$Sign_{owner}[GID||PK_{GID}||CS_{GID}||RepL_{GID}]\} \tag{2}$$

4.2.2   Attribute Certificate Issuance and Revocation

Each node in the blockchain can issue an attribute certificate AC of attribute A to other nodes, and the certificate has a unique identifier $ACID$. The certificate issuer signs the attribute certificate AC for subsequent verification. When a user wants to perform attribute revocation, other nodes are required to issue an attribute revocation certificate ARC for him and sign it. All attribute revocation certificates are indexed by the GID of the certificate issuer and the $ACID$ of the revoked attribute certificate and stored together with their signatures in the attribute revocation list ARL jointly maintained by all nodes. The definitions of attribute certificate AC and attribute revocation certificate ARC are shown in Formula (3) and Formula (4).

$$AC = \left\{GID_{owner}, GID_{issuer}, A, Time, ACID, Sign_{issuer}[GID_{owner}||A||Time]\right\} \tag{3}$$

$$ARC = \left\{GID_{issuer}, ACID, Time, Sign_{issuer}[GID_{issuer}||ACID||Time]\right\} \tag{4}$$

4.2.3   Attribute Certificate Verification

The user submits his attribute certificate AC to the verifier when he applies for the key or data. The certificate verifier obtains his PK_{GID} by looking up the GID of the certificate issuer, verifies the signature of AC to verify the authenticity of AC, and checks whether his attribute is revoked in the ARL list.

4.3   Selection of Attribute Authorities

In the blockchain environment, since there is no centralized institution, consensus mechanisms are needed to select certain nodes among a large number of blockchain nodes to conduct accounting and verify the portion of blocks generated by them to ensure distributed accounting consistency. The common blockchain consensus mechanisms include proof of work (PoW), proof of stake (PoS), proof of authorized shares (DPoS), etc. The PoW mechanism selects the node with the highest arithmetic power through arithmetic competition to perform accounting, which is a completely decentralized consensus mechanism, but it is inefficient and wasteful of resources. The PoS mechanism calculates the product of the number of tokens and the holding time as the equity of the node and selects the node with the highest equity as the billing node. This improves efficiency to some extent, but it may produce the Matthew effect and reduce fairness. In the DPoS mechanism, the nodes holding tokens vote to elect some agents for verification and accounting, which greatly improves the efficiency but has a strong centrality.

Most of the consensus mechanisms mentioned above motivate the node to account by increasing revenue or empowering the node and maintaining the sustainability of the blockchain system, such as the issuance of digital currency. However, the blockchain scenarios used for data privacy protection and access sharing do not require the issuance of digital currency and prefer reputational affirmation. Therefore, this scheme adopts the proof of reputation (PoR) consensus mechanism, uses reputation evaluation to replace digital currency in other consensus mechanisms, and selects a certain number of nodes based on the reputation of the computing node in the network (social relationship, joining time, computing power contribution, etc.). Generally, larger and more reputable companies are more likely to be selected [11]. Once they try to deceive the system, they will face the loss of their own business and reputation, which will produce more benefits than deception for the company. For example, in the banking industry, only banks with the appropriate threshold of reputation are allowed to exchange and share data with other banks, and multiple banks with higher reputations and higher computing power are elected as attribute authorities to initialize the cryptosystem and distribute the attribute key building blocks. A good reputation on the one hand provides moral incentives for banks, and on the other hand, a higher reputation indirectly brings higher economic benefits to their business.

4.3.1   Selection of AAs

Each node issues a verifiable reputation certificate *Rep* to other nodes and uses its own blockchain private key to sign the reputation value *RepValue*, which ensures the integrity of the reputation due to the tamper-evident nature of the blockchain. Each blockchain node receives the *Rep* and stores it to form its own reputation certificate link *RepL* for searching. Reputation certificate *Rep* is defined as shown in Formula (5).

$$Rep = \{GID_{owner}, GID_{issuer}, RepValue, Time, \\ Sign_{issuer}[GID_{owner}||RepValue||Time]\} \tag{5}$$

The reputation level of each node is derived by calculating its own reputation value *RepValue*, and the appropriate number of nodes are elected as AAs through the PoR consensus mechanism.

4.3.2   Entry and Exit of AAs

This scheme supports dynamic changes in attribute authorities. If a node in the system wishes to become a new AA, it must obtain the consent of at least n AAs. When an AA wishes to exit the system, all nodes in the system need to be notified that the AA no longer provides attribute authorization and private key artifact distribution services.

4.4   Hierarchical Linear Secret Sharing Scheme

Data files are often shared with a certain hierarchy. A certain group of files is divided into different access levels for different access objects to access. File encryption can reduce the encryption and storage cost to a certain extent by integrating the hierarchical relationship of different access structures.

For example, as shown in Fig. 3, a set of transaction data is divided into two parts, Data1 and Data2. Data1 contains private data, such as the personal information of customers, which can only be accessed internally by both sides of the transaction, and the access structure is T1. Data2 only contains non-private information, such as transaction amount, which can be accessed by both sides of the transaction as well as the audit unit, and the access structure is T2. Since the access structure of Data2 contains the access structure of the Data1 access structure, the access structure can be set as T3. Encryption and storage are only needed once. If the DV attributes meet all the access structures of these data, all the data can be obtained; if they satisfy only a partial access structure, only partial data are available. Both sides of the transaction can obtain all information Data1+Data2, while the audit unit can only obtain transaction-related information Data2 and cannot obtain customer privacy data Data1.

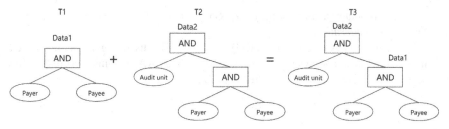

Fig. 3. An access control tree with hierarchical relationships

If the hierarchical structure of access policies is to be utilized, the conventional linear secret sharing scheme (LSSS) needs to be improved. First, the data whose attributes contain hierarchical relationships are integrated into an access control tree, and the access control tree is converted into an access control matrix M by the marking method, whose number of rows l is the total number of attributes involved and whose number of columns n is the total number of thresholds involved. In the traditional LSSS scheme, the secret vector $v = (s, r_2, \ldots, r_n)^T$, only s is the key to be shared, and the remaining $r_2, \ldots, r_n$ are the parameters obtained by random selection. In the improved scheme of this paper, the random parameter part of the secret vector v is utilized. The data are divided into n parts of subdata according to the access hierarchy, and correspondingly n symmetric encryption keys s_j are selected, corresponding to each threshold node in the access policy structure for secret sharing. Therefore, the secret vector is $v = (s_1, s_2, \ldots, s_n)^T$. When secret recovery is performed, the set of visitor attributes is I. If the set of data visitor attributes satisfies this non-leaf node, the constant set $\omega_i \in Z_p (i \in I)$ can be found within polynomial time, satisfying $\sum_{i \in I} \omega_{i,j} M_i = \varepsilon_j$, where ε_j is the row vector with all $n - 1$ elements of 0 except for the j-*th* element which is 1. Then compute the partial key that satisfies this partial policys $s_j = \sum_{i \in I} \omega_{i,j} \lambda_i$. When the data visitor's attributes satisfy the full access structure, all n keys can be obtained to decrypt the full data; if it only satisfies part of the access structure of the hierarchy, only the corresponding partial keys can be obtained to decrypt part of the data.

The cryptographic algorithm in this paper partially utilizes the hierarchy of attributes and improves it on the basis of paper [12] to reduce the encryption and storage costs.

4.5 Blockchain Access Control Algorithm

4.5.1 Password System Initialization

(1) *Initial parameters*

Generate the system public parameters *params* by security parameter λ. Choose two bilinear groups G_1 and G_2 with an order of prime q. There is a bilinear mapping $e: G_1 \times G_1 \rightarrow G_2$ that holds; the generating element of group G_1 is g. The hash function $H: \{0, 1\}^* \rightarrow G_1$. The full set of attributes is U, the number of AAs is *num*, and the threshold value of AAs is t.

We can obtain $params = \{q, g, G_1, G_2, H, U, num, t\}$.

(2) *System key generation*

The distributed key generation (DKG) protocol [13] is used to generate the system master key *SMK* and system public key *SPK*. Each AA$_i$ obtains a corresponding secret share a_i of the system master key *SMK*.

System key generation algorithm

Input: system public parameters $params = \{q, g, G_1, G_2, H, U, num, t\}$

Output: system master key *SMK*, system public key *SPK*, AA$_i$'s secret share a_i

---

①Each AA$_i$ choose a *(t-1)-th* order polynomial $f_i(x)$.
$$f_i(x) = z_{i0} + z_{i1}x + \cdots + z_{i(t-1)}x^{t-1}$$
②Each AA$_i$ calculates and broadcastes Z_{ik} (k=0,...,t-1) separately.
$$Z_{ik} = g^{z_{ik}} \bmod q$$
③Each AA$_i$ calculates the secret value y_{ij} with all other AA$_j$ (j=1,...,num and j≠i) and sends it to the corresponding AA$_j$ respectively.
$$y_{ij} = f_i(AA_j) \bmod q$$
④AA$_j$ verifies AA$_i$ after receiving all Z_{ik} as well as y_{ij}. If AA$_i$ is an honest node, then
$$g^{y_{ij}} = g^{f_i(AA_j)} \bmod q = g^{z_{i0} + z_{i1}AA_j + \cdots + z_{i(t-1)}AA_j^{t-1}} \bmod q$$
$$= \prod_{k=0}^{t-1} g^{z_{ik}AA_j^{k}} \bmod q = \prod_{k=0}^{t-1} Z_{ik}^{AA_j^{k}}$$
Otherwise, AA$_j$ asks AA$_i$ to rebroadcast y_{ij}.

⑤t AAs involved in key generation are denoted as the set S, which generates the system master key *SMK*.
$$SMK = \sum_{i=1}^{num} z_{i0} = \sum_{i=1}^{num} \sum_{j \in S} f_i(j) L_j(0)$$
L_j is the Lagrangian coefficient of AA$_j$. The system master key *SMK* can be reconstructed by t AAs.
⑥Generate the system public key SPK and disclose it in the system.
$$SPK = g^{SMK}$$
⑦AA$_i$'s secret share a_i:
$$a_i = \sum_{j \in S} y_{ji} \bmod q$$

---

(3) *AA initialization*

Each AA$_i$ (with a globally unique identifier of *AID*) generates its own secret key *ASK(AID)*, public key *APK(AID)*, and public key *ATTPK(att$_j$)* of the attribute *att$_j$*.

AA initialization algorithm

Input: AA_i global unique identifier AID, attribute full set U

Output: AA_i secret key $ASK(AID)$, AA_i public key $APK(AID)$, public key $ATTPK(att_j)$ of attribute att_j

---

①Each AA_i randomly selects $\alpha_i, \beta_i, \gamma_i \in Z_q$.

②Each AA_i generates its secret key $ASK(AID)$ and public key $APK(AID)$.

$$ASK(AID) = (\alpha_i, \beta_i, \gamma_i)$$

$$APK(AID) = (e(g,g)^{\alpha_i}, g^{\frac{1}{\beta_i}}, g^{\frac{\gamma_i}{\beta_i}})$$

③Each AA_i randomly selects $v_{att_j} \in Z_q$ and generates the public key $ATTPK(att_j)$ for the attribute att_j.

$$ATTPK(att_j) = (g^{v_{att_j}} H(att_j))^{\gamma_i} a_i = \sum_{j \in S} y_{ji} \; mod \; q$$

---

(4) *User initialization*

Each user (with a globally unique identifier of *UID*) generates a user secret key *USK(UID)* and a user public key *UPK(UID)*.

User initialization algorithm

Input: user global unique identifier *UID*

Output: user secret key *USK(UID)*, user public key *UPK(UID)*

---

①Each user randomly selects $u_m, u_n \in Z_q$.

②Each user generates its own user secret key *USK(UID)* and user public key *UPK(UID)*.

$$USK(UID) = (u_m, u_n)$$
$$UPK(UID) = (g^{u_m}, g^{u_n})$$

4.5.2 Attribute Secret Key Component Generation

The user calculates the attribute private key assignment request $R(UID, att_j)$ and sends it to all AAs. Each AA verifies the user's identity and attribute certificate. Attribute set $UAS(AID, UID)$ is used to represent the set of all attributes of this user that the AA is responsible for. If the verification is passed, the AA generates the user's attribute secret key component $ATTSK(AID, UID)$ and sends it to the user.

Attribute secret key component generation algorithm

Input: system public key *SPK*, attribute att$_j$ public key *ATTPK(att$_j$)*, attribute set *UAS(AID, UID)*

Output: user attribute secret key component *ATTSK(AID, UID)*

---

①user calculates and sends the attribute private key assignment request $R(UID, att_j)$ to AA$_i$ based on the system public key *SPK* as well as the user secret key *USK(UID)*.

$$R(UID, att_j) = (g^{\frac{1}{u_n}}, SPK^{u_m}, AATPK(att_j)^{u_m})$$

②AA$_i$ verifies the identity documents recorded by users on the blockchain, including attribute certificate AC, attribute revocation certificate ARC, etc.

③AA$_i$ generates the user's attribute private key component *ATTSK(AID, UID)* based on $R(UID, att_j)$ as well as the AA$_i$ secret key *ASK(AID)* and sends it to the user.

$$ATTSK(AID, UID) = \{K_{AID,GID} = g^{\frac{\alpha_i}{u_n}} SPK^{u_m}, L_{AID,UID} = g^{\frac{\beta_i}{u_n}},$$
$$\forall att_j \in UAS(AID, UID): Q_{AID,att_j} = g^{\frac{\beta_i \gamma_i}{u_n}} (AATPK(att_j)^{u_m})^{\beta_i}$$

4.5.3 Data Upload

(1) *Data encryption*

For data files with hierarchical relationships of attributes, the DO obtains a lowest-secret hierarchical access control tree covering all access policy attributes by integrating its access control tree. The data are divided into data plaintext sets $F = \{m_1, ..., m_n\}$. Select a symmetric key set $K_{data} = \{K_{data1}, ..., K_{datan}\}$ containing n symmetric encryption keys corresponding to each threshold node in the access policy structure for secret sharing. The data ciphertext set $E_{Kdata}(F) = \{E_{Kdata1}(m_1), ..., E_{Kdatan}(m_n)\}$ is obtained by using the symmetric encryption algorithm and the symmetric key set K_{data} for encryption.

(2) *Key encryption*

Using the improved hierarchical LSSS in Sect. 4.4, the DO performs attribute-based encryption on the symmetric encryption key set K_{data} to generate the key ciphertext set *CT*.

Key encryption algorithm

Input: symmetric encryption key set K_{data}, system public key SPK, access structure T

Output: key ciphertext set CT

---

①DO generates a matrix M (of order $l \times n$) based on the access structure T. The mapping of the k-*th* row M_k of M to the user attributes is noted as $\rho(k)$.

②Randomly selected vectors $v \in Z_q^n$, each key s_m to be shared corresponds to a threshold node in the access policy structure.

$$v = \begin{bmatrix} s_1 \\ s_2 \\ \vdots \\ s_n \end{bmatrix}$$

③Compute the secret share λ_k for each row M_k, $k \in (1, l)$.

$$\lambda_k = M_k v$$

④Randomly select $r_1, \dots, r_l \in Z_q$, and compute the key ciphertext set CT of K_{data}.

$$CT = \{ \forall m = 1, \cdots, n : C_j = K_{datam} \left(\prod_{i=1}^{l} e(g,g)^{\alpha_i} \right)^{s_m} ,$$

$$\forall k = 1, \cdots, l : C_{k,1} = SPK^{\lambda_k} \left(g^{v\rho(k)} H(\rho(k)) \right)^{r_k \gamma_i}, C_{k,2} = g^{\frac{r_k}{\beta_i}}, C_{k,3} = g^{\frac{r_k \gamma_i}{\beta_i}} \}$$

---

(3) *File upload*

The DO uploads the data ciphertext set $E_{KData}(F)$ and the key ciphertext set CT together to the CS.

4.5.4 Data Access

(1) *Decryption token generation*

When a DV wants to access the data file F shared by other users on the cloud server, he first sends his own attribute secret key $ATTSK(AID, UID)$ to the CS. Let $S(UID)$ denote the set of all attributes of the DV. The CS uses them and the DV's public key $UPK(UID)$ to generate the decryption token set DT for the key ciphertext set CT. The use of decryption tokens enables the CS to perform most of the decryption of the key ciphertext, and the user does not have to perform complex computations of bilinear pairs. When the attributes of the DV satisfy part of the access structure, the symmetric key decryption token combined with this part can be obtained, and when the attributes satisfy the whole access structure, the full symmetric key decryption token can be obtained. The CS sends the decryption token set DT, the data ciphertext set $E_{KData}(F)$ and the key ciphertext set CT to the user.

Decryption token generation algorithm

Input: DV attribute secret key $ATTSK(AID,UID)$, DV public key $UPK(UID)$, DV attributes $S(UID)$, matrix M based on the access structure T

Output: decryption token set DT

---

①Access structure determination for each key ciphertext CT_m in the key ciphertext set CT: If the set of attributes of the DV satisfies its corresponding non-leaf node, it is possible to find the constant set $\{\omega_k \in Z_p\}$, $(k \in S(UID))$ in polynomial time, satisfying $\sum_{k\in S(UID)} \omega_{k,m} M_k = \varepsilon_m$, where ε_m is the row vector with all $n\text{-}1$ elements 0 except the $m\text{-}th$ element which is 1.

②The partial key that satisfies this part of the policy:

$$s_m = \sum_{k\in S(UID)} \omega_{k,m}\, \lambda_k$$

③Let $C_m' = g^{s_m}$, compute the corresponding part of the decryption token DT_m.

$$DT_m = \prod_{i=1}^{num}[e(C_m', K_{AID,GID}) \cdot \prod_{j\in UAS(AID,UID)} \left(\frac{e\left(C_{k,2}, Q_{AID,att_j}\right)}{e(C_{k,1}, g^{u_m})e(C_{k,3}, L_{AID,UID})}\right)^{w_j num}]$$

$$= \prod_{i=1}^{num} e(g,g)^{\frac{a_i s_m}{u_n}}$$

④The set of all decryption tokens DT_m that satisfy the corresponding access structure is denoted as DT. The CS sends the decryption token set DT, data ciphertext set $E_{Kdata}(F)$ and key ciphertext set CT to the user.

The derivation of the generation process of decryption token DT_m for each part of the decryption token set DT is shown in Formula (6).

$$DT_m = \prod_{i=1}^{num}[e(C_m', K_{AID,GID}) \cdot \prod_{j\in UAS(AID,UID)} \left(\frac{e\left(C_{k,2}, Q_{AID,att_j}\right)}{e(C_{k,1}, g^{u_m})e(C_{k,3}, L_{AID,UID})}\right)^{w_j num}]$$

$$= \prod_{i=1}^{num}[e\left(g^{s_m}, g^{\frac{a_i}{u_n}}SPK^{u_m}\right) \cdot \prod_{j\in UAS(AID,UID)} \left(\frac{e\left(g^{\frac{r_k}{\beta_i}}, g^{\frac{\beta_i \gamma_i}{u_n}}(g^{v att_j} H(att_j))^{\beta_i \gamma_i u_m}\right)}{e\left(SPK^{\lambda_k}\left(g^{v att_j}H(att_j)\right)^{r_k \gamma_i}, g^{u_m}\right) \cdot e\left(g^{\frac{r_k \gamma_i}{\beta_i}}, g^{\frac{\beta_i}{u_n}}\right)}\right)^{w_j num}]$$

$$= \prod_{i=1}^{num}[e(g,g)^{\frac{a_i s_m}{u_n}} \cdot SPK^{u_m s_m} \cdot \prod_{j\in UAS(AID,UID)} \left(\frac{H(att_j)^{r_k \gamma_i u_m} \cdot e(g,g)^{\frac{r_k \gamma_i}{u_n}} \cdot e(g,g)^{v att_j r_k \gamma_i u_m}}{SPK^{\lambda_k u_m} \cdot H(att_j)^{r_k \gamma_i u_m} \cdot e(g,g)^{v att_j r_k \gamma_i u_m} \cdot e(g,g)^{\frac{r_k \gamma_i}{u_n}}}\right)^{w_j num}]$$

$$= \prod_{i=1}^{num}\left[e(g,g)^{\frac{a_i s_m}{u_n}} \cdot SPK^{u_m s_m} \cdot \prod_{j\in UAS(AID,UID)} \left(\frac{1}{SPK^{\lambda_k u_m}}\right)^{w_j num}\right] = \prod_{i=1}^{num} e(g,g)^{\frac{a_i s_m}{u_n}}$$

$$(6)$$

(2) *File decryption*

The DV receives the decryption token set DT from the CS, decrypts the key ciphertext set CT using his own user secret key $USK(GID)$, obtains all or part of the symmetric cryptographic algorithm key set K_{data}, and then decrypts the data ciphertext set $E_{Kdata}(F)$ using the symmetric key set K_{data} to recover all or part of the data file F.

File decryption algorithm

Input: decryption token set DT, user private key $USK(UID)$, key ciphertext set CT, data cipher set $E_{Kdata}(F)$

Output: data file plaintext F

①DV uses the decryption token set DT and the user private key $USK(UID)$ to decrypt the key ciphertext set CT and obtain all or part of the key set K_{data}' of the symmetric cryptographic algorithm.

$$K_{data}' = \{\forall DT_m \in DT: K_{datam} = \frac{C_m}{DT_m{}^{u_n}} = \frac{K_{datam}(\prod_{i=1}^{num} e(g,g)^{\alpha_i})^{s_m}}{\left(\prod_{i=1}^{l} e(g,g)^{\frac{\alpha_i s_m}{u_n}}\right)^{u_n}}\}$$

②DV uses the symmetric encryption algorithm key set K_{data}' to decrypt the data ciphertext set $E_{Kdata}(F)$ to recover all or part of the data file plaintext F.

---

5 Scheme Analysis

5.1 Security Analysis

In this paper, the MA-ABE scheme is used to protect the privacy and access control of the information on the blockchain. Compared with the traditional ABE scheme, this scheme does not establish a central authorization authority but selects multiple AAs to solve the problem that a single AA is overburdened and must be fully trusted. Using blockchain for autonomous identity management, users on the blockchain manage their own identities and attribute certificates individually, and there is no need to establish a centralized identity authentication authority. In addition, this scheme uses the PoR consensus mechanism to select multiple nodes as AAs, avoiding the waste of arithmetic power under the PoW mechanism and the strong centralization under the PoS and DPoS mechanisms (Table 1).

Table 1. Blockchain ABE access control scheme security comparison

Scheme	Number of AAs	Selection of AA	Identity management	Cryptographic Algorithms
Scheme [14]	Single	Fully credible institutions	Blockchain	Shamir secret sharing scheme
Scheme [15]	Multiple	Direct selection of multiple institutions	Authorization Server	BETHENCOURT CP-ABE algorithm [17]
Scheme [16]	Multiple	DPoS	None	CHASE MA-ABE algorithm [8]
Scheme of this paper	Multiple	PoR	Blockchain	DKG, hierarchical LSSS scheme

The scenario in this paper assumes that each AA and CS are honest and curious imperfectly trusted nodes that honestly follow the protocol and perform each user request but attempt to decrypt the ciphertext. Users are untrustworthy and may conspire with other users to gain unauthorized access. An adversary can breach some of the participants in the system but cannot control all of them.

This paper uses the distributed key generation (DKG) [13] protocol for system initialization, and the cryptographic algorithm is semantically secure against the selection of plaintext attacks. The security proof is based on Sect. 4 of the paper [12]. In addition, the decryption key for data access includes both the attribute private key and the user private key, which can resist the complicity attack of the AAs.

5.2 Comparison of Scheme Cost

The cryptographic algorithm part of this paper is improved on the basis of paper [12]. The hierarchical structure of attributes is used to integrate the data with hierarchical relations into an access control tree, and the hierarchical LSSS scheme is used to reduce the encryption cost and storage cost.

Assume that the plaintext whose attributes have n hierarchical relationships is $M = \{M_1, ..., M_n\}$. The scheme of paper [12] requires n access control policies $T = \{T_1, ..., T_n\}$, where T_1 is the access control policy of plaintext M_1 with the highest confidentiality level and T_n is the access control policy of plaintext M_n with the lowest confidentiality level. The access control policy of the low confidentiality level overrides the access control policy of the high confidentiality level. This improved scheme requires only one access control policy T_n, that is, the access control policy of the lowest confidentiality level.

Let the number of attributes in the access structure attribute set be a, the attributes have n levels of hierarchical relations, and the number of attribute authorization centers be k. The bilinear operation time on group G_1 is D, the power operation time on groups G_1 and G_2 are P_1 and P_2, and the element sizes are l_1 and l_2, respectively. Ignore the operation time of the multiplication operation and hash operation (Table 2).

Table 2. Comparison of scheme cost

Scheme	No hierarchical relationship among attributes	n hierarchy relationships among attributes		
		Key encryption time	Key decryption time	Key ciphertext storage cost
Scheme [12]	The same	$4anP_1 + anP_2$	$nk(3aD + D + aP_2)$	$3nl_1 + anl_2$
Scheme of this paper		$4aP_1 + anP_2$	$k(3aD + D + aP_2)$	$3l_1 + anl_2$

Compared with scheme [12], it can be seen that when there is no hierarchical relationship among attributes, the cost of the two schemes is the same; when there is a hierarchical relationship among attributes, the scheme in this paper has lower encryption and decryption time and storage cost.

6 Conclusion

Blockchain has been widely used in many fields because it can solve the problem of information asymmetry and enable users who do not trust each other to collaborate without the participation of third-party intermediaries. With the increasing demand for applications, encryption technology is needed to protect the security of users' private information while performing data sharing access. This paper improves the existing ABE scheme and completes the design of a blockchain access control scheme based on MA-ABE. The blockchain is used for autonomous identity management, and multiple AAs are selected by the PoR consensus mechanism, which solves the problem that a single AA is overburdened and must be fully trusted. The DKG protocol is used to generate keys, and the traditional LSSS is improved by using the hierarchical relationship among attributes, which reduces the encryption and decryption time and storage cost.

Although this scheme has made certain optimization and improvement in security and efficiency compared with other corresponding schemes, the multiple AAs established in this scheme have caused high costs in their mutual communication while ensuring the decentralization of the system, and the workload of system reinitialization when an AA withdraws or fails is large, which will be the direction for further optimization of this scheme in the future.

Acknowledgment. This work is supported by the Emerging Interdisciplinary Project of CUFE, the National Natural Science Foundation of China (No. 61906220) and Ministry of Education of Humanities and Social Science project (No. 19YJCZH178).

References

1. He, P., Yu, G., Zhang, Y., Bao, Y.: Survey on blockchain technology and its application prospect. Comput. Sci. **44**(04), 1–7+15 (2017)
2. Zhu, Y., Gan, G., Deng, D., Ji, F.F., Chen, A.: Security architecture and key technologies of blockchain. J. Inf. Secur. Res. **2**(12), 1090–1097 (2016)
3. Fiat, A., Naor, M.: Broadcast encryption. In: Stinson, D.R. (ed.) CRYPTO 1993. LNCS, vol. 773, pp. 480–491. Springer, Heidelberg (1994). https://doi.org/10.1007/3-540-48329-2_40
4. Shamir, A.: Identity-based cryptosystems and signature schemes. In: Blakley, G.R., Chaum, D. (eds.) CRYPTO 1984. LNCS, vol. 196, pp. 47–53. Springer, Heidelberg (1985). https://doi.org/10.1007/3-540-39568-7_5
5. Sahai, A., Waters, B.: Fuzzy identity-based encryption. In: Cramer, R. (ed.) EUROCRYPT 2005. LNCS, vol. 3494, pp. 457–473. Springer, Heidelberg (2005). https://doi.org/10.1007/11426639_27
6. Su, J.S., Cao, D., Wang, X.F., Sun, Y.P., Hu, Q.L.: Attribute-based encryption schemes. J. Softw. **22**(6), 1299–1315 (2011)

7. Rahulamathavan, Y., Phan, R.C.W., Misra, S., et al.: Privacy-preserving blockchain-based IoT ecosystem using attribute-based encryption. In: IEEE International Conference on Advanced Networks and Telecommunications Systems(ANTS), Bhubaneswar, India. NJ, 17–20 December 2017, pp. 1–6. IEEE (2017)

8. Chase, M.: Multi-authority attribute based encryption. In: Vadhan, S.P. (eds.) TCC 2007. LNCS, vol. 4392, pp. 515–534. Springer, Heidelberg (2007). https://doi.org/10.1007/978-3-540-70936-7_28

9. Lewko, A., Waters, B.: Decentralizing attribute-based encryption. In: Paterson, K.G. (ed.) EUROCRYPT 2011. LNCS, vol. 6632, pp. 568–588. Springer, Heidelberg (2011). https://doi.org/10.1007/978-3-642-20465-4_31

10. Lewko, A., Waters, B.: Decentralizing attribute-based encryption. In: Paterson, K.G. (eds.) EUROCRYPT 2011. LNCS, vol. 6632, pp. 568–588. Springer, Heidelberg (2011). https://doi.org/10.1007/978-3-642-20465-4_31

11. Ma, Z.: Research on Distributed Authentication and Access Control Based on Blockchain. Chongqing University of Posts and Telecommunications (2020)

12. Yang, X., Zhou, Q., Yang, M., Liu, T., Wang, C.: Muti-authority ABE without central authority for access control scheme in cloud storage. J. Chin. Comput. Syst. **38**(04), 826–829 (2017)

13. Gennaro, R., Jarecki, S., Krawczyk, H., Rabin, T.: Secure distributed key generation for discrete-log based cryptosystems. J. Cryptol. **20**(1), 51–83 (2007)

14. Pan, Q.: Research on Shared Data Access Control Based on Blockchain. Nanjing University of Posts and Telecommunications (2020)

15. Lu, X., Fu, S.: A trusted data access control scheme combining attribute-based encryption and blockchain. Netinfo Secur. **21**(03), 7–14 (2021)

16. Wang, J., Xie, Y., Wang, G., Li, Y.: A methond of privacy preserving and access control in blockchain based on attribute-based encryption. Netinfo Secur. **20**(09), 47–51 (2020)

17. Bethencourt, J., Sahai, A., Waters, B.: Ciphertext-policy attribute-based encryption. In: 2007 IEEE Symposium on Security and Privacy, Berkeley, USA. New York, 20 May 2007, pp. 321–334. IEEE (2007)

18. Li, J., Huang, Q., Chen, X., et al.: Multi-authority ciphertext-policy attribute-based encryption with accountability. In: Proceedings of the 6th ACM Symposium on Information, Computer and Communications Security, pp. 386–390 (2011)

19. Li, F., Rahulamathavan, Y., Rajarajan, M., et al.: Low complexity multi-authority attribute based encryption scheme for mobile cloud computing. In: 2013 IEEE Seventh International Symposium on Service-Oriented System Engineering, pp. 573–577. IEEE (2013)

20. Rouselakis, Y., Waters, B.: Efficient statically-secure large-universe multi-authority attribute-based encryption. In: Böhme, R., Okamoto, T. (eds.) FC 2015. LNCS, vol. 8975, pp. 315–332. Springer, Heidelberg (2015). https://doi.org/10.1007/978-3-662-47854-7_19

21. Zhong, H., Zhu, W., Xu, Y., et al.: Multi-authority attribute-based encryption access control scheme with policy hidden for cloud storage. Soft. Comput. **22**(1), 243–251 (2018)

Applications of Data Science

Study on the Intelligent Control Model of a Greenhouse Flower Growing Environment

Jinyang Zhen[1], Rui Xu[1], Jian Li[1], Shiming Shen[2(✉)], and Jianhui Wen[3]

[1] School of Computer Science and Information Security, Guilin University of Electronic Technology, Guilin 541004, China
[2] Satellite Navigation Positioning and Location Service Engineering Research Center, Guilin University of Electronic Technology, Guilin 541004, China
mcdull_s@163.com
[3] Ecological and Environmental Monitoring Center of Guangxi, Guilin 541002, China

Abstract. Intelligent control of the greenhouse planting environment plays an important role in improving planting efficiency and guaranteeing the quality of precious flowers. Among them, how to adapt the air humidity, temperature and light intensity in greenhouses to the different needs of the flower growth cycle is the key problem of intelligent control. Therefore, an intelligent flower planting environment monitoring and control system model (named) based on the Internet of Things and fuzzy-GRU network adaptive learning is proposed. The above three parameters in the greenhouse were used as model input parameters. The optimal growth humidity, temperature and illumination intensity of flowers are determined by the model, and the output temperature, humidity and illumination intensity act on the executing organ of the greenhouse room by the single-chip microcomputer. The model was evaluated using field greenhouse crops. The results show that the performance of this model is better than that of the PID model and fuzzy control model in simulation experiments and actual scene control. Compared with the flowers in the natural state, the plants of the flowers under systematic control were approximately 6 cm higher than those in the natural state on average, the blooming time of the flowers was approximately two days longer than that in the natural state, and the quality of the flowers was stable.

Keywords: Internet of Things · Intelligent agriculture · AI · Greenhouse cultivation · Real-time control

1 Introduction

At present, most farmers use artificial management to detect and control the carbon dioxide concentration, temperature and humidity of greenhouses for flower cultivation. The manual management mode increases the production cost, wastes human resources, and has low measurement and control accuracy and untimely measurement and control, which makes it difficult to achieve the desired effect and easily causes irreparable economic losses [1]. Greenhouse climate is one of the key factors affecting plant production and is influenced by a variety of factors, such as external weather, driving factors and crops themselves [2–4]. Therefore, how to combine the physical model in the greenhouse with the growth model of flowers to control the concentration of carbon dioxide, temperature and humidity intelligently will greatly benefit the planting and quality improvement of flowers.

Existing control mostly adopts the proportional-integral-derivative (PID) algorithm, which is the earliest and most widely used control algorithm and can eliminate the error between the target value and the actual value to carry out set value tracking control [2]. For example, to improve the accuracy of a greenhouse controller, a control method using the Levenberg–Marquardt (LM) algorithm to adjust PID parameters was proposed [3]. Aiming at the nonlinearity and time-variation of greenhouse temperature control, a temperature PID controller based on a Kalman filter was designed for greenhouse temperature control [4]. To optimize the error of the greenhouse data acquisition system and eliminate the corresponding noise and interference, the PID control method based on particle filter optimization technology is designed [5]. However, when the PID model is used for greenhouse control, a large initial error easily causes overshoot, and the introduction of error integral feedback produces more side effects, which is a limitation of PID.

To overcome the limitation of PID control, researchers began to pay attention to the fuzzy control model. A fuzzy control model based on a switching mechanism is proposed, which is a complex and variable parameter analysis model. This model can combine the estimated values of various parameters in greenhouses to conduct intelligent management of crop growth [6]. Fuzzy associative memory (FAM) was used to control the greenhouse antifreeze irrigation system, which has five outputs to control the presence of climatic frost: nonfrost (NF), possible frost (PF), light frost (MF), severe frost (SF) and severe frost (HF) [7]. Shenan et al. used a fuzzy inference machine to monitor and manage the system and realized a refrigeration/heating system, irrigation control system and light control system, which were managed by a fuzzy controller [8]. When the control variables increase, the fuzzy control rules will grow exponentially, and the rules are completely acquired by the operator's experience and expert knowledge, which cannot guarantee the optimal or suboptimal rules [9].

At present, neural network models have been widely used in modeling and parameter optimization, prediction and control of greenhouse environments because of the characteristics of neural networks, such as self-organization, self-learning and self-adaptive identification [10–12]. For example, a deep neural network model time series analysis model RNN-LSTM was proposed for greenhouse gas forecasting, which can predict the temperature, humidity, and CO2 concentration in the greenhouse [10]. Using the linear-nonlinear structure network of the radial basis function neural network, a method based on online learning was established to predict greenhouse temperature [11]. NamDS et al., using an artificial neural network (ANN) through the air and root zone environment as well as the growth factor in a greenhouse, the hydroponic chili at a specific time [12]. Jung D H et al. developed a new general risk control logic using an output feedback neural network prediction and optimization method and applied it to a strawberry multiwindow greenhouse. Compared with the traditional general risk control system, the improved control performance has been confirmed [13]. These studies prove that the neural network model has excellent performance in greenhouse environmental control compared with other intelligent control algorithms because it can better capture time sequence information in greenhouse environmental control. However, because its own algorithm easily falls into a local minimum, the convergence speed is slow, and the execution speed is low.

In the field of flower greenhouse planting intelligent control, the neural network model can realize an arbitrary complex nonlinear mapping problem, which is widely used in the parameter optimization, prediction and control of greenhouse environments. However, due to the limitations of its own algorithm, it easily falls into local minima and has a slow convergence speed and low execution speed. Therefore, the execution speed and accuracy of the model can be further improved. By combining fuzzy control with the GRU network model, the multi-input multioutput greenhouse system can achieve the optimal control effect.

2 Problem Scenario

2.1 A Floral Growth Factor Analysis

The environmental conditions that affect the growth and development of flowers mainly include light, water, soil, air and fertilizer, which are the basic conditions for the growth of flowers. This paper takes Monthly rose as an example. Monthly rose likes warmth and is afraid of heat. The optimum temperature for the growth of monthly rose is 22–25 °C. The daytime temperature is high, approximately 15–26 °C, and the nighttime temperature is low, approximately 10–15 °C. The high temperature in summer is not conducive to the flowering of monthly rose. If the high temperature lasts for more than 30 °C, the flowering will decrease, the quality of monthly rose will decrease, and it will enter the semidormancy state. Monthly rose prefers a sunny, well-ventilated, well-drained and sheltered environment. Especially in the middle of the summer period, the need for appropriate shade to monthly rose; otherwise, flowers are easy to dry out. Monthly rose flowers like moist air, and the relative humidity of the air should be between 75% RH and 80% RH. It can be seen that temperature and humidity are the decisive factors affecting the growth of monthly rose.

The following table shows the growth habits of Monthly rose (Table 1):

Table 1. Growth characteristics of Monthly rose

Florescence	Peak flowering stage	Optimum temperature		Air humidity	Illumination time
May–November	May	Daytime 15–26 °C	Night 10–15 °C	75%–80%	Per day ≥ 6 h

Therefore, temperature, humidity and light intensity are taken as control objects in system regulation, the temperature threshold is set as [15, 26], the humidity threshold is set as [0.75, 0.80] and the light intensity threshold is set as [1000, 10000], and a fuzzy GRU neural network with multiple inputs is selected as the controller of the system.

2.2 Fuzzy Neural Network

The fuzzy-GRU neural network model is adopted as the controller in the greenhouse environmental regulation, and e model is assumed to have M-dimensional input, N-dimensional output and K rules. First, the model fuzzifies the input variables by a Gaussian membership function:

$$\mu(x) = e^{(-((x-b)/c))^2} \tag{1}$$

Then, the fuzzized data are deduced by a fuzzy multiplication operation:

$$F(\mu_k) = \Pi_{i=1}^{m}\mu_k(x), \quad k = 1, 2 \cdots H \tag{2}$$

The triggering intensity of each rule is obtained, and then the triggering intensity of each rule input at this moment and at the last moment is calculated recursively through the GRU cycle network:

Update the door:

$$z_t = \sigma(w_z X_t + U_Z h_{t-1}) \tag{3}$$

Reset the door:

$$r_t = \sigma(w_t X_t + U_t h_{t-1}) \tag{4}$$

Candidate hidden state:

$$\tilde{h} = tanh(wX_t + U(r_t h_{t-1})) \tag{5}$$

The hidden state:

$$h_t = \left(1 - z_t h_{t-1} + z_t \tilde{h}\right) \tag{6}$$

Finally, the obtained results are defuzzified by the center of gravity method:

$$y = \frac{\sum_{k=1}^{H} \mu_k v_k}{\sum_{k=1}^{H} \mu_k} \tag{7}$$

Obtain the exact system output value.

2.3 Practice Site and Flowers

The experimental site was selected in Daxu Village, Lingchuan County, Guangxi, and monthly rose was selected as the experimental flower. Dawei Village, located in southeastern Lingchuan County, between 25°05"–25°19" north latitude and 110°21"–110°33" east longitude, is located in the middle of the subtropical monsoon climate zone and has four distinct seasons: abundant rainfall, sunshine, abundant heat, long summer and short winter, and rain and heat in the same season. The annual average sunshine duration is 1614.7 h, with a daily illumination rate of 36%. The sunshine duration is 1607.7 h, accounting for 99% of the annual sunshine duration, and 1505.2 h, accounting for 93% of the annual sunshine duration, when the sunshine duration is greater than 5 °C. The sunshine duration is 1354 h, accounting for 84% of the annual sunshine duration when the sunshine duration is greater than 10 °C. The maximum frost-free period of the year is 349 days, and the minimum is 256 days. The annual average rainfall is 1941.5 mm, the maximum annual rainfall is 2460.7 mm, and the maximum juvenile rainfall is 1543.2 mm. The rainfall is mainly concentrated in the first half of the year. The rainy season is from March to August, the rainfall is greater from April to July, and the peak rainfall is from May to June. After September, large-scale rainfall decreased, and most of them were local hot thunderstorms. The annual average relative humidity was 73–79%. The annual average temperature is 18.7 °C, the coldest in January, the monthly average temperature is 6.8 °C–8.4 °C, the hottest in July, the monthly average temperature is 27.0 °C–28.6 °C, the extreme maximum temperature is 39.5 °C, and the extreme minimum temperature is −5.1 °C.

The following are the experimental greenhouse and experimental flowers (Figs. 1 and 2):

Fig. 1. The experimental

Fig. 2. Experimental greenhouse room

3 Methods

3.1 Framework

Combined with the climatic characteristics of the test site and the growth habit of rose flowers, the overall framework of the fuzzy neural network greenhouse control system design is proposed, as shown in the figure (Fig. 3):

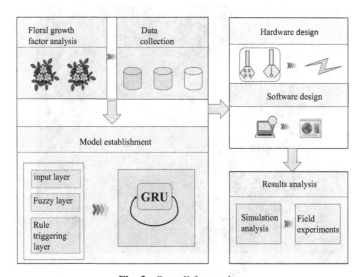

Fig. 3. Overall frame chart

First, the growth habit of rose flowers was qualitatively analyzed, and the mathematical model was established through the fuzzy-GRU network according to the optimal growth environment parameters. Second, the light, temperature and humidity in the greenhouse environment were collected through sensors deployed in multiple locations, and mean value processing was used to eliminate errors in the collection process. Then, in the analysis of the results, the PID model and fuzzy control model are constructed, and the three models are simulated and compared by using the collected greenhouse environmental data to verify the reliability of the fuzzy-GRU network model. Finally, the hardware and software system platform is deployed in the field scene. The PID model, fuzzy control model and fuzzy-GRU network model are applied to the field scene to regulate the greenhouse. The superiority, accuracy, uniformity and effectiveness of the fuzzy-GRU network model in greenhouse regulation are verified again.

3.2 Fuzzy Neural Network Model

The fuzzy neural network structure adopted by the system is shown in the following figure (Fig. 4):

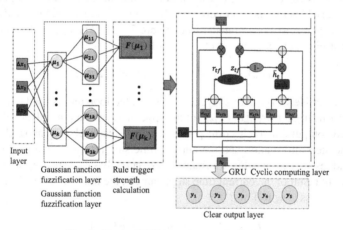

Fig. 4. Fuzzy-GRU neural network diagram

The following is the structure of each layer of the fuzzy-GRU network:

The determination of input and output variables and their fuzzy sets: we define the input variable as Δx_m and the output variable as y_n. Let the system have t fuzzy rules. The greenhouse environment variables that we need to control in the greenhouse are temperature, humidity and light intensity. Therefore, the temperature error, humidity error and light intensity error are regarded as the input variables of the system, which are defined as Δx_1, Δx_2 and Δx_3. The control mechanism of these three variables in the greenhouse is sunshade net, fan, spray humidifier, heater and LED light intensity supplement lamp. These 5 mechanisms are used as output variables and are expressed as y_1, y_2, y_3, y_4 and y_5. Therefore, the input variable is 3, and the output variable is 5.

$$\Delta_{x_m,\ldots}m = 1, 2, 3 \tag{8}$$

The Second Layer Network: This layer uses the Gauss function as the membership function of the fuzzy layer, and the variables of the input layer are fuzzy. Each rule corresponds to 3 input variables of each group, namely, $[\Delta x_1, \Delta x_2, \Delta x_3]$; then, this layer produces 3 * k neural nodes. The membership degree of rule k corresponding to the mth input variable is calculated as follows:

$$\mu_{mk} = e^{\left(-\left((\Delta x_m - b_m^k)/\sigma_m^k\right)\right)^2} \tag{9}$$

b_m^k and σ_m^k represent the mean and standard deviation. The membership degrees of the rules corresponding to temperature, humidity and light intensity are 7. Because there are three input variables corresponding to each rule, there are 3 * 7 * 7 * 7 neuron nodes in this layer.

The Third Layer Network: This layer calculates the rule trigger intensity of each set of input variables corresponding to each premise rule through fuzzy multiplication. A set of input variables in this layer corresponds to a rule trigger strength, and each rule strength corresponds to a neural node, so there are 7 * 7 * 7 neuron nodes in this layer. The trigger strength of rule k is calculated as follows:

$$F(\mu_k) = \mu_{1k} * \mu_{2k} * \mu_{3k}, k = 1, 2 \cdots k \tag{10}$$

Fourth Layer Network: In this layer, the GRU gated loop unit is used to recursively calculate the rule trigger strength $F(\mu_k)$ of the upper network output and the rule strength h_t of the GRU output. Each rule corresponds to a GRU feedback loop network. In GRU, the last time step information and the current time step information are combined by updating and reset gates, and this information is selectively saved and forgotten to obtain the final output result of the current time step, namely, the final strength of the rule. The calculation process is as follows:

$$r_{tf} = \sigma\left(w_{rtf} \cdot F(\mu_k) + w_{rfh}h_{t-1}\right) \tag{11}$$

$$z_{tf} = \sigma\left(w_{ztf} \cdot F(\mu_k) + w_{zfh}h_{t-1}\right) \tag{12}$$

$$\tilde{h} = tanh\left(w_{htf} \cdot F(\mu_k) + w_{htf} \cdot \left(r_{tf} \circ h_{t-1}\right)\right) \tag{13}$$

$$h_t = z_{tf} \circ h_{t-1} + \left(1 - z_{tf}\right) \circ \tilde{h} \tag{14}$$

The Fifth Layer Network: this layer clarifies the output of the fuzzy results and uses the gravity center method to weighted average the intensity of each rule output by the fourth layer network:

$$y = \frac{\sum_{k=1}^{H} h_{tk} w_k(\mu)}{\sum_{k=1}^{H} h_{tk}} \tag{15}$$

$w_k(\mu)$ is the weight value of rule k, and the output is the control state of 5 actuators. The logarithmic S-type activation function is used to limit the output control to 0–1. When the output is greater than 0.5, the actuator works and is closed to less than 0.5. There are 5 neuron nodes in this layer, corresponding to 5 actuators.

3.3 Design Implementation

Hardware Design. Only two levels of headings should be numbered. Lower level headings remain unnumbered; they are formatted as run-in headings (Fig. 5).

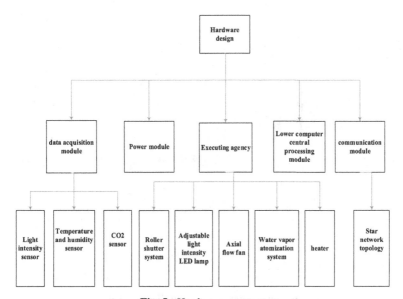

Fig. 5. Hardware structure

System hardware design mainly includes two parts: detection node design and control node design. This system uses STM32F407 as the control core of the lower machine, and the data acquisition module and execution module of the intelligent greenhouse are concentrated here. The detection node sensors are arranged in multiple points. The DHT12 sensor is used as the air temperature and humidity sensor. SGP30 is used as the CO_2 concentration sensor. The CO_2 concentration is only displayed in the model without data processing. BH1750 is used as the light intensity sensor. The CC2530 integrated chip is selected as the MCU of the ZigBee node, and the network structure adopts a star network topology.

The following is the schematic diagram of the sensor and the circuit schematic diagram of the central minimum system of the control node (Fig. 6):

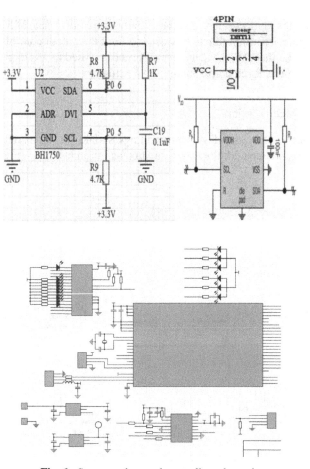

Fig. 6. Sensor and central controller schematic

Software Function Design. The software module program includes the lower computer software program, communication protocol program, and upper computer software program.

Software Program Design of Lower Computer: Lower computer is mainly divided into data acquisition module and actuator module. In the working process of the system, the lower computer mainly completes the work of data acquisition, data storage and control of the actuator. The collected indoor and outdoor temperature, humidity and light intensity parameters are saved. On the other hand, according to the control instructions sent by the upper computer, the actuator in the greenhouse is controlled to carry out the corresponding work, and finally, the temperature, humidity and light intensity in the greenhouse are controlled to keep them within the required range.

PC Software Program Design: PC is mainly divided into an intelligent control module and a data visualization module. The work of the host computer is to receive and visualize the data sent by the lower computer. After the received data are processed and analyzed by the intelligent system, the execution instructions are sent to the lower computer

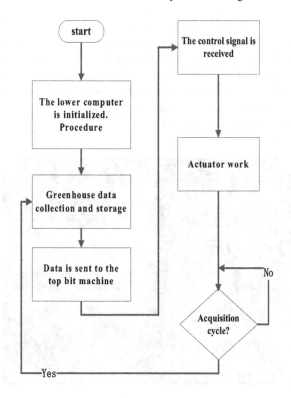

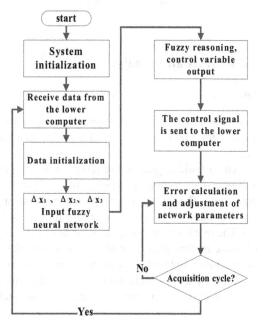

Fig. 7. Procedure execution flow of the lower computer

through the communication program. The upper computer first completes the adaptive adjustment of the parameters of the intelligent network through system initialization and then receives the data sent by the lower computer. After data processing, the light intensity error, temperature error and humidity error are transmitted to the intelligent control module. After processing and analysis by a fuzzy neural network, the control signal of the actuator of the lower computer is obtained, and finally, the control signal is sent to the lower computer (Fig. 7 and 8).

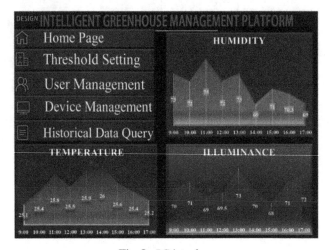

Fig. 8. PC interface

4   Experimental Results and Analysis

4.1 Simulation Analysis

For the proposed fuzzy-GUR network model, three reference models are established to evaluate its performance: the PID control model, fuzzy PID control model and fuzzy-RNN network model.

MATLAB was used to build a greenhouse space with a space of 2000 m³. The greenhouse has a rolling system, a cooling and humidifying system, a lighting auxiliary system, and a heating system. The shutter system can reduce the solar radiation energy by 79%. The cooling and humidification system is divided into an atomized water sprayer and an axial flow fan. The maximum water spray volume of the atomized water sprayer is 15 g $[H_2O]$/min * cubic meters, the maximum air volume of the axial flow fan is set to 22.2 m³/s, and the maximum light intensity of the illumination auxiliary system is set to Lux. Because the data used in the simulation process are the climate data of Guilin in August, the heater control signal is 0 in the simulation process, and the heater heating is not set.

The environmental setting scenario of the simulation experiment is the climate and environment data collected in Guilin from 6:00 a.m. to 8:00 p.m. on July 23. During the experiment, the response ability of this scheme to the setting point under the premise of continuous changes in the external environment is proven by changing the indoor temperature, humidity and illumination settings. The indoor temperature response is shown in Fig. 10, the indoor relative humidity response is shown in Fig. 11, and the illumination response is shown in Fig. 9.

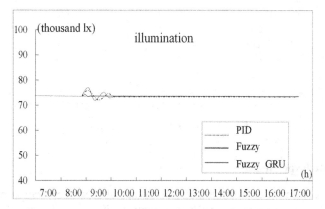

Fig. 9. Illumination response

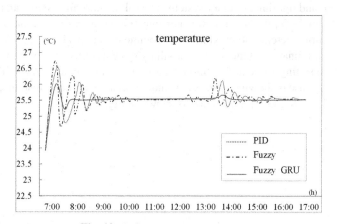

Fig. 10. Indoor temperature response

By comparing the simulation results from Fig. 9 to Fig. 10, it can be seen that the conventional PID control has a large overshoot, long adjustment time and poor anti-interference ability. Fuzzy control has a small overshoot, fast control response, short adjustment time, and certain anti-interference ability. However, because it reflects the error quantity through the membership function, what is found in the fuzzy rule table is a fuzzy quantity, and thus the steady-state error cannot be eliminated. The fuzzy-GRU model overcomes the above shortcomings, which not only has small overshoot, fast response, no oscillation, and good stability but also has small steady-state error. This shows that the dynamic and static characteristics of the fuzzy-GRU model are superior.

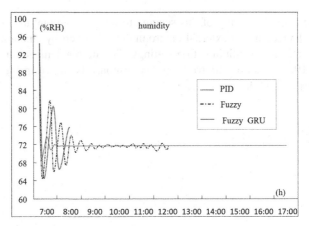

Fig. 11. Indoor humidity response

4.2 Field Site Experiments

To verify the control effect of the model in the actual scene application process, the system is deployed in a room in Lingchuan County, Guilin City. The greenhouse area is $300\,m^2$, and the height is 2.5 m. The greenhouse has a rolling curtain system, humidifying cooling system and lighting auxiliary system. The rolling curtain system can reduce the solar radiation by 70%, the maximum water injection volume of the atomizer is 16 g $[H_2O]$/min * cubic meters, the maximum air volume of the axial fan is 24.3 m^3/s, and the maximum illumination of the lighting auxiliary system is Lux. In the control process, the temperature setting value and humidity setting value are changed once. The indoor temperature response is shown in Fig. 12, the indoor relative humidity response is shown in Fig. 13, and the illumination response is shown in Fig. 14.

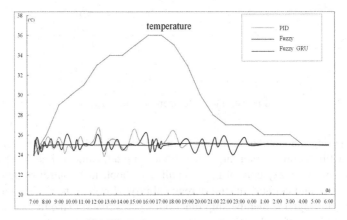

Fig. 12. Indoor temperature response

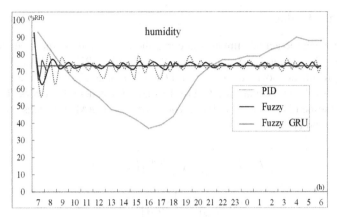

Fig. 13. Indoor relative humidity response

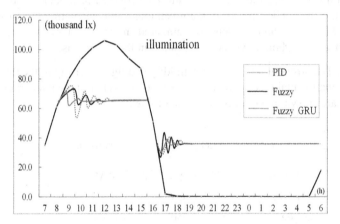

Fig. 14. Illumination response

From the actual operation situation, the control effect of the field experiment is slightly inferior to that of the simulation experiment. This is because there is a certain gap between the tightness of the greenhouse in the field experiment and the ideal state in the simulation laboratory, which leads to the transfer of indoor and outdoor temperature and humidity and produces certain interference in the process of temperature and humidity control. Due to the influence of machine construction error and signal interference between circuits, the water output of the atomizing water injector and axial flow fan cannot reach the ideal state in the simulation experiment. Due to the loss of energy conversion, the illumination adjustment process has a certain delay lag effect.

However, on the whole, the fuzzy-GRU model can better meet the control requirements compared with the other three models and is superior to the other three models in temperature, humidity and illumination control regardless of time or overshoot. Therefore, the energy consumption in the actual experiment is also lower, which meets the demand of production efficiency.

4.3 Model Evaluation

The error between the actual climate value and the target value in the greenhouse controlled by the PID model, fuzzy control model and fuzzy-GRU model is calculated and compared by using the root mean square error (RMSE) and average absolute error (MAE) to objectively evaluate the control effect of the fuzzy-GRU neural network. The smaller the error is, the better the control effect. The RMSE and MAE formulas are shown in formulas (16) and (17).

$$RMSE_{(y',y)} = \sqrt{\frac{1}{n}\Sigma_{i=1}^{n}\left(y_i' - y_i\right)^2} \qquad (16)$$

$$RAE_{(y',y)} = \frac{1}{n}\Sigma_{i=1}^{n}\left|y_i' - y_i\right| \qquad (17)$$

By randomly selecting the three-day control data of the PID model, the fuzzy control model, and the fuzzy-GRU model in the greenhouse and calculating the error through RMSE and MAE, the three models are compared in terms of temperature, humidity, light intensity, and carbon dioxide concentration in the greenhouse.

Error Value. The errors of temperature, humidity and light intensity of the three models were calculated to obtain the average errors of the three environmental parameters. The final results are shown in the following table (Table 2):

Table 2. Model control effect comparison.

Method	RMSE	MAE
Fuzzy	0.424	0.368
PID	0.350	0.319
Fuzzy-GRU	0.292	0.243

5 Conclusion and Discussion

A greenhouse control system based on a fuzzy control GRU neural network is proposed. The GRU network is used to optimize the parameter weights of the fuzzy rules. The control effects of the PID model, fuzzy control model and fuzzy neural network are compared by simulation software. The results show that the fuzzy neural network has higher control accuracy and less control time for greenhouses.

At present, the PID control model mostly remains in the single factor adjustment of environmental parameters. The fuzzy control model cannot guarantee the optimal or suboptimal rule. Neural network models, because of their own algorithm limitations, easily fall into local minima. The fuzzy-GRU neural network model obtains the rule strength of each parameter through a fuzzy rule operation for multiple input coupling factors. Using the memory function of the GRU update gate and reset gate to selectively save and forget the rule intensity of these parameters can effectively reduce the optimization time. The problem of weight oscillation caused by gradient explosion and gradient disappearance can be avoided by combining the memory information of the previous time with the current input information. Therefore, this model solves the problems that the parameters are difficult to fit and the regulation time is too long due to the complex nonlinear time-varying and multi-input coupling characteristics in the greenhouse intelligent control.

After the field test, it was found that the control effect of the system on the temperature, humidity and light intensity in the greenhouse in spring and summer was relatively good. However, after entering late autumn and winter, due to the large temperature and humidity difference between the indoor and outdoor environments, the climate was harsh, which had a certain impact on the temperature and humidity control in the greenhouse. Due to the complex aerodynamic problems involved in the control process, it was difficult to grasp the air convection in the winter ventilation process.

At present, the integrated system is only in a relatively friendly external environment, and the control effect is good, but the greenhouse system experiences deep winters and other adverse environmental climates. In the future, if the aerodynamic model can be considered in the regulation of the system, it will achieve a better regulation effect in harsh environmental climates.

Acknowledgments. This work was supported by the Guangxi Key Research and Development Program [Grant no: AB21196063]; Major Achievement Transformation Foundation of Guilin [Grant No. 20192013-1]; Innovation and Entrepreneurship Training Program for College Students of Guilin University of Electronic Technology [Grant No. 202010595031].

References

1. Park, S.H., Moon, J.P., Kim, J.K., Kim, S.H.: Development of fog cooling control system and cooling effect in greenhouse. Protected Hortic. Plant Factory **29**(3), 265–276 (2020)
2. Somefun, O.A., Akingbade, K., Dahunsi, F.: The dilemma of PID tuning. Annu. Rev. Control **52**, 65–74 (2021)
3. Su, Y., Yu, Q., Zeng, L.: Parameter self-tuning PID control for greenhouse climate control problem. IEEE Access **8**, 186157–186171 (2020)
4. Gao, Z., He, L., Yue, X.: Design of PID controller for greenhouse temperature based on Kalman. In: Proceedings of the 3rd International Conference on Intelligent Information Processing, pp. 1–4 (2018)
5. Wang, Z.: Greenhouse data acquisition system based on ZigBee wireless sensor network to promote the development of agricultural economy. Environ. Technol. Innov. **24**, 101689 (2021)
6. Wang, L., Wang, B., Zhu, M.: Multi-model adaptive fuzzy control system based on switch mechanism in a greenhouse. Appl. Eng. Agric. **36**(4), 549–556 (2020)
7. Castañeda-Miranda, A., Castaño-Meneses, V.M.: Internet of things for smart farming and frost intelligent control in greenhouses. Comput. Electron. Agric. **176**, 105614 (2020)
8. Shenan, Z.F., Marhoon, A.F., Jasim, A.A.: IoT based intelligent greenhouse monitoring and control system. Basrah J. Eng. Sci. **1**(17), 61–69 (2017)

9. Revathi, S., Sivakumaran, N.: Fuzzy based temperature control of greenhouse. IFAC-PapersOnLine **49**(1), 549–554 (2016)
10. Jung, D.H., Kim, H.S., Jhin, C., et al.: Time-serial analysis of deep neural network models for prediction of climatic conditions inside a greenhouse. Comput. Electron. Agric. **173**, 105402 (2020)
11. Hongkang, W., Li, L., Yong, W., et al.: Recurrent neural network model for prediction of microclimate in solar greenhouse. IFAC-PapersOnLine **51**(17), 790–795 (2018)
12. Nam, D.S., Moon, T., Lee, J.W., et al.: Estimating transpiration rates of hydroponically - grown paprika via an artificial neural network using aerial and root-zone environments and growth factors in greenhouses. Hortic. Environ. Biotechnol. **60**(6), 913–923 (2019)
13. Jung, D.-H., Kim, H.-J., Kim, J.Y., Lee, T.S., Park, S.H.: Model predictive control via output feedback neural network for improved multi-window greenhouse ventilation control. Sensors **20**(6), 1756 (2020)
14. Wang, G., Wu, J., Zeng, B., et al.: A nonlinear model predictive tracking control strategy for modular high-temperature gas-cooled reactors. Ann. Nucl. Energy **122**, 229–240 (2018)
15. Escamilla-García, A., Soto-Zarazúa, G.M., Toledano-Ayala, M., et al.: Applications of artificial neural networks in greenhouse technology and overview for smart agriculture development. Appl. Sci. **10**(11), 3835 (2020)
16. Huang, H., Zhang, S., Yang, Z., et al.: Modified Smith fuzzy PID temperature control in an oil-replenishing device for deep-sea hydraulic system. Ocean Eng. **149**, 14–22 (2018)
17. Subahi, A.F., Bouazza, K.E.: An intelligent IoT-based system design for controlling and monitoring greenhouse temperature. IEEE Access **8**, 125488–125500 (2020)
18. Li, Z., Wang, J., Higgs, R., et al.: Design of an intelligent management system for agricultural greenhouses based on the internet of things. In: 2017 IEEE International Conference on Computational Science and Engineering (CSE) and IEEE International Conference on Embedded and Ubiquitous Computing (EUC), vol. 2, pp. 154–160. IEEE (2017)
19. Riahi, J., Vergura, S., Mezghani, D., et al.: Intelligent control of the microclimate of an agricultural greenhouse powered by a supporting PV system. Appl. Sci. **10**(4), 1350 (2020)
20. Li, L., Cheng, K.W.E., Pan, J.F.: Design and application of intelligent control system for greenhouse environment. In: 2017 7th International Conference on Power Electronics Systems and Applications-Smart Mobility, Power Transfer & Security (PESA), pp. 1–5. IEEE (2017)
21. Alaviyan, Y., Aghaseyedabdollah, M.H., Sadafi, M.H., et al.: Design and manufacture of a smart greenhouse with supervisory control of environmental parameters using fuzzy inference controller. In: 2020 6th Iranian Conference on Signal Processing and Intelligent Systems (ICSPIS), pp. 1–6. IEEE (2020)
22. Iddio, E., Wang, L., Thomas, Y., et al.: Energy efficient operation and modeling for greenhouses: a literature review. Renew. Sustain. Energy Rev. **117**, 109480 (2020)
23. Moon, T.W., Jung, D.H., Chang, S.H., et al.: Estimation of greenhouse CO_2 concentration via an artificial neural network that uses environmental factors. Hortic. Environ. Biotechnol. **59**(1), 45–50 (2018)
24. Zhao, H., Kong, D.: The design and realization of intelligent greenhouse control system based on cloud integration. J. Phys. Conf. Ser. **1646**(1), 012113 (2020)
25. Guo, Y., Zhao, H., Zhang, S., et al.: Modeling and optimization of environment in agricultural greenhouses for improving cleaner and sustainable crop production. J. Clean. Prod. **285**, 124843 (2021)
26. Blondin, M.J., Sáez, J.S., Pardalos, P.M.: Control engineering from classical to intelligent control theory—an overview. In: Blondin, M.J., Pardalos, P.M., Sáez, J.S. (eds.) Computational Intelligence and Optimization Methods for Control Engineering, pp. 1–30. Springer, Cham (2019). https://doi.org/10.1007/978-3-030-25446-9_1
27. Qin, H., Wang, X.: A multi-discipline predictive intelligent control method for maintaining the thermal comfort on indoor environment. Appl. Soft Comput. **116**, 108299 (2022)

28. Cao, L., Li, H., Zhou, Q.: Adaptive intelligent control for nonlinear strict-feedback systems with virtual control coefficients and uncertain disturbances based on event-triggered mechanism. IEEE Trans. Cybern. **48**(12), 3390–3402 (2018)
29. Wang, B., Jahanshahi, H., Dutta, H., et al.: Incorporating fast and intelligent control technique into ecology: a Chebyshev neural network-based terminal sliding mode approach for fractional chaotic ecological systems. Ecol. Complex. **47**, 100943 (2021)
30. Sun, Q., Zhang, M., Mujumdar, A.S.: Evaluation of potential application of artificial intelligent control aided by LF-NMR in drying of carrot as model material. Drying Technol. **39**(9), 1149–1157 (2021)
31. Hadipour, M., Derakhshandeh, J.F., Shiran, M.A.: An experimental setup of multi-intelligent control system (MICS) of water management using the Internet of Things (IoT). ISA Trans. **96**, 309–326 (2020)
32. Sagdatullin, A.: Development of an intelligent control system based on a fuzzy logic controller for multidimensional control of a pumping station. In: Hu, Z., Petoukhov, S., He, M. (eds.) CSDEIS 2019. AISC, vol. 1127, pp. 76–85. Springer, Cham (2020). https://doi.org/10.1007/978-3-030-39216-1_8
33. He, C., Shen, M., Liu, L.S., et al.: Design and realization of a greenhouse temperature intelligent control system based on NB-IoT. J. South Chin. Agric. Univ. **39**(2), 117–124 (2018)
34. Liu, J.: Intelligent Control Design and MATLAB Simulation. Springer, Singapore (2018). https://doi.org/10.1007/978-981-10-5263-7
35. Borase, R.P., Maghade, D.K., Sondkar, S.Y., et al.: A review of PID control, tuning methods and applications. Int. J. Dyn. Control **9**(2), 818–827 (2021)
36. Mu, S., Shibata, S., Lu, H., Yamamoto, T., Nakashima, S., Tanaka, K.: Study on the learning in intelligent control using neural networks based on back-propagation and differential evolution. In: Mu, S., Yujie, Li., Lu, H. (eds.) 4th EAI International Conference on Robotic Sensor Networks. EICC, pp. 17–29. Springer, Cham (2022). https://doi.org/10.1007/978-3-030-70451-3_2
37. Data Science: 6th International Conference of Pioneering Computer Scientists, Engineers and Educators, ICPCSEE 2020, Taiyuan, China, 18–21 September 2020, Proceedings, Part II. Springer Nature (2020)

A Multi-event Extraction Model for Nursing Records

Ruoyu Song[1], Lan Wei[2], and Yuhang Guo[1(✉)]

[1] School of Computer Science and Technology, Beijing Institute of Technology,
Beijing 100081, China
{songruoyu,guoyuhang}@bit.edu.cn
[2] Xuanwu Hospital Capital Medical University, Beijing 100053, China

Abstract. Nursing records contain information on patients' treatment processes, which reflect the changes in patients' conditions and have legal effects. However, some of the written records of intensive care unit (ICU) nurses are incomplete according to our observations. This paper proposes an approach extracting structured nursing events from unstructured nursing records for detecting the missing items automatically. According to the PIO (problem, intervention, outcome) principle in the field of medical care, we propose event schemas for nursing records and annotate a Chinese nursing event extraction dataset (CNEED) on ICU nursing records. We find that several events may occur in a nursing record. Therefore, we present a multi-event extraction model for the nursing records. The experimental results demonstrate that our model achieves good results on CNEED and outperforms competitive methods on the multi-event argument attribution problem. By observing the results of automatic event extraction by our model, we detect missing items in the existing nursing records. This proves that our model can be used to help nurses check and improve the method of recording nursing processes.

Keywords: Event extraction · Nursing records · Multi-event

1 Introduction

Nursing records include the observation of the patient's condition by the nursing staff and the implementation of nursing measures, which are an important part of clinical nursing work. At present, most hospitals have realized the importance of nursing records and carry out statistics, analysis, and sorting of daily nursing records. However, the nursing records are manually recorded, which has unstructured problems, such as incomplete descriptions and missing items. A survey shows that among the 600 nursing documents counted by a hospital, there are 229 documents with writing problems [1]. Therefore, it is necessary to apply information extraction technology to nursing records.

Event extraction (EE) uses the predefined event schemas to detect whether events are included in texts. If an event exists, it identifies the event type and extracts all its attributes [2]. Event schemas define several event types and their corresponding event attributes. As defined by ACE, the event extraction task can be divided into two subtasks, i.e.,

© The Author(s), under exclusive license to Springer Nature Singapore Pte Ltd. 2022
Y. Wang et al. (Eds.): ICPCSEE 2022, CCIS 1629, pp. 146–158, 2022.
https://doi.org/10.1007/978-981-19-5209-8_10

event detection (trigger detection and trigger classification) and argument extraction (argument detection and argument classification). The results of event extraction are helpful for tasks such as question answering and knowledge graphs.

Due to the complexity of event extraction tasks, the difficulty of constructing domain data, the high cost of labeling, and the lack of evaluation indicators for labeling data, it is difficult to design nursing event schema and labeling specifications.

Generally, the argument roles of different events defined by the event schema are different, but because of the characteristics of nursing records, the argument under the event template is the same. For example, Fig. 1(a) presents an event mention of type *Use of Pump* by "pumped in". There are several arguments, such as "nicardipine" playing the *object* role in the event.

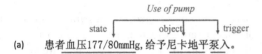

(a)

The patient's **blood pressure was 177/80mmHg**, and **nicardipine** was **pumped in.**

The **body temperature** was **re-measured at 38.7°C**, and **25mg of indomethacin suppository** was given to the patient by **anal infusion** as directed by the doctor.

Fig. 1. Examples of nursing events

However, events often appear in sentences complicatedly, where one sentence may contain multiple events. When the event template is the same, it is difficult to determine which event the argument belongs to. Sometimes even triggers and arguments may have overlaps in a sentence. For example, Fig. 1(b) shows two event mentions of type Check index interpretation and Specific organ operation. "Body temperature" and "25 mg of indomethacin suppository" separately play the object role in the Check index interpretation event and Specific organ operation event. They are all labeled as objects to which we need to distinguish which event they belong to.

In this paper, our challenges can be summarized as follows:

- There is no standard definition of nursing events, and we need to design appropriate nursing event schemas and annotation rules.
- We need to present an event extraction model for solving the multi-event argument attribution problem of our nursing event extraction dataset.

2 Related Work

Over the past decade, information extraction from electronic medical records has been studied [3]. However, it mostly focuses on named entity recognition and relation extraction [4–7]. In the past two years, there have been studies on medication change events, and the Contextualized Medication Event Dataset (CMED) has been presented which consists of 9,013 medication mentions annotated over 500 clinical notes [8]. There are also studies that extracted tumor events from Chinese liver cancer surgery records in 2014. It can be seen that there are few public nursing event corpora at home and abroad. In addition, nursing records contain detailed nursing operations, focusing on more accurate content.

Regarding the issues above, extracting structured nursing events from unstructured nursing records is necessary, which can detect writing errors and help organize nursing records.

Event extraction mostly experiments on the ACE 2005 dataset [9], the TAC-KBP corpus [10] and the TDT corpus, where the annotated text comes from news, knowledge, etc. Some biomedical domain event corpora have also been established and released, such as BioNLP [11].

The two subtasks of event extraction can be executed in a pipeline manner, where trigger is extracted first and then arguments are extracted. The two subtasks can also be executed in a joint manner, where the results of the two subtasks are directly output. The pipeline event extraction approach may cause error propagation and the joint event extraction approach is regarded as a sequence labeling task [12] where the arguments of the same role for multiple events in the sentence have the same label.

3   Method

3.1   Dataset Annotation

We define a nursing event as a nursing operation in the nursing record. Triggers are core nursing actions, and arguments are attributes of those nursing actions with the roles. A record can contain multiple events or no events. We collected the nursing records from 2018 to 2019, which only involved the nursing content and did not involve the patient's personal information, and there was no security problem of patient privacy.

Rule-Based Trigger Selection. After counting the most frequently appearing verbs in the 290k nursing records through Jieba [13], we filter out many verbs via rules: (1) filter out verbs connected with the trigger words; (2 filter out verbs that have nothing to do with nursing operations; (3) filter out verbs that are used as nouns in nursing records; (4) filter out verbs that are used as adjectives in nursing records; and (5) filter out verbs contained in arguments. Some cases are shown in Table 1.

According to the ICU nursing technology category [14], we assigned the remaining verbs to 32 nursing event subtypes under two categories of recording and operation.

Argument Role Design. Nursing records reflect dynamic changes and record problems, measures, and results. According to the PIO principle [15], we design 32 types of nursing events with three argument roles: (1) state; (2) object; and (3) result.

Because arguments tend to have fuzzy boundaries, and ambiguities in argument role classification, we make the argument labeling rules including the following: (1) The state is before the trigger, and the result is after the trigger; (2) Arguments include modifiers and contain as much information as possible; (3) The medical name and dosage

Table 1. Cases of high-frequency verbs that are deprecated as triggers

RuleID	Verb	Case
1	给予(give)	给予物理降温 (Give physical cooling)
2	通知 (notify) 告知 (notify)	告知医生 (Notify the doctor) 通知出院 (Notify the discharge)
3	观察 (observation) 治疗 (therapy) 渗出 (exudate), 渗血(bleed)	医学观察 (Medical observation) 泵入治疗 (Pump in therapy) 无渗血及渗出 (No bleeding and exudation)
4	悬浮 (suspend), 输血 (transfuse blood)	悬浮红细胞, 未有输血反应发生 (Suspended red blood cells, no blood transfusion reaction occurred)
5	未诉 (Not inform)	现未诉恶心, 本班入量225 ml, 症状有所缓解 (No complaints of nausea, the class has taken 225 ml, and the symptoms have eased)

RuleID : 1,2

Text : 停止连续性床旁血液净化治疗, 患者现HR100次/分。

(The continuous bedside blood purification treatment was stopped, the patient now had an HR of 100 beats/min.)

Trigger : 停止 (stopped)

State : None

Object : 连续性床旁血液净化治疗 (the continuous bedside blood purification treatment)

Result : 患者现HR100次/分 (the patient now had an HR of 100 beats/min)

RuleID : 3

Text : 给予去甲肾上腺素6mg+0.9%氯化钠47ml以10ml/h微量泵入。

(Give norepinephrine 6mg+0.9% sodium chloride 47ml with 10ml/h micro-pumping.)

Trigger : 泵入 (pump)

State: None

Object : 去甲肾上腺素6mg+0.9%氯化钠47ml (norepinephrine 6mg+0.9% sodium chloride 47ml)

Result : 10ml/h

RuleID : 4

Text : 患者神志为睁眼昏迷、吞咽、言语、肌力查体不合作, 胰岛素以4iu/h静脉泵入, HR122-126次/分, 已通知医生。

(The patient was in a coma with his eyes open, uncooperative in swallowing, speech, and muscle strength examination. Insulin was pumped intravenously at 4iu/h, HR122-126 beats/min, and the doctor was notified.)

Trigger : 泵入 (pumped)

State : 神志为睁眼昏迷、吞咽、言语、肌力查体不合

(in a coma with his eyes open, uncooperative in swallowing, speech, and muscle strength examination)

Object : 胰岛素 (Insulin)

Result : HR122-126 beats/min

Fig. 2. Cases of ambiguous arguments

belong to the object, and the flow rate belongs to the result; (4) When a record contains multiple events, the arguments of the two events do not overlap; (5) There can be multiple argument mentions under one argument role. Some cases are shown in Fig. 2.

3.2 Dataset Analysis

CNEED consists of 4758 event mentions annotated over 4,000 nursing records. The event types are shown in Table 2.

The role of the argument in each event schema is the same, but each role is not necessary. There may be multiple or zero nursing incidents in each sentence.

Table 2. Event types and subtypes in CNEED

Category	Event type
Recording	Report of laboratory results, Check index interpretation, Scoring, Vital Sign Summary, Death diagnosis, Intracranial Pressure Monitoring
Operation	Drainage tube care, Intravenous Indwelling Needle Infusion Technique, Compression Bandaging, Medical Technology Cooperation, Nasal Feeding, Oxygen method, External use, Use of pump, Intubation Care, Sputum Suction, Infusion, Anal Exhaust, Pressure Sore Prevention, Rewarming/Insulation Technology, Instrument usage, Perineal irrigation, Continuous blood purification, Specific organ operation, Fluid aspiration, Enema, Posture, Ventilator ultrasonic atomization, Expectoration, CPR, Basic adjustment operation, Basic deactivation operation

3.3 Model

Given an input sentence, event extraction aims to extract triggers with their event types and arguments with their roles, where arguments may have the same role but belong to different events. To solve this problem, we consider event extraction as a two-stage task, which includes trigger extraction and argument extraction, and propose a conditional pipeline event extraction model (CPEE), where the argument targets are extracted under the condition of the specific trigger. Figure 3 shows the structure of our model. It consists of two parts, the trigger extractor and the argument extractor, both of which rely on the feature representation of pre-trained model BERT [16].

Trigger Extractor. The trigger extractor predicts the triggers with their event types in the sentence and the number of triggers and length of trigger spans are unknown. Therefore, we formulate trigger extraction as a sequence labeling task, where each token x_i will be annotated with a label y_i to show whether it is related to the trigger and if it is, what type of event it represent.

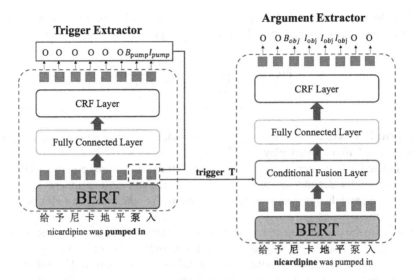

Fig. 3. Architecture of our model CPEE model

We apply the BIO annotation schema to assign trigger label t_i to each token x_i, which contains not only the position information of the current token in the trigger span but also the event type. Finally, the trigger extractor has 65 tags for 32 types of nursing events formulated.

We adopt BERT encoder to capture textual features and the input of the trigger extractor follows the BERT. BERT [16] is a bi-directional language model based on transformer architecture [17], which generates textual representations containing semantic information based on context. Let $X = (x_1, x_2, \cdots, x_n)$ be a sentence of length n where x_i is the i-th token. We input the tokens into BERT, and then obtain the hidden states $H = (h_1, h_2, \cdots, h_n)$ as the token representations for the following fully connected layer.

The fully connected layer converts each token x_i from a high-dimensional vector h_i to a low-dimensional vector g_i as Eq. (1), in preparation for decoding.

$$g_i = \delta(Wh_i + b) \tag{1}$$

Modeling the relationship between labels jointly rather than separately improves the model's performance (e.g., the tag "I-state" should not follow "B-object"). We model labels jointly using a conditional random field (CRF). The output $G = (g_1, g_2, \cdots, g_i)$ of the fully connected layer is input to CRF for decoding. In the CRF layer, we have the matrix B of n× tag_size, where $B_{i,j}$ is the score of the tag j of the i-th token in the sequence. For a tag sequence $y_T = (y_1, \cdots, y_n)$, we have the score for the sequence tag pair as Eq. (2).

$$\text{Score}(X, y_T) = \sum_{i=0}^{n} A_{y_i, y_{i+1}} + \sum_{i=0}^{n} B_{i, y_i} \tag{2}$$

A is the transition matrix of scores, where $A_{i,j}$ represents the score of a transition from the tag i to tag j. For each possible trigger tag sequence, the probability is as Eq. (3).

$$P(y_T|X) = \frac{e^{score(X,y_T)}}{\sum_{y_T' \in Y_X} e^{score(X,y_T')}} \tag{3}$$

Argument Extractor. Given the trigger, the argument extractor aims to extract related arguments and all roles they play.

Compared with trigger extraction, argument extraction is more complicated because of three issues: the dependency of arguments on the trigger, most arguments being long noun phrases, and the argument attribution problem. We take exactly a series of actions to address these issues.

Similar to the trigger extractor, the argument extractor also inputs the token sequence $X = (x_1, x_2, \cdots, x_n)$ into BERT and then obtains the hidden states $G = (g_1, g_2, \cdots, g_n)$ as the token representations. It is worth noting that G is different from H obtained in the trigger extractor on account of the separate training of the trigger extractor and argument extractor.

To capture the dependencies between triggers and arguments, we use conditional layer normalization (CLN) [18] to fuse trigger embedding T which is decoded in the trigger extractor and token embedding g_i as Eq. (4)

$$e_i = \alpha_T \odot \frac{g_i - \mu}{\sigma} + \beta_T$$

$$\alpha_T = W_\alpha[h_s; h_e] + b_\alpha$$

$$\beta_T = W_\beta[h_s; h_e] + b_\beta \tag{4}$$

where $\odot$ is the element-wise product, and h_s, h_e is the embedding of BERT at the start and end position of the trigger T.

Since the argument span is relatively large, in order to better capture the long-range dependence, we also regard it as a sequence labeling task and use CRF to learn the relationship between tags. Similar to the trigger extractor, the fusion embedding is input to the fully connected layer and CRF layer. There are four tags for the argument extractor in the BIO schema. For each possible argument tag sequence, the probability is given by Eq. (5).

$$P(y_A|X) = \frac{e^{score(X,y_A)}}{\sum_{y_A' \in Y_X} e^{score(X,y_A')}}$$

$$Score(X, y_A) = \sum_{i=0}^{n} R_{y_i,y_{i+1}} + \sum_{i=0}^{n} Q_{i,y_i} \tag{5}$$

where R is the transition matrix of scores, $R_{i,j}$ represents the score of a transition from the tag i to tag j and $Q_{i,j}$ is the score of the tag j of the i-th token in the sequence.

Loss Function. Since we want to predict the arguments when a specific and precise trigger is known, it cannot be regarded as a multi-task learning. We need to train the two extractors separately, and the decoding result of the trigger is used when the argument is decoded. To train the networks, we maximize the probability of sequence labels $P(y|X)$, that is, minimize two extractors' negative log-likelihood loss functions respectively. The loss function $J_1(\theta)$ of the trigger extractor and loss function $J_2(\theta)$ of the argument extractor are shown in Eq. (6), where P is shown in Eq. (3) and Eq. (5).

$$J_1(\theta) = -\log(P(y_T|X))$$

$$J_2(\theta) = -\log(P(y_A|X)) \tag{6}$$

4 Experiments

4.1 Experiment Settings

Dataset. We evaluate our model on CNEED and split the dataset into training set, validation set and test set by 6:2:2. Table 3 shows more details, and column denotes the number.

Table 3. Statistics of the CNEED dataset.

	Sentence	Event	Argument	Argument in multi-event cases
Training set	2418	2877	5848	2052
Validation set	806	935	1904	645
Testing set	814	946	1920	652
All	4038	4758	9672	3349

Evaluation Metric. For evaluation, we follow the four traditional evaluation metrics [19] and two specific evaluation metrics.

1. Trigger Identification (TI): If the predicted trigger span matches with a golden span, it is a correct answer and correctly identified;
2. Trigger Classification (TC): If the predicted trigger is correctly identified and assigned to the golden event type, it is a correct answer and correctly classified;
3. Argument Identification (AI): If the predicted argument span matches with a golden span, it is a correct answer and correctly identified.
4. Argument Classification (AC): If the predicted argument is correctly identified and the predicted role matches with a golden role, it is a correct answer and correctly classified;

5. Argument Classification with Event Type (ACET): an argument is correctly classified with event type if it is correctly classified and the event type predicted by the trigger it belongs to matches with a golden event type.
6. Event Detection (ED): an event is correctly detected if all triggers and arguments in the event are correctly classified.

We report Precision (P), Recall (R) and F1 measure (F1) for each of the six metrics.

Hyperparameter Setting. For all the experiments below, we adopt BERT-Base-Chinese model as encoder. The batch size in our experiments is 64. The learning rate is tuned in $5e-5$ for the trigger extractor and $3e-5$ for the argument extractor. We utilize a maximum length $n = 128$ of sentences in the experiments by padding shorter sentences and cutting off longer ones. In the fully connected layer, we use ReLU as the activation function and dropout is set to 0.5. For all the models, we minimize the negative log-likelihood loss function and use the Adam optimizer.

4.2 Comparison Methods

We compared three joint methods, BERT-softmax [16] directly uses the last layer hidden states of BERT to classify each token into triggers and arguments. BERT-CRF [20] model outputs the hidden layer representation of BERT as the inputs of CRF for joint trigger and argument classification. Although CRF appears in both BERT-CRF and CPEE, the former jointly models the relationship of triggers and arguments, and the latter is in a pipeline manner. In contrast to our pipeline method, CasEE [21] divides the task into three stages in a cascade manner, which can also improve the argument attribution problem. It first identifies the event type, then extracts triggers based on the event type, and finally merges the event type and trigger to extract arguments.

4.3 Main Results

The overall performance in Trigger Identification, Trigger Classification, Argument Identification, Argument Classification is shown in Table 4.

Table 4. Overall results in traditional metrics

	TI (%)			TC (%)			AI (%)			AC (%)		
	P	R	F1	P	R	F1	P	R	F1	P	R	F1
BERT-softmax	94.4	93.6	94.0	93.5	92.7	93.1	89.0	90.3	89.6	88.5	89.9	89.2
BERT-CRF	94.9	**94.5**	**94.7**	94.2	**93.8**	94.0	89.9	**91.6**	**90.8**	89.6	**91.3**	**90.5**
CasEE	93.9	92.5	93.2	92.3	92.5	92.4	88.0	88.8	88.4	87.7	88.5	88.1
Ours	**95.5**	93.8	94.6	**95.3**	93.6	**94.4**	**91.5**	88.7	90.1	**91.2**	88.5	89.8

Although the joint model often outperforms the pipeline model in event extraction tasks, our method still outperforms these models on the Precision score. The results show that ours achieves 1% and 2% improvements on Precision score of TI, TC and AI, AC over BERT-CRF, which prove that extracting triggers independently is purer, and extracting triggers and arguments jointly is more complicated and not easy to train.

It can also be found that our model outperforms CasEE on all scores of all subtasks. While it can also solve the problems specific to the CNEED dataset, our model performs better on the traditional four tasks for the separate training way in two specific extractors.

4.4   Analysis on the Multi-event Argument Attribution Problem

Since arguments of the same role for multiple events in a sentence have the same label, we need to pay attention to the event type predicted by the trigger that the argument belongs to.

Therefore, we use evaluation metrics (5) and (6) to evaluate the argument attribution problem.

We selected all the arguments in the multi-event sentences to test and evaluate their prediction results with the metric (5) ACET. The performance is shown in Table 5.

Table 5. Results in argument classification with event type

	P(%)	R(%)	F1(%)
BERT-softmax	56.0	41.7	47.8
BERT-CRF	36.11	52.7	43.86
CasEE	91.1	**93.7**	92.3
Ours	**95.1**	90.8	**92.9**

In these examples, multiple events are to be extracted. However, BERT-softmax and BERT-CRF cannot match the arguments to their corresponding events, leading to the huge drop between the results of Table 4 and Table 5. Our model achieves 4% improvements on Precision score but drops 3% on Recall score. Due to the pipeline method, triggers need to be predicted first, so that the number of arguments we predicted is approximately 200 less than that of CasEE. However, this also ensures that our accuracy rate is higher. The Precision score is obviously more important than the Recall score on this issue, because when there are fewer cases, we can roughly assign arguments to all events, or the event triggered by the nearest trigger.

In addition, we also use the evaluation metric (6) to evaluate the test set in the unit of event as shown in Table 6.

This method not only solves the argument attribution problem, but also has an intuitive understanding of the overall performance. Our model performs better than the other models. Observing the prediction results, we conclude that the main reason for the poor performance of the event-based evaluation is that the errors are distributed under different tasks in different sentences, and the errors of trigger extraction and argument extraction are usually not concentrated in one sentence.

Table 6. Results in event detection

	P (%)	R (%)	F1 (%)
BERT-softmax	46.81	42.81	44.75
BERT-CRF	51.98	51.27	51.62
CasEE	76.44	76.36	76.40
Ours	**79.9**	**78.4**	**79.1**

5 Discussion

5.1 Analysis on High Score Performance

It can be seen from the above results that all other evaluation scores are very high except for event detection. We observe the CNEED dataset and believe that this is because of the corpus characteristics of the nursing field, and the distinction between ordinary texts and numbers. Medical nouns such as drugs are relatively large, so the boundary in triggers and arguments is easy to determine.

5.2 Application of Missing Item Detection

We selected a high-quality corpus and constructed CNEED, which is standardized in writing. We count the argument roles of each event type in CNEED. For example, there are 104 sentences with Report of laboratory results event type, and 99% of the sentences have arguments with the role of result. Therefore, we regard the ratio of the result role in the report of laboratory results event as 0.99. We assume that this argument role is necessary for this event if the role's ratio reaches 0.9.

Table 7. Comparison of the argument role ratios of event types with missing items

	Result	
	CNEED	Sampled records
Report of laboratory results	0.99	0.88
Intravenous indwelling needle infusion technique	1.0	0.88
Continuous blood purification	1.0	0.82
Death diagnosis	1.0	0.82

CNEED is standardized in writing, but there are still nonstandard nursing records that lack the argument items with some roles. Nursing records from March to May 2019 were sampled, and nursing events were automatically extracted. It can be found in Table 7 that the argument role result's ratio is 0.99 of Report of laboratory results in CNEED, but only 0.88 in sampled records, which indicates records are lacking the role of result. This shows that there are nonstandard and incomplete nursing records, and our method can automatically detect these missing items.

6   Conclusion

This paper proposes the task of event extraction on nursing records first and designs appropriate nursing event schemas and annotation rules. To solve the multi-event argument attribution problem of our CNEED dataset, we present a conditional pipeline pre-trained event extraction model. Compared with other methods, the experimental results show that our model not only has better overall performance, but can also solve the argument attribution problem. Finally, we extracted structured events to detect the missing items in nursing records, confirming that it is meaningful to study the nursing event extraction.

Acknowledgments. This work is supported by the National Key R&D Program of China (No. 2020AAA0106600).

References

1. Wang, Y.: Analysis of writing quality of 600 nursing documents and intervention counter-measures. Chin. J. Sch. Doct. **35**(1), 60–62 (2021)
2. Xiang, W., Wang, B.: A survey of event extraction from text. IEEE Access **7**, 173111–173137 (2019). https://doi.org/10.1109/ACCESS.2019.2956831
3. Demner-Fushman, D., Chapman, W.W., Mcdonald, C.J.: What can natural language processing do for clinical decision support? J. Biomed. Inform. **42**(5), 760–772 (2009)
4. Özlem, U., Luo, Y., Peter, S.: Evaluating the state-of-the-art in automatic de-identification. J. Am. Med. Inform. Assoc. **14**(5), 550–563 (2007)
5. Özlem, U., Ira, G., Luo, Y., Isaac, K.: Identifying patient smoking status from medical discharge records. J. Am. Med. Inform. Assoc. **15**, 14–24 (2008)
6. Uzuner, Ö., Solti, I., Cadag, E.: Extracting medication information from clinical text. J. Am. Med. Inform. Assoc. Jamia **17**(5), 514–518 (2010)
7. Sameer, P., et al.: Evaluating the state of the art in disorder recognition and normalization of the clinical narrative. J. Am. Med. Inf. Assoc. Jamia **22**(1), 143–154 (2015)
8. Mahajan, D., Liang, J.J., Tsou, C.H.: Toward understanding clinical context of medication change events in clinical narratives. arXiv preprint arXiv:2011.08835 (2020)
9. https://www.ldc.upenn.edu/language-resources/data/obtaining
10. https://tac.nist.gov/2017/KBP/Event/index.html
11. Nédellec, C., et al.: Overview of BioNLP shared task 2013. In: Proceedings of the BioNLP Shared Task 2013 Workshop, pp. 1–7 (2013)
12. Nguyen, T.H., Cho, K., Grishman, R.: Joint event extraction via recurrent neural networks. In: Proceedings of the 2016 Conference of the North American Chapter of the Association for Computational Linguistics: Human Language Technologies, pp. 300–309 (2016)
13. https://github.com/fxsjy/jieba
14. Nuo, Z.,: Investigation and research on the admission of nursing technology in intensive care unit. Ph.D. thesis, Chinese People's Liberation Army Military Medical Training College (2009)
15. Monsen, K.A.: Problem-intervention-outcome meta-model (PIO MM): a conceptual meta model for intervention effectiveness research, quality improvement activities, and program evaluation. In: Monsen, K.A. (ed.) Intervention Effectiveness Research: Quality Improvement and Program Evaluation, pp. 17–28. Springer, Cham (2018). https://doi.org/10.1007/978-3-319-61246-1_2

16. Devlin, J., Chang, M.W., Lee, K., et al.: BERT: pre-training of deep bidirectional transformers for language understanding. arXiv preprint arXiv:1810.04805 (2018)
17. Vaswani, A., Shazeer, N., Parmar, N., et al.: Attention is all you need. In: Advances in Neural Information Processing Systems, pp. 5998–6008 (2017)
18. Su, J.: Conditional text generation based on conditional layer normalization (2019)
19. Chen, Y., Xu, L., Liu, K., Zeng, D., Zhao, J.: Event extraction via dynamic multipooling convolutional neural networks. In: Proceedings of the 53rd Annual Meeting of the Association for Computational Linguistics and the 7th International Joint Conference on Natural Language Processing of the Asian Federation of Natural Language Processing, pp. 167–176 (2015)
20. Du, X., Cardie, C.: Document-level event role filler extraction using multi-granularity contextualized encoding. In: Proceedings of the 58th Annual Meeting of the Association for Computational Linguistics, pp. 8010–8020 (2020)
21. Sheng, J., Guo, S., Yu, B., et al.: CasEE: a joint learning framework with cascade decoding for overlapping event extraction. arXiv preprint arXiv:2107.01583 (2021)

Cuffless Blood Pressure Estimation Based on Both Artificial and Data-Driven Features from Plethysmography

Huan Li$^{(\boxtimes)}$, Yue Wang, and Yunpeng Guo

College of Software Engineering, Sichuan University, Chengdu 610065, China
`lihuan@stu.scu.edu.cn`

Abstract. Blood pressure (BP) is an important indicator of individuals' health conditions for the prevention or treatment of cardiovascular disease. However, conventional measurements require inconvenient cuff-based instruments and are not able to detect continuous blood pressure. Advanced methods utilize machine learning to estimate BP by constructing artificial features in plethysmography (PPG) or using an end-to-end deep learning framework to estimate BP directly. Empirical features are limited by current research on cardiovascular disease and are not sufficient to express BP variability, while data-driven approaches neglect expert knowledge and lack interpretability. To address this issue, in this paper we propose a method for continuous BP estimation that extracts both artificial and data-driven features from PPG to take advantage of expert knowledge and deep learning at the same time. Then a deep residual neural network is designed to reduce information redundancy in the gathered features and refine high-level features for BP estimation. The results show that our proposed methods outperforms the compared methods in three commonly used metrics.

Keywords: Blood pressure estimation · Plethysmography · Artificial features · Data-driven features

1 Introduction

As the World Health Organization reports, an increasing number of people today are suffering from cardiovascular diseases [1]. Blood pressure(BP) is one of the most critical physiological factors of human health and is necessary for the accurate diagnosis of hypertension [2]. However, traditional BP measurements are not convenient and cannot detect continuous BP. They can be classified into invasive and noninvasive groups. In hospitals, invasive methods such as arterial puncture measurements are widely used in the intensive care unit (ICU). Although carefully prepared, they still carry a risk to patients not to mention their comfort [8]. For noninvasive methods, cuff-required devices such as sphygmomanometers are the most commonly used. However, the cuff must be inflated during measurement, which causes discomfort and does not allow for continuous

Y. Wang et al. (Eds.): ICPCSEE 2022, CCIS 1629, pp. 159–171, 2022.
https://doi.org/10.1007/978-981-19-5209-8_11

BP. In addition, this method requires advanced training and is therefore not quite friendly to elderly individuals, who need BP monitoring the most [9].

To solve these problems, many studies have been conducted to obtain accurate continuous BP via other noninvasive convenient-detection physiological signals. Pulse waves (PWs) are periodic blood ripples that travel along the arteries produced by heartbeat [3]. It can be easily measured on the finger end or radial artery by the plethysmography (PPG) method [5]. Based on the wave propagation theory of fluids in elastic pipes, two main kinds of BP measurements using PW have been developed [6]. The first group utilizes pulse wave velocity (PWV), which uses multichannel PPG signals to calculate the velocity of the pressure wave propagation in the vessels. Nevertheless, the PWV-based method requires complex calibration procedures for individual physiological parameters which are not practical yet. Another kind of method is based on the analysis of a single PPG signal. Researchers have focused on the morphological features of PPG [10]. They apply several analysis schemes to the original PPG signal or its derivatives to construct better artificial features. However, this kind of empirical feature may be limited by expert knowledge and cannot sufficiently express BP variability. Thus, some deep learning models are also applied to extract features from the PPG signal directly. However, these data-driven estimation methods neglect the current cardiovascular research and therefore have poor medical interpretability.

Inspired by the latest work of Hong [7] and Deng [4], this paper proposes a method for BP estimation based on a single PPG signal. To express more information in features and improve the robustness of estimation, we utilize existing knowledge and a deep learning model to extract empirical and data-driven features at the same time. Specifically, we gather features from two designed branches: one branch constructs morphological features artificially based on current research while the other branch uses an LSTM network to extract data-driven features. Despite the information redundancy in those features, an efficient residual deep learning model is introduced to filter and refine valuable features in an end-to-end model. The experimental results confirm the validity of each module in our method and indicate that it surpasses the compared methods.

The rest of this paper is organized as follows. Section 2 presents the current works on the empirical features or deep learning frameworks. Section 3 describes each module of the proposed method in detail. Section 4 compares our model with other methods, while Sect. 5 concludes the paper.

2   Related Work

In this section, we briefly introduce some related works that implement continuous BP estimation by extracting artificial features from PPG signals or using deep learning techniques. Kachuee extracts features from PPG and electrocardiograms (ECGs), and collects additional personal information(age, sex, etc.) for calibration, then various conventional machine learning methods are used for estimation [11]. Xie extracted time features, area features, and ratio features from the preprocessed PPG signal and then uses a regression model method to estimate BP [12]. Aman Gaurav extracted 46 features from PPG and its derivatives while the target BP is estimated with an artificial neural network (ANN) [9]. The ANN is designed to reduce the information redundancy in the extracted features. These methods merely extract features artificially, which might neglect some implicit information and are not sufficient enough to express BP variability.

In addition to traditional machine learning methods based on artificial features, more end-to-end deep learning techniques are applied to extract features through a data-driven approach. Su built a sequence-to-sequence deep recurrent neural network to directly estimate arterial blood pressure (ABP) [13]. Shimazaki used a deep convolutional neural network (CNN) to estimate BP from PPG and its third and fourth order derivatives, and personal information (age, sex, height, weight, etc.) is also utilized [14]. Despite the enrichment of the input data, his method still merely utilizes a data-driven feature extraction approach and neglects expert knowledge. Thus, in this paper, we explicitly take advantage of both artificial and data-driven features to capture more information from the PPG signal and make our method more robust.

3   Proposed Model

In general, the model we proposed is a combination of an LSTM branch and an MLP-included feature engineering branch embedded in a deep neural network. These two branches are responsible for extracting artificial features and data-driven features respectively. The final module gathers the extracted features, and refines the information for BP estimation. Since systolic blood pressure (SBP) and diastolic blood pressure (DBP) are both required in diagnosing hypertension, the targets of our model are SBP and DBP. The overview of our model is illustrated in Fig. 1.

To obtain the SBP and DBP at time t, the empirical feature extraction branch utilizes the last valid cycle in the PPG signal before t to construct 21 artificially set features and obtains high-level patterns with a simple MLP. The data-driven feature extracting branch uses a single layer LSTM to learn the implicit features from the input PPG sequence of length L before moment t. The feature gathering and multichannel output module concatenates the features extracted by two branches and captures deeper information from the fusion of those features in a deep learning framework. These two branches and feature gathering modules comprise our complete end-to-end model to estimate SBP and DBP simultaneously.

$$SBP = Maximum(ABP \ in \ the \ following \ minute) \qquad (1)$$
$$DBP = Minimum(ABP \ in \ the \ following \ minute) \qquad (2)$$

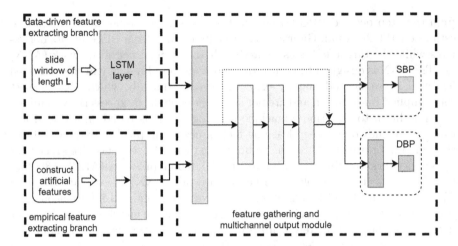

Fig. 1. The architecture of the proposed method

3.1 Empirical Feature Extracting Branch

This branch includes two main processes. We first capture the last complete valid PPG cycle before moment t from the original PPG sequence. Then 21 artificially set features will be constructed from that cycle. Finally, the constructed features are fed into a simple MLP to obtain high-level empirical patterns.

Segmenting the Last Valid Cycle. Figure 2 shows an example fluctuation of a normal PPG cycle, which can be broadly divided into systolic and diastolic states. During diastole, the aortic valve closes causing some blood to rush back, which results in a repulse wave in the diastolic part called the dicrotic wave.

Although the pulse wave has the same periodicity as the heartbeat, the actual fluctuations of the PPG signal are rather complex because of the volatile heart rate and external interference. To guarantee the value of artificially constructed features, we need to filter those abnormal cycles by the following rules:

1. The deviation between A and A' should be less than 1/10 of the main peak amplitude. The heights of A and A' represent the arterial pressure in the heartbeat interval, which should not be mutated. The sudden change is usually caused by external disturbances, for example, the movement of the device or the subject under measurement.
2. The height of both the dicrotic notch and dicrotic crest should be between 25% and 75% of the main peak amplitude. The dicrotic wave happens due to a small amount of refluxed blood after the aortic valve closes during diastole. Therefore, those dicrotic waves that happen too early or too late or too drastically mean that there is some error in the measurement process.

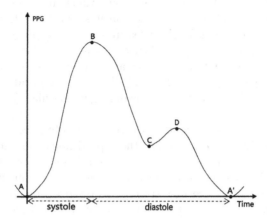

Fig. 2. An example waveform of one PPG cycle. Points A and A' are the beginning and end of one PPG cycle respectively; B is the systolic peak; C is the dicrotic notch and D is the dicrotic peak also called the diastolic peak.

Table 1. Further information of artificially constructed features.

Category	Id	Description
Magnitude features	1	The height of point B
	2	The height of point C
	3	The height of point A
	4	The height deviation between point E and D
Temporal features	5	The total time of the systole
	6	The time interval from A to reach 25% height of systole amplitude
	7	The time interval from A to reach 50% height of systole amplitude
	8	The time interval from A to reach 75% height of systole amplitude
	9	The total time of the diastole
	10	The time interval from B to reach 75% height of diastole amplitude
	11	The time interval from B to reach 50% height of diastole amplitude
	12	The time interval from B to reach 25% height of diastole amplitude
	13	The time interval between B and C
	14	The time interval between C and D
	15	The time interval from C to reach 50% height of dicrotch pulse amplitude
	16	The time interval from D to reach the height of C
	17	The time interval from D to reach 50% height of dicrotch pulse amplitude
Area features	18	The integral from A to B
	19	The integral from B to C
	20	The integral from C to D
	21	The integral from D to A'

Artificially Construct Features. After analyzing the PPG curve, we design 21 simple features including magnitude features, temporal features, and area features. Further descriptions of these features are shown in Table 1.

Each kind of feature has its physiological significance. The magnitude features have an inner correlation with age and the arterial characteristic information of a subject which greatly affects the SBP and DBP. The temporal and area features correspond to the force and blood volume during ventricular systole and diastole [15].

Then a neural layer is designed to turn the 21 obtained artificial features into a more informative form. We use the typical linear neural layer with the ReLU function to perform primary information fusion among those empirical features. Although further feature gathering will be subsequently carried out, this fusion process is still necessary since it can discover some inner correlations among artificial features without the interference of data-driven features.

3.2 Data-Driven Feature Extracting Branch Based on LSTM

To extract data-driven features, long short-term memory (LSTM) [16] is utilized. The excellent capability of LSTM to extract information from long sequences is achieved by the ingenious cooperation of its cell and input gate, output gate, and forget gate [18]. The cell is designed to remember valuable information over arbitrary time intervals, while the three gates regulate the update of the cell state.

With the excellent power, LSTM could effectively capture the implicit features from the input PPG sequence which reflects the softness of blood vessels or the condition of the subject's heart. Since this process is data-driven, LSTM may extract some hidden features that are neglected in the empirical branch.

3.3 Feature Gathering and Multichannel Output Module

This module is designed to gather the features from two previous branches and to estimate both SBP and DBP. The processes of obtaining empirical features and data-driven features are isolated, so that some information redundancy may exist.

To reduce the information redundancy and capture some high-level implicit patterns, the gathered features are fed forward into a simple MLP. However, those additional layers make our model much deeper, which could lead to gradient vanishing or exploding problems during the end-to-end training process. To overcome this issue, we add a residual connection [22], which is a sort of skip-connection that learns residual parameters with reference to the layer inputs. By adding the shortcut, the gathering module can obtain some new complex information without missing existing valuable features.

Although SBP and DBP both reflect the human cardiovascular status which means that the features they depend on are quite similar, the mapping function might be rather different so it is necessary to build an output channel for each estimation target. After obtaining the condensed features, we build two isolated channels for SBP and DBP estimation. This mode not only reinforces the representation of shared patterns of different targets but also maintains the task-oriented patterns for each task. The evaluation indicator Eq. 3 is used to analyze the estimation accuracy.

$$MAE = \frac{1}{n} \sum_{i=1}^{n} \left| S\hat{B}P_i - SBP_i \right| + \left| D\hat{B}P_i - DBP_i \right| \tag{3}$$

4   Experiments

4.1   Experimental Setup

Dataset and Data Preprocessing. In this paper, we use the Multiparameter Intelligent Monitoring in Intensive Care (MIMIC) dataset [19], which is publicly available and commonly utilized in biomedical and health research. This dataset provides various clinical physiological signals of thousands of intensive care unit (ICU) patients over weeks. The signals required in this paper, arterial blood pressure (ABP) and PPG are both sampled 125 Hz.

For our experimentation, we filtered the subjects for lack of ABP or PPG and randomly selected 100 subjects. Then the records of the beginning and end of 3 h are deleted because of null values and strong fluctuation. The input data consist of two parts, the L-second-long segmentation of PPG before t and the 21 artificial features extracted from the last PPG cycle. To obtain the target output of time t, we pick the 60-second-long segmentation of ABP before t and then analyze its maximum and minimum as SBP and DBP respectively. The proportions of SBP and DBP are shown in Fig. 3. To ensure the robustness of the model, we select 8000 samples from each subject.

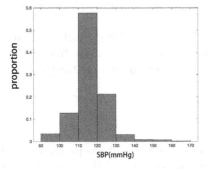 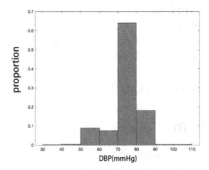

Fig. 3. The proportion of target SBP and DBP of our samples

Basic Training Settings. The samples are split into three subsets, training set, validation set, and testing set with a ratio of 6:2:2. The obtained PPG segmentation data and target BP data are not normalized but the artificial feature data are normalized to $[0, 1]$ via the feature-specific scope obtained from the training set.

Performance Metrics. We choose three widely used metrics to evaluate the performance of each estimation method. Mean Absolute Percentage Error (MAPE), Mean Absolute Error (MAE), Root Mean Squared Error (RMSE). The definitions follow:

$$MAPE = \frac{100\%}{n} \sum_{i=1}^{n} \left| \frac{\hat{y}_i - y_i}{y_i} \right| \tag{4}$$

$$MAE = \frac{1}{n} \sum_{i=1}^{n} |\hat{y}_i - y_i| \tag{5}$$

$$RMSE = \sqrt{\frac{1}{n} \sum_{i=1}^{n} (\hat{y}_i - y_i)^2} \tag{6}$$

4.2 Competing Methods

The methods chosen in the comparison experiments cover classical machine learning techniques and some state-of-the-art approaches.

These chosen machine learning methods are the most typical schemes and perform well in many aspects, including multilayer perceptron (MLP) [17], convolution neural network (CNN) [20], and support vector machine (SVM) [21]. The CNN only uses the original PPG segmentation, while MLP and SVM merely utilize the normalized artificial features. All competing models have a similar number of parameters as our proposed model. The SVM method does not need the validation set, so it uses both the training set and the validation set for training.

We also compare our method with 3 current related works, including the FFT-based neural network by Xing, X. [23], a calibration-free CNN by Schlesinger, O. [24], and a fully convolutional network using only the PPG signal proposed by Baek, S. [25]

4.3 Results and Analysis

Main Results. Table 2 shows the results of our model and competing methods. Comparing the results, we can find that except for the RMSE when estimating the DBP, our proposed model all obtains the best score, which demonstrates that our model surpasses the other methods. Meanwhile, to check whether CNN could have better capability to extract data-driven features, we also build a complete model with CNN. However, its performance is close to that of CNN-only model.

Table 2. Estimation results of the different methods

	Results for SBP			Results for DBP		
	MAPE	MAE	RMSE	MAPE	MAE	RMSE
SVM	5.109%	6.075	8.855	8.103%	5.427	8.080
MLP	3.617%	4.283	7.927	5.032%	3.508	6.074
CNN	3.684%	4.380	7.478	4.796%	3.315	5.945
Xing [23]	3.344%	4.053	7.175	4.872%	3.377	6.185
Schlesinger [24]	3.266%	3.979	7.035	4.784%	3.301	6.044
Baek [25]	3.248%	3.913	6.910	4.759%	3.298	**5.917**
Proposed	**3.243%**	**3.869**	**6.886**	**4.755%**	**3.292**	6.059

Ablation Study. To verify the contribution of each module in our proposed method, we conduct a clear ablation study. We remove the empirical feature extracting branch, the data-driven feature extracting branch, and the multi-channel output mode in the gathering learning module with the same hyper-parameters to form three new models: 1) model A utilizes the PPG sequence with only the LSTM branch; 2) model B uses artificial features only; 3) model C uses both empirical and data-driven features but does not use the multichannel output mode, which means there are two models for SBP and DBP estimation respectively. The results are shown in Table 3. In general, the performance of our hybrid-feature model is significantly better than other incomplete models, indicating that each module in our proposed framework contributes to the improvement of measuring accuracy and model robustness.

Table 3. Results of the ablation study

	Results for SBP			Results for DBP		
	MAPE	MAE	RMSE	MAPE	MAE	RMSE
A	3.897%	4.601	7.071	5.474%	3.635	6.240
B	3.492%	4.148	7.151	4.990%	3.512	6.161
C	3.573%	4.247	7.082	4.787%	3.363	6.256
Complete	**3.243%**	**3.869**	**6.886**	**4.755%**	**3.292**	**6.059**

By comparing models A, B, and the complete model, we can conclude that both empirical features and data-driven features contain valuable information for BP estimation while the latter includes some implicit patterns neglected by the former. As presented in the table, the performance of model B is better than model A, which reveals that the 21 artificially constructed explicit features express more hidden information than data-driven features. However, the complete model combining two branches of features obtains the best score. It not only proves that the LSTM can learn some implicit patterns neglected in artificial features but also verifies the effectiveness of our gathering module to blend and refine information.

The experimentation also verifies the effect of the multichannel output mode. To test the performance of model C, we have to train two task-oriented models for SBP and DBP, which is less efficient. As the results show, the complete model with multichannel output mode improves the performance for both SBP and DBP estimation, but the change of the former is much more significant compared to the latter. The reason behind this might be that some valuable hidden patterns are easier to learn from the DBP channel while the multichannel mode makes it possible for the SBP channel to eavesdrop on those shared patterns.

Parameter Study. There are two the most significant hyperparameters in our proposed model, the input PPG sequence length (L) of the data-driven feature extraction branch based on LSTM, and the output feature size (H) of the two feature extraction branches. Considering both medical practicality, we set 4 groups of experiments to investigate the impact of L on the measuring results, $L = 10$ s, $L = 20$ s, $L = 30$ s, and $L = 40$ s, while H is set to 16, 32, 64, and 128. The other parameters remain the same as the proposed method and the experiment results are shown in Figs. 4 and 5 respectively.

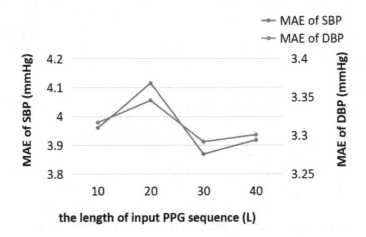

Fig. 4. Model performance with different input lengths of PPG (L)

L determines the input length of the data-driven pattern learning module, which represents the temporal range for LSTM to extract hidden patterns. Intuitively, the longer the signal we feed into the LSTM, the more information it can learn. However, the results of experiments show that the performance of LSTM is neither linearly related to the input length nor does it simply improve before it decreases because of the underfitting or overfitting. The $L = 30$ group has the best performance while the $L = 10$ group performs better than the $L = 20$ and $L = 40$ groups. For this, we speculate that the PPG signals in the range of 10 to 20 s ahead or 30 to 40 s ahead contain less extra valuable information but some interference factors.

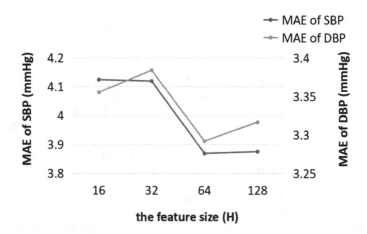

Fig. 5. Model performance with different feature sizes of the middle layer (H)

H is the hyper-parameter determining the capability of both the empirical pattern learning module and data-driven pattern learning module to represent a high level hidden pattern. When the pattern size is too small to pass all useful information to the deeper layers, an excessive size of patterns will not only lead to more parameters but also sparse the pattern and result in poor performance. Via experiments, we have found that the best value of H is 64.

5   Conclusions

In this paper, we proposed a cuffless BP estimation method using both artificial and data-driven features in PPG. The LSTM discovers some implicit features neglected by artificial features, while the empirical features utilize expert knowledge to make the end-to-end model more interpretable. A deep residual network is designed to reduce information redundancy in the gathered features and construct some higher-level features. Meanwhile, the multichannel output mode we applied to estimate SBP and DBP simultaneously can not only learn the task-oriented estimation function efficiently but also reinforce the shared features. We consider introducing some state-of-the-art empirical features and applying efficient noise filtering methods to the original PPG sequence in future works.

Acknowledgments. We would like to thank Prof. Yingjie Zhou at Sichuan University who provided us with valuable suggestions and feedback that improved the quality of this manuscript.

References

1. World Health Organization: World health statistics 2015. World Health Organization (2015)
2. W. WHO: A global brief on hypertension: silent killer, global public health crisis (2013)
3. Van de Vosse, F.N., Stergiopulos, N.: Pulse wave propagation in the arterial tree. Annu. Rev. Fluid Mech. **43**, 467–499 (2011)
4. Deng, Y., Zhou, Y., Zhang, Z.: Short-long correlation based graph neural networks for residential load forecasting. In: Mantoro, T., Lee, M., Ayu, M.A., Wong, K.W., Hidayanto, A.N. (eds.) ICONIP 2021. LNCS, vol. 13109, pp. 428–438. Springer, Cham (2021). https://doi.org/10.1007/978-3-030-92270-2_37
5. Parreira, V.F., Vieira, D.S., Myrrha, M.A., Pessoa, I.M., Lage, S.M., Britto, R.R.: Optoelectronic plethysmography: a review of the literature. Braz. J. Phys. Ther. **16**, 439–453 (2012)
6. Chung, E., Chen, G., Alexander, B., Cannesson, M.: Non-invasive continuous blood pressure monitoring: a review of current applications. Front. Med. **7**(1), 91–101 (2013)
7. Hong, Y., Zhou, Y., Li, Q., Xu, W., Zheng, X.: A deep learning method for short-term residential load forecasting in smart grid. IEEE Access **8**, 55785–55797 (2020)
8. Sorvoja, H., Myllyla, R.: Noninvasive blood pressure measurement methods. Mol. Quantum Acoust. **27**, 239–264 (2006)
9. Gaurav, A., Maheedhar, M., Tiwari, V.N., Narayanan, R.: Cuff-less PPG based continuous blood pressure monitoring—a smartphone based approach. In: 2016 38th Annual International Conference of the IEEE Engineering in Medicine and Biology Society (EMBC), pp. 607–610. IEEE (2016)
10. Korpas, D., Halek, J., Doležal, L.: Parameters describing the pulse wave. PhysioEl. Res. **58**(4), 473-479 (2009)
11. Kachuee, M., Kiani, M.M., Mohammadzade, H., Shabany, M.: Cuffless blood pressure estimation algorithms for continuous health-care monitoring. IEEE Trans. Biomed. Eng. **64**(4), 859–869 (2016)
12. Xie, Q., Wang, G., Peng, Z., Lian, Y.: Machine learning methods for real-time blood pressure measurement based on photoplethysmography. In: 2018 IEEE 23rd International Conference on Digital Signal Processing (DSP), pp. 1–5. IEEE (2018)
13. Su, P., Ding, X.-R., Zhang, Y.-T., Liu, J., Miao, F., Zhao, N.: Long-term blood pressure prediction with deep recurrent neural networks. In: 2018 IEEE EMBS International Conference on Biomedical & Health Informatics (BHI), pp. 323–328. IEEE (2018)
14. Shimazaki, S., Kawanaka, H., Ishikawa, H., Inoue, K., Oguri, K.: Cuffless blood pressure estimation from only the waveform of photoplethysmography using CNN. In: 2019 41st Annual International Conference of the IEEE Engineering in Medicine and Biology Society (EMBC), pp. 5042–5045. IEEE (2019)
15. Elgendi, M.: On the analysis of fingertip photoplethysmogram signals. Curr. Cardiol. Rev. **8**(1), 14–25 (2012)
16. Hochreiter, S., Schmidhuber, J.: Long short-term memory. Neural Comput. **9**(8), 1735–1780 (1997)
17. Ramchoun, H., Ghanou, Y., Ettaouil, M., Janati Idrissi, M.A.: Multilayer perceptron: architecture optimization and training. Int. J. Interact. Multimedia Artif. Intell. **4**(1), 26–30 (2016)

18. Zhao, R., Wang, J., Yan, R., Mao, K.: Machine health monitoring with LSTM networks. In: 2016 10th International Conference on Sensing Technology (ICST), pp. 1–6. IEEE (2016)
19. Johnson, A.E.: MIMIC-III, a freely accessible critical care database. Sci. Data **3**(1), 1–9 (2016)
20. Li, Z., Liu, F., Yang, W., Peng, S., Zhou, J.: A survey of convolutional neural networks: analysis, applications, and prospects. IEEE Trans. Neural Networks Learn. Syst. 1–21 (2021)
21. Sapankevych, N.I., Sankar, R.: Time series prediction using support vector machines: a survey. IEEE Comput. Intell. Mag. **4**(2), 24–38 (2009)
22. Alaeddine, H., Jihene, M.: Deep residual network in network. Comput. Intell. Neurosci. 1–9 (2021)
23. Xing, X., Sun, M.: Optical blood pressure estimation with photoplethysmography and FFT-based neural networks. Biomed. Opt. Express **7**(8), 3007–3020 (2016)
24. Schlesinger, O., Vigderhouse, N., Eytan, D., Moshe, Y.: Blood pressure estimation from PPG signals using convolutional neural networks and Siamese network. In: ICASSP 2020–2020 IEEE International Conference on Acoustics, Speech and Signal Processing (ICASSP), pp. 1135–1139. IEEE (2020)
25. Baek, S., Jang, J., Yoon, S.: End-to-end blood pressure prediction via fully convolutional networks. IEEE Access **7**, 185458–185468 (2019)

User Attribute Prediction Method Based on Stacking Multimodel Fusion

Qiuhong Chen, Caimao Li[(✉)], Hao Lin, Hao Li, and Yuquan Hou

School of Computer Science and Technology, Hainan University, Haikou 570228, China
{20085400210006,20085400210028,20085400210039,
20085400210018}@hainanu.edu.cn, lcaim@126.com

Abstract. The user's age and gender play a vital role within the user portrait. In view of the lack of basic attribute information, such as the age and gender of users, this paper constructs an attribute prediction method based on stacking multimodel integration. The user's browsing and clicking history is analyzed to predict the user's basic attributes. First, LR, RF, XGBoost, and ExtraTree were selected as the base classifiers for the first layer of the stacking framework, and the training results of the first layer were input as new training data into the second layer LightGBM for training. Experiments show that the proposed model can improve the accuracy of prediction results.

Keywords: Machine learning · Attribute prediction · Model fusion · LightGBM

1 Introduction

Advertising on the Internet has the advantages of faster dissemination, wider dissemination, and higher efficiency than offline advertising. Therefore, the Internet has gradually become the main position for business promotion. In advertising targeting, the user's search content, browsing history and basic attributes play an important role, among which the basic attributes gender and age are very important, but not all users are willing to disclose their age and gender information, so it will lead to users' basic attributes. If attribute data are missing, it is necessary to use existing data and related algorithms to make predictions. Hu et al. [1] used to the user to browse web information, and combined it with Bayesian network algorithm to predict the gender and age of the user. Bock et al. [2] predict the user' gender, age, education level and other attributes. Lu Xun [3] took Weibo users as the research object and predicted the user's gender, age distribution and education level based on user nicknames, tags, Weibo text, etc. Liu Baoqin [4] trained a support vector machine based on emotional features to predict the gender of Chinese microblog users, and the recognition accuracy was 73.6%. Wang Man et al. [5] integrated LightGBM and FM, analyzed the installation and usage of mobile apps, and predicted the basic attributes of users. The accuracy of gender prediction was 67.65%.

At present, the prediction of the gender and age of advertising users is still in its infancy, and the classification effect still has room for growth. Consequently, this paper proposes an advertising user attribute prediction method based on stacking multimodel

Y. Wang et al. (Eds.): ICPCSEE 2022, CCIS 1629, pp. 172–184, 2022.
https://doi.org/10.1007/978-981-19-5209-8_12

fusion. Applying the idea of model fusion of stacking [16, 17], integrating LR [18], RF [19], XGB [20], ET [21] and LGB [22, 23] algorithms, using the training method of fivefold cross-validation to predict the age and gender of users, achieved good results. The accuracy of dataset 1 is 77.97% for gender and 53.15% for age. To verify the effectiveness of the model, another dataset is used for training; the gender accuracy of dataset 2 is 72.95%, and the age accuracy is 61.28%.

2 Related Machine Learning Algorithms

In the field of machine learning algorithms, the two most commonly used algorithms in supervised learning algorithms are regression algorithms and classification algorithms [6]. The process of predicting the age and gender of a user is a typical classification problem. Therefore, a classification algorithmic algorithm can be used to accurately predict the age and gender of a user. Commonly used classification algorithms are logistic regression, random forest, XGBoost, ExtraTree and LightGBM in the integrated learning boosting algorithm.

Logistic regression (LR) [24, 25] is a generalized linear regression model. The LR algorithm has the benefits of low computational cost, low storage resources, and convenient model updating during classification. Of course, there are also some shortcomings, such as when the feature space is massive, its performance is not terribly good; it is simple to underfit and the accuracy is not terribly high.

Random forest (RF) [15] is a joint prediction model integrated by many decision trees. It divides the sample set based on the bootstrap method, and uses the CART method to construct a decision tree algorithm for each subsample set. The random forest algorithm has better accuracy in classifying and predicting the results to be tested. In contrast, random forests also have some shortcomings. When too many basic learners are established in random forests, the time and space complexity of model training will increase.

XGBoost [9] uses sparse perception to deal with missing values. On the basis of GBDT, the concept of the stacking tree model is proposed, adding regularization and improving the selection of the loss function, and optimizing the objective function. The objective loss function after XGBoost optimization is:

$$J(f_t) = -\frac{1}{2} \sum_{j=1}^{T} \frac{G_j^2}{H_j + \lambda} + \gamma \cdot T_t \qquad (1)$$

In the formula, G_j is the first-order statistic of the loss function; H_j is the second-order statistic of the loss function; and λ is the penalty term coefficient.

Using a differentiable convex arithmetic function to perform empirical risk prediction and adding a penalty term to perform structural risk prediction will help reduce the probability of overfitting.

ExtraTree, the algorithm principle of [14] is generally similar to that of random forest. It is based on a decision tree as the base learner, and the algorithm model is established through an ensemble. The difference is that during the model training process, the extreme random tree uses all the samples to build a decision tree, and there is no random sample. Moreover, in the process of node splitting, the extreme random tree is divided by random attribute splitting, which is inaccurate for the result of a single tree, but it will achieve better results in the integration process of multiple trees.

LightGBM [7, 8] is an algorithm in boosting, and similar to XGBoost, it is an optimization algorithm of GBDT. There are still large differences between the two in many aspects. For example, in the segmentation algorithm, LightGBM adopts the histogram algorithm; in the decision tree growth strategy, LightGBM uses the leafwise growth method. When the leaf starts to split, it can reduce more errors and have higher efficiency when the number of splits is the same.

3   Multimodel–LightGBM

To boost the accuracy of the model, this paper adopts stacking model to construct a multimodel-LightGBM machine learning algorithm to predict the basic attributes of advertising users' age and gender by using the cross-validation training method.

3.1   Stacking Algorithm

The stacking [11–13] algorithm is an ensemble learning framework for model fusion. It refers to training a model structure with multiple stages, training multiple different models in the first layer, and combining the obtained prediction results into a new feature set, which is used as the input variable of stage 2, and then the model of stage 2 is used. Carry out training, down in turn, and finally get an output result. The principle of this algorithm is to combine several models with different operating angles and learn from each other to improve the overall prediction accuracy.

This paper constructs an advertising user attribute prediction model based on the idea of stacking, and adopts the 5-fold cross-validation method to train the model. Stacking is two layers, and its implementation process is shown in Fig. 1.

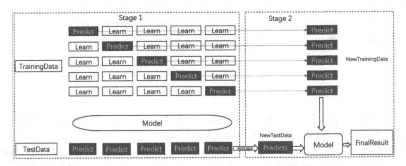

Fig. 1. Stacking structure

As shown in Fig. 1, 5-fold cross-validation means that the training set is split into five parts, four of which are used as the training set, and the other is used as the test set. The training repeated five times to obtain the predicted result of the same length as the original training set and then used as the training set of the stage 2 model. Similarly, the test set uses the trained Stage 1 model to repeat the prediction five times, the predicted value is averaged as the test data of stage 2, and the final result is obtained after the prediction of the stage 2 model. It can also be seen from the framework process that the stacking algorithm is not essentially the result of simple model stacking, but improves the accuracy of the result by using the different feature extraction capabilities of different models.

3.2 Multimodel–LightGBM

This paper adopts the idea of the stacking algorithm and combines LR, RF, XGBoost, ExtraTree, LightGBM and other algorithms to construct an advertising user age and gender prediction model. The model structure is shown in Fig. 2.

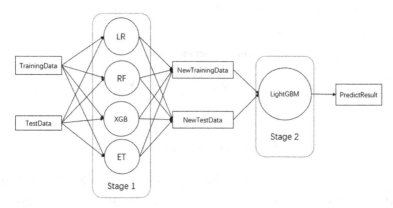

Fig. 2. Multimodel–LightGBM

As shown in Fig. 2, in the first stage of the prediction model, four machine learning algorithms, LR, RF, XGB, and ET, were selected, and the LightGBM algorithm was selected in stage 2. The training process of the model is as follows: first, the dataset is divided into a training set and a test set, and then the training set is input into the algorithm of the first stage. The training results obtained in the first stage are saved and input into the algorithm of the second stage as a new training set for training. The test set obtained from the first-stage training is averaged to obtain a new test set, which is input into the second-stage model as a new test set, and finally, the prediction result is obtained.

4   Experiment and Result Analysis

The overall training process of the model algorithm proposed in this paper can be split into five steps: data collection, feature engineering, model training, cross-validation and accuracy evaluation. To verify the effectiveness of the proposed model, ablation experiments are also performed.

4.1 Dataset

The dataset used for the experiment in this paper is the 91-day advertisement click history of users disclosed by Tencent in 2020, which is organized into three tables: user click log, user information, and clicked advertisement information. Advertisement, click time (from 1 to 91), and the number of times the user clicked the advertisement on the day, as shown in Table 1, a total of 30 million; the user information table includes user id, age and gender, as shown in Table 2, a total of 900,000; the clicked advertisement information includes the creative id, advertisement id, productid, product category, advertiser id, industry id, etc., as shown in Table 3, a total of 2.4 million records.

Table 1. User click log

Time	User_id	Click_times	Creative_id
23	34042	2	325532

Table 2. User information

User_id	Age	Gender
59352	3	1

Table 3. Advertising information

Creative_id	Ad_id	Product_id	Product_category	Advertiser_id	Industry
325532	1	34647	5	312	267

Male users and female users are represented as 1 and 2 respectively. The user's age is divided into 5 years into a total of 10 categories. For example, the number of 16-years-old corresponds to 2, and the specific age group numbering method is shown in Table 4.

Table 4. Age map

Age	<15	16–20	20–25	26–30	31–35	36–40	41–45	46–50	51–55	>56
Number	1	2	3	4	5	6	7	8	9	10

4.2 Feature Engineering

The importance of feature engineering in machine learning is self-evident. In this process, it is necessary to screen out data features that are more suitable for the task. The study of feature engineering will affect the prediction effect of the model. Therefore, we first conduct big data mining analysis on the training dataset with 900,000 users.

4.2.1 Data Preprocessing

First, data preprocessing is performed on the original dataset. When the missing rate of a variable is high (greater than 80%), its variable is deleted, and the rest is filled with the median, that is, missing value processing.

4.2.2 Feature Selection

There may be a high degree of correlation between the features of the original data, which is prone to overfitting and is not conducive to the training of the model. Therefore, it is necessary to perform feature selection. In feature selection, this paper first performs correlation analysis on the dataset, and then combines the embedded feature selection method to sort and filter the sample features. This paper uses the heatmap to analyze the correlation of the data. The feature importance of the XGBoost algorithm belongs to the embedded feature selection. Therefore, the XGBoost algorithm is used to obtain the feature importance ranking, and the results of the two are combined to eliminate the features with high correlation.

In this paper, the correlation between each feature is represented in the form of a heatmap, as shown in Fig. 3.

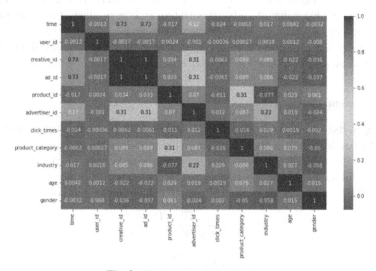

Fig. 3. Feature correlation heatmap

As shown, the larger the value on the color block is, the higher the correlation, and the correlation between the advertisement id and the creative id is as high as 1.0. The correlation between time and advertisement id is high; the correlation between creative id and time is high; the correlation between advertiser id and creative id is high; the correlation between advertiser id and advertisement id is high; and product category id and product id are highly correlated.

In addition, the XGBoost algorithm is used for feature importance ranking, as shown in Fig. 4.

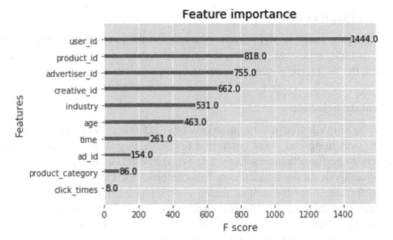

Fig. 4. Feature importance ranking

As seen from Fig. 4, among the indicators with high correlation in the heatmap, the advertiser id ranks higher than the creative id, the advertisement id is higher than the time, the creative id is higher than the advertisement id, and the product id is higher than the product category id. Therefore, this paper removes 4 features: creative id, advertisement id, time and product category id. Finally, 7 characteristics, such as user id, product id, advertiser id, advertiser industry id, age, click times and gender, are used for training.

4.3 Evaluation Metrics

In this paper, the precision (P), the recall (R), the F1-score and the accuracy (acc) values are used to evaluate the prediction effect of the model on the age and gender of the user's basic attributes. The accuracy rate indicates the percentage of correctly classified samples to the total number of samples. The larger the accuracy rate value, the better the classifier. The calculation method of each evaluation index is as follows:

$$acc = \frac{TP + TN}{TP + FP + TN + FN} \tag{2}$$

$$P = \frac{TP}{TP + FP} \tag{3}$$

$$R = \frac{TP}{TP + FN} \tag{4}$$

$$F1 = \frac{2 * P * R}{P + R} \tag{5}$$

TP represents the total number of correct predictions; TN means predicting the negative class as the number of negative classes; FN is the total number that has not been predicted; FP represents the total number of mispredictions.

4.4 Experimental Results

This paper studies the differences and characteristics of five commonly used machine learning algorithms including LR, RF, ExtraTree, XGBOOST and LightGBM, and integrates various machine learning algorithms based on stacking ideas. Model training. Data processing and model training were implemented through Python. During training, the method of 5-fold cross-validation is employed to optimize the different weights among the features in the iterative classification process of the algorithm in this paper.

4.4.1 Ablation Experiment

To verify the prediction effect of this model, three groups of ablation experiments are designed in this paper:

The first group: The first stage of stacking uses a machine learning algorithm, and the second stage uses the LGB algorithm. There are 4 model combinations, including LR-LGB, RF-LGB, ET-LGB, and XGB-LGB;

The second group: The first stage of stacking uses two machine learning algorithms, and the second stage uses the LGB algorithm. There are 6 model combinations, including RF-XGB-LGB, ET-XGB-LGB, ET-LR-LGB, RF-ET-LGB, RF-LR-LGB and LR-XGB-LGB;

The third group: The first stage of stacking uses three machine learning algorithms, and the second stage uses the LGB algorithm. There are 4 model combinations, including RF-ET-LR-LGB, RF-ET-XGB-LGB, LR-ET-XGB-LGB and LR-RF-XGB-LGB.

The experimental results of the above 15 models are compared comprehensively from the four aspects of precision, recall, F1-score and accuracy and are ranked according to the accuracy of each model. The gender prediction results are shown in Table 5, and the age prediction results are shown in Table 6.

Table 5. Gender prediction results

Model	Precision%	Recall%	F1-score%	Accuracy%
LR-RF-ET-XGB-LGB	**78.47**	**77.97**	**76.54**	**77.97**
RF-ET-LR-LGB	76.67	76.87	75.79	76.87
RF-XGB-LGB	76.55	76.80	75.77	76.80
ET-XGB-LGB	77.19	76.73	75.09	76.73
RF-ET-XGB-LGB	77.34	76.60	74.74	76.60
ET-LGB	76.92	76.53	74.88	76.53
ET-LR-LGB	76.89	76.46	74.77	76.46
LR-ET-XGB-LGB	76.93	76.39	74.61	76.39
LR-RF-XGB-LGB	75.75	75.63	73.98	75.63

(*continued*)

Table 5. (*continued*)

Model	Precision%	Recall%	F1-score%	Accuracy%
RF-ET-LGB	75.37	75.57	74.20	75.57
XGB-LGB	75.12	75.36	74.00	75.36
RF-LR-LGB	74.98	75.15	73.65	75.15
RF-LGB	75.03	75.09	73.45	75.09
LR-XGB-LGB	72.27	72.68	70.68	72.68
LR-LGB	68.99	69.87	67.11	69.87

Table 6. Age prediction results

Model	Precision%	Recall%	F1-score%	Accuracy%
LR-RF-ET-XGB-LGB	**53.35**	**53.12**	**52.61**	**53.15**
ET-LR-LGB	52.50	52.51	51.74	52.51
LR-ET-XGB-LGB	52.15	52.23	51.54	52.23
LR-RF-XGB-LGB	51.32	51.96	51.11	51.96
RF-ET-LR-LGB	51.46	51.82	51.18	51.82
RF-LGB	50.97	51.54	50.71	51.75
ET-XGB-LGB	51.52	51.68	51.12	51.68
ET-LGB	50.91	51.61	50.46	51.61
XGB-LGB	51.11	51.34	50.53	51.33
LR-XGB-LGB	51.30	50.72	50.04	50.72
RF-ET-XGB-LGB	50.90	50.51	49.80	50.51
RF-XGB-LGB	49.20	50.24	48.94	50.24
RF-ET-LGB	49.68	49.97	49.53	49.97
RF-LR-LGB	51.54	51.75	50.73	49.69
LR-LGB	49.74	49.62	48.90	49.62

According to the results in Table 5, in the gender prediction task, the multimodel LightGBM model has the best prediction effect, with an accuracy of 77.97%. Accuracy, recall rate and F1 value are all the highest among the 15 models. The LR-LGB model has the lowest accuracy of 69.87%.

Table 6 shows that in the age prediction tasks, the multimodel LightGBM model prediction effect is best, with an accuracy of 53.15%, and the LR-LGB model has the lowest accuracy of 49.62%.

From the model prediction results of gender and age, it can be seen that the expressiveness of a single model is not as good as the combination of multiple models, and the prediction effect of LR is the worst. For binary classification problems such as gender and multiclassification problems such as age, the model constructed in this paper has the best effect compared with the control experiment.

4.4.2 Effectiveness Experiment

To verify the validity of this model, an additional dataset is used to verify the model, which is the advertising behavior log of Taobao users, with a total of 21 features and a total of 900,000 pieces of data. After data preprocessing, feature selection and other feature engineering, 11 features, including adgroup_id, time_stamp, nonclk, clk, cate_id, shopping_level, occupation, price, cms_segid, age and gender, are used to predict the age and gender of users. The rest of the operations are the same as those for dataset 1. The experimental results of the models are comprehensively compared from the four aspects of precision rate, recall rate, F1 value and accuracy rate and are ranked according to the accuracy of each model. The gender prediction results are shown in Table 7, and the age prediction results are shown in Table 8.

Table 7. Gender prediction results

Model	Precision%	Recall%	F1-score%	Accuracy%
LR-RF-ET-XGB-LGB	**72.73**	**68.09**	**68.74**	**72.95**
RF-XGB-LGB	73.03	67.61	68.22	72.84
LR-RF-XGB-LGB	72.88	67.63	68.24	72.79
ET-XGB-LGB	73.19	67.25	67.81	72.71
RF-ET-XGB-LGB	73.03	67.35	67.93	72.71
XGB-LGB	73.02	67.04	67.57	72.56
LR-ET-XGB-LGB	73.23	66.87	67.37	72.53
ET-LR-LGB	73.20	66.86	67.35	72.52
LR-XGB-LGB	73.20	66.84	67.33	72.51
RF-LGB	71.66	65.26	65.51	71.22
ET-LGB	71.22	64.64	64.77	70.78
LR-LGB	70.16	65.36	65.73	70.74
RF-ET-LR-LGB	69.38	62.81	62.58	69.31
RF-ET-LGB	69.93	62.06	61.47	69.10
RF-LR-LGB	69.53	61.97	61.40	68.95

It can be seen from the above results that the accuracy of gender prediction for dataset 2 is 72.95%, and the accuracy of age prediction is 61.28%. Experiments show that the prediction performance of the model proposed in this paper is universal.

Table 8. Age prediction results

Model	Precision%	Recall%	F1-score%	Accuracy%
LR-RF-ET-XGB-LGB	**61.10**	**65.49**	**61.10**	**61.28**
LR-RF-XGB-LGB	65.12	60.99	61.22	60.99
RF-XGB-LGB	64.71	60.96	61.17	60.96
ET-XGB-LGB	65.14	60.96	61.21	60.96
LR-ET-XGB-LGB	64.88	60.91	61.16	60.91
LR-XGB-LGB	65.00	60.89	61.07	60.89
RF-ET-XGB-LGB	68.77	60.72	60.79	60.72
XGB-LGB	66.24	60.49	60.85	60.49
RF-ET-LR-LGB	65.84	60.29	60.26	60.29
RF-LGB	66.18	60.19	60.47	60.19
RF-ET-LGB	64.74	59.96	60.03	59.96
ET-LGB	65.99	59.94	60.12	59.94
RF-LR-LGB	64.34	59.69	59.92	59.69
ET-LR-LGB	64.35	59.30	59.42	59.30
LR-LGB	32.51	32.23	31.47	32.23

5 Conclusion

To improve the performance of advertising user attribute prediction, this paper proposes a multimodel blending method. The model integrates multiple machine learning algorithms based on stacking ideas. The fusion of multiple models can make up for the deficiency of different algorithms and improve the prediction accuracy by using the difference in feature extraction of algorithms. The experimental result demonstrates that the multimodel LightGBM model has a great prediction result on the gender and age of users. In a follow-up study, if more features are added, the prediction effect of the model may be better, which can help advertisers have a more comprehensive understanding of users so that they can more accurately push advertisements according to the characteristics of users, improve the efficiency of advertising, and maximize their interests.

Acknowledgment. This work is supported by Hainan Province Science and Technology Special Fund, which is Research and Application of Intelligent Recommendation Technology Based on Knowledge Graph and User Portrait (No. ZDYF2020039). Thanks to Professor CaiMao Li, the correspondent of this paper.

References

1. Hu, J., Zeng, H.-J., Li, H., Niu, C., Chen, Z.: Demographic prediction based on user's browsing behavior. In: Proceedings of the 16th International Conference on World Wide Web, pp. 151–160 (2007)
2. De Bock, K., Van den Poel, D.: Predicting website audience demographics for web advertising targeting using multiwebsite clickstream data. Fund. Inform. **98**(1), 49–70 (2010)
3. Lu, X.: Research and analysis of user attribute inference of Chinese social networking websites based on WEB. Ph.D. thesis, Nanjing University of Aeronautics and Astronautics (2017)
4. Liu, B.Q., Niu, Y.: Gender recognition of chinese microblog users based on emotion characteristics. Compt. Eng. Sci. **38**(9), 7 (2016)
5. Wang, M., Cao, Q., Sun, J., Zhang, Q., Xu, F.: A user based attribute prediction method based on ensemble learning. Small Microcomput. Syst. **41**(12), 7 (2020)
6. Gao, J., Zhang, T., Cheng, X., Jian, G.: An age and gender prediction method based on lightGBM machine learning algorithm. Post Telecommun. Des. Technol. **9**, 4 (2019)
7. Ke, G., et al.: LightGBM: a highly efficient gradient boosting decision tree. In: Advances in Neural Information Processing Systems, vol. 30 (2017)
8. Yun, J., Sun, G., Chen, Q., Zhang, M., Zhu, H., Rehman, M.U.: A model combining convolutional neural network and lightGBM algorithm for ultra-short-term wind power forecasting. IEEE Access **7**, 28309–28318 (2019)
9. Pan, B.: Application of XGBoost algorithm in hourly PM2.5 concentration prediction. IOP Conf. Ser. Earth Environ. Sci. **113**, 012127 (2018)
10. Chen, T., Guestrin, C.: XGBoost: a scalable tree boosting system. In: The 22nd ACM SIGKDD International Conference (2016)
11. Yin, Q.: Research on intelligent stock selection in biomedical sector based on stacking fusion model. Ph.D. thesis, Shanghai Normal University (2020)
12. Zhang, J., Huo, J.: Research on purchasing behavior prediction based on stacking model integration. Shanghai Manage. Sci. **43**(1), 12–19 (2021)
13. Xu, G., Shen, Y.: Multiclassification detection method for malicious programs based on XGBoost and stacking model. Inf. Netw. Secur. **21**(6), 11 (2021)
14. Wang, H.: Research on old vehicle valuation model based on characteristic price theory and CatBoost. Ph.D. thesis, Tianjin University of Commerce (2019)
15. Hang, Q., Jinghui, Y.: Application status of machine learning stochastic forest algorithm. Electron. Technol. Softw. Eng. **24**, 3 (2018)
16. Pu, Y., Xingfu, L.: Application of stacking ensemble learning in sales forecasting. Softw. Guide **21**(4), 6 (2022)
17. Yangheran, P., Junling, R.: An integrated detection method of malicious web pages based on stacking. Comput. Appl. **39**(4), 8 (2019)
18. Xuemei, H., Ying, X., Huifeng, J.: Breast cancer prediction based on penalized logistic regression. Data Collect. Process. **36**(6), 13 (2021)
19. Jiaqing, W., Zhe, W., Taipeng, Z., Jinyu, Y., Xin, Z., Yunxiang, M.: Research on breast cancer prediction model based on random forest algorithm. Chin. Med. Equip. **19**, 119–123 (2022)
20. Aihong, W.: Research on the prediction of railway passenger refund rate based on XGBoost algorithm. Railw. J. **41**(12), 7 (2019)
21. Hongyan, H., Guoyan, H., Bing, Z., Damiao, J.: Anomaly detection model based on limit tree feature recursive elimination and lightGBM. Inf. Netw. Secur. **1**, 8 (2022)
22. Tong, G., Guoliang, X., Wanlin, L., Jiahao, L.: An intelligent evaluation model for housing prices based on integrating lightGBM and Bayesian optimization strategies. Comput. Appl. **40**(9), 6 (2020)

23. Xuan, W., Xiangwei, L.: A classification model that integrates feature selection algorithms and lightGBM fusion. Fujian Comput. **38**(4), 4 (2022)
24. Cong, Z.: Research on MOOC dropout prediction based on feature engineering. Ph.D. thesis, Jiangxi University of Finance and Economics (2020)
25. Wei, Y., Fengli, Y., Jing, J., Meng, Y.: A method for online parameter tuning of models based on improved logistic regression algorithm. Commun. Technol. **53**(8), 5 (2020)

How is the Power of the Baidu Index for Forecasting Hotel Guest Arrivals? –A Case Study of Guilin

Haitao Yu[1,2], Le Liu[1], Zhongjun Wu[1], and Yajun Jiang[1(✉)]

[1] Guilin University of Technology, No. 12, Jiangan Street, Qixing District, Guilin 541004, Guangxi, China
jyjall@163.com

[2] Guangxi Tourism Industry Research Institute, No. 12, Jiangan Street, Qixing District, Guilin 541004, Guangxi, China

Abstract. The prediction of the Baidu index for tourism demand has been increasingly focused on by scholars. However, few studies have evaluated the predictive power of the Baidu index for hotel guest arrivals in fine granularity at the micro level. Taking Guilin as a case study, we use the OLS regression method to quantitatively investigate the forecasting power of the Baidu index for daily hotel guest arrivals and to comprehensively evaluate the performance of the forecasting model and to optimize the forecasting model by deeply mining the hidden characteristics of tourism flow in a special case study. The contributions of this paper mainly have threefold: first, to the best of our knowledge, based on the actual full-example of daily hotel guest check-in data in fine granularity, we evaluated the predictive power of the Baidu index by comparison of 5 forecasting models for the first time. Second, we proposed two metrics for forecasting: the trend forecasting index and the forecasting stability index. Finally, we introduce a kind of punishment strategy to optimize forecasting models based on the potential pattern of research objects.

Keywords: Tourism big data · Baidu index · Hotel guest arrivals · Performance optimization · Forecasting power

1 Introduction

With the advancement of new-generation information technologies, such as the Internet of Things (IoT), various social media and mobile Internet, cloud computing and big data, have become available in various industries. Big data technology has moved to various industries from its initial concept, and we are truly in the real era of big data [1], for example enterprise marketing focuses on customers instead of product center, and customer relationship management has become the core issue [2].

In the tourism industry, various types of information are generated in real time, such as pretrip plans, information searching, reservation and booking, information sharing and transactions during a trip, posttrip experience sharing, interaction activities, recommendations, or trace trajectories. These massive, unstructured and various tourism

Y. Wang et al. (Eds.): ICPCSEE 2022, CCIS 1629, pp. 185–211, 2022.
https://doi.org/10.1007/978-981-19-5209-8_13

data can be further deeply analyzed and mined to seek the hidden regular patterns and correlations behind the data for the purpose of management, marketing, product design and service quality improvement and operation optimization.

In tourism big data research, the forecasting of hotel guest arrivals is widely focused on based on the Baidu index or Google Trends due to the availability and timeliness of query data of search engines. However, most of the current studies based on search indexes are carried out in coarse granularity, such as weekly or monthly scales or official statistical data, and few studies are based on full-sample data of daily real-time check-in data to evaluate the predictive power of Internet search indexes. **To fill this research gap, based on 16 months of full-sample data of the daily check-in of hotels in Guilin tourism destination, we explore the forecasting powers of Baidu search indexes for actual daily hotel guest arrivals. To the best of our knowledge, this paper is the first to assess the predictive power of the Baidu index using a full sample in the hospitality industry.** Because guests generally use the Baidu search engine to search for information on tourism destinations prior to travel, there is a certain time difference between the Baidu index and the check-in of guests. The phenomenon of time difference is called the lag effect, which is meaningful for the prediction of tourism demands. The value of the time difference is called the lag period. In addition, the number of guests in the hotel also has a lag effect, which means that the data of the guest number of the last period have an effect on the data of the following periods. Therefore, the current research considered the lag effects of both the Baidu index and the number of guests. The current study has two advantages. Practically, the prediction based on full-example fine granularity plays an important role in tourism flow security prewarning and on-time schedules of human power, various tourism resources and facilities. Theoretically, the study expands the literature knowledge of tourist forecasting and provides a new perspective for tourism prediction research.

The research questions are as follows:

1. **How do Baidu indexes forecast daily hotel guest arrivals in tourist destinations quantitatively?**
2. **What are the effective lag periods of Baidu indexes and hotel guest arrivals for forecasting models?**
3. **How does the forecasting performance of the model be improved effectively by means of deep data exploration and pertinent mechanism design?**
4. **How is the forecasting performance of models evaluated comprehensively and objectively?**

The main contributions of this study lie in the following: first, based on municipal administrative-level tourism destinations in daily fine granularity using actual check-in full-sample big data of a tourism destination, we quantitatively conducted empirical research modeling hotel guest arrivals to explore the complicated relationship between actual hotel guest arrivals and Baidu indexes. Second, a punishment mechanism is designed to effectively improve the forecasting performance of models. Third, we proposed two metrics for measuring the trend forecasting power and the forecasting robustness, which are capable of quantitatively measuring the trend consistency between the forecasting result and actual trend and the forecasting power stability. Finally, we make a detailed analysis and comparison of models in terms of forecasting accuracy, trend forecasting power and the stability and robustness of forecasting power.

The rest of this paper is organized as follows: the literature relevant to the Baidu index and tourism demand forecasting are overviewed briefly in Sect. 2. Data description and preparation are provided in Sect. 3. Variable tests and metrics are described in Sect. 4. An empirical study of forecasting models is presented in detail in Sect. 5. Finally, Sect. 6 concludes this paper and presents limitations and future research.

2 Literature Review

2.1 Baidu Index and Tourism Research

With the popularity of the Internet worldwide, network search has become an important daily information acquisition mode due to the available information on the Internet. One can understand and forecast social and economic behaviors according to the search content and the corresponding volume. Search engine volumes, such as Google trend analytic and Baidu index, are used to predict illnesses, such as Dengue Outbreak [3], diagnosed cases of HIV/AIDS [4], AIDS Epidemics [5], COVID-19 [6], human brucellosis [7] and hand-foot-mouth disease [8], and economic phenomena, such as volatility of China's stock markets [9], mobile phone sales [10], and Chinese stock returns [11].

As the largest Chinese search engine worldwide, Baidu is most popular for network users in China. Until December 2015, the user penetration rate of Baidu search accounted for 91.3% of all search engines based on official statistical data. The tourism research with the Baidu index is mainly classified into 2 categories: network attention of tourism and tourism forecasting. Research on the network attention of tourism has focused on temporal and spatial distribution features. Based on Baidu index data in 31 provinces, autonomous regions and municipalities in China, Lin et al. (2014) analyzed the characteristics of the regional disparity of the national tourism security network attention [12]. Based on an index of the cyberspace attention of 53 5A-level tourist attractions in China, LI Shan et al. (2008) revealed the temporal distribution and precursor effect of tourism network attention [13]. Taking Tianmu Lake as a case study, Liu P et al. examined the relationship between daily tourist arrivals and search indexes from 13 cites to test their spatial and temporal correlations using the impulse function in a vector autoregressive mode [14]. Huang, Xiankai et al. (2016) used the Baidu index to predict tourism flow in the Forbidden City [15]. Zhang Xing et al. (2017) conducted an empirical study on the relationship between the Baidu index and trading volume on a P2P platform [16]. Yongwei Liu et al. (2021) used the Baidu index to analyze the spatial characteristics of tourism flows in China [17].

The existing study used the Baidu index to make trend predictions at the macro level, and most of the data sources are statistical results. Few studies adopt a full sample to predict tourism demand at the micro level. Therefore, it is necessary to use full sample data to evaluate the predictive power of the Baidu index and optimize the current regression method according to the characteristics of tourism destinations and tourism resources.

2.2 Tourism Forecasting with Big Data

Currently, research on tourism forecasting for tourism flow or demand with big data is widely carried out based on different space dimensions from a tourism attraction to the European Union, including many countries, different sampling periods from several weeks to several years, and different forecasting methods, including traditional statistics, machine learning, deep learning and econometrics, most of which are based on the search volume of keywords on search engines used as the independent variables, such as the Baidu index and Google trend analysis.

Based on website traffic data and Google Analytics indicators, Gunter U et al. constructed a novel Bayesian FAVAR for predicting monthly tourist arrivals to Vietnam [18]. Liu Y Y et al. selected Miao village in Guizhou Province in China as the study case to explore the correlation between weather, temperatures, weekends and public holidays with tourism destination arrivals and web search queries using the VAR (vector autoregressive model) method based on Baidu search index data [19]. Based on the Baidu index and Google trend, Sun S et al. (2019) used a kernel extreme learning method to model the monthly volume tourist arrivals to Beijing [20]. Li X et al. (2017) adopted a generalized dynamic factor model (GDFM) to predict monthly tourist volumes to Beijing based on the Baidu index [21]. Xie, G et al. (2021) utilized a least squares support vector regression model with a gravitational search algorithm to forecast Chinese cruise demand based on search query data (SQD) from Baidu and economic indexes [22]. Silva E S et al. (2019) introduced a hybrid singular spectrum analysis and neural network model to forecast monthly tourism demand for 10 European countries [23]. Wen L et al. proposed a new hybrid model that combines the linear and nonlinear features of component model-based search query data [24].

Based on Internet search data from Google Trends, Park S et al. constructed short-term forecasts for the inflow of Japanese tourists to South Korea, which perform much better than the standard time-series models in terms of short-term forecasting accuracy [25]. Tang J et al. constructed a forecasting model of the inflow index of Chinese tourists to Thailand based on Baidu trends and internet big data [26]. Camacho M et al. utilized a dynamic factor approach to forecast the checked-in and overnight stays of travelers in Spain based on Google's search volume indexes and a real-time database of vintages [27]. Li S et al. proposed the PCA-ADE-BPNN method, a combination of a dimensional reduction algorithm, an optimization algorithm, and a neural network, for tourist volumes in Beijing city and Hainan Province, China [28]. Önder I compared the forecasting accuracy of cities and countries using Google Trends Web and image indexes regarding 2 cities (Vienna and Barcelona) and 2 countries (Austria and Belgium) [29]. Pan B et al. constructed an accurate time-series forecasting model of weekly hotel occupancy for a destination by incorporating several tourism big data sources, including search engine queries, website traffic, and weekly weather information [30].

Volchek E et al. investigated the relationships between Google search queries for the most popular London museums and actual visits to these attractions [31]. Xie G et al. proposed models with KPCA-based (kernel principal component analysis) web search indexes for tourist arrivals at Hong Kong [32]. Zhang B et al. constructed a deep learning framework to forecast the overnight passenger flows for hotel accommodations in Hainan Province, China, using an internet search index [33]. Taking short-haul travel from Hong Kong to Macau as the study object, Hu M et al. examined whether combining causal variables with search engine data can further improve the forecasting performance of search engine data models [34]. Taking Jiuzhaigou, a popular tourist spot in China, as a case study for empirical analysis, Zhang B et al. used a long short-term memory (LSTM) network to forecast daily tourist flow [35].

These studies verified the validation in forecasting the social and economic phenomenon in tourism and hospitality with the combination of tourism big data and web search data, which demonstrates that web search data are a powerful predictor. However, the current literature on tourism forecasting based on the network search index has the following limitations. In terms of predictive granularity, most are based on weekly or monthly periods, and few studies are conducted based on real-time full-sample tourism data in fine granularity. In terms of research methods, most are based on the existing algorithms, and few optimize the forecasting methods according to the detailed features of the specific data set. In terms of evaluation indexes, most are based on forecasting error and MAPE, which cannot measure the forecasting trend consistency and the robustness and stability of forecasting methods. Most importantly, there is a lack of comprehensive and quantitative research on the evaluation of the predictive power of the Baidu index.

3 Data

3.1 Data Collection

We selected Guilin, Guangxi Zhuang Autonomous Region of China, as the study case for the following reasons. As a famous international tourist attraction, Guilin is a famous cultural city with 2000 years of history that has both world-level natural scenery heritage and splendid cultural heritage. Until 2017, Guilin has owned 60 A-level scenic spots, including 4 5A-level scenic spots, 25 4A-level scenic spots and 31 3A-level scenic spots. It is a cultural tourist destination with abundant natural tourism sources of Li-River and Karst mountains, historical culture and simple folk customs and unique regional culture of Zhuang, Miao, Dong, Yi nationalities and so on, such as farming, nomadism, festivals, clothing. Each year Guilin attracts a large number of tourists at home and abroad, and the tourism arrivals of Guilin destination exceeded 80 million in 2017 according to official statistics data.

The authors are authorized to access hotel guest arrival data from Guilin Culture, Tourism and Broadcast Administration for the purpose of the current research. Each record in the data stored in the oracle database management system includes fields: hotel code, check-in date and time, check-out date and time, the top six number of identity cards of guests, and the bottom two number of identity cards of customers for Chinese citizens. The top six identity cards indicate the province code, city code and county code of the identity registration place, and the bottom six identity cards indicate the gender codes for Chinese residents born in the People's Republic of China. The period of data is from January 2015 to April 2016. The data are divided into in-sample subsets and out-sample subsets. The in-sample subset was used for training the prediction model with data from January 1, 2015, to December 31, 2015, and the out-sample subset was used for testing the prediction capability of the trained model with data from January 1, 2016, to April 30, 2016.

3.2   Keyword Selection for the Baidu Search

To reveal the relationship between the Baidu index and daily hotel guest arrivals, relatively accurate search keywords about Guilin Hotel should be found first. This is because the effectiveness and validity of prediction models mainly depend on the selection accuracy of search keywords in addition to the factors of the models. However, there are currently no effective theories or methods for improving the selection accuracy of search keywords for the Baidu Index. In most of the current literature, researchers usually make use of Baidu demand mapping to determine the most concerned and frequently searched keywords about destinations. In this study, the process of keyword selection is divided into three steps.

In the first step, we first input Guilin Hotel as a keyword to the Baidu index search and then seek the keywords with a strong correlation with the Guilin hotel keyword. In the second step, we filtered some keywords rarely used by tourists by means of surveys and interviews with tourist experts. In the last step, using SPSS statistical software, we calculate the correlation between the keywords kept in the second step with Guilin hotel guest arrival data. The correlation coefficients of the Baidu indexes of Guilin Hotel and Guilin Accommodation are higher than 0.6, which have a high correlation with the Guilin accommodation number. The correlation coefficients of Guilin Tourism and Guilin weather are below 0.6, which has a low correlation with the Guilin accommodation number. Finally, we keep four keywords: 'Guilin hotel' (**GH**), 'Guilin weather' (**GW**), 'Guilin accommodation' (**GA**) and 'Guilin tourism' (**GT**). Table 1 shows the correlation among the selected Baidu indexes, where **AAN** denotes the actual daily accommodation number.

Table 1. Pearson correlation test between Baidu indexes.

	AAN	GH	GW	GA	GT
AAN	1	0.686^{**}	0.177^{**}	0.625^{**}	0.304^{**}
GH	0.686^{**}	1	0.078	0.857^{**}	0.656^{**}
GW	0.177^{**}	0.078	1	0.036	-0.155^{**}
GA	0.625^{**}	0.857^{**}	0.036	1	0.645^{**}
GT	0.304^{**}	0.656^{**}	-0.155^{**}	0.645^{**}	1

Note: Pearson correlation test is used by SPSS software

3.3   Data Preprocessing

Data Preprocessing: Guilin hotel guest arrival data contain not only domestic tourists but also inbound tourists. Therefore, we need to remove the accommodation data of inbound tourists from the database according to the top six numbers of IDs. Finally, we kept the accommodation data of tourists from 32 provincial administrative units (excluding Taiwan, Hong Kong and Macao region) of Guilin, approximately 2950 hotels registered in industrial and commercial administration.

The Variables: In this study, *AAN* (actual accommodation number) is the dependent variable. In addition to the independent variables *GH*, *GW*, *GA* and *GT*, to study the effect of public holidays on the prediction model, we set public holidays as an independent variable *HOL*. The variables *GH*, *GW*, *GT*, and *GA* are numerical data. We set nonnumerical data *HOL* as a dummy variable. If the date is a public holiday or weekend, *HOL* is equal to 1; otherwise, *HOL* is 0. In this study, public holidays also include New Year's Day, Tomb Sweeping Day, Dragon Boat Festival holiday, golden week of Labor Day, golden week of National Day and the Spring Festival in addition to known weekends. In hospitality, there are more guests on Friday or Saturday than on other days in a week, and guests who check in on Sundays begin to decrease obviously. If dates are Friday or Saturday, *HOL* is 1.

Missing or Abnormal Data: To eliminate invalid data or abnormal data, we performed an exploratory analysis of the descriptive statistics of the Baidu indexes of four search keywords using the SPSS statistical tool. The boxplots of the Baidu indexes are shown in Fig. 1, where the Y-axis denotes the daily distribution of the Baidu index in a given period. As shown in Fig. 1, the Baidu indexes of Guilin Tourism keyword from February 6, 2015, to February 7, 2015, are rather abnormal for unknown reasons. It can be found that the Baidu indexes of the three days are equal to 326236 and 319829, respectively, which are far higher than the corresponding mean of 4843 in 2015. By analyzing the trend of the Baidu index, except on holidays and weekends, the Baidu index approximately shows a linear trend on adjacent days. At the same time, the Baidu index will show a similar trend with the period of one week. To guarantee the validity of the prediction model, we use the linear interpolation method to fill the missing data by weighting the adjacent data and the corresponding data of the last week and next week.

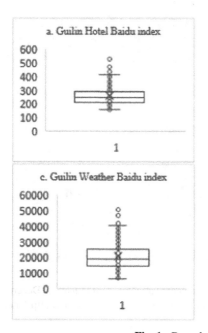 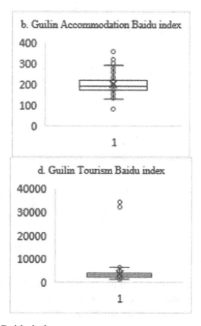

Fig. 1. Box plot of Baidu indexes.

After the preprocessing operation, the trends of the dependent variable and independent variables in 2015 are shown in Fig. 2, where the X-axis denotes the date from January 2015 to April 2016 and the Y-axis denotes the daily Guilin hotel guest arrivals and selected Baidu indexes. Because the value ranges of ANN, GH, GW, GA and GT are very different from each other, it is not suitable to place them in the same coordinate system. Otherwise, the trend of the variable with the small value range will be shielled by the one with the large value range. It is clearly shown that there are consistent relationships between the trend of the dependent variable *AAN* and the independent variables *GH*, *GW*, *GA* and *GT*, especially *GH* and *GT*.

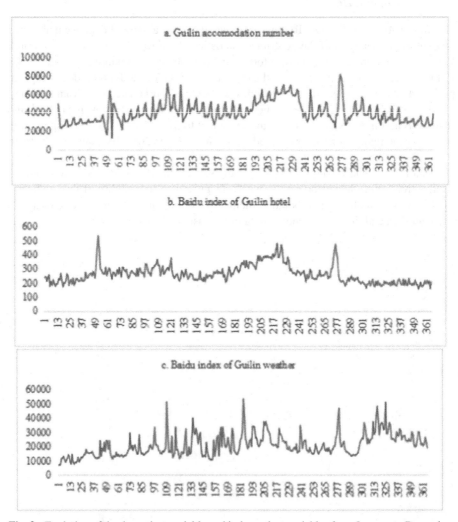

Fig. 2. Evolution of the dependent variable and independent variables from January to December 2015. Source: Guilin Tourism Culture, Tourism and Broadcast Administration and http://index. baidu.com/v2/index.html#

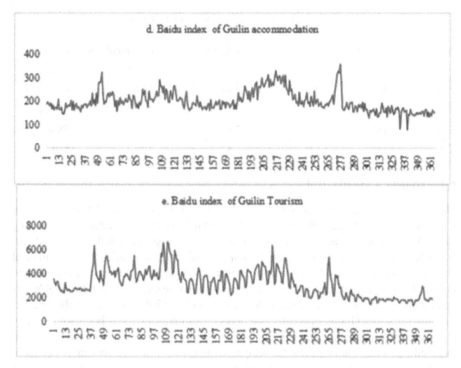

Fig. 2. continued

4   Variable Test and Metrics

4.1   Unit Root

The unit root test is a stochastic process problem that is used to test whether there exists a unit root in time series data. If a unit root exists, it means that time series data are not stationary. Using nonstationary time data, pesu-regression is generated, which cannot truly express the relationship between the dependent variable and independent variables. Therefore, a unit root test is necessary for regression analysis. The variables *AAN*, *GH*, *GW*, *GA* and *GT* are time-series data. Therefore, prior to constructing the econometric prediction model, we need to make a unit test for variables to guarantee the stationary time series, thus avoiding spurious regression. In this study, we adopt the ADF method to perform a unit root test. The results of the unit root of variables are shown in Table 2. The test results showed that all the variables are a stationary sequence with a zero-order single whole due to no existing unit root. Due to the stationary characteristic of the variables, we can directly make a Granger casualty test between the dependent variable and independent variables with no need for a cointegration test.

Table 2. Results of the variable unit root test.

Variable	Test statistic	1% critical value	5% critical value	10% critical value	Conclusion
AAN	−7.448	−3.441	−2.871	−2.570	Stationary
GH	−5.441	−3.441	−2.871	−2.570	Stationary
GW	−8.755	−3.441	−2.871	−2.570	Stationary
GA	−7.196	−3.441	−2.871	−2.570	Stationary
GT	−4.885	−3.441	−2.871	−2.570	Stationary

4.2 Granger Casualty Test

The Granger casualty test was created by Clive W. J. Granger Nobel Prize winner in economics, which is capable of analyzing the Granger casualty relationship among economic variables to account for the prediction power of independent variables for dependent variables. As a kind of prediction for stationary time series data in statistical essence, a Granger casualty test is only applicable to variable prediction in the econometrics field instead of being used as the basis of the true causality test. If variable x is helpful to explain the future change of variable y, it can be believed that x is the Granger casualty for y. If a Granger casualty test between variables exists, it is shown that they have predictive powers upon each other. The purpose of the Granger casualty test in this study is to test whether there are reciprocal predictive powers between independent variables and dependent variables. Because a Granger causality test is sensitive to lag selection, this study identified the best lag based on the Akaike information criterion (**AIC**) and Schwarz criterion (**SC**). The results of the Granger causality test between the dependent variable and independent variables are shown in Table 3. It can be seen from Table 3 that *AAN* has causality with *GH*, *GW*, *GA* and *GT*. Bidirectional causality relationships existed between *AAN* and *GH*, *AAN* and *GT*, and *AAN* and GA. However, an omni-directional casualty relationship exists between *AAN* and *GW*.

Table 3. Results of the Granger causality test.

Lag	Null hypothesis	F value	Prob > F	chi2 value	Prob > chi2	Conclusion
1	GH **NOT** AAN	55.72	0.000*	56.06	0.000*	Refuse
7	AAN **NOT** GH	1.92	0.065***	13.87	0.054***	Refuse
2	GW **NOT** AAN	4.24	0.015**	8.57	0.014**	Refuse
8	AAN **NOT** GW	1.60	0.123	13.25	0.104	Non refuse
2	GA **NOT** AAN	36.99	0.000*	74.76	0.000*	Refuse
1	AAN **NOT** GA	3.21	0.074***	3.23	0.072***	Refuse
1	GT **NOT** AAN	29.92	0.000***	30.11	0.000*	Refuse
1	AAN **NOT** GT	3.01	0.083***	3.03	0.082***	Refuse

(***), (**) and (*) denote statistical significance levels at 1%, 5% and 10%, respectively.
NOT: does not Granger-cause.

The findings can be found in Table 3: 1) **GH** causes **AAN** and vice versa; 2) **GT** causes **AAN** and vice versa; 3) **GA** causes **AAN** and vice versa; 4) **GW** has the predictive power for **AAN**, but **AAN** does not have the predictive power of **GW**, which is consistent with common sense about tourism behaviors. In previous trips, tourists commonly paid more attention to accommodations than to weather. Therefore, it is feasible to construct prediction models for forecasting **AAN** using **GH, GT, GW** and **GA**.

4.3 Metrics

The forecasting error index **MAPE** is defined as follows, which is used for evaluating the performance of forecasting models as formula (1).

$$\text{MAPE} = \frac{1}{n} \sum\nolimits_{t=1}^{n} \left| \frac{\hat{y}_t - y_t}{y_t} \right| \times 100\% \tag{1}$$

where $\hat{y}_t$ denotes the forecasting value of hotel guest arrivals at time t, y_t denotes the actual value of hotel guest arrivals at time t and n denotes the number of test samples in out-sample subsets. When MAPE equals 0, a model is perfect.

However, MAPE only denotes the average forecasting errors between actual values and forecasting values, which is not capable of measuring the trend forecasting capacity and the stability of forecasting models. In practice, the trend forecasting and stability of forecasting performance are also important. Therefore, we proposed two metrics for measuring trend forecasting capacity and the stability and robustness of forecasting models. This is one of the main novelties of our contributions.

Define 1: The trend forecast index (**TFI**) is defined as formula (2):

$$TFI = \frac{1}{n-1} \sum_{i=2}^{n} \text{sgn}(\hat{y}_i - \hat{y}_{i-1}) \oplus \text{sgn}(y_i - y_{i-1}) * 100\% \tag{2}$$

where n denotes the number of forecast samples; a symbol function $\oplus$ denotes an exclusive OR operator; $\hat{y}_i$ denotes the forecast value at time i; and y_i denotes the actual value at time i. **TFI** denotes the consistency between the actual trend and forecasting trend. The larger the **TFI** is, the more similar the forecasting trend is to the actual trend.

Define 2: Forecasting stability and robustness index **STD** is defined as formula (3)

$$STD = \sqrt{\frac{1}{N} \sum_{i=1}^{N} (x_i - \bar{x})^2} \tag{3}$$

where x_i denotes the forecasting error of the ith test sample and $\bar{x}$ denotes the mean of forecasting errors. **STD** is used to quantitatively evaluate the robustness and stability of the forecasting performance of models. A lower **STD** indicates a smaller fluctuation in forecasting error.

By means of **MAPE, TFI** and **STD**, we can comprehensively evaluate the forecasting performance of models in terms of number error, trend error and stability.

5 Model Fitting and Performance Evaluation

5.1 Forecasting Model Without Any Baidu Index

Model Fitting: The estimation results are shown as Eq. (4) without considering public holidays and with public holidays, which are abbreviated as **AR** (autoregression) and **AR+HOL**, respectively. The adjusted goddesses of fitting of Eq. (4) are 0.64 and 0.75. Equation (4) as well as all the following equations hereinafter are fitted by OLS linear regression (**least square method**) using Eviews statistical software. The root mean square errors within the sample period are 6990 and 6108. All the regression variables and intercept item c in the two equations can pass the statistical significance test at a probability level of 1%. The estimation results show that both the first- and third-order lags of *AAN* have a significantly positive correlation with its current order lag at the 1% probability level. In contrast, the second-order lag of *AAN* has a significantly negative correlation with its current order lag at the 1% probability level.

According to the first item in Eq. (4), without considering public holidays, when keeping other variables constant, the current order item will increase by 0.902 units if the first order lag item increases by 1 unit. That is, the hotel guests will increase by 90.2 on average if hotel guest arrivals yesterday increase by 100. When keeping other variables constant, the hotel guests will decrease by 25.7 on average if hotel guest arrivals the day before yesterday increase 100. Similarly, the hotel guests will decrease by 16.2 on average if hotel guest arrivals three days ago increase 100. The interpreting powers of the first lag, second lag and third lag are 0.902, −0.257 and 0.162 for hotel guests forecasting in the case of not considering public holidays.

According to the second item of equation group (5), *HOL* has a significant positive correlation with its current order lag at the 1% probability level. Due to the introduction of the dummy variable HOL, the prediction powers of the first-order lag and the second lag are weakened, but the prediction power of the third-order lag of *AAN* conversely is enhanced. It can be concluded that hotel guest arrivals increase by approximately 7900 on public holidays compared to working days on average.

$$
\begin{aligned}
AAN(t) &= \underset{(0.0000*)}{7824} + \underset{(0.0000*)}{0.902* AAN(t-1)} - \underset{(0.0002*)}{0.257 * AAN(t-2)} \\
&\quad + \underset{(0.0000*)}{0.162 * AAN(t-3)} + \varepsilon \\
AAN(t) &= \underset{(0.000*)}{5793} + \underset{(0.000*)}{0.731 * AAN(t-1)} - \underset{(0.017*)}{0.148 * AAN(t-2)} + \underset{(0.0000*)}{0.214 * AAN(t-3)} \\
&\quad + \underset{(0.000*)}{7923 * HOL} + \varepsilon
\end{aligned}
\tag{4}
$$

Foresting Evaluation: To evaluate the forecasting performance of the two models, equation group (4) is used to forecast hotel guest arrivals with 120 observation samples in the out-sample subset spanning from January 1, 2016, to April 30, 2016. The forecasting results are shown in Fig. 3 and Fig. 4. Due to lag order = 3, there are 117 forecasting cases in the figures. It can be seen from Fig. 3 that either AR or AR+HOL models can show good performance in forecasting the trend of hotel guest arrivals, but the AR+HOL model shows better trend forecasting performance than the AR model. In Fig. 4, the error stationarity of the AR+HOL model is better than that of the AR model overall.

In the box plot in Fig. 5, the abnormal data denote the number of cases. Figure 5 shows that the mean *MAPE* of AR+HOL is 9.0%, which is 1.3% lower than that of the AR model, with a mean forecasting error of 10.6%. In addition, on the one hand, the STD of AR+HOL is 9.48, which is 2.0% lower than that of AR. On the other hand, it is shown that there are 7 outliers in the AR model and that there are no outliers in AR+HOL. We can demonstrate that AR+HOL has better forecasting stationarity than AR. Finally, it can be found that 11 forecasting errors are larger than 30% in AR, which accounts for 9.4% of all forecasting samples, and only 2 forecasting errors are beyond 30% in AR+HOL. Twenty-five forecasting errors in AR are larger than 20%, which accounts for 21% of all forecasting samples, but there are only 12 forecasting errors larger than 20%. Therefore, the introduction of the dummy variable *HOL* to the model can improve the forecasting performance in terms of accuracy and stationarity according to the comparison of the mean, standard error and outlier distribution.

In Fig. 4, it can be seen that the forecasting errors of AR+HOL at 3 dashed circle locations decrease drastically compared with the AR model. The dates at the 3 dashed circles are 2 days before the Spring Festival, the Lantern Festival and Tomb Sweeping Day. Through the analysis of raw data, it can be found that hotel guest arrivals at 3 days decrease drastically compared with the day before the corresponding dates. As usual, the percentage of decrease in the day before the Spring Festival, in the Lantern Festival and in the Tomb Sweeping Day is 22%, 24% and 33%, respectively. By consulting the experts of Guilin engaging in tourism industries, we determine the causes hidden behind such phenomena as follows: according to traditions and customs, Chinese people usually do not select tourism activity in different places from their hometowns or usual residences crossing New Year's Eve, especially in domestic tourism destination cities. The Lantern Festival in China means family reunion, so Chinese usually stay at home with family in the festival instead of going out in most cases. On Tomb Sweeping Day, Chinese commemorate ancestors at hometowns, especially in southern China. Therefore, a drastic decrease in hotel guest arrivals occurs in these 3 cases.

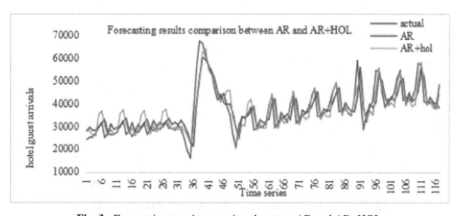

Fig. 3. Forecasting trend comparison between AR and AR+HOL.

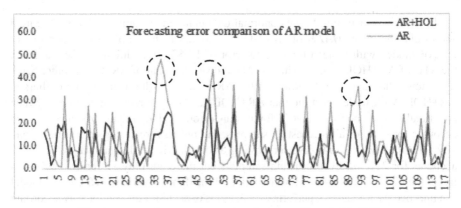

Fig. 4. Forecasting error comparison between AR and AR+HOL.

To reduce the forecasting errors in such special cases, a novel punishment mechanism is introduced in the AR+HOL model, called the improved AR+HOL model. The
regression equation with variable **HOL** is below:

$$
\begin{aligned}
\text{AAN}(t) = \underset{(0.000^*)}{5793} &+ \underset{(0.000^*)}{0.731 * \text{AAN}(t-1)} - \underset{(0.017^*)}{0.148 * \text{AAN}(t-2)} \\
&+ \underset{(0.0000^*)}{0.214 * \text{AAN}(t-3)} + \underset{(0.000^*)}{7923 * \text{HOL}} * P + \varepsilon
\end{aligned}
\tag{5}
$$

where variable P denotes a kind of punishment factor. In such special days, the value
of p is set to -1; otherwise, it is set to 1. Through the mechanism, the forecasting error
can decrease by 15%. MAPE can be reduced to 9.0% due to the introduction of the
punishment mechanism. The foresting error distribution of AR with the punishment
mechanism is shown in Fig. 6. The forecasting errors are greatly reduced compared with
AR.

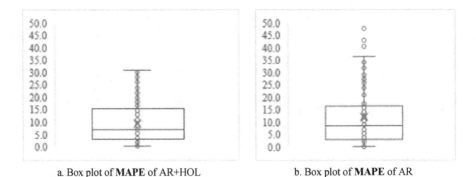

a. Box plot of **MAPE** of AR+HOL   b. Box plot of **MAPE** of AR

Fig. 5. Forecasting results comparison between AR and AR+HOL.

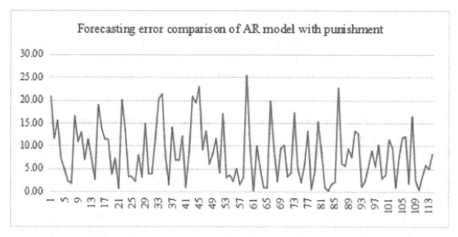

Fig. 6. Forecasting error distribution of AR with a punishment mechanism

5.2 Autoregressive Distributed Lag Model Establishment with Univariate Baidu Index

To verify the predictive power of the Baidu index of a single search keyword for hotel guest arrivals, the combination of various lags of a single search keyword about Guilin Hotel is used as an independent variable as well as the dummy variable *HOL* to construct regression equations. Then, the lag of the *AAN* together with the combination of various lags of a single search keyword are used as independent variables to construct another kind of regression equation to examine whether the combination of the *AAN* and a single Baidu index has better predictive power than the Baidu index of a single search keyword. We only adopt the continuous lag of *AAN* and 4 Baidu indexes in view of the time efficiency of the data.

Model Fitting of Univariate GH, GA, GT and GW: We estimated the regression equations with various continuous lags of **GH**, *GA*, *GT* and *GW,* which can pass the statistical significance test at the probability 1% level, as shown in Eq. (6). Here, we only present the regression equations of *GH*, *GA* and *GT* with the lowest forecasting errors. The regression equation of univariate *GW* cannot pass the statistical significance test at the probability level of 1%. The estimation results show that *AAN* has a significantly positive correlation with *GH*, *GA* and *GT*, which is consistent with the analysis results of selecting search keywords.

In addition, in univariate Baidu index models of *GH* and **GA**, the first order lags have greater forecasting power for *AAN* compared with the second lag or the third order lag because the coefficient of the first order lag is obviously higher than that of the second order or the third order lag. However, in the *GT* model, the second lag has greater forecasting than the first lag.

$$AAN(t) = 689 + 81.8 * GA(t-1) + 50.8 * GA(t-2) + 53.5 * GA(t-3) + 8839 * HOL * P + e$$
$$AAN(t) = 6281 + 84.8 * GH(t-1) + 35.5 * GH(t-2) + 8606 * HOL * P + e$$
$$AAN(t) = 23930 + 1.56 * GT(t-1) + 2.68 * GT(t-2) + 10343 * HOL * P + e$$

$$(6)$$

Model Fitting of the Combination of AAN Lag and GH, GA, GT and GW: AAN lag as well as univariate *GH*, *GA*, *GT* and *GW* are used for equation estimation to further improve the forecasting performance. The corresponding regression equations with the lowest forecasting errors are shown as equation group (7). In the following analysis, the corresponding equations are abbreviated as GH+AAN, GA+AAN, GT+AAN and GW+AAN.

By comparing equation groups (6) and (7), the combination regression equations have smaller coefficients of the Baidu indexes than the univariate regression equations. This is because the introduction of the *AAN* lag weakens the forecasting power of the Baidu index. In addition, we can observe that the coefficient of *AAN* in AAN+GW is the largest, and AAN+GA has the smallest *AAN* coefficient when considering the various orders of magnitude of the Baidu indexes. This indicates that *GA* has the highest forecasting power and *GW has* the lowest forecasting power among the 4 Baidu indexes.

$$AAN(t) = 4570 + 0.576 * AAN(t-1) + 23.52 * GH(t-2) + 16.06 * GH(t-3)$$
$$+ 7491 * HOL * P + \varepsilon$$

$$AAN(t) = 2155 + 0.55 * AAN(t-1) + 41.6 * GA(t-2) + 27.3 * GA(t-3)$$
$$\quad\quad\quad\quad\quad\quad [0.000]^{***} \quad\quad\quad\quad [0.000]^{***} \quad\quad\quad [0.002]^{**}$$
$$+ 7566 * \underset{[0.000]^{***}}{HOL} * P + \varepsilon$$

$$AAN(t) = 5298 + 0.69 * AAN(t-1) + 2.91 * GT(t-1) - 1.27 * GT(t-2)$$
$$\quad\quad\quad\quad\quad\quad [0.000]^{***} \quad\quad\quad\quad [0.000]^{***} \quad\quad\quad [0.02]^{**}$$
$$+ 7707 * \underset{[0.000]^{***}}{HOL} * P + \varepsilon$$

$$AAN(t) = 8863 + 0.73 * AAN(t-1) - 0.129 * GW(t-2) + 0.130 * GW(t-3)$$
$$\quad\quad\quad\quad\quad\quad [0.000]^{***} \quad\quad\quad\quad [0.0243]^{**} \quad\quad\quad [0.0211]^{**}$$
$$+ 7664 * \underset{[0.000]^{***}}{HOL} * P + \varepsilon$$

$$(7)$$

Except for *GT*, the second and third orders of the Baidu indexes are statistically significant for the regression equations. The forecasting power of first order for hotel guest arrivals is replaced. Interestingly, by the sum of coefficients of each Baidu index, the coefficient sum of *GW* is close to 0 ($-0.129 + 0.130 \approx 0$), and the coefficient sum of *GA* is still largest and 42% higher than GH. Therefore, we can conclude that *GW* still shows far low forecasting power for hotel guest arrivals even in the combination equations, and *GA* has the most powerful forecasting power.

Forecasting Performance Evaluation: The *MAPE*s of univariate models and combination models are shown in Fig. 7. Based on the comparison and analysis of Fig. 7, the

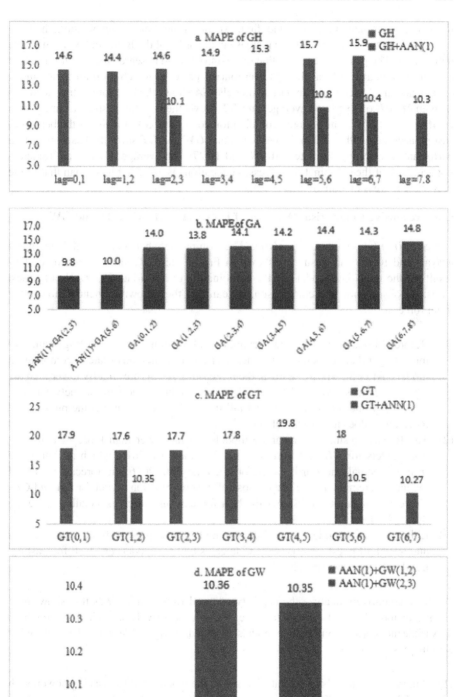

Fig. 7. Comparison of **MAPE** at various lags of Baidu indexes and auto regression.

following phenomenon can be found: first, except for the 0-order combination, the forecasting accuracy decreases with the increase in the order of the Baidu index in univariate regressions. The possible reason is that more tourists tend to search tourism information prior to the departure of travel, and fewer tourists prefer to search for information many days in advance. Second, in univariate models, *GA* shows the best forecasting accuracy with an *MAPE* of 13.8% on average, and *GT* shows the best forecasting accuracy with an *MAPE* of 17.6%. Third, in the combination equations, *GA* still shows the best forecasting accuracy with 9.8% in the best case, and *GW* and *GT* show the best forecasting performance with approximately 10.1% and 10.27% on average. Finally, the forecasting accuracy of the combination regression equations is 4–7.5% higher than that of the univariate regression equation on average.

Comprehensive Comparison Analysis of Univariate GH, GA, GT and GW

Forecast Trend Comparison Analysis: The comparison between actual hotel guest arrivals and forecasting results is shown in Fig. 8. Here, we present the forecasting results of the univariate model and the combination of univariate and *AAN* in the best cases. By observation and comparison, we can find the following phenomenon in the trend forecast:

(1) Before February 10, 2016, the Spring Festival of China, the forecasting results are much larger than the actual hotel guest arrivals for either univariate or combination models. After the Spring Festival, the forecasting results are slightly lower than the actual hotel guest arrivals. The possible reason is that the guest arrivals between New Year's Day and the Spring Festival fall into the slack season, but the punishment mechanism does not go into effect.

(2) The forecast results of univariate models are far lower than those of combination models for *GH*, *GA* and *GT* after February 10, 2016, which deviate more from the actual guest arrivals. Before February 10, 2016, the forecast results of the univariate model are close to those of the combination model for *GA* and *GH*. However, univariate *GT* shows the best forecasting power, especially during the Spring Festival.

(3) Either univariate models or combination models are capable of forecasting the crest and trough of guest arrivals, which indicates that the models can well forecast the trend of hotel guest arrivals.

To quantitatively analyze the capacity of trend forecasting, *TFIs* that show trend change of today's guest arrivals to yesterday's guest arrivals are calculated for both univariate models and combination models, as shown in Fig. 9. We can find the following conclusions:

(1) Among the 7 models, univariate *GT* shows the worst trend forecasting power, with a *TFI* of only 52.9%, but the *TFIs* of the others are above 66.4%, which agrees with the assumption that the correlation between *GT* and Guilin hotel guest arrival is lower than that of *GT* and *GA*.

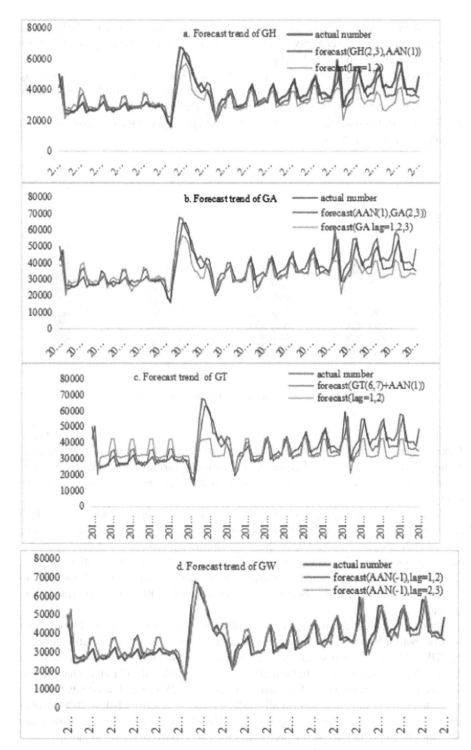

Fig. 8. Comparison of trend forecasts in the best cases.

(2) On the whole, either the univariate model or the combination model of **GH** shows a better trend forecast capacity. The possible reason is that the meaning of GH (Guilin hotel) is closest to Guilin hotel guest arrival.

(3) Interestingly, the combination models of **GW** and **GT** show better trend forecasting power with the help of **AAN** lag, although univariate models of **GT** and **GW** are inferior to **GA** and **GH** in terms of forecasting error.

(4) In terms of trend forecasting, **GH** is rather similar to GA based on a comparison of Fig. 8(a) and Fig. 8(b).

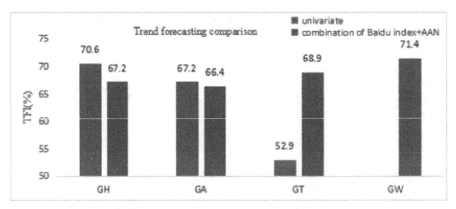

Fig. 9. Comparison of **TFI** in the best cases.

Forecast Error Distribution Comparison Analysis: To quantitatively analyze the forecasting power, we calculated a frequency distribution of various forecasting error ranges for univariate models and combination models in the respective best 3 cases, as shown in Fig. 10.

In univariate models, the **GT** model has 33 forecasting errors above 25%, which accounts for 27.7% of the total forecasting samples; the **GA** model has 15 errors above 25%, which accounts for 12.6% of the total forecasting samples; and the **GA** model has 16 errors above 25%, which accounts for 13.4% of the total forecasting samples. Therefore, **GT** shows the worst forecasting performance among **GT**, **GA** and **GH**. In all models, forecasting errors below 20% account for 60–70% of the total forecasting samples.

By observation, it can be found that combination models show more obvious long-tail distribution features compared with univariate models. In the combination models, the **GW** model has the worst forecasting performance, with 10 forecasting errors above 25%, which accounts for 8.4% of the total forecasting samples; the **GA** model has the best forecasting errors, with only 3 forecasting errors above 25%, which accounts for 2.5% of the total forecasting samples. In all models, forecasting errors below 20% account for 83–89% of the total forecasting samples. Therefore, the combination models can substantially improve forecasting performance compared to univariate models.

Forecast Error Means and Standard Error Comparison Analysis: To quantitatively analyze the stability of forecasting performance, we calculated the statistics of the standard errors of forecasting error of both univariate models and combination models, as shown in Fig. 11. It can be seen from Fig. 11 that combination models are approximately decreased by 30% compared with the respective corresponding univariate model in terms of standard errors and mean. Therefore, the forecasting performance of the combined model is superior to that of the univariate model in view of either forecasting accuracy or forecasting stability. In the 4 combination models, GA+AAN shows the best performance with a standard error of 7.15%, followed by GH+AAN with 7.4%, and GW+AAN shows the worst performance with a standard error of 9.2%, followed by GT+AAN. In general, GA+AAN still shows the best forecasting performance compared to the other models.

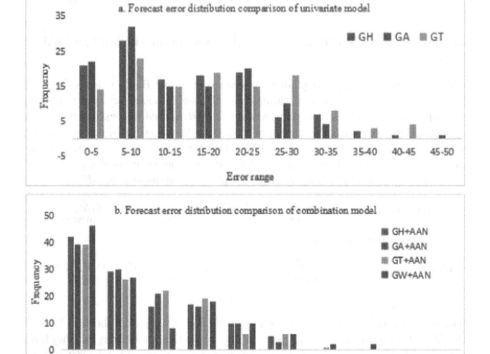

Fig. 10. Plot of Forecast error distribution comparison.

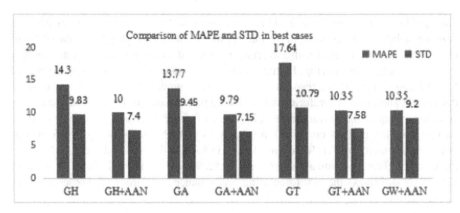

Fig. 11. Comparison of **MAPE** and **STD** of forecasting errors in best cases.

5.3 Autoregressive Distributed Lag Model Establishment with Multiple Baidu Indexes

Model Fitting: To verify the predictive power of multiple Baidu indexes for hotel guest arrivals, the combination of various lags of a single search keyword is used as an independent variable, and the dummy variable **HOL** *is used* to construct regression equations. Here, we present the regression equations when showing the best forecasting performance of the model with multiple Baidu indexes and with the corresponding combination form as equation group (8). Because other combinations of Baidu indexes cannot pass the significance test, we only presented the combination of **GH** and **GA**.

$$AAN(t) = 1352 + 0.52 * \underset{[0.000]^{***}}{AAN(t-1)} + 21.57 * \underset{[0.023]^{**}}{GH(t-1)} + 50.69 * \underset{[0.000]^{***}}{GA(t-1)}$$

$$+ 7491 * \underset{[0.000]^{***}}{HOL} * P + \varepsilon$$

$$AAN(t) = 4282 + 74.4 * \underset{[0.000]^{***}}{GH(t-1)} + 70.72 * \underset{[0.000]^{***}}{GA(t-1)} + 8742 * \underset{[0.000]^{***}}{HOL} * P + \varepsilon$$

$$(8)$$

Performance Evaluation: The forecasting results of each sample are shown in Fig. 12. The **TFIs** of equation group (8) are 73% and 78%, respectively, indicating that they have better trend forecasting capability. In general, compared with actual hotel guest arrivals, the forecasting results of the model with **GT** and **GA** are much lower than those of the model with **GT**, **GA** and **AAN** lag.

The mean and standard error of forecasting errors of models capable of passing the statistical significance test at the probability level of 0.01 are shown in Fig. 12. For the model with the combination of **GH** and **GA**, the forecasting performance decreases with increasing lag order according to the results of the mean and standard error of forecasting error.

With comparison with the best results in Fig. 11 and Fig. 13, the combination of **GH** and **GA** is not capable of improving the forecasting performance compared with the univariate **GH** or **GA** model in terms of either the mean or standard error of forecasting error. In addition, the combination of *GH*, *GA* and *GT* with a forecasting accuracy of 18.1% and forecasting error standard error of 10.8% is even worse than the combination of *GH* and *GA*, which is not superior to the univariate *GT* model.

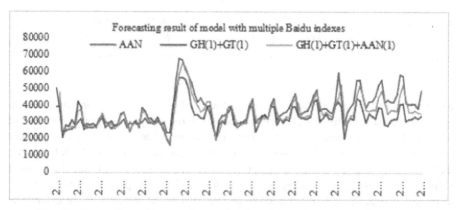

Fig. 12. Comparison of forecasting results of models with multiple Baidu indexes.

However, the model with *GH*, *GA* and *AAN* lag obviously can improve the forecasting performance with a forecasting error of 9.6% and a standard error of forecasting error of only 6.5%, which is superior to the model with the combination of univariate Baidu index and *AAN* lag order 1, especially the forecasting error stability.

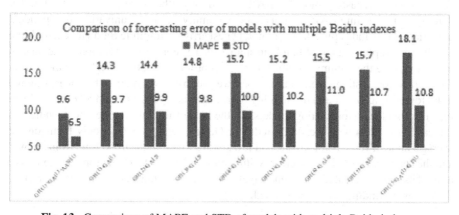

Fig. 13. Comparison of MAPE and STD of models with multiple Baidu indexes.

5.4 Summary

To compare all the models, the forecasting performance of the models is summarized in Table 4. In Table 4, due to more than one forecasting regression model in the case of the

Table. 4. Performance comparison of various forecasting models.

	AR	AR+HOL	Improved AR+HOL	Uni-Baidu GA	Uni-Baidu + ANN GA+AAN	Multi-Baidu GA + GH	Multi-Baidu + AAN GH(1)+GA(1)+AAN(1)
MAPE	11.9%	10.6%	9.0%	13.77%	9.79%	14.3%	9.6%
TFI	75%	81%	84.0%	67.2%	66.4%	73%	78%
STD	11.5	9.48	8.16	9.45	7.15	9.7	6.5

univariate Baidu index, univariate Baidu index + ANN, multi-Baidu and multi-Baidu + ANN, we list only the models showing the best forecasting performance. In terms of forecasting error and trend forecasting capacity, the improved AR+HOL shows the best forecasting performance. In terms of stability, the multi-Baidu index + AAN lag shows the best forecasting performance. Therefore, the forecasting capacity of the Baidu index is limited. In this paper, we adopt daily check-in data with a period of only one year to train the forecasting models, and the MAPE is reduced to 9.0%. If we extend the period of training data and optimize the models by means of mining the potential laws or patterns behind data, the forecasting error should be further improved.

6   Conclusions, Limitations, and Future Work

6.1   Conclusions

We construct forecasting models of AR without any other variables, AR plus holiday dummy variables, improved AR plus holiday dummy variables, univariate Baidu index, univariate Baidu index plus the lag of hotel guest arrivals, multiple Baidu index, multiple Baidu indexes plus the lag of hotel guest arrivals. Except for the AR model, the other 5 models include the dummy variable HOL. Except for the metric *MAPE* for measuring the average forecasting error, in this paper, we proposed two metrics for measuring the trend forecasting power and the stability of forecasting errors, *TFI* and *STD,* to evaluate the performance of forecasting models. To the best of my knowledge, most studies on forecasting models with the Baidu index or Google search analysis only focus on the average value of forecasting errors, and few studies focus on the trend forecasting power and the stability of forecasting error. With metrics *TFI* and *STD,* the forecasting power of models can be revealed comprehensively. According to the analysis results in Sect. 5, the following finding can be concluded:

Forecasting Power of the Baidu Index: (1) The Baidu index has the forecasting power of hotel guest arrivals, especially in trend forecasting. According to the comparison and analysis of the *TFI* index in the models, the Baidu index shows close forecasting performance in the trend forecasting of hotel guest arrivals in all fitting models in this study, which empirically demonstrates that the Baidu index can show tourism demand again. (2) Among the 4 Baidu indexes, *GA* and *GH* have the highest forecasting power of Guilin hotel guest arrivals, and *GT* has a relatively lower forecasting power than *GH*. Even univariate *GW* cannot forecast hotel guest arrivals effectively.

Effect of Lag Period: (1) When the lag periods are set to 1, 2, 3, the Baidu index has the best forecasting power for hotel guest arrivals. (2) The first-order lag of the independent variable plays the most important role in forecasting the performance of hotel guest arrivals.

About Forecasting Performance Comparison Among Models: (1) The combination of univariate Baidu index and lag of hotel guest arrivals has the higher forecasting accuracy than univariate Baidu index, which can increase by 2%–3% absolute value of forecasting errors. (2) The combination of Baidu indexes cannot effectively improve the forecasting accuracy compared with the univariate Baidu index due to the collinearity issue of keywords. (3) Interestingly, the improved AR+ HOL mode shows the best forecasting performance, with an *MAPE* of 9.0% among the 6 kinds of forecasting models, which indicates the limited forecasting power of the Baidu index.

Model Optimization: (1) In constructing a forecasting model, forecasting performance can be improved effectively only when some punishment mechanism or award mechanism should be introduced in terms of special situations of the study case. For example, we introduce a kind of punishment mechanism in terms of the rapid drop in hotel guest arrivals. (2) Public holidays and weekends show a significant positive correlation with actual hotel guest arrivals to Guilin. By introducing the dummy variable *HOL*, the forecast performance can be improved significantly.

6.2 Limitations

The limitation of this research lies in the following: first, the limited time range of samples is used for fitting the forecasting model due to the privacy nature of hotel accommodation data, although we construct a forecasting model of daily hotel guest arrivals with full-sample data, and we only chose 365-day check-in big data in 2015 as the training set of models and 121-day check-in data as the test set. Therefore, the possible long-term lag length is not shown. Second, by means of expert interviews, only 4 Baidu indexes related to hotel accommodations are selected as forecasting indicators. The Baidu indexes of other unobvious search keywords related to hotels or accommodations may further contribute to the improvement of forecasting performance. Finally, this study only OLS—a common linear regression method to explore the predictive power of the Baidu index for hotel guest arrivals.

6.3 Future Work

In the future, with the improvement of sharing mechanisms and corresponding laws and regulations of big data utilization, more big data sources may be available in the hospitality and tourism industry. First, using a multisource larger sample volume with a longer time dimension, we can further investigate the forecasting power of search queries with the Baidu search engine for hotel guest arrivals and design a fine-granularity adaptable reward or punishment mechanism for drastic changes in guest arrivals. Second, although Baidu search accounts for most of the market share in China, other search engines, such as Sogou and 360, should not be neglected. To improve the forecasting performance, the combination of the Baidu index and Sogou index can be used as an independent variable for constructing a fine-granularity model according to the geographic area features. Finally, we will further investigate the predictive power of the Baidu index using other regression models and make a performance comparison among different models.

Acknowledgement. We would like to extend our gratitude to the Guilin Tourism and Development Committee for the support of this study. We acknowledge the anonymous reviewers and the editors of Journal. This study is supported by the research on key technology of tourism destination safety warning and its application demonstration granted by No. Guike AB17195028, technology development of tourist safety warning system for smart scenic spot and virtual spatiotemporal reconstruction of special culture and its application demonstration granted by No. 20170220, and research on sustainable utility technology integration of Longji terrace landscape resources and tourism industry demonstration granted by No. 20180102-2, Guangxi natural science fund by No. 2018GXNSFAA138209.

References

1. Miller, E.: Community cleverness needed. Nature **455**, 1 (2008)
2. Tao, Y.: Analysis method for customer value of aviation big data based on LRFMC model. In: Zeng, J., Jing, W., Song, X., Lu, Z. (eds.) ICPCSEE 2020. CCIS, vol. 1257, pp. 89–100. Springer, Singapore (2020). https://doi.org/10.1007/978-981-15-7981-3_7
3. Liu, K., et al.: Using Baidu search index to predict dengue outbreak in China. Sci. Rep. **6**(1), 38040 (2016). https://doi.org/10.1038/srep38040
4. Huang, R., et al.: Using Baidu search index to monitor and predict newly diagnosed cases of HIV/AIDS, syphilis and gonorrhea in China: estimates from a vector autoregressive (VAR) model. BMJ Open **10**(3), e036098 (2020). https://doi.org/10.1136/bmjopen-2019-036098
5. Li, K., Liu, M., Feng, Y., et al.: Using Baidu search engine to monitor AIDS epidemics inform for targeted intervention of HIV/AIDS in China. Sci. Rep. **9**(1), 320 (2019). https://doi.org/10.1038/s41598-018-35685
6. Tu, B., Wei, L., Jia, Y., et al.: Using Baidu search values to monitor and predict the confirmed cases of COVID-19 in China: – evidence from Baidu index. BMC Infect. Dis. **21**, 98 (2021). https://doi.org/10.1186/s12879-020-05740-x
7. Zhao, C., Yang, Y., et al.: Search trends and prediction of human brucellosis using Baidu index data from 2011 to 2018 in China. Sci Rep **10**(1), 5896 (2020). https://doi.org/10.1038/s41598-020-62517-7
8. Zhao, Y., Xu, Q., Chen, Y., Tsui, K.L.: Using Baidu index to nowcast hand-foot-mouth disease in China: a meta learning approach. BMC Infect. Dis. **18**(1), 398 (2018). https://doi.org/10.1186/s12879-018-3285-4
9. Fang, J., Gozgor, G., Lau, C.-K., Zhou, L.: The impact of Baidu Index sentiment on the volatility of China's stock markets. Finan. Res. Lett. **32**, 101099 (2020)
10. Fang, J., Wu, W., Lu, Z., Cho, E.: Using Baidu index to nowcast mobile phone sales in China. Singap. Econ. Rev. **64**(01), 83–96 (2019)
11. Shen, D., Zhang, Y., Xiong, X., Zhang, W.: Baidu index and predictability of Chinese stock returns. Finan. Innov. **3**(1), 1–8 (2017). https://doi.org/10.1186/s40854-017-0053-1
12. Lin, W., Zou, Y., Zheng, X.: Study on the regional disparity in the network attention of China tourism security: based on the Baidu index of tourism security in 31 provinces. Hum. Geogr. **6**, 154–160 (2014)

13. Li, S., Qiu, R., Chen, L.: Cyberspace attention of tourist attractions based on Baidu index: temporal distribution and precursor effect. Geogr. Geo-Inf. Sci. **6**, 102–107 (2008)
14. Liu, P., Zhang, H., Zhang, J., et al.: Spatial-temporal response patterns of tourist flow under impulse pretrip information search: from online to arrival. Tour. Manage. **73**, 105–114 (2019)
15. Huang, X., Zhang, L., Ding, Y.: The Baidu index: uses in predicting tourism flows – a case study of the forbidden city. Tour. Manage. **58**, 301–306 (2017). https://doi.org/10.1016/j.tourman.2016.03.015
16. Xing, Z.: An empirical analysis and forecast study between the Baidu index based on the amount of web search and trading volume on the platform of P2P. Sci. Innov. **5**(5), 256–262 (2017). https://doi.org/10.11648/j.si.20170505.12
17. Liu, Y., Liao, W.: Spatial characteristics of the tourism flows in China: a study based on the Baidu index. ISPRS Int. J. Geo-Inf. **10**(6), 378 (2021). https://doi.org/10.3390/ijgi10060378
18. Gunter, U., Önder, I.: Forecasting city arrivals with Google analytics. Ann. Tour. Res. **61**, 199–212 (2016)
19. Liu, Y.Y., Tseng, F.M., Tseng, Y.H.: Big Data analytics for forecasting tourism destination arrivals with the applied vector autoregression model. Technol. Forecast. Soc. Chang. **130**, 123–134 (2018)
20. Sun, S., Wei, Y., Tsui, K.L., et al.: Forecasting tourist arrivals with machine learning and internet search index. Tour. Manage. **70**, 1–10 (2019)
21. Li, X., Pan, B., Law, R., et al.: Forecasting tourism demand with composite search index. Tour. Manage. **59**, 57–66 (2017)
22. Xie, G., Qian, Y., Wang, S.: Forecasting Chinese cruise tourism demand with big data: an optimized machine learning approach. Tour. Manage. **82**, 104208 (2021). https://doi.org/10.1016/j.tourman.2020.104208
23. Silva, E.S., Hassani, H., Heravi, S., et al.: Forecasting tourism demand with denoised neural networks. Ann. Tour. Res. **74**, 134–154 (2019)
24. Wen, L., Liu, C., Song, H.: Forecasting tourism demand using search query data: a hybrid modeling approach. Tour. Econ. **25**(3), 309–329 (2019)
25. Park, S., Lee, J., Song, W.: Short-term forecasting of Japanese tourist inflow to South Korea using Google trends data. J. Travel Tour. Mark. **34**(3), 357–368 (2017)
26. Tang, J.: Evaluation of the forecast models of Chinese tourists to Thailand based on search engine attention: a case study of Baidu. Wirel. Pers. Commun. **102**(4), 3825–3833 (2018)
27. Camacho, M., Pacce, M.J.: Forecasting travelers in Spain with Google's search volume indexes. Tour. Econ. **24**(4), 434–448 (2018)
28. Li, S., Chen, T., Wang, L., et al.: Effective tourist volume forecasting supported by PCA and improved BPNN using Baidu index. Tour. Manage. **68**, 116–126 (2018)
29. Önder, I.: Forecasting tourism demand with Google trends: accuracy comparison of countries versus cities. Int. J. Tour. Res. **19**(6), 648–660 (2017)
30. Pan, B., Yang, Y.: Forecasting destination weekly hotel occupancy with big data. J. Travel Res. **56**(7), 957–970 (2017)
31. Volchek, E., Song, H., Law, R., et al.: Forecasting London museum visitors using Google trends data. e-Rev. Tour. Res. **11**, 447–475 (2018)
32. Xie, G., Li, X., Qian, Y., Wang, S.: Forecasting tourism demand with KPCA-based web search indexes. Tour. Econ. **27**(4), 721–743 (2020)
33. Zhang, B., Pu, Y., Wang, Y., et al.: Forecasting hotel accommodation demand based on LSTM model incorporating internet search index. Sustainability **11**(17), 4708 (2019)
34. Mingming, H., Song, H.: Data source combination for tourism demand forecasting. Tour. Econ. **26**(7), 1248–1265 (2019). https://doi.org/10.1177/1354816619872592
35. Zhang, B., Li, N., Shi, F., et al.: A deep learning approach for daily tourist flow forecasting with consumer search data. Asia Pac. J. Tour. Res. **25**(3), 323–339 (2020)

A Facial Size Automatic Measurement and Analysis Technology

Dongliang Yang[✉], Changjiang Song, Hantao Zhao, and Tongjun Liu

Intelligent Manufacturing Institute, Heilongjiang Academy of Sciences, Harbin 150090, China
yangdongliang4825@dingtalk.com

Abstract. Facial measurement and analysis is an important part of anthropometry, which provides data support for the design of facial protective equipment. To overcome the inconveniences, low efficiency and poor measurement accuracy of facial size parameters measured and analyzed by manual contact, a method of automatic measurement and analysis of face size parameters is proposed. First, the automatic marking method of faces based on deep learning can improve the efficiency of measuring facial parameters. Then, facial parameters, including nose middle width, nose width, face width and eye width, can be measured. Finally, the data set of face size parameters is classified and counted based on fuzzy clustering analysis. Sixty-five groups of Han youth facial data are collected for measurement and analysis, and compared with the existing algorithms, the facial morphology analysis system presented in this paper has higher measurement accuracy.

Keywords: Facial measurement · Facial morphology analysis · Convolutional neural network · Clustering analysis

1 Introduction

Facial measurement and analysis is an important part of anthropometry, which provides a data reference for the plastic surgery department and the design of facial protective equipment. In recent years, many researchers have measured and analyzed the facial parameters of people of different ages in different regions, such as Hebei, Henan and Xi'an [1]. In the literature, face parameters are measured in 19 regions of Han nationality using facial contact measurements. However, direct contact measurement is time-consuming and labor-consuming, which has certain requirements for surveyors.

Literature [2] uses two-dimensional images for noncontact measurement, but two-dimensional images are projections of faces in a certain direction, and there is a large error between measured and real data. In reference [3], the 3D face model is used for measurement, which can obtain high-precision measurement results, but the method manually marks points in the 3D face model, and the workload is very large. Traditional faces need to manually use the 3D model for measuring the parameters of choosing and then measuring, but when dealing with a large amount of facial data, efficiency can be slow. With deep learning widely used in the field of computer vision and using artificial intelligence technology, automatic access landmark facial models and automatic measurement and analysis are of great significance. To the best of my knowledge, there is no such automated face measurement system at present.

The face morphology analysis system classifies and analyses face data by analyzing a large amount of facial morphology data. Given the three-dimensional face database, we measure the facial parameters of each facial data, and according to these facial parameters, the 3D face database is classified.

The traditional measurement method requires manual selection of points on the 3D model and measurement. To overcome the inconvenience of measuring facial parameters by manual contact, we transform the measurement problem of facial parameters into the recognition problem of facial key points. The deep learning algorithm is used to automatically mark and measure key points of facial features, which can achieve the effect of automatically measuring facial parameters.

Based on fuzzy clustering analysis, the data set of facial parameter values is classified and counted. According to the positioning of facial key points, facial length parameters, including nose width, face width and eye width, can be calculated. Then, multivariate data analysis technology was used to study facial morphology. The fuzzy clustering algorithm is used to classify and calculate the face data according to these parameters.

This paper is divided into five parts. Section 2 describes the collection of 3D facial data. Section 3 details our proposed approach. Section 4 presents the experiments and results. The last section summarizes this paper.

2 Facial Data Acquisition

Three-dimensional facial data acquisition is the basis of facial measurement and analysis. Noncontact 3D model acquisition technology mainly includes binocular 3D reconstruction technology and structured light reconstruction technology. Binocular vision technology requires the matching of corresponding points between two views. However, matching is a difficult task, and the matching error will lead to the low accuracy of 3D reconstruction. Structural light 3D reconstruction technology is fast, stable and more suitable for our system. First, the projector projects interference streaks onto the face. The camera then captures patterns that adjust the stripes according to the surface of the face. Finally, according to the geometric relationship between the projector, camera and face, the relative distance between the face and the datum plane is calculated. Through this acquisition system, we can obtain two-dimensional face images and corresponding three-dimensional face data at the same time. For detailed principles, please refer to the literature [4] for more details.

3 Facial Morphology Analysis System

The system is mainly composed of three modules, as shown in Fig. 1.

(1) Face measurement module. The cascaded convolutional neural network model is constructed to obtain the coordinates of feature points in facial images.

Through the mapping function, the spatial coordinates of three-dimensional facial feature points are obtained, so the facial parameters are measured.

(2) Facial morphology analysis module. FCM is used for fuzzy clustering analysis of facial parameters.

(3) Visual output module. Show the dimension parameters of each face data and the clustering center of all face data.

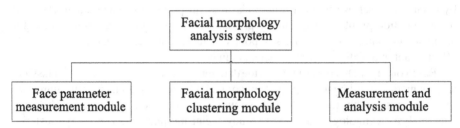

Fig. 1 Schematic diagram of the facial morphology analysis system structure

3.1 Feature Point Recognition

The CNN network model is used to obtain facial feature points. The CNN network model includes a convolution layer, a pooling layer and a fully connected layer. This CNN network structure can achieve better positioning accuracy [5, 6].

Fig. 2. Network structure diagram

In our facial key point recognition, we use the convolutional layer and the pooling layer of VGG16 as the feature extraction module of the face [7, 8]. However, the training will not start directly from the first layer. Instead, the pretrained VGG model is used to fix the parameters of the previous convolutional layer and fine-tune the data when applying it to the recognition of facial key points [9, 10]. The so-called fine-tuning means that we update the weights of the models trained for a similar task, such as VGG-16 and ResNet, through the face training data sets. The regression module is followed by three fully connected layers, and the ReLU activation function is followed by FC1 and FC2, as shown in Fig. 2. The whole model is optimized using the L2 loss function. Finally, we obtain the coordinate position of our facial feature points.

The training data set used in our network is 5000 face images obtained from the network. Each image has 10 key points, including the left and right corners of the eyes, corner of the nose, corner of the mouth, middle nose and tip of the nose. These key points are manually marked. We enhanced the facial training data by translating and rotating.

3.2 Location of Three-Dimensional Facial Feature Points

The location of 3D facial feature points is the basis of measuring facial parameters. To solve this problem, facial image feature points can be obtained through the CNN model.

There is a mapping relationship between two-dimensional image data $U = \{u_i \in R^2 : i = 1, 2, \cdots, N\}$ in our face database and three-dimensional face data $V = \{v_j = (x_j, y_j, z_j) \in R^3 : j = 1, 2, \cdots, M\}$, denoted as $\phi : u_i \to v_j, i = 1, 2, \cdots N, v_j \notin 0$. u_i is a pixel in the two-dimensional image, and v_j is a spatial point in three-dimensional facial data. According to this mapping function, the spatial coordinates of three-dimensional facial feature points can be obtained, as shown in Fig. 3.

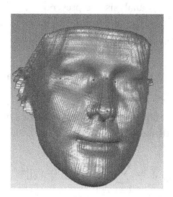

Fig. 3. 3D facial marker location

3.3 Facial Morphology Analysis

The facial morphology analysis module first inputs face data into the database. The feature point position of the human face is obtained by the above algorithm. The spatial coordinates of three-dimensional feature points are obtained by combining the mapping relationship, and the facial parameters are calculated. Determine whether to traverse the entire face database. When the whole face database is measured, the fuzzy C-means algorithm is used to obtain the clustering center and display it. The flow of this module is shown in Fig. 4.

Compared with simple statistical analysis, it is more scientific and theoretical to use a clustering analysis algorithm to analyze facial morphology categories [11, 12]. Fuzzy clustering analysis (FCM) [13] in clustering analysis is used here, which is a multivariate statistical technique, and its idea is to divide the whole data set according

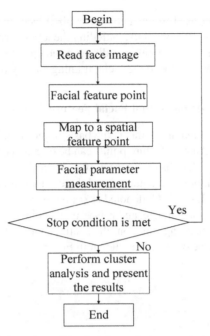

Fig. 4. Flow chart of the facial morphology analysis module

to the approximation degree of the attribute characteristics of each clustering center [14, 15]. The objective function of the FCM algorithm is

$$J = \sum_{k=1}^{c} \sum_{i=1}^{N} u_{ki}^{m} ||x_i - v_k||^2 \tag{1}$$

where N is the number of samples, x_i is the ith sample, c is the preset number of clusters, v_k is the kth clustering center, m is the fuzzy weighting parameter, and u_{ki} is the membership function of the ith sample in the kth class. For any pixel point i and category k, there are the following constraints:

$$\sum_{k=1}^{c} u_{ki} = 1, 0 \leq u_{ki} \leq 1 \tag{2}$$

3.4 Output Module

The output module of the system can display the measurement results of the dimension parameters of each facial data, as shown in Table 1. At the same time, the face data set is divided into three categories, and the center and variance of each category are obtained, as shown in Table 2. Table 2 shows the size parameter range of each category obtained by clustering. Three categories are displayed using three colors, with each dot representing a parameter of facial data, as shown in Fig. 5.

Table 1. Measurement results of facial parameters

	Measured value (mm)
Face length	165.08
Long nose	58.10
Nose width	39.87
Mouth wide	91.21
Face width	211.07

Table 2. Cluster analysis results

	The first cluster (mm)	The second cluster (mm)	The third cluster (mm)
Face length	156.81 ± 9.88	174.45 ± 8.60	132.19 ± 13.33
Long nose	53.67 ± 3.32	61.12 ± 3.57	45.80 ± 4.94
Nose width	33.37 ± 2.18	39.43 ± 2.53	28.35 ± 3.28
Mouth wide	69.64 ± 7.06	83.87 ± 9.21	59.80 ± 11.21
Face width	194.89 ± 9.72	223.11 ± 9.45	168.99 ± 14.23

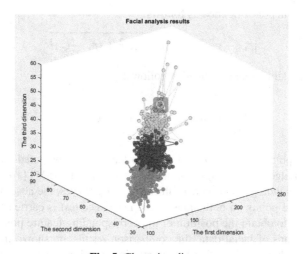

Fig. 5 Clustering diagram

4 Experimental Comparison

The data for statistical analysis were collected by the abovementioned equipment. A total of 65 sets of data were collected, including 50 males and 15 females between 25 and 40 years old. Face point cloud data contain approximately 45,000 points, and the size of texture images is 256*256, BMP format. Sixty-five groups of 3D face data are collected to establish the facial data set. We obtain the facial parameters for each group of facial data and display the interface of the facial morphology analysis system. The software is developed under the i7 processor, 18 GB memory, 1 G video memory and Windows 10 operating system.

The performance of the proposed algorithm is measured by calculating the error rate. The error rate is defined as follows:

$$ErrorRate_i = \frac{\frac{1}{M}\sum_{j=1}^{M}|p_{i,j} - g_{i,j}|}{r_i} \tag{3}$$

Therefore, the average error rate is:

$$ErrorRate = \frac{1}{N}\sum_{i=1}^{N}ErrorRate_i \tag{4}$$

where $ErrorRate_i$ represents the error rate of the ith picture and $ErrorRate$ is the average error of all pictures. N indicates the number of images in the data set. M represents the number of key points in the face. $p_{i,j}$ is the calculated coordinate of the jth feature point of the ith data, and $g_{i,j}$ is the actual coordinate of the jth feature point of the ith data. r_i is the width of the face, and $p_x p_y$ represent $p_{i,j}$ the coordinates, and g_x and g_y represent $g_{i,j}$ the coordinates.

The metric function is defined as follows:

$$|p_{i,j} - g_{i,j}| = \sqrt{(p_x - g_x)^2 + (p_y - g_y)^2} \tag{5}$$

The average failure rate is defined as follows:

$$FailureRate = \frac{1}{N}\sum_{i=1}^{N}1_{\{ErrorRate_i>0.1\}} \tag{6}$$

The feature point algorithm based on deep learning is compared with the feature point extraction algorithm based on an active shape model (ASM) [16]. We extracted feature points, including the corners of the left and right eyes, the nose tip, the corners of the mouth, and points on both sides of the nose. The test is carried out on 65 sets of collected data to obtain the position of the corresponding feature points of a certain face. For 65 groups of facial data, the comparison results are shown in Tables 3 and 4. It can be seen that our algorithm has greatly improved the accuracy and failure rate compared with the traditional algorithm. Compared with the active shape model, the network model has better learning ability, and the error rate is greatly reduced.

Table 3. Comparison table of average error rates of experimental results

Average error rate (%)	ASM	CNN
Left eye	5.44	3.46
Right eye	4.26	3.41
Tip of nose	5.39	4.21
Left mouth	4.51	3.61
Right mouth	4.73	3.72

Table 4. Comparison of the failure rates of the experimental results

Average failure rate (%)	ASM	CNN
Left eye	3.65	0.52
Right eye	2.26	0.51
Tip of nose	5.31	0.52
Left mouth	6.31	0.79
Right mouth	4.55	0.78

5 Conclusion

Aiming at the problems of low efficiency and poor accuracy of traditional facial measurement methods, an automatic facial measurement method based on deep learning was proposed. Then, based on fuzzy clustering analysis, the facial data sets are classified and counted. Finally, 65 groups of facial data were collected for facial data measurement and statistics. Compared with existing algorithms, our method has an average error rate of less than 0.8% and an average failure rate of less than 4.5% in feature point positioning, which greatly improves the performance compared with the active shape model method. Due to the use of the deep neural network, rich facial information is better utilized to achieve a better detection effect.

Acknowledgements. This work was supported in part by the Natural Science Foundation of Heilongjiang Province (No: LH2020F049).

References

1. Feng, X., Han, M., Wang, X., et al.: Head and facial measurement and analysis of young male of HAN nationality in Henan Province. J. Zhongyuan Univ. Technol. **30**(5), 12–17, 64 (2019)
2. Ren, J.: Facial Soft tissue Aesthetic Measurement and Analysis of Han Young Women in Henan Province. Zhengzhou University, Henan (2018)
3. Wang, Y.: Facial 3D image measurement of Han Young Adults in northern China. Peking Union Medical College, Beijing (2017)

4. Donlic, M., Petkovic, Y., Pribanic, T.: On tablet 3D structured light reconstruction and registration. In: IEEE International Conference on Proceedings of Computer Vision Workshop (ICCVW), pp. 2462–2471 (2017)
5. Ma, X.: Fatigue driving state detection based on facial features and deep learning. J. Electr. Test. **30**(5), 12–17, 64 (2019)
6. Guo, X.: Adaptive deep convolution neural network and its application in human face recognition. Tech. Autom. Appl. **36**(07), 72–77 (2017)
7. Li, S., Deng, W.: Deep facial expression recognition: a survey. IEEE Trans. Affect. Comput. **3**, 1–10 (2020)
8. Wu, Y., Ji, Q.: Facial landmark detection: a literature survey. Int. J. Comput. Vis. **127**, 115–142 (2019)
9. Sadiq, M., Shi, D.: Facial landmark detection via attention-adaptive deep network. IEEE Access **7**(11), 180141–181050 (2020)
10. Wu, Y., Hong, C., Chen, L., Zeng, Z.: Regression-based face pose estimation with deep multi-modal feature loss. In: ICPCSEE (2020)
11. Jiang, Y., Zhao, K.: A novel distributed multitask fuzzy clustering algorithm for automatic MR brain image segmentation. J. Med. Syst. **43**, 118–128 (2019)
12. Mishro, P.K., Agrawal, S., Panda, R.: A novel type-2 fuzzy C-means clustering for brain MR image segmentation. IEEE Trans. Cybern. **51**(8), 3901–3912 (2021)
13. Wang, Y.: Research and application of knowledge discovery method based on fuzzy theory. Dalian University of Technology, Dalian (2018)
14. Lei, T., Liu, P.: Automatic fuzzy clustering framework for image segmentation. IEEE Trans. Fuzzy Syst. **28**(9), 2078–2092 (2020)
15. Hoang, V., Mumtaz, A.: A novel approach for fuzzy clustering based on neutrosophic association matrix. Comput. Ind. Eng. **127**, 687–697 (2019)
16. Liu, L.: Multifeature Fusion driver Fatigue Detection Algorithm based on Active Shape Model. Hunan University, Changsha (2017)

Intelligent Industrial Auxiliary System Based on AR Technology

Tao Wang[1], Xinqi Shen[2], Junpei Ma[3], Zhuorui Chang[3], and Linyan Guo[1](✉)

[1] School of Geophysics and Information Technology, China University of Geosciences,
Beijing, China
guoly@cugb.edu.cn
[2] China Academy of Information and Communications Technology, Beijing, China
[3] School of Information Engineering, China University of Geosciences, Beijing, China

Abstract. The traditional industrial assembly method is inefficient, and it is difficult to meet the needs of rapid production in modern society. Although many virtual assembly systems based on augmented reality technology have appeared in recent years, most of these assembly systems are noninteractive, nonintelligent, and inefficient. In response to this problem, we combined AR technology and AI technology to design and implement a strong interactive, high-intelligence virtual assembly guidance system in four levels. Finally, we took a UAV assembly as an example to show our design results. Operators can use this system to quickly master the assembly process. Finally, according to some problems encountered in the process of system implementation, the improvement direction of the virtual assembly system is proposed, and the future development of the AR auxiliary industry is proposed. This article describes the functions and implementation process of the four levels of the AR assembly system in detail, which can help readers quickly understand the connection between AR technology and AI technology and understand the principle of the virtual assembly guidance system.

Keywords: Augmented reality · Artificial intelligence · Virtual assembly

1 Introduction

Augmented reality technology is a technology that skillfully integrates virtual information with the real world. After virtual information such as 3D models and videos is simulated and applied to the real world, the two kinds of information complement each other, thereby realizing the "enhancement" of the real world [1].

In recent years, due to the rapid development of deep learning and other fields, AR technology has also achieved rapid development. AR technology has not only won people's love in life and entertainment with its novel and convenient functions but also brought great reforms to industry.

In most manufacturing industries, there is a need for part assembly. In traditional maintenance and assembly tasks, it is difficult for workers with short working hours to complete assembly or maintenance tasks by themselves and often require the guidance of one or more staff with maintenance experience, which is a rather long learning

Y. Wang et al. (Eds.): ICPCSEE 2022, CCIS 1629, pp. 221–230, 2022.
https://doi.org/10.1007/978-981-19-5209-8_15

process [2]. With the continuous progress and development of society, the interactive virtual assembly system based on agile manufacturing technology has developed rapidly, bringing new technical support to the development of various industries. With the continuous optimization and upgrading of various products, the difficulty of assembling the components of these products has also increased. The assembly control capability of the traditional virtual assembly system has been unable to assemble more refined items in detail and cannot meet the needs of enterprises [3].

Therefore, we designed a virtual assembly guidance system based on AR+AI technology, which has a highly intelligent perception and understanding ability and decision-making function and can integrate various elements in the scene in real time to provide operators with multimodal and multidimensional immersion. Guidance, error correction, and early warning quickly help operators learn the assembly process and assist assemblers in assembly in real time.

2 System Design and Implementation

We plan to implement an auxiliary assembly system based on AR equipment. In this system, people can assemble and match industrial parts according to the assembly route planned by the system, which ensures the accuracy of the assembly process. Compared with other virtual assembly systems, our virtual assembly system not only realizes the display of a fixed process but also uses the latest artificial intelligence technology to make the system more intelligent and efficient to complete various tasks. At the same time, with the help of spatial positioning, 3D reconstruction [4] and other technologies, the system can better understand the surrounding environment and achieve function switching with various scenes. Not only can it be suitable for a variety of industrial scenarios, but the algorithm is also versatile for various assembly environments, which greatly improves the migration ability of the system between various scenarios. The Microsoft HoloLens2 glasses we selected are one of the most advanced AR glasses at present. The development tools it provides can greatly reduce the development difficulty and make the system more user-friendly.

2.1 Real-Time Display

To realize the display function through AR equipment, that is, to project the virtual graphics generated by the computer onto the real-world image seen by the human eye from the correct viewpoint, which can visually provide people with supplementary information. This can not only improve work efficiency but also greatly improve the accuracy of assembly. In the specific assembly work, the operator can realize the scan code confirmation function by wearing AR glasses [5] and simultaneously intuitively observe the progress of the current work task.

To realize this function more effectively, we plan to select Microsoft's holoLens2 holographic glasses as the carrier of the system. This is because glasses use optical waveguide technology to achieve a holographic image display with a resolution of 2k, which can effectively alleviate dizziness caused by low resolution and support 3D display optimization based on eye position. In addition, the glasses integrate a 5-channel microphone array, and spatial sound can realize visual and auditory immersive guidance for the operator. Glasses do not need to connect other equipment through cables, which effectively enhances mobility and wearing comfort [6]. In addition, we plan to develop on the Unity platform because the Unity platform provides a complete set of software solutions that can be used to create any real-time interactive 2D and 3D content [7, 8].

2.2 Human-Computer Interaction

Operators can interact with virtual objects through gestures, such as selecting, moving, rotating, and zooming, and can operate virtual panels through gestures, enabling natural interaction between operators and material models.

In order to achieve better human-computer interaction capabilities, AR devices not only need to have a good understanding of the environment, but also must be able to better understand the instructions issued by humans. By combining instructions with the environment, a more friendly human-computer interaction experience is achieved. Therefore, to achieve better interactive functions, multiple high-precision sensors of AR glasses are required, including RGB cameras, depth cameras, VLC cameras, and IMUs. At the same time, we use the Hololens2 mixed reality development kit MRTK [6] officially provided by Microsoft, which can help effectively implement gesture tracking interaction, eye tracking interaction, shaders, bounding boxes, hand menus, lists, sliders, and rendering, spatial awareness, scenes understanding and other functions (see Fig. 1) [9]. Through MRTK, we can quickly design a user-friendly interface. We only need to open the palm to call up the menu, and click to make a selection.

Fig. 1. Example scene provided by MRTK

2.3 Perception and Positioning

In the whole system, to achieve an ideal recognition ability, it can identify whether the parts taken by the assembler are required for the current step, whether the assembly sequence is appropriate, whether the assembly method is correct, etc. This requires the help of computer vision and spatial positioning technology. At the same time, the system needs to perceive and locate the materials and objects in the actual work, display the material and object models established in advance in the prescribed way, and achieve the effect of real-time updates. In terms of perception, we employ a deep learning-based instance segmentation network BlendMask [10] to determine the location, class and mask of objects of interest in the image. This method needs to obtain the labeled samples of the objects in advance and send them to the neural network for training. BlendMask can be divided into two branches, namely, the semantic segmentation branch and the target detection branch. The semantic segmentation branch obtains the mask, and the target detection [11] branch realizes the classification and recognition effect (see Fig. 2).

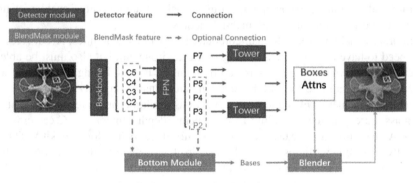

Fig. 2. BlendMask algorithm flow chart

Traditional dense segmentation networks mainly include top-down segmentation networks and top-down segmentation networks [12]. Taking DeepMask [13] as an example, the top-down dense segmentation network mainly predicts a mask proposal on each spatial region through the sliding window method [14]. This will cause many problems, such as loss of feature consistency and redundant feature extraction. The bottom-up dense instance segmentation network generally generates per-pixel embedding features [15] and then uses postprocessing methods such as clustering and graph theory to group them. Although this method maintains better underlying features, it also has many problems, such as high requirements for segmentation quality and poor generalization ability. BlendMask combines two methods, uses the high-dimensional information of the instance level generated by the top-down method, fuses the per-pixel prediction generated by the bottom-up method, and proposes a Blender module, which can better integrate the instance-level information. The global information and low-level features provide details and location information.

In terms of spatial scene positioning, we use the high-precision sensors of AR glasses and the perfect SLAM algorithm [16] based on deep learning. In the 3D scene reconstruction stage, we use the instance segmentation and target detection network in the previous perception algorithm to obtain the mask of the object and convert it into the coordinates and normal vector coordinates of the point cloud voxel. Then, we use the SuperPoint [17] deep neural network to extract the 3D key points on the object and calculate the position and attitude of the current camera according to the point cloud of the current frame and the point cloud predicted by the previous frame through the ICP algorithm [18]. In essence, the optimal registration method based on the least squares method repeatedly matches the corresponding points and calculates the optimal rigid body transformation until the convergence accuracy requirements of the correct registration are met. Then, the TSDF value is updated according to the camera position and pose, the point cloud is fused, and the surface is estimated according to the TSDF value. As shown in Algorithm 1, we provide pseudocode for perceptual localization.

Algorithm 1 Perceptual Positioning

1:    **Input**: a real-time image X={imgl,img2,...imgn}, from the image captured by AR glasses.
2:    **Objective** find the target position, return the target position coordinate y.
3:    BlendMask (img) means using BlendMask to process the image, PCD (mask) means converting the mask into a point cloud, SuperPoint (pcd) means using the SuperPoint algorithm to describe t he key points, ICP (keypoint, **pcd'**, pcd) means using the ICP algorithm for registration.
4:    **for** step $\in$ X **do**
5:        mask = BlendMask （frame）
6:        **pcd'** = PCD （mask）
7:        keypoint = SuperPoint （pcd）
8:        **for** step $\in$ {1...∞} **do**
9:            **y'** = ICP （keypoint，**pcd'**，pcd）
10:           **pcd'** = pcd （）
11:           **If** Registration succeeded **then**
12:               y =y'
13:               break
14:           **end if**
15:       **end for**
16:   **end for**

To achieve this function more effectively, we use AR glasses to collect image information and transmit it to the edge server for positioning and perception. The edge server returns the processed results to the AR glasses and then guides the operator to assemble (see Fig. 3).

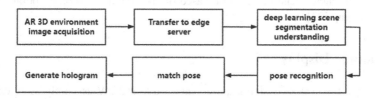

Fig. 3. Visual information processing flow

The core technology of this part lies in the recognition, understanding and spatial positioning of objects in the scene [19], and it is also the core level of edge computing in this project. Compared with ordinary AR applications that can only provide instructive content, we plan to combine AI to recognize and understand objects in the assembly process to truly generate a more realistic interaction with assembly personnel. This can not only guide the assembler to carry out the assembly operation step by step but also when the assembler makes a wrong operation, the system will remind the assembler in real time and give the correct operation process guidance. While recognizing the object, the system will calculate the 3D pose of the object in real time, generate accurate 3D guidance animation to guide the assembler to carry out the next operation [20], and make the virtual 3D model superimposed on the real object accurately.

2.4 Scene Switch

Edge AI analyzes the current assembly scene in real time, matches the corresponding process data, model, assembly process and other information from the database, and automatically realizes scene switching and process guidance corresponding to the scene information. When the assembly scene changes, the assembly guide also changes accordingly, and when the assembly object position shifts, the position indicator changes accordingly.

Among them, for spatial scene understanding, we leverage the existing powerful capabilities of HoloLens2 glasses to achieve highly accurate but less structured spatial mapping and new AI drivers. Three distinct but related objects are generated using this technique, namely, a simplified watertight environment mesh, a planar area for placement, and a spatial mapping mesh snapshot. Holographic images of walls, bases, objects, etc., can be automatically constructed. By combining these techniques, scene understanding produces a representation similar to the 3D environment used in the Unity framework [21].

Spatial positioning points are used to track objects and scenes [22]. Spatial anchors represent important points that the system tracks over a period of time. Each anchor point has an adjustable coordinate system based on other anchor points or reference frames to ensure that the anchored hologram remains accurate. By rendering the hologram in the coordinate system of the locator point, it is possible to precisely locate the hologram at any given time. At the same time, edge-side AI is used to analyze the objects in front of the operator in real time to determine the type of object and the assembly process. If the operator is found to have changed the type or assembly process of the object, it will reconstruct the new scene information based on the spatial positioning point. Realize scene switching [23].

3 Example Display

According to the system design requirements, we took UAV assembly as an example to realize various functions of the virtual assembly system and achieved good results. Microsoft HOLOLENS2 AR glasses, a custom vision server, and an H3C GR-5400AX enterprise wireless router were used. Table 1 below lists the details of our experimental equipment.

Table 1. Experimental equipment information

device name	Model	Introduce
AR glasses	Microsoft HOLOLENS2	Microsoft's most advanced AR glasses
vision server	Custom CASIA-CV	8 cores and 16 threads, NVIDIA high-performance graphics card
Industrial Wireless Router	H3C GR-5400AX	High-performance industrial wireless router
Visualization screen	AOC 24B2XHM	1920 * 1080 resolution 23.8 inches

We designed three simple assembly scenarios, namely, quadcopter drone, fat airplane, and thin airplane (see Fig. 4). First, the workbench is designed according to the tools, equipment and parts used in the assembly of the drone. The parts and tools of the four-wing drone are placed in the blue box, and the tools of other aircraft are placed on the left side of the workbench. Inside the red box. The color of the table top of the workbench is green and should be quite different from the color of the drone. This is to facilitate the machine in capturing the position of the drone more quickly and accurately (see Fig. 5).

Fig. 4. Shows the assembly model. The fat plane is on the left, the thin plane is in the middle, and the quadcopter drone is on the right.

Fig. 5. The layout of the workbench. The picture on the left is the model of the workbench that is conceived, and the picture on the right is the actual workbench assembled by the drone.

Taking the assembly of four-wing drones as an example, after the virtual assembly system is turned on, the operator can choose three modes. The first mode is the AR demonstration assembly, that is, the real installation steps are presented through the method of holographic image projection so that the operator can intuitively watch the whole process of the installation. The second mode is manual virtual assembly; that is, the operator can manually select virtual parts and perform virtual assembly by dragging and other methods. The third mode is physical-guided assembly; that is, the operator directly assembles the real UAV through virtual guidance [24].

In the object-guided assembly mode, the system will guide the operator to assemble according to the prescribed steps. If the operator has a wrong step or wrong handling during the assembly process, the system will issue an installation error prompt (see Fig. 6). If the assembly is correct, the next assembly work will continue until the entire assembly work is completed. During assembly, the system will track the position of the drone in real time and respond to changes in the position of the drone at any time. When the position of the drone changes, the position indicator also moves with the drone (see Fig. 7).

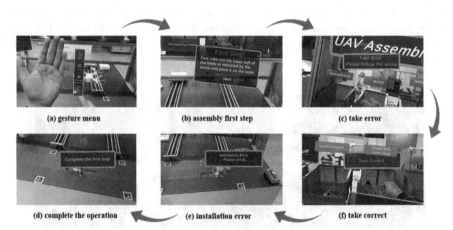

Fig. 6. Demonstration diagram of experimental results. In the figure, we can see that the system guides the operator to assemble according to the prescribed steps, raises an error warning when taking and installing errors, and prompts the next step when the installation is correct.

Fig. 7. Visual positioning effect diagram. We can see that when the drone is removed, the red position indicator repoints to the position where the drone moved. (Color figure online)

4 Summary and Outlook

This research takes AR technology and AI technology as the core, combines augmented reality and artificial intelligence, transforms and upgrades traditional industries, and creates a more modern, intelligent, networked, and digital industrial auxiliary system. In this system, we have a clear and friendly user interface, which can intelligently perceive the position of assembly objects and realize switching between different assembly scenes.

Although the assembly system designed this time has been able to complete the general assembly tasks well, there is still much room for improvement. For example, the AR remote guidance function can be added at the same time as assembly to guide workers more effectively in assembly work [25]. For assembly tasks with a large amount of data, transformer-based [26] segmentation and detection algorithms can be used to improve the accuracy of perception. This research uses artificial intelligence instance segmentation at the 2D image level and then transforms it into 3D space according to the camera pinhole model. No artificial intelligence algorithm is applied at the 3D registration level, so algorithms such as Pointnet++ [27] can be directly used at the 3D point cloud level. Obtain richer information more efficiently.

In addition, we hope to build an AR auxiliary framework that serves the industry, linking the underlying technology with the realization functions and industrial scenarios, and helping to quickly develop AR auxiliary products suitable for different scenarios and functions. In this framework, the underlying technology can be linked to the implementation function with less code, greatly reducing development time.

References

1. Chen, Y., Wang, Q., Chen, H.: An overview of augmented reality technology. J. Phys: Conf. Ser. **1237**, 022082 (2019)
2. Feng, S., He, W., Zhang, S.: Seeing is believing: AR-assisted blind area assembly to support hand–eye coordination. Int. J. Adv. Manufact. Technol. 1–10 (2022)
3. Wang, Z., et al.: User-oriented AR assembly guideline: a new classification method of assembly instruction for user cognition. Int. J. Adv. Manufact. Technol. **112**(1–2), 41–59 (2020). https://doi.org/10.1007/s00170-020-06291-w
4. Osorio Quero, C.A., Durini, D., Rangel-Magdaleno, J.: Single-pixel imaging: an overview of different methods to be used for 3D space reconstruction in harsh environments. Rev. Sci. Instrum. **92**(11), 111501 (2021)
5. Kipper, G., Rampolla, J.: Augmented Reality: An Emerging Technologies Guide to AR. Elsevier (2012)
6. Zhang, W., Chen, J.: Research on experiential design of RV interior using HoloLens2. In: 2020 International Conference on Innovation Design and Digital Technology (ICIDDT). IEEE, pp. 202–206 (2020)
7. Qin, L., Si, Z.: Visual information transfer design of packaging product instructions based on unity platform. In: Zhao, P., Ye, Z., Xu, M., Li., Yang, Zhang, L., Zhu, R. (eds.) Advances in Graphic Communication, Printing and Packaging Technology and Materials. LNEE, vol. 754, pp. 257–262. Springer, Singapore (2021). https://doi.org/10.1007/978-981-16-0503-1_38
8. Sarosa, M., Chalim, A., Suhari, S.: Developing augmented reality based application for character education using unity with Vuforia SDK. J. Phys: Conf. Ser. **1375**(1), 012035 (2019)

9. Huang, W., Alem, L., Tecchia, F., Duh, H.-L.: Augmented 3D hands: a gesture-based mixed reality system for distributed collaboration. J. Multimodal User Interf. **12**(2), 77–89 (2017). https://doi.org/10.1007/s12193-017-0250-2

10. Chen, H., Sun, K., Tian, Z.: BlendMask: top-down meets bottom-up for instance segmentation. In: Proceedings of the IEEE/CVF Conference on Computer Vision and Pattern Recognition, pp. 8573–8581 (2020)

11. Tian, Z., Shen, C., Chen, H.: FCOS: fully convolutional one-stage object detection. In: Proceedings of the IEEE/CVF International Conference on Computer Vision, pp. 9627–9636 (2019)

12. Mo, Y., Wu, Y., Yang, X.: Review the state-of-the-art technologies of semantic segmentation based on deep learning. Neurocomputing **493**, 626–646 (2022)

13. Xu, K., Guan, K., Peng, J.: DeepMask: an algorithm for cloud and cloud shadow detection in optical satellite remote sensing images using deep residual network. arXiv preprint arXiv:1911, 03607 (2019)

14. O Pinheiro, P O., Collobert, R., Dollár, P.: Learning to segment object candidates. In: Advances in Neural Information Processing Systems, vol. 28 (2015)

15. Chen, S., Fang, J., Zhang, Q.: Hierarchical aggregation for 3D instance segmentation. In: Proceedings of the IEEE/CVF International Conference on Computer Vision, pp. 15467–15476 (2021)

16. Song, J., Kook, J.: Visual SLAM based spatial recognition and visualization method for mobile AR systems. Appl. Syst. Innov. **5**(1), 11 (2022)

17. DeTone, D., Malisiewicz, T., Rabinovich, A.: Superpoint: self-supervised interest point detection and description. In: Proceedings of the IEEE Conference on Computer Vision and Pattern Recognition Workshops, pp. 224–236 (2018)

18. Wang, X., Li, Y., Peng, Y.: A coarse-to-fine generalized-ICP algorithm with trimmed strategy. IEEE Access **8**, 40692–40703 (2020)

19. Hoque, S., Arafat, M.D.Y., Xu, S.: A comprehensive review on 3D object detection and 6D pose estimation with deep learning. IEEE Access (2021)

20. Hou, L., Wang, X., Bernold, L.: Using animated augmented reality to cognitively guide assembly. J. Comput. Civ. Eng. **27**(5), 439–451 (2013)

21. Neb, A., Brandt, D., Awad, R.: Usability study of a user-friendly AR assembly assistance. Procedia CIRP **104**, 74–79 (2021)

22. Mo, Y., Zhang, Z., Meng, W.: A spatial division clustering method and low dimensional feature extraction technique based indoor positioning system. Sensors **14**(1), 1850–1876 (2014)

23. Marino, E., Barbieri, L., Colacino, B.: An Augmented Reality inspection tool to support workers in Industry 4.0 environments. Comput. Ind. **127**, 103412 (2021)

24. Gors, D., Birem, M., De Geest, R.: An adaptable framework to provide AR-based work instructions and assembly state tracking using an ISA-95 ontology. Procedia CIRP **104**, 714–719 (2021)

25. Wang, P., Zhang, S., Bai, X.: A gesture-and head-based multimodal interaction platform for MR remote collaboration. Int. J. Adv. Manufact. Technol. **105**(7), 3031–3043 (2019)

26. Dosovitskiy, A., Beyer, L., Kolesnikov, A.T.: An image is worth 16x16 words: transformers for image recognition at scale. arXiv preprint arXiv:2010.11929 (2020)

27. Qi C.R., Yi, L., Su, H.: PointNet++: deep hierarchical feature learning on point sets in a metric space. In: Advances in Neural Information Processing Systems, vol. 30 (2017)

Infrastructure for Data Science

Industry-Oriented Cloud Edge Intelligent Assembly Guidance System

Difei Liu[1], Zhuorui Chang[2]([✉]), Junpei Ma[2], Tao Wang[3], and Mei Li[2]

[1] China Academy of Information and Communications Technology, Beijing, China
[2] School of Information Engineering, China University of Geosciences, Beijing, China
2104210029@email.cugb.edu.cn
[3] School of Geophysics and Information Technology, China University of Geosciences, Beijing, China

Abstract. In view of the low efficiency of production and assembly in traditional industries, we use AR technology to replace traditional assembly instructions and design an industry-oriented cloud-edge intelligent assembly guidance system. Since the computing power of AR glasses cannot meet the high complexity requirements of scene understanding, we adopt the joint solution of Cloud-Edge. First, the sensor data collected by the AR glasses are streamed to the edge server using high-speed and low-latency wireless interconnection technology. Then, the product artifacts in the data scene are identified and understood through the instance segmentation network BlendMask based on deep learning. Then, the 3D pose of the object is calculated in real time by combining pose estimation and 3D reconstruction. Furthermore, an accurate 3D guidance animation is generated, and the virtual 3D model in the AR glasses is accurately superimposed on the real object to determine whether the assembly is correct in real time. Experiments show that the system effectively combines artificial intelligence and intelligent manufacturing, integrates various elements in the scene in real time to provide operators with multimodal and multidimensional immersive guidance, and corrects in time when assembly errors occur. It can not only quickly guide the operator to complete the learning of the assembly process but also assist the staff in the assembly in real time. Ultimately, it improves assembly speed and accuracy, which in turn improves enterprise productivity.

Keywords: Mixed reality · Instance segmentation · Pose estimation · Cloud edge · Edge computing · Smart assembly

1 Introduction

With the new round of scientific and technological revolution and industrial transformation, the pace of quality and efficiency improvement of intelligent manufacturing [1, 2] has been accelerated, and various products have been continuously optimized and upgraded, developing in the direction of complexity, miniaturization and high precision. In the manufacturing industry, the assembly density and precision of these products are increasing, and the difficulty for workers to assemble the components is also increasing.

The problems of difficult assembly of complex products, low assembly efficiency, and large employee memory and cognitive load during the assembly process are becoming increasingly prominent on the factory floor [3, 4]. The traditional assembly instructions can not only assemble more refined items in detail but also make it difficult to correct errors and give early warnings in the assembly process, which can no longer meet the needs of enterprises [5]. However, with the continuous maturity of mixed reality technology, to solve these problems in the traditional assembly process, the interactive virtual assembly system based on manufacturing technology has developed rapidly, bringing new technical support to the development of various industries.

To overcome the problems of complex assembly and low production and assembly efficiency, this paper provides a cloud-edge intelligent assembly guidance system for the new generation of industrial Internet and intelligent manufacturing industries. The visual point of view is projected onto the real-world image seen by the human eye, providing supplementary information for people visually, using AR technology to replace the traditional assembly instruction book, realizing real-time intelligent guidance of assembly and data analysis, and real-time synthesis of various elements in the scene. The operator provides multimodal and multidimensional immersive guidance, error correction and early warning, quickly helps the operator to complete the learning of the assembly process, and assists the assembler in assembly in real time. At the same time, it can cooperate with unmanned AGVs, robotic arms, etc. Intelligent. Therefore, strengthening the assembly speed, accuracy and real-time performance of physical object assembly effectively improves the production efficiency of enterprises and realizes the effective combination of artificial intelligence and intelligent manufacturing. The main contributions of this work are as follows:

1. Apply the real-time single-stage segmentation method of Blender Mask [6] to industrial scenarios to realize real-time image detection and segmentation of assembly products;
2. The pose estimation and completion detection parts are integrated into an AR assembly guidance system to realize the recognition and understanding of items in the scene and provide effective and accurate guidance.
3. Using the combined solution of Cloud-Edge [7, 8], cloud edge technology is applied to the industrial environment to solve the problem that the computing power of the AR glasses themselves cannot understand the AI scene with high complexity.

This paper is organized as follows: We first introduce the methods we use in Sect. 2. In the Sect. 3, experiments and results analysis are carried out. Section 4 gives some conclusions and outlook.

2   Method

We input the sensor data and image information collected by industrial cameras and AR glasses into the edge server and use deep learning algorithms such as BlendMask instance segmentation and point cloud pose estimation to extract the product workpiece information and combine pose estimation. Realize scene understanding and determine the 3D coordinate information of the workpiece. The relative position relationship between the standard point cloud stored offline and the point cloud collected by the depth camera is determined by pose estimation and point cloud matching, and the position of the work-piece under the field of view is obtained through coordinate transformation. Combined with returning its information to the AR glasses for display in the unity system for the next guide operation.

2.1 Instance Segmentation

The deep learning network for UAV (Unmanned Aerial Vehicle) workpiece recognition and pose estimation is as follows: The instance segmentation network BlendMask based on deep learning determines the position, category and mask of the object of interest in the image. This method needs to obtain the labeled samples of objects in advance and send them to the neural network for training (Fig. 1).

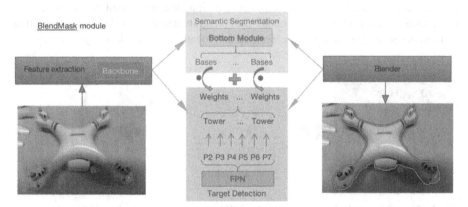

Fig. 1. This is BlendMask's network. After the network uses Backbone to extract features from the image, it is divided into two branches, namely, the semantic segmentation branch and the target detection branch. The semantic segmentation branch uses the bottom module to semantically segment images to obtain bases. The target detection branch imitates the FCOS (fully convolutional one-stage object detection) method, using FPN (feature pyramid network) and Tower to perform object classification and bounding box regression for each pixel and predict the centrality of the pixel and the attention weight of the corresponding bases. Use centrality and classification confidence to perform nonmaximum suppression on the results of each pixel regression and classification to obtain the bounding box and classification confidence of the object. Finally, the attention weight of bases is used to weight and sum the bases to obtain the masks of each object.

The core idea of FCOS target detection [9] is to predict the target category and target box to which each point in the input image belongs. It retains the anchor-free mechanism, introduces three strategies of pixel-by-pixel regression prediction, multiscale features and center-ness [6, 9], and finally realizes that the effect can be comparable to various mainstream anchor-based target detection algorithms without anchor boxes. By eliminating predefined anchor boxes, complex calculations related to anchor boxes are avoided. In addition, it avoids setting all hyperparameters related to anchor boxes that are very sensitive to the final detection performance. It achieves better results than previous anchor box-based one-stage detection algorithms [10]. The detection problem of the FCOS detection framework can be unified to other problems that have been solved by FCN, such as semantic segmentation, which can reuse the ideas of other tasks and does not need to adjust many parameters like the algorithm based on anchor boxes, which makes training simpler, does not need to calculate IOU, saves a lot of computing power and memory.

After the target detection is completed by the FCOS method, an attention mechanism is added to the detected region candidate frame for instance segmentation. Combining the top-down and bottom-up ideas, using the instance-level high-dimensional information (such as bbox) generated by the top-down method, the prediction information of each pixel generated by the bottom-up method is used. fusion. The Blender module normalizes the feature map, multiplies the k bbox-sized masks and the corresponding attention, and then superimposes them by channel to obtain the final mask [6]. To obtain multiple masks of the entire graph, the performance of each mask is not necessarily good, but some have overall information, and some have edge information, which are combined to generate a perfect mask. This Blender module can better integrate instance-level global information and low-level features that provide detailed and location information (Fig. 2).

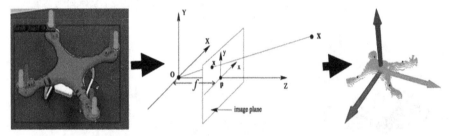

Fig. 2. The left is the pixel coordinate result of 2D image segmentation, the middle is the camera pinhole model, and the right is the converted 3D point cloud model. First, image segmentation is performed on a scene containing a certain target component to obtain the mask of the target component, and the depth map is combined to obtain the target component point cloud under the scene frame and the camera coordinate system. Second, the cam2world matrix that comes with HoloLens is used to convert the center coordinates of the main body to the world coordinate system and judge whether it is still at the corresponding position.

2.2 Pose Estimation

The workpiece positioning and recognition is divided into two stages: 3D reconstruction and real-time pose estimation [11]. In the 3D reconstruction stage, the depth camera is used to shoot the object. Section 2.1 uses BlendMask to obtain the object mask and extracts its corresponding depth image and uses the SuperPoint deep neural network [12, 13] to extract the 3D key points on the object.

The basic process of the real-time pose estimation stage is to use the FPFH [14] point cloud descriptor to extract the local features of the point cloud and then use ransac [15] to perform the maximum consistent set matching to complete the rough matching. Finally, the ICP algorithm [16] is used for fine matching to make the results more accurate. It needs to know the pose of the object and the camera, shoot the object from multiple angles, and obtain 3D key points and object pose information from multiple angles to reconstruct the sparse 3D key point map of the object. In the real-time pose estimation stage, the image is obtained through the depth camera on the AR glasses and is sent back to the server for pose matching in real time. The specific method is also BlendMask + SuperPoint [13] to obtain the 3D key points on the object, match it with the sparse 3D key point map of the object reconstructed in 3D, and then obtain the pose of the object through the Ransac+least squares method.

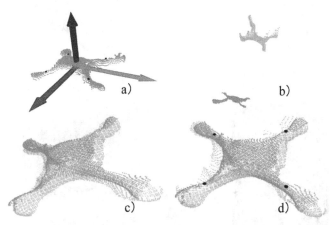

Fig. 3. (a) is the workpiece coordinates of the standard point cloud prepared offline, (b) is the point cloud collected by the online depth camera, (c) is the transformation matrix between the two point clouds obtained by registration, and (d) is the workpiece coordinate multiplied by the same transformation matrix.

First, the scene frame containing a certain subject is segmented to obtain the mask of the subject, and the depth map is combined to obtain the subject point cloud under the scene and the camera coordinate system. Second, the standard point cloud stored offline in advance is used to estimate the pose of the subject point cloud to obtain the pose transformation matrix. Third, the coordinates of the component under the standard point cloud are multiplied by the pose matrix to obtain the installation target position coordinates of the component under the camera coordinate system under the current scene frame. Finally, the workpiece installation target position coordinates under the camera system are converted to the HoloLens coordinate system using the calibration relationship between HoloLens and the world coordinate system for display in the Unity system (see Fig. 3).

2.3 Cloud-Edge Joint Technology

Due to the complexity of AI scene understanding, the computing power of AR glasses itself cannot meet the demand. Therefore, the cloud-edge joint solution is adopted. The solution uses high-speed and low-latency wireless interconnection technology to stream sensor data collected by AR glasses to edge computing for visual processing. within the server[7, 8]. The edge computing vision processing server runs the deep learning algorithm to analyze and understand the data and then returns the results to the AR glasses for further guidance (see Fig. 4).

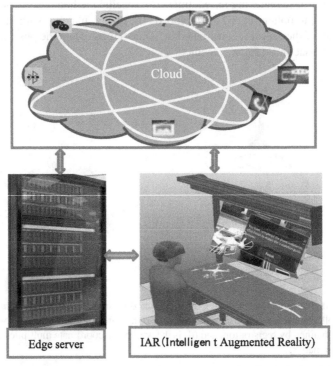

Fig. 4. This part reflects the application of cloud edge technology in the industrial environment and is mainly divided into three parts: (1) Cloud: used to deploy process database, model database, etc.; (2) Edge: used to realize real-time interactive tasks such as AI reasoning and IAR interaction; (3) AR: The user wears the device, which is used for visual perception and information display.

3   Experiment and Analysis

3.1   Experimental Setup

Our experiments are run in an environment where the edge server is deployed locally and combined with a local area network. The edge server uses an Intel server, an NVIDIA RTX3090 graphics card, and an Ubuntu 64-bit, 8 GPU operating system environment. Using the PyTorch deep learning framework, the compilation environment is Torch1.3, CloudPickle, Matplotlib, and Python 3.7 language programming. We use the MSCOCO2017 instance segmentation dataset as the standard to create a quadrotor UAV dataset, and the training and testing experiments are performed on this dataset. After the instance segmentation of the target workpiece is realized, the result is stored offline as a standard point cloud model and the point cloud model collected by the depth camera in combination with the pose estimation technology for real-time registration to obtain the position of the workpiece in the field of view. Finally, combined with cloud edge technology, it is returned to the AR glasses, and after a series of coordinate transformations, it is used for the final Unity display. To realize the function of assembly guidance and error correction.

3.2 Experimental Results

When the operator wears the HoloLens2 mixed reality glasses at the quadrotor assembly station, the industrial wireless router combines cloud edge technology to establish the link and data transmission between the HoloLens2 glasses, the vision processing server, and the visualization screen. The artificial intelligence AI program running on the edge computing vision processing server can judge the current station and assembly scene in real time and use computer vision and neural network assistance to judge the current quadrotor assembly station and production line production products through target detection. The neural network determines the stage process of the production line. The operator can observe the composition of the virtual quadrotor in real time, observe it from different perspectives, understand the structure of the quadrotor, and interact with it through gestures. At the same time, various operations on the holographic control panel and holographic images, such as clicking, dragging, zooming, etc., can be completed through touch gestures and visual feedback of the UI interface of gesture actions.

When assembling the body, HoloLens2 mixed reality glasses can give corresponding guidance and animation prompts according to the current workpiece, install screws and batteries, connect the battery wiring to the socket, and then connect the camera wiring to the specified socket. Install the bracket with the wing to the designated position and the rotor, and finally fix the rotor with screws. Through real-time animation, the entire installation process is demonstrated, including the objects installed in each step, with text and animation prompts, making the assembly personnel more familiar with it more quickly. The installer can install the virtual quadrotor object, and all the parts can be manipulated directly through gesture interaction. Each part will display its name and the position where it needs to be installed when the hand touches it, which is convenient for the installer to learn quickly. At the same time, the artificial intelligence AI program running in the edge computing vision processing server can perform intelligent interaction tasks such as object recognition, object positioning, process judgment, operation guidance, natural gesture recognition interaction, and abnormal judgment in real time. When assembly personnel make mistakes, the AI assistant reminds them in real time and gives guidance on the correct operation process (see Figs. 5 and 6).

According to the running performance test, our system supports a deep learning instance segmentation network and multiobject recognition. The delay is less than 0.2 s, the recognition rate is not less than 90%, it supports multiobject pose estimation [17], the processing frame rate is not less than 5 frames, and it supports recognition and understanding. Objects in the assembly process can guide the assembler to perform the assembly operation step by step. When the assembler makes a wrong operation, the system will remind him/her in real time and give the correct operation process guidance.

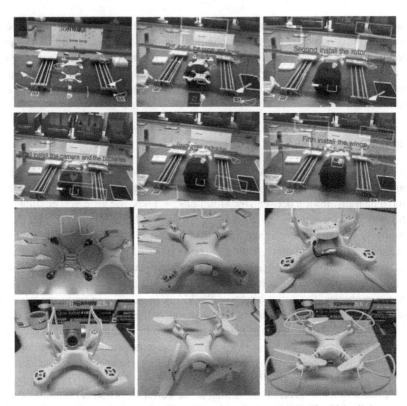

Fig. 5. The picture shows the process of installing a quadcopter drone. First, install the upper and lower body, then install the rotor, then install the camera and battery, then install the bracket, and finally install the wing.

Fig. 6. In the figure, when an installation error occurs, a corresponding alarm prompt will be given, and a mark will also be given when the installation is completed. After the installation is complete, if you want to choose to install other models, you can select the aircraft model as shown in the lower left picture and another thin aircraft model as shown in the lower right picture.

4 Conclusion

At present, the related applications of mixed reality assembly guidance mainly focus on aircraft assembly skills training, on-site assembly guidance, and assembly result inspection [7]. Most of the research on mixed reality assembly guidance technology focuses on mixed reality technology. There are problems such as insufficient system construction and weak human-computer interaction capabilities, and the existing mixed reality assembly guidance systems are generally based on specific scenarios. Customized development lacks system applications for cable assembly of electronic equipment, and there are still problems such as the inability to solve the problem of incompetent cable assembly process information and the inability to provide effective support for process design and assembly [18, 19]. In addition, there are problems of product complexity, product assembly accuracy and low assembly efficiency in the traditional assembly process, and the traditional AR assembly guidance and inherent procedural guidance cannot meet the suddenness and optimization capabilities of the assembly process and lack sufficient flexibility.

Therefore, it provides a highly intelligent perception and understanding ability and decision-making function, which can integrate various elements in the scene in real time to provide operators with multimodal [20] and multidimensional immersive guidance, error correction and early warning and can be used for new cloud-edge intelligent assembly guidance systems for the generation of industrial Internet and intelligent manufacturing industries.

In summary, our system is AR, computer vision, and deep learning working together to realize real-time monitoring of the assembly process, provide real-time prompts and guidance (including arrows, texts, animations, etc.) to the installation personnel's assembly behavior and cooperate with animation demonstrations. The system can perceive the installation position of the workpiece in real time and give arrow prompts. The right side shows the processing and display of environmental objects by server-side computer vision and artificial intelligence. During the process, the system runs smoothly, the interface is simple and beautiful, and the functions are stable. The industry-oriented cloud-edge intelligent assembly guidance system uses AR technology to replace traditional assembly instructions, which can not only improve the accuracy of assembly but also improve production efficiency, which is of great significance to the development of intelligent manufacturing.

References

1. Hořejší, P., Novikov, K., Šimon, M.: A smart factory in a smart city: virtual and augmented reality in a smart assembly line. IEEE Access **8**(8), 94330–94340 (2020)
2. Deb, M., Kannan, P.: Assembly and installation guidance by augmented reality. In: CIRED 2021-The 26th International Conference and Exhibition on Electricity Distribution, pp. 1317–1320 (2021)
3. Jianjun, T., Bo, Y., Junhao, G.: Exploration and practice of AR intelligent guidance technology for aircraft assembly operations. Aviat. Manufact. Technol. **62**(08), 22–27 (2019)
4. Yue, W.: Research on Augmented Reality Fusion Technology for Product Assembly Guidance (2018)

5. Liu, R., Fan, X., Yin, X., et al.: Estimation method of matrix and part pose combination in AR-assisted assembly. Mech. Des. Res. **34**(06), 119–125+137 (2018)
6. Chen, H., Sun, K., et al.: BlendMask: top-down meets bottom-up for instance segmentation. In: IEEE/CVF Conference on Computer Vision and Pattern Recognition, pp. 8570–8578. IEEE, New York (2020)
7. Liu, Q., Wang, Y., et al.: Remote collaborative assembly and maintenance guidance based on mixed reality. Chin. J. Graph. **42**(02), 216–221 (2021)
8. Varghese, B., Reaño, C., et al.: Accelerator virtualization in fog computing: moving from the cloud to the edge. Cloud Computing, IEEE, vol. 5, pp. 28–37. IEEE (2018)
9. Tian, Z., Shen, C., et al.: FCOS: fully convolutional one-stage object detection. In: International Conference on Computer Vision, ICCV, pp. 9626–9635. IEEE/CVF (2019)
10. Su, L., Sun, Y., et al.: A review of instance segmentation based on deep learning. J. Intell. Syst. **17**(01), 16–31 (2022)
11. Liu, R.: Research on the pose estimation and state detection method of base parts in AR-assisted assembly. Shanghai Jiaotong University (2018)
12. Hui, L., Yuan, J., et al.: Superpoint network for point cloud oversegmentation. In: International Conference on Computer Vision 2021, pp. 5490–5499. IEEE/CVF (2021)
13. Landrieu, L., Simonovsky, M.: Large-scale point cloud semantic segmentation with superpoint graphs. In: Conference on Computer Vision and Pattern Recognition 2018, CVPR, pp. 4558–4567. IEEE/CVF, Salt Lake City (2018)
14. Rusu, R., Blodow, N., et al.: Fast Point Feature Histograms (FPFH) for 3D registration. In: International Conference on Robotics and Automation, pp. 3212–3217. IEEE (2009)
15. Peng, C.: K-means based RANSAC algorithm for ICP registration of 3D point cloud with dense outliers. In: International Conference on Consumer Electronics-Taiwan, ICCE-TW, pp. 1–2. IEEE (2021)
16. Li, P., Wang, R., et al.: Evaluation of the ICP algorithm in 3D point cloud registration. IEEE Access **8**, 68030–68048 (2020)
17. Sock J., Garcia-Hernando G., et al.: Active 6D multi-object pose estimation in cluttered scenarios with deep reinforcement learning. In: International Conference on Intelligent Robots and Systems (IROS), pp. 10564–10571. IEEE/RSJ (2020)
18. Agati, S., Bauer, R., et al.: Augmented reality for manual assembly in Industry 4.0: gathering guidelines. In: Symposium on Virtual and Augmented Reality 2020, SVR, pp. 179–188. SVR (2020)
19. Bauer R., Agati S., et al.: Manual PCB assembly using augmented reality towards total quality. In: Symposium on Virtual and Augmented Reality, pp. 189–198. SVR (2020)
20. Liu, S., Li, M.: Multimodal GAN for energy efficiency and cloud classification in Internet of Things. Internet of Things Journal **6**(4), 6034–6041 (2019)

An Intelligent Data Routing Scheme for Multi-UAV Avionics System Based on Integrated Communication Effectiveness

Yan Zou[1], Meng Wu[2], Shaoqing Zhang[1,3], Feiyan Li[1,3], Jiarun Chen[1,4], Yazhuo Wang[1], Chuan Lin[2,4], and Guangjie Han[3(✉)]

[1] Shenyang Aircraft Design and Research Institute AVIC, Shenyang, China
[2] Daqing Oilfield Powerlift Pump Industry Co., Ltd., Daqing, China
[3] Hohai University, Changzhou, China
hanguangjie@gmail.com
[4] Northeastern University, Shenyang, China

Abstract. The rapid development of information technology promotes the transformation and development of future air combat, from mechanization to informatization, intelligence, and multiplatform integration. For the multiplatform avionics system in the unmanned aerial vehicle (UAV)-based network, we aim to address the data routing and sharing issues and propose an integrated communication effectiveness metric. The proposed integrated communication effectiveness is a hierarchical metric consisting of link effectiveness, node effectiveness, and data effectiveness. The link quality, link stability, node honesty, node ability, and data value are concurrently taken into account. We give the normal mathematical expression for the integrated communication effectiveness. We propose a hop-by-hop routing scheme based on a Q-learning algorithm considering the proposed effectiveness metric. Simulation results demonstrate that the proposed scheme is able to find the most efficient routing in the UAV network.

Keywords: Multiplatform avionics system · UAV · Integrated communication effectiveness · Q-learning

1 Introduction

To adapt to the modern battlefield, the air combat mode is developing from traditional single-platform combat to multiplatform combat, especially for multi-UAV cooperation combat [1–8]. As the foundation of multi-UAV combat, multiplatform avionics systems may produce error information due to accidental failure or enemy invasion, affecting the combat results [9–15]. Therefore, it is very important to evaluate the effectiveness of the multiplatform avionics system, especially when the UAVs perform data routing for sharing information and determining cooperative combat [16–20] (Fig. 1 displays a schematic for multi-UAV avionics system). By evaluating the communication effectiveness of the multiplatform avionics system, the past or current state of the entire system

Y. Wang et al. (Eds.): ICPCSEE 2022, CCIS 1629, pp. 243–252, 2022.
https://doi.org/10.1007/978-981-19-5209-8_17

can be determined, and the future state can be predicted. Thus, the unavailable or untrustworthy information or platform can be isolated [21–23]. Hence, it is indispensable to determine the effectiveness evaluation mechanism to evaluate the communication of the data routing in the UAV network [24–26].

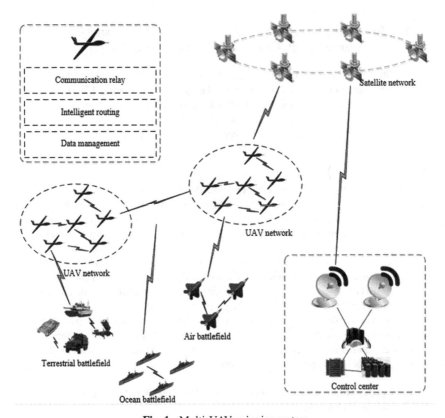

Fig. 1. Multi-UAV avionics system

For this purpose, in this paper, we propose integrated communication effectiveness consisting of link effectiveness, node effectiveness, and data effectiveness. Based on the proposed effectiveness metric, we propose a Q-learning-enabled routing scheme for the UAV network.

The rest of the paper is summarized as follows. Section 2 presents the integrated communication effectiveness metric. In Sect. 3, we introduce the proposed routing scheme. Section 4 presents the evaluation results. The conclusion is drawn in Sect. 5.

2 Proposed Integrated Communication Effectiveness Metric

This paper proposes an integrated communication effectiveness metric to evaluate the data routing among the UAVs in the multi-UAV network. The proposed integrated communication effectiveness is a hierarchical metric including link effectiveness, node effectiveness, and data effectiveness, which can be computed as follows:

2.1 Link Effectiveness

In the paper, the link effectiveness contains the "link quality" and "link stability".

Link Quality: In a real battlefield, the complicated electromagnetic environment will affect the data communication among the UAVs. This results in a serious packet error rate that can be computed by Eq. 1.

$$r_p = 1 - (1 - B)^L \tag{1}$$

where B is the bit error rate. In a real multi-UAV network with variable network topology, the network links frequently fail and are rebuilt. Thus, the duration $T_{a,b}$ of a network link $<a, b>$ (connecting UAVs a and b) is critical for performing data routing. Normally, $T_{a,b}$ should satisfy the following equation.

$$R = \sqrt{\left((x_b + s_{a,b}T_{a,b}) - x_a\right)^2 + \left((y_b + s_{a,b}T_{a,b}) - y_a\right)^2} \tag{2}$$

where R is the communication range of each UAV, and (x_a, y_a) and (x_b, y_b) are the coordinates of UAVs a and b, respectively. In particular, in Eq. (2), $s_{a,b}$ is the relative speed between UAVs a and b.

In a multi-UAV network, the data communication delay due to a network link $<a, b>$ is similar to the delay computation in computer networks, which can be expressed by the following Eq. (3).

$$T_{\text{total}} = T_b + T_p + T_c \tag{3}$$

where T_p is the data propagation delay, which is related to the distance between UAVs a and b, T_b is the delay resulting from the bandwidth $B_{a,b}$ of the network link $<a, b>$, and T_c is the data processing delay it takes to handle the data at node a. Therefore, it can be inferred that a network link $<a, b>$ is available to deliver the data only when the following is satisfied.

$$T_{a,b} > T_{total} \tag{4}$$

Furthermore, to standardize the link evaluation, we give the normalized expression for $T_{a,b}$ as follows:

$$T_{a,b}^{Nor} = \frac{T_{a,b}}{T_{a,b}^{\max}} \tag{5}$$

where $T_{a,b}^{\max}$ is the maximum delay that requires the data to traverse link $<a, b>$.

Based on Eqs. (3)–(5), we have the following equation to evaluate the link quality of a network link $<a, b>$.

$$Q_{a,b} = T_{a,b}^{Nor} \cdot k + r_p \cdot (1 - k), k < 1 \tag{6}$$

Link Stability: For a network link <a, b> connecting two UAVs a and b, with a period t, the relative displacement $D_{a,b}$ of UAVs a and b is utilized to evaluate the link stability of link <a, b>, which can be expressed as follows.

$$S_{a,b} = 1 - \frac{l_{a,b}^{t+1} - l_{a,b}^{t}}{R} \tag{7}$$

where $l_{a,b}^{t+1}$ and $l_{a,b}^{t}$ represent the length of the link <a, b> at time t and $t + 1$, respectively.

Thus, the link effectiveness of a network link <a, b> can be expressed by the following equation:

$$LE_{a,b} = \mathcal{Q}_{a,b} \cdot S_{a,b} \tag{8}$$

2.2 Node Effectiveness

In general, the node effectiveness of a network node is computed based on the observation of the sensor node and the recommendation from the third-party referees. Since collecting and managing the recommendations from the third-party referees consumes considerable energy, only the direct node effectiveness is taken into account, neglecting the recommended node effectiveness computed by the third-party referees. In this paper, the direct node effectiveness consists of node honesty and node ability.

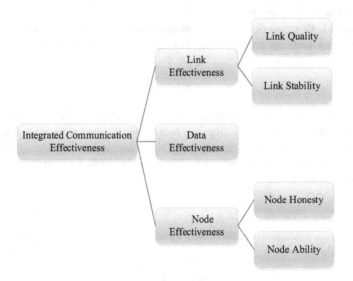

Fig. 2. Proposed integrated communication effectiveness

Node Honesty: The node honesty of a node a represents the intuitive judgment of a node. The node honesty NH(a) of node a can be computed by the following

$$NH(a) = \frac{o(a) + 2 \cdot z(a)}{2} \tag{9}$$

where $z(a)$ and $o(a)$ can be, respectively expressed by the following

$$z(a) = \frac{u(a)}{u(a) + f(a) + 1} \tag{10}$$

$$o(a) = \frac{1}{u(a) + f(a) + 1} \tag{11}$$

where $u(a)$ and $f(a)$ represent UAV a's number of successful and unsuccessful communications.

Node Ability: Assume that all the UAVs have the same initial energy level and energy consumption rate. However, malicious UAVs always consume too much energy to launch attacks. For example, the avionics system of a UAV under a high electromagnetic environment consumes much more energy than a normal UAV. The energy consumption is utilized to quantify the effectiveness of a node or whether the node is under a high adversarial environment. Thus, the node ability of node a can be computed by the following:

1) First, define the lowest energy threshold η. When the residual energy E_l of the UAV is below η, the UAV is considered to have insufficient energy to perform its desired task. At this point, the capability of the UAV is specified to be 0.

2) Otherwise, the node ability is computed based on the energy consumption rate $r_e \in [0, 1]$. The larger r_e is, the lower the residual energy of the UAV. This means that UAVs are less capable of performing sensing, data computation, and communication tasks. Thus, the node ability of a UAV is considered to be smaller. The node ability $NA(a)$ of a node a can be computed by

$$NA(a) = \begin{cases} 1 - r_e, E_L \geq \eta \\ 0, E_L < \eta \end{cases} \tag{12}$$

In total, the node effectiveness $NE(a)$ of a node a can be expressed by the following

$$NE(a) = \begin{cases} 0.5 + (NH(a) - 0.5) \cdot NA(a), NH(a) \geq 0.5 \\ NA(a) \cdot NH(a), NH(a) < 0.5 \end{cases} \tag{13}$$

2.3 Data Effectiveness

Data effectiveness is defined as effectiveness evaluation toward fault tolerance and data consistency. In general, the data sent from UAVs are time-spatially variable. That is, in a specified period, the sensing data from two closer UAVs are similar. The effectiveness of the sensing data follows a normal distribution. In this paper, we assume that the sensing data from each UAV follow a normal distribution with probability density function $f(x) = \frac{1}{\sigma\sqrt{2\pi}} e^{-\frac{(s-\mu)^2}{2\sigma^2}}$ where μ and σ denote the mean and variance of the sensing data, respectively, and s is the sensing data. In this paper, the mean of the sensing data is selected as the reference of the sensing sample because the mean of the data represents the trend of the sample. Namely, the closer the data are to the mean, the greater the effectiveness of the value is. Hence, the data effectiveness $DE(s,a)$ of the data s from node a can be computed by the following:

$$DE(s, a) = 2 \cdot \int_{|s|}^{\infty} f(s)ds \qquad (14)$$

where $|s|$ is the value of s.

Thus, following the concept in Fig. 2, the integrated effectiveness $IE_{a,b}$ of the communication tunnel $<a, b>$ can be computed by the following:

$$IE_{a,b} = LE_{a,b} \cdot NE(a) \cdot NE(b) \cdot DE(s, a) \cdot DE(s, b) \qquad (15)$$

For instance, in Fig. 3, the path $<u, v, b>$ can be selected as the potential route to deliver the data between UAVs u and b.

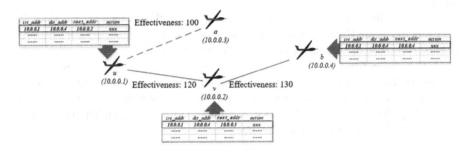

Fig. 3. Integrated communication effectiveness-based data routing

3 Proposed Data Routing Scheme

In this paper, we proposed a hop-by-hop heuristic algorithm based on the proposed integrated communication effectiveness. Assuming the distance among the UAVs can be acquired at any time, the data are delivered to its one-hop neighbor UAV with its communication range, when the energy efficiency and the communication effectiveness are concurrently taken into account. Each UAV a is associated with a residual energy $RE(a)$, the distance $Dist(a, t)$ between UAV a and destination UAV t, and its set $S_n(a)$ of its one-hop neighbor with its communication range R. The routing is in the direction where $Dist(a, t)$ can decrease. For each hop, the UAV in $S_n(a)$ is ranked in descending order of the distance between a and $\alpha \in S_n(a)$. Note that if $t \in S_n(a)$, the data are directly sent to the sink UAV t. The candidate neighbor set $C_n(a)$ is determined as follows: 1) the routing distance between the neighbor and the sink node plus the distance between the neighbor and the current node is smaller than the routing distance between the current node and the sink node; 2) the integrated communication effectiveness value is larger than a given threshold θ. A UAV can be added in $C_n(a)$ only when the above two requirements are satisfied. Based on the above ideas, we summarize our proposed routing algorithm in Algorithm 1. Note that the Q-learning algorithm is utilized to output the optimal routing from a to destination t.

Algorithm 1 Optimal Data Routing

Input: a, $S_n(a)$, $C_n(a)$, IE, θ

Output: Optimal routing from a to the destination t.

1   For each $\alpha \in S_n(a)$

2   if α is sink

3   add α to $C_n(a)$

4   elseif Dist(a, t)> Dist(α, t) && $IE(a, \alpha)>\theta$

5   add α to $C_n(a)$

6   End for

7   Evoking Q-learning to compute the reward of the action, the data delivery between any α and the UAV $\beta \in S_n(\alpha)$. (p.s. the reward is specified as the integrated communication effectiveness)

8   Output the optimal routing from a to t.

---

4 Evaluation

This paper utilizes MATLAB to simulate the proposed integrated communication effectiveness-based routing scheme. Note that we assume that all the UAVs in the UAV network are static and follow a prespecified formation. This paper calculates the integrated communication effectiveness value by multiplying the link effectiveness, data effectiveness, and node effectiveness with different weights. The simulation results are shown in Fig. 4, where the proposed Q-learning-based routing scheme performs better than the normal approaches. This results from the Q table recording the routing strategy learned by the current node, and every time the node is selected, the routing algorithm based on Q-learning will be selected according to the Q table. Therefore, we can infer that our proposed Q-learning-based scheme can perform stable and secure data routing in the UAV network.

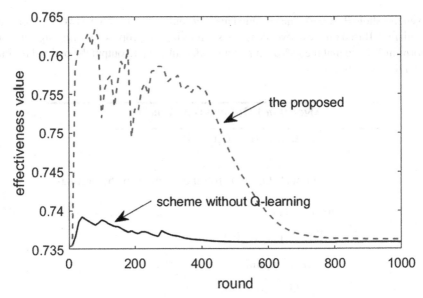

Fig. 4. Evaluation result

5   Conclusion

As an important part of modern space combat, the avionics system integrating navigation, search, and communication functions, especially the multiplatform avionics system,
realizes the effective coordination and deployment of complex air combat tasks by integrating the equipment and electronic system of each platform and making full use of
the characteristics of each platform. To guarantee secure and stable data communication in the multiplatform avionics system of the UAV network, we take link stability,
link quality, node quality, node honesty, and data availability into account and propose
an integrated communication effectiveness metric to quantify the communication links
among the UAV network. We propose a routing scheme based on hop-by-hop routing and Q-learning based on the proposed effectiveness metric. The evaluation results
demonstrate that our proposed scheme can support dependable data delivery in the UAV
network.

References

1. Sun, W., Yuan, X., Wang, J., Li, Q., Chen, L., Mu, D.: End-to-end data delivery reliability
   model for estimating and optimizing the link quality of industrial WSNs. IEEE Trans. Autom.
   Sci. Eng. **15**(3), 1127–1137 (2018)
2. Wen, J., Dargie, W.: Evaluation of the quality of aerial links in low-power wireless sensor
   networks. IEEE Sens. J. **21**(12), 13924–13934 (2021)
3. Dabiri, M.T., Rezaee, M., Yazdanian, V., Maham, B., Saad, W., Hong, C.S.: 3D channel
   characterization and performance analysis of UAV-assisted millimeter wave links. IEEE Trans.
   Wireless Commun. **20**(1), 110–125 (2021)

4. Zhao, Z., et al.: Exploiting link diversity for performance-aware and repeatable simulation in low-power wireless networks. IEEE/ACM Trans. Networking **28**(6), 2545–2558 (2020)
5. Xu, C., Xiong, Z., Han, Z., Zhao, G., Yu, S.: Link Reliability-based adaptive routing for multilevel vehicular networks. IEEE Trans. Veh. Technol. **69**(10), 11771–11785 (2020)
6. Xia, H., Zhang, S., Li, Y., Pan, Z., Peng, X., Cheng, X.: An attack-resistant trust inference model for securing routing in vehicular ad hoc networks. IEEE Trans. Veh. Technol. **68**(7), 7108–7120 (2019)
7. Li, F., Guo, Z., Zhang, C., Li, W., Wang, Y.: ATM: an active-detection trust mechanism for VANETs based on blockchain. IEEE Trans. Veh. Technol. **70**(5), 4011–4021 (2021)
8. Huang, M., Liu, A., Xiong, N.N., Wu, J.: A UAV-assisted ubiquitous trust communication system in 5G and beyond networks. IEEE J. Sel. Areas Commun. **39**(11), 3444–3458 (2021)
9. Du, J., Han, G., Lin, C., Martínez-García, M.: LTrust: an adaptive trust model based on LSTM for underwater acoustic sensor networks. IEEE Trans. Wirel. Commun. (2022). https://doi.org/10.1109/TWC.2022.3157621
10. Du, J., Han, G., Lin, C., Martínez-García, M.: ITrust: an anomaly- resilient trust model based on isolation forest for underwater acoustic sensor networks. IEEE Trans. Mob. Comput. **21**(5), 1684–1696 (2022)
11. Cui, S., Wang, H., Xie, Y., et al.: Intelligent storage system of machine learning model based on task similarity. In: Zeng, J., Qin, P., Jing, W., Song, X., Lu, Z. (eds.) International Conference of Pioneering Computer Scientists, Engineers and Educators, vol. 1451, pp. 119–124. Springer, Singapore (2021). https://doi.org/10.1007/978-981-16-5940-9_9
12. Shu, Y., Ma, Z., Liu, H., et al.: An analysis and validation toolkit to support the undergraduate course of computer organization and architecture. In: Zeng, J., Qin, P., Jing, W., Song, X., Lu, Z. (eds.) International Conference of Pioneering Computer Scientists, Engineers and Educators, vol. 1452, pp. 465–474. Springer, Singapore (2021). https://doi.org/10.1007/978-981-16-5943-0_38
13. Wu, Y., Li, Z., Li, Y., et al.: Teaching reform and research of data structure course based on BOPPPS model and rain classroom. In: Zeng, J., Qin, P., Jing, W., Song, X., Lu, Z. (eds.) International Conference of Pioneering Computer Scientists, Engineers and Educators, ICPCSEE 2021, pp. 410–418. Springer, Singapore (2021). https://doi.org/10.1007/978-981-16-5943-0_33
14. Sheng, R., Wang, Y., Huang, L.: Intelligent service robot for high-speed railway passengers. In: Zeng, J., Qin, P., Jing, W., Song, X., Lu, Z. (eds.) International Conference of Pioneering Computer Scientists, Engineers and Educators, vol. 1452, pp. 263–271. Springer, Singapore (2021). https://doi.org/10.1007/978-981-16-5943-0_21
15. Zhao, T., Jin, L., Jia, Y.: Prediction of enzyme species by graph neural network. In: Zeng, J., Qin, P., Jing, W., Song, X., Lu, Z. (eds.) International Conference of Pioneering Computer Scientists, Engineers and Educators, pp. 283–292. Springer, Singapore (2021). https://doi.org/10.1007/978-981-16-5943-0_23
16. Lin, C., Han, G., Shah, S.B.H., et al.: Integrating mobile edge computing into unmanned aerial vehicle networks: an SDN-enabled architecture. IEEE Internet Things Mag. **4**(4), 18–23 (2021)
17. Qin, Z., Liu, Z., Han, G., et al.: Distributed UAV-BSs trajectory optimization for user-level fair communication service with multi-agent deep reinforcement learning. IEEE Trans. Veh. Technol. **70**(12), 12290–12301 (2021)
18. Liu, X., Lai, B., Gou, L., et al.: Joint resource optimization for UAV-enabled multichannel Internet of Things based on intelligent fog computing. IEEE Trans. Netw. Sci. Eng. **8**(4), 2814–2824 (2020)
19. Jiang, J., Han, G.: Routing protocols for unmanned aerial vehicles. IEEE Commun. Mag. **56**(1), 58–63 (2018)

20. Lin, C., Han, G., Qi, X., et al.: Energy-optimal data collection for unmanned aerial vehicle-aided industrial wireless sensor network-based agricultural monitoring system: a clustering compressed sampling approach. IEEE Trans. Industr. Inf. 17(6), 4411–4420 (2020)
21. Chen, X., Bi, Y., Han, G., et al.: Distributed computation offloading and trajectory optimization in Multi-UAV-enabled edge computing. IEEE Internet Things J. 1 (2022)
22. Osco, L.P., Junior, J.M., Ramos, A.P.M., et al.: A review on deep learning in UAV remote sensing. Int. J. Appl. Earth Obs. Geoinf. 102, 102456 (2021)
23. Pham, Q.V., Zeng, M., Ruby, R., et al.: UAV communications for sustainable federated learning. IEEE Trans. Veh. Technol. 70(4), 3944–3948 (2021)
24. Wang, Z., Zhou, W., Chen, L., et al.: An adaptive deep learning-based UAV receiver design for coded MIMO with correlated noise. Phys. Commun. 47, 101365 (2021)
25. Tang, S., Zhou, W., Chen, L., et al.: Battery-constrained federated edge learning in UAV-enabled IoT for B5G/6G networks. Phys. Commun. 47, 101381 (2021)
26. Fu, S., Tang, Y., Wu, Y., et al.: Energy-efficient UAV-enabled data collection via wireless charging: a reinforcement learning approach. IEEE Internet Things J. 8(12), 10209–10219 (2021)

Intelligent Scheduling Strategies for Computing Power Resources in Heterogeneous Edge Networks

Zhixiang Ji[1(✉)], Jie Zhang[2], and Xiaohui Wang[1]

[1] China Electric Power Research Institute Co. Ltd., Beijing 100192, China
jizhixiang@epri.sgcc.com.cn
[2] Sichuan Power Research Institute SGCC, Chengdu 610041, China

Abstract. The edge computing model enables real-time and low-power processing of data, while contributing to data security and privacy protection. However, the heterogeneity and diversity of edge computing devices pose a great challenge to task scheduling and migration. Most of the existing studies only consider the allocation of computational resources, but lack comprehensive consideration of data resources, storage space, etc. In this paper, we proposed intelligent scheduling strategies for computing power resources in heterogeneous edge networks. We define the relevant models and construct a comprehensive matching matrix in terms of task matching with computing resources, data resources, storage resources, load balancing of computing devices and storage space matching, and design an intelligent scheduling algorithm based on iteration and load balancing according to the matching degree of tasks and computing devices in the heterogeneous edge network environment. The iterative and load-balanced scheduling algorithm is based on the least-cost flow solution scheduling strategy, which effectively reduces the task computation response time and improves the computation and storage resource utilization of computing devices. Experimental validation of the proposed intelligent scheduling strategy is carried out based on a simulation environment. The experimental results show that the proposed intelligent scheduling strategy has obvious advantages over random scheduling methods in terms of task processing delay, computing power resource utilization and number of satisfactory tasks.

Keywords: Edge computing · Resource heterogeneity · Task scheduling · System simulation

1 Introduction

With the rapid development of mobile Internet, an increasing number of smart devices constitute edge networks and generate huge amounts of data [1]. If all the data generated by edge devices are centralized and run in the center of cloud servers, not only does it take more time to transfer the data, it also generates great load pressure on the cloud servers, and there are also data security and privacy leakage problems when the data are transmitted in the network [2], while the proposal of edge computing presents an optional solution to solve the above problems [3]. Edge computing makes full use of edge devices distributed in different places to schedule computational tasks to the immediately adjacent computing devices for execution, which on the one hand reduces the load pressure on cloud servers and on the other hand improves task processing time and facilitates data security.

In an edge network, the time overhead of task execution depends on two main aspects: the time overhead of data transmission on the one hand, and the time overhead of task execution on the edge computing devices on the other. With the massive deployment of edge devices and smart terminals, although the load pressure on cloud-centric servers and the network bandwidth shortage problem have been alleviated to some extent, there is an inevitable tendency for computing resources to be deployed in a panoptic manner, leading to the emergence of the computing power silo effect. On the one hand, although some edge computing nodes have a high load, other computing nodes in the same edge network may have a low load or even be idle, resulting in underutilization of computing resources. On the other hand, some computing tasks, especially computationally intensive tasks, usually require high computing power, and it is difficult for a single computing node to meet the computing power requirements of the task, while the current edge network computing devices usually suffer from a lack of collaboration.

In the edge computing environment, how to perform efficient scheduling of computational tasks is a popular research problem in the field of edge computing and an urgent problem to be solved. Resources in edge networks can be classified into three types, namely communication resources, storage resources (also known as cache resources) and computational resources [4–6].

In edge computing networks, edge computing devices have obvious disadvantages compared to cloud computing central servers in terms of computing power, storage space, network bandwidth, etc. How to effectively schedule the limited resources to improve the utilization of computing devices is the key to the smooth execution of tasks in edge computing environments. In an edge computing environment, there are usually large differences in the hardware and performance of computing devices, such as processors involving different types of CPUs and GPUs, while there are also large differences in the number of cores, processing power and cache of processors [7–9], which also makes the edge computing network usually a heterogeneous computing network. With the diversification of computing tasks, the execution of a computing task may require multiple computing resources of different types and processing powers, which are often difficult to satisfy by a single edge computing device. Thus, the heterogeneity of edge computing devices, the limited computing resources, and the high demand of computing tasks on computing resources make the task scheduling of computing resources in heterogeneous edge networks challenging.

Some current studies model the edge computing network task scheduling problem as a multitask multiobjective optimization problem, and use linear programming [10, 11], graph theory [12, 13], machine learning [14, 15] and other methods to solve the problem. Although the utilization efficiency of computing resources is somehow improved, the global consideration of computing device resources is usually lacking, and the main consideration is the utilization improvement of computing capacity, while the space resource utilization of computing devices and network resource utilization are not considered.

The paper proposes an intelligent task scheduling strategy that takes into account the utilization of computing resources, space resources and network resources of edge computing devices, defines a model and constructs an evaluation matrix for the matching degree of tasks and computing resources, data resources and storage resources, load balancing of computing devices and matching degree of storage space, and constructs an optimization model for the computing resources of heterogeneous edge network environments. The iterative and load balancing-based intelligent scheduling algorithm is designed based on the matching degree of tasks and computing devices in the heterogeneous edge network environment, which effectively reduces the task computation response time and improves the computation and storage resource utilization of computing devices.

2 Modeling of Intelligent Scheduling of Algorithmic Networks

Unlike general computing resource scheduling, the most important feature of heterogeneous edge computing network arithmetic scheduling is how to highly match diverse computing tasks with heterogeneous computing resources to maximize the utilization of computing resources and thus reduce the task response time. The arithmetic network intelligent scheduling algorithm proposed in this section takes into account the arithmetic resources, storage resources and data resources of edge computing devices, matches the computing devices that meet the task requirements according to the task requirements for arithmetic resources, storage resources and data resources, and optimizes the bandwidth allocation among devices to improve data transmission efficiency by taking into account the distribution of arithmetic network communication bandwidth.

2.1 Task Model

Assume that the heterogeneous edge arithmetic network has U user tasks, and the set of user tasks is denoted by Ψ, $\Psi = \{1,2,...,U\}$, with $u \in \Psi$ being a task.

Each task u of the heterogeneous edge computing network is represented by the quaternion $u(M_u, N_u, D_u, V_u)$, where Mu denotes the demand matrix of the floating-point operations of different types of computing resources by task u, N_u denotes the demand for the number of cores of different types of computing resources by task u, D_u denotes the scale matrix of the data demanded by task u for the data owned by each device, and V_u denotes the scale matrix of the data to be transmitted by the end device initiating task u to V_u denotes the data size matrix to be transferred from the end device initiating task u to each computing device.

2.2 Computing Resource Model

The computing resources of heterogeneous edge computing networks involve heterogeneous hardware devices such as CPUs and GPUs. When scheduling tasks, it is important to consider whether the CPUs, GPUs and other computing resources of each computing device meet the task requirements, and if they do, the task can be scheduled to that computing device; otherwise, the task cannot be scheduled to that device.

In edge computing scenarios, many computing tasks will involve tasks such as artificial intelligence and scientific computing, where the computation time of the task is directly related to the processor's floating-point computing processing power. In this paper, the floating-point operations (FLOP) of a task on a single core are used as the computational demand metric, considering that the operation of a task may require multiple cores; therefore, the demand of a task on the number of cores of different types of processors should also be considered. The matrices Mu and Nu are used to represent the demand matrix of task u for different types of computing resources for floating-point operations and the demand for different types of computing resources for the number of cores, respectively, as follows:

$$M_u = [m_{u1}, m_{u2}, \cdots, m_{uk}] \tag{1}$$

$$N_u = [n_{u1}, n_{u2}, \cdots, n_{uk}] \tag{2}$$

where k is the number of heterogeneous processor types, m_{uj} is the FLOPs demanded by task u on class j computing resources, and n_{uj} is the maximum number of cores demanded by task u on class j computing resources. If task u does not require class j resources, both m_{uj} and n_{uj} values are zero.

In this paper, we consider the abstraction of heterogeneous computing devices and use floating-point operations per second (FLOPs) as the performance metric for a single core of the processor. This metric, which can mask the differences between different hardware devices, is applicable to heterogeneous edge computing environments. For each edge computing device, its performance is determined by the FLOPs and the number of cores of each type of processor on the device, and is represented by the matrices P_h and Q_h, respectively.

$$P_h = [p_{h1}, p_{h2}, \cdots, p_{hk}] \tag{3}$$

$$Q_h = [q_{h1}, q_{h2}, \cdots, q_{hk}] \tag{4}$$

where k is the number of computing resources of each type of computing device, p_{hj} represents the FLOPs of computing resources of type j on computing device h, and q_{hj} represents the number of cores of computing resources of type j on device h. If there are no computing resources of type j on computing device h, both p_{hj} and q_{hj} values are zero.

The number of cores currently available on device h is represented by the matrix Q_h', which, according to Eqs. (2) and (4), is expressed as follows:

$$Q_h' = Q_h - \sum_{u \in J_h} N_u \tag{5}$$

where J_h is the set of tasks running on the current device h and $\sum_{u \in J_h} N_u$ represents the set of occupied cores on the current device h.

When a certain computational resource currently available on device h cannot satisfy task u's demand for that resource, the computational response time of task u on device h is considered to be infinite when $N_u > Q_h'$ in real time. In this case, task u is not suitable for scheduling to device h. The ratio of the computation FLOP of task u to its occupied arithmetic resources is used as a prediction of the computation time of task u on target device h. The occupied arithmetic resources are obtained by multiplying the matrix N_u of the maximum number of cores occupied by task u with the matrix Ph of core arithmetic FLOPs owned by device h. The estimated computation time of task u on target device h, $t_{cal}(u,h)$, is expressed as follows.

$$t_{cal}(u, h) = \begin{cases} \frac{M_u}{N_u * P_h}, N_u \leq Q_{h'} \\ \infty, N_u > Q_h' \end{cases} \tag{6}$$

According to Eq. (6), the estimated computation time matrix T_{cal} for each task on the target device can be obtained.

2.3 Storage Resource Model

When the data required for task u are not distributed on any of the current edge computing devices, the data required for the computation must be transferred from the end node that initiated the task to the computing device. The storage space of each computing device is represented by the matrix S. The matrix S is represented as follows.

$$S = [s_1, s_2, \cdots, s_m] \tag{7}$$

where m is the number of computing devices within the edge computing environment and s_j denotes the storage space of the jth device. Considering that each device storage space may be partially occupied, the size of the storage space available for task u for each device is represented by the matrix S_u', which is expressed as follows:

$$S_u' = [S_{u1}, S_{u2}, \cdots, S_{um}] \tag{8}$$

where s_{uj} denotes the size of the storage space available to device j for task u.

For $\forall j \in E$, where E is the set of edge computing devices, $s_{uj} \leq s_j$. Assume that the size of the data to be transmitted to each computing device by the end device initiating task u is represented by the matrix V_u, where the matrix V_u is represented as follows:

$$V_u = [V_{u1}, V_{u2}, \cdots, V_{um}] \tag{9}$$

where v_{uj} denotes the size of the data transmitted by task u to device j.

Denote the communication capability matrix of the end device initiating task u by C_u, which is expressed as follows:

$$C_u = [C_{u1}, C_{u2}, \cdots, C_{um}] \tag{10}$$

where c_{uj} denotes the effective communication bandwidth between the terminal device initiating task u and the jth computing device. The data transmission response time t_{com_s} of the terminal device initiating task u to each computing device is the maximum value of the data transmission response time of this terminal device to each computing device. t_{com_s} is expressed as follows:

$$
\begin{aligned}
t_{\text{com}_s}(u, h, w') &= \max \frac{V_u}{W'} \\
&= \max_{j' \in E} \frac{V_{uj'}}{C_{uj'}}
\end{aligned}
\tag{11}
$$

where W' is the effective communication bandwidth matrix between the terminal device initiating task u and each computing device, c_{uj}' denotes the effective communication bandwidth between the terminal device initiating task u and the j'th computing device, and v_{uj}' denotes the size of the data transmission from the terminal device initiating task u to the j'th computing device.

2.4 Data Resource Model

In a heterogeneous edge computing scenario, the data required for a task may be scattered across different computing devices and require data transfer, and the time consumed for data transfer affects the total response time of the task. The data requirements of task u are represented by the matrix D_u, as follows:

$$D_u = [d_{u1}, d_{u2}, \cdots, d_{um}] \tag{12}$$

where m is the number of computing devices within the heterogeneous computing environment and d_{uj} indicates the size of the storage on the jth computing device for the data required by task u. If no data required by task u is stored on the jth device, the value of d_{uj} is 0.

In a heterogeneous computing environment, the communication bandwidth varies greatly between different devices, and the thesis uses the actual effective bandwidth of communication between devices as a measure of interdevice communication capability. The communication capability of device h with adjacent devices is represented by the matrix C_h as follows:

$$C_h = [C_{h1}, C_{h2}, \cdots, C_{hm}] \tag{13}$$

where c_{hj} denotes the current effective bandwidth of device h to the jth device. If device h is not adjacent to the jth device, the value of c_{hj} is 0. The bandwidth of device h to itself is represented by c_{hh}, which is expressed as the I/O rate of device h.

The time required for the data to be transferred from other edge computing devices to the current device while task u is executing on device h is the maximum time required for the data to be transferred from other edge computing devices to the current device while task u is executing on device h, expressed as follows:

$$t_{com\_d}(u, h, w) = \max Du/W \tag{14}$$

$$= \max_{j' \in E} \frac{d_{uj'}}{C_{hj'}}$$

where W is the effective bandwidth allocated by device h to task u for data transmission with other devices, $c_{hj'}$ is the effective bandwidth between device h and device j', and d_{hj}' is the storage size of the data required by task u on the j'th device. According to Eq. (14), the estimated data processing time for different tasks on each computing device can be calculated, which in turn leads to the estimated data response time matrix T_{com_d}.

2.5 Resource Matching Model

According to Eq. (6), the estimated data processing time of different tasks on each computing device can be calculated, which leads to the estimated processing response time matrix T_{cal}. According to Eq. (11), the data transfer response time from the end device initiating different tasks to each computing device can be calculated, which leads to the estimated data response time matrix T_{com_s}. According to Eq. (14), the data transfer time from other edge computing devices to the current device required for the execution of different tasks can be calculated T_{com_d}. The smaller the values of T_{cal}, T_{com_s} and T_{com_d} are, the better the resource matching between the task and the device. Normalize T_{cal}, T_{com_s} and T_{com_d} to obtain Γ_{cal}, Γ_{com_s} and Γ_{com_d} respectively, and calculate the difference between 1 and Γ_{cal}, Γ_{com_s} and Γ_{com_d} to obtain the normalized computational resource matching degree matrices A_{cal}, A_{com_s} and A_{com_d}, denoted as follows:

$$A_{cal} = 1 - \Gamma_{cal} \tag{15}$$

$$A_{com_s} = 1 - \Gamma com\_s \tag{16}$$

$$Acom_{d_{com}} = 1 - \Gamma com\_d \tag{17}$$

Each element in A_{cal} indicates the matching degree of computing resources between task u and computing device h, each element in A_{com_s} indicates the matching degree of spatial resources between task u and computing device h, and each element in A_{com_d} indicates the matching degree of data resources between task u and computing device h. The value magnitude of the elements in A_{cal}, A_{com_s} and A_{com_d} reflects the correlation between tasks and resources. Smaller element values indicate a better match between resources and tasks.

2.6 Equilibrium Model

In a heterogeneous computing environment, there may be a situation where some of the computing resources of a computing device are overoccupied and the remaining resources caused by other idle resources are greatly wasted. To avoid this situation, this paper measures the load balance of edge computing devices based on idea of balanced resource allocation and based on the computational resource utilization. It is known that task u occupies a matrix of computing resource cores as N_u, and the task on the current device h occupies a matrix of computing resource cores as denoted by N, where N' is denoted as follows:

$$N' = \sum_{u \in J_h} N_{u'} \tag{18}$$

Here J_h is the set of all tasks on device h. For device h, the resource utilization ξ after assigning task u to h is expressed as follows:

$$\xi = \frac{N_u + N'}{Q_h} \tag{19}$$

The variance ψ is found for the resource utilization rate ξ and is calculated as follows:

$$\Psi(u, h) = \begin{cases} D(\xi), N_u \leq Q_{h'} \\ \infty, N_u > Q'_h \end{cases} \tag{20}$$

The variance of the resource utilization of each device after being assigned to different tasks is calculated using Eq. (20) to obtain the variance matrix, denoted by Ψ. The normalization of Ψ is carried out and the normalization result Ψ' is made to differ from 1 to obtain the load balancing degree evaluation matrix $\mathcal{M}$, expressed as follows:

$$\Phi = 1 - \Psi' \tag{21}$$

Each element ϕ_{uh} in Φ indicates the degree of load matching between task u and device h computing resources, and a larger value represents a more balanced load after task u migrates to h.

2.7 Storage Space Matching Model

Resource matching and load balancing do not focus on the impact of available storage space on the edge computing device on task execution. In cases where tasks need to migrate data to the computing device, the relationship between the amount of storage space available on the computing device and the amount of space required by the task on the computing device needs to be considered, and the amount of space required by the task on the computing device cannot exceed the amount of storage space currently available on the computing device.

According to Eqs. (8) and (9), for $\forall j \in E$, if $v_{uj} \leq s_{uj}$, there is enough storage space on computing device j to store the data migrated from task u to device j and device j is available for task u storage space; otherwise device j is not available for task u storage space. Denote by B the task to computing the device storage space adaptation matrix, expressed as follows:

$$B = [b_{u1}, b_{u2}, \cdots, b_{um}] \tag{22}$$

Here, a b_{uj} of 1 indicates that computing device j can meet the storage space requirement of task u; otherwise, b_{uj} is 0. Considering the different remaining space sizes of different devices, tasks are dispatched to nodes with abundant storage space in priority provided that the devices meet the computing resources of task u. The ratio of the remaining available storage space of each edge computing device to the space required by task u, θ_{uj} is calculated using Eqs. (8) and (9) to construct a storage space availability matrix Θ. Normalizing Θ, a storage space matching degree matrix $\mathbb{Z}$ is constructed, expressed as follows:

$$\mathbb{Z} = \Theta * B \tag{23}$$

2.8 Integrated Matching Model

Different weights are assigned to the above five assessment matrices and the combined matching matrix $\mathcal{L}$ is obtained by weighted summation, as expressed below:

$$\mathcal{L} = \alpha A_{cal} + \beta A_{\mathrm{com}_s} + \gamma A_{\mathrm{com}_d} + \delta \Phi + \varepsilon \mathbb{Z} \tag{24}$$

Here, $\alpha + \beta + \gamma + \delta + \varepsilon = 1$, and α, β, γ, δ and ε are the weight coefficients for calculating the resource matching degree, data resource matching degree, storage resource matching degree, device load balancing degree and storage space matching degree respectively. The initial values of α, β, γ, δ and ε are flexibly selected for different application scenarios, and the above weight coefficients can be dynamically adjusted in the course of operation according to changes in load, storage space remaining, etc. The element l_{uj} in the integrated matching degree matrix $\mathcal{L}$ indicates the matching situation between task u and device j. The larger its cvalue is, the higher the integrated relevance of the task and device.

3 Intelligent Scheduling Algorithms for Algorithmic Networks

3.1 Intelligent Scheduling Optimization Model

On the basis of the above storage resource model and data resource model, the data transfer time $t_{com}(u,h)$ from task u to computing device h is expressed as follows:

$$t_{\mathrm{com}}(u, h) = t_{\mathrm{com}_d}(u, h, w) + t_{\mathrm{com}_s}\left(u, h, w'\right) \tag{25}$$

The total time overhead $t(u,h)$ for completing task u on computing device h is expressed as follows:

$$t(u, h) = t_{com}(u, h) + t_{cal}(u, h) \tag{26}$$

With the objective of minimizing the total time overhead to complete the task, the final task scheduling optimization problem is modeled in this paper as follows:

$$\min \frac{1}{U} \sum_{U \in \Psi} t(u, h) \tag{27}$$

$$\text{s.t.} \sum_{j' \in \Psi, j' \neq j} d_{uj'} + \sum_{u \in J_h} V_{uh} \leq s_h$$

$$\sum_{u \in J_h} n_{uj} \leq q_{hj}, j \in [1, k]$$

where J_h is the set of all tasks scheduled to computing device h, v_{uh} is the size of the data transferred to device h by the end device initiating task u, s_h denotes the size of the storage space available to device h, $d_{uj}{'}$ denotes the size of the data required by task u to be stored on the j'th computing device, and q_{hj} denotes the FLOPs of the jth class of computing resources on device h.

3.2 Intelligent Scheduling Algorithm

According to Eq. (26), to minimize the total time overhead of completing a task, the data transfer time and task computation response time need to be optimized. In this paper, we consider the optimization of bandwidth allocation based on an iterative approach to reduce data transfer time; and the scheduling of tasks based on the idea of load balancing to reduce task computation response time.

Based on the amount of data migration D_u required by task u, the maximum time t_{com_max} taken to transfer data to device j and the data transfer path r_{max} are calculated iteratively, where t_{com_max} is the sum of the maximum values of t_{com_d} and t_{com_s}. W_{used} is further optimally adjusted to reduce t_{com_max}, and an approximate optimal solution for bandwidth allocation is obtained when the difference between t_{com_max} and the optimal solution approaches a set threshold value δ. The intelligent scheduling algorithm proposed in this paper is specified as shown in Algorithm 1.

Algorithm 1. Intelligent scheduling algorithm based on iteration and load balancing.

Input: Set of tasks to be scheduled Ψ, set of edge computing devices E, arithmetic network communication topology G, used bandwidth matrix W_{used}, unused bandwidth matrix W_{idle}

Output: Task allocation set $\mathcal{U}$

1.　**for each** $u \in \Psi$ **do**

2.　　　Calculate t_{com_max} from W_{used};

3.　　　**if** $t_{com\_max} - t_{com\_min} < \delta$/* The threshold δ is a sufficiently small number */

4.　　　　　Break

5.　　　Update the W_{used} matrix;

6.　　　$t_{com\_min} \leftarrow t_{com\_max}$;

7.　**end for**

8.　Update the W matrix based on the W_{used} matrix; /* The W matrix is the communication bandwidth matrix between device j running task u and the other computing devices */

9.　Update the W_{idle} matrix based on the arithmetic network communication topology G and the W_{used} matrix;

10.　**for each** $h \in E$ **do**

11.　　　$Q_{h\_used} \leftarrow Q_h - Q_h'$;/* Obtain the matrix of used cores */

12.　　　**for each** $u \in \Psi$ **do**

13.　　　　　**If** $N_u < Q_h'$

14.　　　　　　$T_{cal}(u,h) \leftarrow M_u/(N_u * P_h)$; /* Task u calculates the elapsed time on device h */

15.　　　　　　$T_{com\_d}(u,h) \leftarrow D_u/W$; /* Time consuming transfer of data resources */

16.　　　　　　$T_{com\_s}(u,h) \leftarrow V_u/W$; /* Time consumption of data transfer at the terminal */

17.　　　　　　$$\psi(u,h) \leftarrow D(\frac{N_u + N'}{Q_h})$$

18.   ELSE

19. $T_{com\_d}(u,h)$, $T_{com\_s}(u,h)$, $\psi(u,h) \leftarrow \infty$;

19.   **end for**;

20. $A_{cal} \leftarrow 1 - \Gamma_{cal}$;

21. $A_{com\_s} \leftarrow 1 - \Gamma_{com\_s}$;

22. $A_{com\_d} \leftarrow 1 - \Gamma_{com\_d}$;

23. $\Phi \leftarrow 1 - \Psi'$;

24. $\mathbb{Z} \leftarrow \Theta * B$;

25. $\mathcal{L} = \alpha A_{cal} + \beta A_{com\_s} + \gamma A_{com\_d} + \delta \Phi + \varepsilon \mathbb{Z}$;

26. **Constructing a network flow diagram $\mathcal{F}$ based on $\mathcal{L}$**;

27. Calculating preliminary set $\mathfrak{R}$ of scheduling policies based on minimum cost flow on $\mathcal{F}$;

28. **for each** $<u,h> \in \mathfrak{R}$ **do**

29.   **if** $<u,h>$ and $\mho$ are not conflict

30. $\mho \leftarrow \mho \cup <u,h>$;

31. **end for**.

This algorithm first calculates t_{com_max} based on W_{used} for each task in the to-be-scheduled task set Ψ, updates the bandwidth matrix W_{used}, further obtains the communication bandwidth matrix between device j running task u and other computing devices, and then updates the W_{idle} matrix based on the arithmetic network communication topology graph G and the W_{used} matrix. Next, for each task-device combination, the integrated matching degree matrix $\mathcal{L}$ is computed, and finally, the current task-device combination is judged to be in conflict with the already assigned ones in the task assignment set $\mho$. If not, the task-device combination is added to the task assignment set $\mho$. After all the verification, the resulting $\mho$ is the set of task assignments.

4 Simulation Experiments

4.1 Simulation Scenarios and Parameter Settings

In the simulation experiments, an edge computing network consisting of end devices, edge computing devices and network routing is considered, as shown in Fig. 1. Among the simulated edge computing network, three terminal nodes, T1, T2 and T3, seven routing nodes, R1, R2, R3, R4, R5, R6 and R7, and four edge computing nodes, C1,

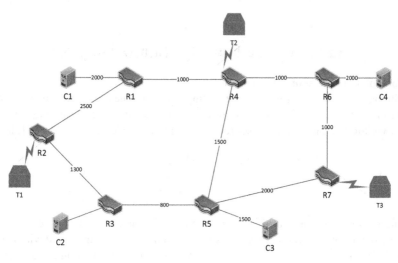

Fig. 1. Simulation scenes

C2, C3 and C4, are included. The connection between the devices is the communication bandwidth in Mbps.

Table 1 shows the specifications of the equipment parameters.

Table 1. Equipment parameter specification table

Device	Core category	Single core FLOPs	Number of cores
C1	CPU	2.1 G	6
	GPU	1.73 G	2560
C2	CPU	1.48 G	2
C3	CPU	2.56 G	4
C4	CPU	2.37 G	4
	GPU	1.5 G	1536

4.2 Evaluation Indicators

The method proposed in the thesis is mainly evaluated in terms of task processing latency, computing power resource utilization and number of satisfied users.

(1) Task processing latency.
Task processing latency includes data transmission latency and task computation latency, which are expressed as follows:

$$t(u,h) = t_{com_d} d(u,h,w) + t_{com_s}\left(u,h,w'\right) + t_{cal}(u,h) \tag{28}$$

(2) Computational Resource Utilization

The Storage resource utilization of computing nodes in a heterogeneous edge network is the ratio of the storage resources used for the task execution of computing devices in the network to the total storage resources of computing devices, expressed as follows:

$$S_E = \frac{\sum\limits_{u \in \Psi} (V_u + D_u)}{\sum\limits_{j \in [1,m]} s_j} \tag{29}$$

The compute resource utilization of a computing node in a heterogeneous edge network is the ratio of the compute resources used by the computing devices in the network for task execution to the total compute resources of the computing devices, expressed as follows:

$$C_E = \frac{\sum\limits_{u \in \Psi} (N_u)}{\sum\limits_{h \in E} Q_h} \tag{30}$$

(3) Number of satisfactory tasks

If the task obtains the required arithmetic resources and the required storage resources, the task is considered to be processed successfully; otherwise, it is considered to be processed unsuccessfully. This is expressed as follows:

$$I(u_2) = \begin{cases} 1, & \text{If the required storage resources are acquired} \\ 0 & \text{If the required storage resources are not acquired} \end{cases} \tag{31}$$

$$I(u) = I(u_1)I(u_2) \tag{32}$$

The number of satisfactory assignments is expressed as follows:

$$I_{satisfy} = \sum_{u \in \Psi} I(u) \tag{33}$$

4.3   Simulation Results and Analysis

In the experiments, the scheduling method proposed in the thesis, represented as ITLB, is analyzed in comparison with random scheduling considering load balancing and random scheduling without considering load balancing in different scenarios.

(1) Task processing time delay

The relationship between task processing latency and the total number of tasks in a heterogeneous edge computing network is shown in Fig. 2. As the number of tasks increases, the average latency of the three scheduling methods, ITLB scheduling, random scheduling considering load balancing, and random scheduling without considering load balancing, all show an increasing trend, and the greater the number of tasks, the greater their processing elapsed time keeps increasing. It is easy to see that with a fixed number of tasks, the average latency of the ICLB method proposed in the thesis is significantly shorter than that of the random scheduling method. As the number of tasks increases, the advantage of the ICLB method in terms of processing latency becomes more obvious. The random scheduling method, which considers load balancing, has an advantage in terms of processing latency compared to the method that does not consider load balancing.

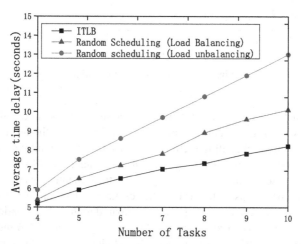

Fig. 2. Task processing latency versus number of tasks

(2) Computational resource utilization
The relationship between storage resource utilization and the number of tasks is shown in Fig. 3. It is easy to see that the storage resource utilization of the three scheduling methods, ITLB scheduling, random scheduling with load balancing and random scheduling without load balancing, all show an increasing trend as the number of tasks increases, mainly because the increase in tasks requires more task data to be migrated to the computing device, increasing the storage space occupation of the computing device. In the case of a fixed number of tasks, the ICLB method proposed in the thesis shows a significant increase in storage resource utilization compared to the random scheduling method. As the number of tasks increases, the advantage of the ICLB method in terms of storage resource utilization becomes more obvious. Among the random scheduling methods, the method that considers load balancing has an advantage in storage resource utilization compared to the method that does not consider load balancing.

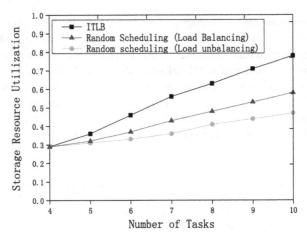

Fig. 3. Storage resource utilization

The relationship between the utilization of computing resources and the number of tasks is shown in Fig. 4. As the number of tasks increases, the arithmetic resource utilization of the three scheduling methods, ITLB scheduling, random scheduling with load balancing and random scheduling without load balancing, all show an increasing trend, mainly because the increase in tasks increases the utilization of computing devices. With a fixed number of tasks, the arithmetic resource utilization of the ICLB method proposed in the thesis increases significantly compared to that of the random scheduling method. As the number of tasks increases, the advantage of the ICLB method in terms of arithmetic resource utilization becomes more obvious. Among the random scheduling methods, the method that considers load balancing has an advantage in arithmetic resource utilization compared to the method that does not consider load balancing.

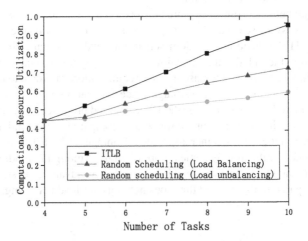

Fig. 4. Computational resource utilization

(3) Number of satisfactory tasks

The relationship between the number of satisfactory tasks and the total number of tasks is shown in Fig. 5. It can be seen that in terms of the number of satisfactory tasks, the ITLB method proposed in the thesis is able to perform the tasks better than the random scheduling method. As the number of tasks increases, the ITLB method proposed in the thesis is able to process the tasks better, while the random scheduling method is unable to process more tasks better due to the possibility of falling into resource exhaustion of some computing devices.

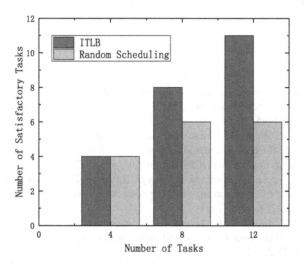

Fig. 5. The relationship between the number of satisfactory tasks and the total number of tasks

5 Conclusion

In this paper, we study the arithmetic resource scheduling problem of a heterogeneous edge network environment, define the relevant model and construct the relevant evaluation matrix in terms of task and arithmetic resources, data resources, storage resource matching degree, computing device load balancing and storage space matching degree, calculate the comprehensive matching degree matrix by setting weight coefficients, construct an optimization model for the arithmetic resource scheduling problem of a heterogeneous edge network environment, and design an intelligent scheduling algorithm based on iteration and load balancing according to the matching degree of heterogeneous edges. The intelligent scheduling algorithm based on iteration and load balancing is designed based on the matching degree of tasks and computing devices in the heterogeneous edge network environment, and the scheduling strategy is solved based on the minimum cost flow, which effectively reduces the response time of task computation and improves the utilization of computing and storage resources of computing devices. Experimental validation of the intelligent scheduling strategy proposed in the thesis is carried out based on a simulation environment. The experimental results show that the intelligent scheduling strategy proposed in the thesis has obvious advantages in terms of task processing delay, computing power resource utilization and number of satisfactory tasks compared with random scheduling methods. The next step is to consider further research on the setting of matching degree weight coefficients and task scheduling fairness for different application scenarios.

Acknowledgments. This work was supported by the Science and Technology Project of State Grid Corporation "Research on Key Technologies of Power Artificial Intelligence Open Platform" (5700-202155260A-0-0-00).

References

1. Naveen, S., Kounte, M.R.: Key technologies and challenges in IoT edge computing. In: 2019 Third International conference on I-SMAC (IoT in Social, Mobile, Analytics and Cloud) (I-SMAC), pp. 61–65 (2019)
2. Yu, R, Zhang, X., Zhang, M.: Smart home security analysis system based on the Internet of Things. In: 2021 IEEE 2nd International Conference on Big Data, Artificial Intelligence and Internet of Things Engineering (ICBAIE), pp. 596–599 (2021)
3. Goudarzi, M., Wu, H., Palaniswami, M., Buyya, R.: An application placement technique for concurrent IoT applications in edge and fog computing environments. IEEE Trans. Mobile Comput. **20**(4), 1298–1311 (2021)
4. Tan, Z., Yu, F.R., Li, X., Ji, H., Leung, V.C.: Virtual resource allocation for heterogeneous services in full duplex-enabled scans with mobile edge computing and caching. IEEE Trans. Veh. Technol. **67**(2), 1794–1808 (2017)
5. Wang, P., Yao, C., Zheng, Z., Sun, G., Song, L.: Joint task assignment, transmission, and computing resource allocation in multilayer mobile edge computing systems. IEEE Internet Things J. **6**(2), 2872–2884 (2018)
6. Tran, T.X., Pompili, D.: Joint task offloading and resource allocation for multiserver mobile-edge computing networks. IEEE Trans. Veh. Technol. **68**(1), 856–868 (2019)
7. Auluck, N., Azim, A., Fizza, K.: Improving the schedulability of real-time tasks using fog computing. IEEE Trans. Serv. Comput. (2019)
8. Mehrabi, M., You, D., Latzko, V., Salah, H., Reisslein, M., Fitzek, F.H.P.: De-vice-enhanced MEC: multiaccess edge computing (MEC) aided by end device computation and caching: a survey. IEEE Access **7**, 166079–166108 (2019)
9. Chen, W., Wang, D., Li, K.: Multiuser multitask computation offloading in green mobile edge cloud computing. IEEE Trans. Serv. Comput. **12**(5), 726–738 (2019)
10. Balcazar, E.H., Cerda, J., Avalos, A.: A validation method to integrate non linear non convex constraints into linear programs. In: 2018 IEEE International Autumn Meeting on Power, Electronics and Computing (ROPEC), pp. 1–7 (2018)
11. Kuendee, P., Janjarassuk, U.: A comparative study of mixed-integer linear pro-gramming and genetic algorithms for solving binary problems. In: 2018 5th International Conference on Industrial Engineering and Applications (ICIEA), pp. 284–288 (2018)
12. Mohammad, C.W., Shahid, M., Husain, S.Z.: A graph theory based algorithm for the computation of cyclomatic complexity of software requirements. In: 2017 International Conference on Computing, Communication and Automation (ICCCA), pp. 881–886 (2017)

13. Susymary, J., Lawrance, R.: Graph theory analysis of protein-protein interaction network and graph based clustering of proteins linked with Zika virus using MCL algorithm. In: 2017 International Conference on Circuit, Power and Computing Technologies (ICCPCT), pp. 1–7 (2017)
14. Lu, M., Li, F.: Survey on lie group machine learning. Big Data Mining Analyt. 3(4), 235–258 (2020)
15. Kalinina, E.A., Khitrov, G.M.: A linear algebra approach to some problems of graph theory. Comput. Sci. Inf. Technol. **2017**, 5–8 (2017)

A Survey of Detection Methods for Software Use-After-Free Vulnerability

Faming Lu, Mengfan Tang, Yunxia Bao[✉], and Xiaoyu Wang

Shandong University of Science and Technology, Qingdao 266590, Shandong, China
baoyunxia98@163.com

Abstract. Due to the absence of validity detection on pointers and automatic memory rubbish reclaim mechanisms in programming languages such as the C/C++ language, software developed in these languages may have many memory safety vulnerabilities, such as Use-After-Free (UAF) vulnerability. An UAF vulnerability occurs when a memory object has been freed, but it can still be accessed through a dangling pointer that points to the object before it is reclaimed. Since UAF vulnerabilities are frequently exploited by malware which may lead to memory data leakage or corruption, much research work has been carried out to detect UAF vulnerabilities. This paper investigates existing UAF detection methods. After comparing and categorizing these methods, an outlook on the future development of UAF detection methods is provided. This has an important reference value for subsequent research on UAF detection.

Keywords: Memory safety · Use-after-free vulnerability · Dangling pointer · Software concurrency defect

1 Introduction

In recent years, with the increasing development of IT technology, a growing number of software systems have been developed and applied. C/C++ is the mainstream language for developing system software due to its high execution efficiency, flexibility and compactness. For example, software such as FireFox, Google Chrome and Microsoft Windows are all developed in the C/C++ language. However, with the continuous expansion of the software scale and the increase in software development complexity, some drawbacks of the C/C++ language have been gradually revealed. For example, C/C++ lacks pointer validity detection and an automatic memory rubbish reclaim mechanism, making programmers prone to vulnerabilities related to memory security, such as Use-After-Free (UAF) vulnerabilities.

Spported by the Project supported by the National Natural Science Foundation, China (61602279), the Taishan Scholars Program of Shandong Province (No. ts20190936), the Excellent Youth Innovation Team Foundation of Shandong Higher School (2019KJN024), the Postdoctoral Innovation Foundation of Shandong Province (201603056), the Open Foundation of First Institute of Oceanography, China (2018002), and the Distinguished Teachers Training Plan Program of Shandong University of Science and Technology.

Y. Wang et al. (Eds.): ICPCSEE 2022, CCIS 1629, pp. 272–297, 2022.
https://doi.org/10.1007/978-981-19-5209-8_19

An UAF vulnerability arises, when a program accesses already freed memory through a dangling pointer, where the dangling pointer points to the object before it is reclaimed but is not modified or assigned a *null* value when the object is freed [7]. If an operating system reallocates the freed memory to another process, the original program re-refers to the current dangling pointer. Access to the memory object through the dangling pointer will corrupted or leaked the memory. For example, Chrome 78.0.3904.87, an update to Google's Chrome browser released in 2019, was targeted for two UAF vulnerabilities. One of them has been heavily used by hackers to hack and hijack computers. In fact, a month before that, Google had also released an emergency security update to fix another four UAF vulnerabilities. The most serious among them could even allow hackers to fully control the infected computer. As a result, UAF vulnerability has become one of the most common security vulnerabilities in Chrome. According to the CVE (Common Vulnerabilities and Exposures) database [15], the statistics of UAF vulnerabilities from 2006 to present are shown in Fig. 1. Obviously, the number of UAF vulnerabilities has grown rapidly and has remained at a high level since 2010.

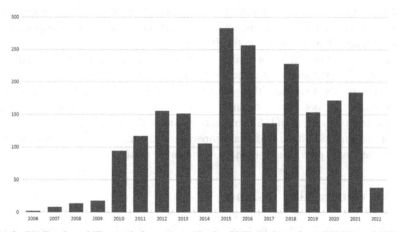

Fig. 1. UAF vulnerability statistics reported in the CVE database from 200601 to 202203.

It is difficult to detect UAF vulnerabilities because the *allocation, release,* and *use* of a memory object may be in completely different program modules. Some UAF vulnerabilities may only be triggered during specific thread scheduling processes. In view of the highly dangerous nature of UAF vulnerabilities and the difficulty of detection, various detection methods have been developed. Generally, UAF vulnerability detection methods are divided into three main categories: static detection methods [3, 10, 19, 32, 36, 39, 52], dynamic detection methods [22, 28, 30, 33, 41, 45, 50], and a combination of static and dynamic detection methods [6, 16, 21, 22].

Static detection methods take the software source code as input, which does not need to execute the program. Dynamic methods usually take the program behavior information captured during program execution as input. Their combined methods usually use static analysis to detect potential vulnerabilities at first and then verify the authenticity of the potential vulnerability through dynamic replay techniques. Each type of UAF vulnerability detection method has its own advantages and limitations. Static ones, such as the GUEB (Graph of Use-After-Free to Exploit Binary) [19] tool, can detect UAF vulnerabilities in large-scale and complex code with very few false negatives. However their lack of program runtime information causes many false positives. In contrast, dynamic detection techniques, whether based on taint tracking technology [12, 14, 17, 35, 37] or DSE (dynamic symbolic execution) [7–9, 20, 39, 40, 47], make full use of the program's runtime information and thus have few false positives. However, a single execution of a program can only reveal a small part of its behaviors, which usually leads to false negatives. For instance, AddressSanitizer [41] and KASAN (Kernel Address Sanitizer) [45] are representative of dynamic UAF vulnerability detection tools. The combination methods aim to combine both of their advantages. For example, "Detect use-after-free vulnerabilities in binaries" [23] and "Finding the Needle in the Heap" [18]. In the static detection section, literature [23] applies inter-procedural [2, 38] and points-to [36] based on static analysis in the static detection section, and literature [18] uses the GUEB static detection tool to detect potential vulnerabilities to reduce false negatives. In the dynamic detection part, both literatures use DSE for dynamic replay to identify real vulnerabilities to reduce false positives. However, the combined methods still need to reduce false positives, improve detection efficiency at the static analysis stage, and reduce false negatives at the dynamic stage.

To better carry out UAF vulnerability detection work, this paper provides an overview for existing UAF vulnerability detection methods, as well as their shortcomings and development trends. To achieve this goal, Sect. 2 presents an UAF vulnerability example and illustrates the root causes of such vulnerabilities. Section 3 summarizes typical existing detection methods of UAF vulnerability and analyzes their respective advantages and shortcomings. Section 4 makes a comparison and points to the directions for future research. Section 5 provides a summary of the entire thesis.

2   UAF Vulnerability Instance

Figure 2 shows a program containing a UAF vulnerability. It has two threads: *Thread1* and *Thread2*, one shared variable *x*, one lock object *l* and two pointers *p* and *q*. *Thread1* allocates memory for pointers *p* and *q* at first. Then, it starts *Thread2* and acquires lock *l* twice. For the first time, *Thread1* acquires lock *l* and releases the memory object referenced by *q*. For the second time, it acquires lock *l* and modifies the value of variable *x* to 1. After *Thread1* releases the lock *l* and releases memory object referenced by *p*, then it waits for *Thread2* to end. *Thread2* acquires lock *l* and writes a value to the memory object referenced by *q*. After *Thread2* writes a value to the memory object referenced by *p* when the value of *x* is 0. Finally, *Thread2* releases *l* and ends the thread.

The program has three different execution scenarios as follows. The first two trigger a UAF vulnerability. While the last one does not.

In the first scenario, *Thread1* first acquires lock *l* and releases the memory object referenced by *q*. Then, *Thread2* acquires *l* and assigns a value to the already released memory object through *q*. Thereby, a UAF vulnerability is triggered, which is constituted by the memory release operation on line 15 and the write operation to that object on line 26.

In the second scenario, *Thread1* acquires lock *l* for the first time and releases the memory object referenced by *q*. Afterwards *Thread1* acquires lock *l* for the second time, modifies the value of *x* to 1 and releases the memory object referenced by *p*. Then, *Thread2* acquires lock *l* and assigns a value to the already released memory object through *q*. Thereby, the UAF vulnerability identical to the first scenario is triggered. Finally, *Thread1* executes the judgment statement; at this time, the value of *x* has been modified to 1 by *Thread1*, so *Thread1* does not to execute the write operation on line 28, and there is no UAF vulnerability in this scenario. Then, the thread ends.

```
1  #inclode <stdio.h>
2  #include <stdlib.h>
3  #include <pthread.h>
4  #include <unistd.h>
5  int x=0;
6  int *p=NULL;
7  int *q=NULL;
8  pthread_mutex_t l=PTHREAD_MUTEX_INITIALIZER;
9  void *Thread1(void *arg){
10   p=(int*)malloc(sizeof(int));
11   q=(int*)malloc(sizeof(int));
12   pthread_t tid2;
13   pthread_create(&tid2,NULL,Thread2,NULL);
14   pthread_mutex_lock(&l);
15   free(q);
16   pthread_mutex_unlock(&l);
17   pthread_mutex_lock(&l);
18   x=4;
19   pthread_mutex_unlock(&l);
20   free(p);
21   pthread_join(tid2,NULL);
22   return 0;
23  }
24  void *Thread2(void *arg){
25   pthread_mutex_lock(&l);
26   *q=8;
27   if(x==0){
28     *p=6;
29   }
30   pthread_mutex_unlock(&l);
31   return 0;
32  }
```

Fig. 2. Program 1, which contains a UAF vulnerability.

In the third scenario, *Thread2* first acquires lock *l* and assigns a value to the memory object through *q*. Then, *Thread2* executes a judgment statement; at this point, the value of *x* is 0. Afterwards, *Thread2* executes the write operation on line 28 and *Thread2* ends. Then, *Thread1* acquires lock *l* and sequentially executes until to end. At this time, all the release operations in *Thread1* occur after the write operations in *Thread2*, so this scenario will have no UAF vulnerability.

For the above program, the next section illustrates the detection principles and results, points out their respective advantages and shortcomings, and gives directions for further research in combination with several detection methods.

3 Analysis of Existing UAF Detection Methods

This section introduces common UAF detection methods in combination with the program shown in Sect. 2. Several representative static methods, dynamic methods, and combined methods are analyzed in this paper.

3.1 Static UAF Detection Methods

Static detection methods [3] focus on the analysis of program source code through lexical, syntax, control flow and data flow analysis. They do not actually run the code. Static detection methods allow for a more comprehensive and thorough inspection of the program code, improving code coverage and providing all possible executions of the

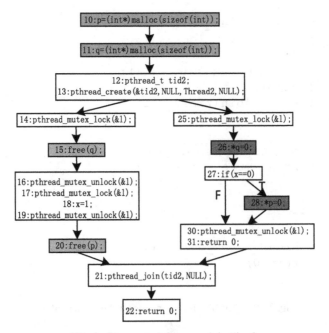

Fig. 3. The control-flow graph in Fig. 2.

program. For example, the static detection tool GUEB first uses dedicated value analysis to track heap operations and address transfers, taking aliases into account while doing so. Second, it uses these results to statically identify vulnerabilities in the UAF. Finally, it creates dataflow subgraphs for each UAF, describing in turn the situation where dangling pointers are created, released and used.

Figure 3 shows the CFG (control-flow graph) extracted from the program in Fig. 2. First, the program allocates memory chunks to pointers p and q on lines 10 and 11, respectively. Then, the pointers q and p are freed on lines 15 and 20. Finally, q and p are used on lines 26 and 28.

GUEB can identify the static paths in the CFG above and is able to continuously *allocate, free* and *use* the same memory object. First, it uses value set analysis (VSA) [4, 15] to locate the program point where heap elements are allocated or freed. In the static detection tool GUEB, *HE* is defined as the set of all heap elements. An element in the set *HE* is called a chunk denoted by *(base, size)*, where *base* is the allocation identifier of the heap element and *size* is the allocation size. For example, the allocation pointer p on line 10 is denoted by $(p, chunk_0)$. *PC* is defined as the set of all program points, and *HA* is defined as the set of all allocated elements at each program point. *HF* is the set of all freed elements, and *AbsEnv* holds a set of values associated with each memory address that may be contained at that address.

Table 1 shows the results of the analysis of the example in Fig. 2 (Scenario 2). In the first two allocations, two new chunks, $chunk_0$ and $chunk_1$, are created on lines 10 and 11. Then, two chunks are added to the set *HA*, and at this time, *HE* is empty. Then, the program executes the *free* operation on lines 15 and 20, and $chunk_0$ and $chunk_1$ are freed; thus, pointers p and q both become dangling pointers. Two chunks are removed from the set *HA* and added to the set *HF*. Finally, the program assigns a value to $chunk_1$ through q on line 26. The *write* operation on line 28 is not performed because $x = 1$ on line 18 is executed.

Table 1. VSA detection results.

Code	AbsEnv	Heap
10: p = (int*) malloc(sizeof(int));	$\{(p, chunk_0)\}$	HA = $\{chunk_0\}$; HF = $\emptyset$
11: q = (int*) malloc(sizeof(int));	$\{(q, chunk_1)\}$	HA = $\{chunk_0, chunk_1\}$; HF = $\emptyset$
15: free(q);	$\{(q, chunk_1)\}$	HA = $\{chunk_0\}$; HF = $\{chunk_1\}$
20: free(p);	$\{(p, chunk_0)\}$	HA = $\emptyset$; HF = $\{chunk_1, chunk_0\}$
26: *q = 0;	$\{(q, chunk_1)\}$	HA = $\emptyset$; HF = $\{chunk_1, chunk_0\}$

Based on the above results, GUEB defines *AccessHeap(pc)* to store all elements of the set *HE* that are accessed to program point *pc*. The implementation of GUEB takes REIL [16] as an intermediate representation. In REIL, two instructions are defined specifically for memory accesses:

– *LDM ad,, reg*, load the value Mem(ad) on the address to the register
– *STM reg,, ad*, store the value of the register to address Mem(ad)

As a result, GUEB defines *AccessHeap* as:

$$AccessHeap(LDMad, , reg) = AbsEnv(ad) \cap HE$$

$$AccessHeap(STMreg, , ad) = AbsEnv(ad) \cap HE$$

Finally, GUEB defines a detection formula $UafSet = \{(pc, chunk)|chunk \in (AccessHeap(pc) \cap HF(pc))\}$ to detect UAF vulnerabilities. For example, the program executes the *write* operation on line 26; this time, $AccessHeap(26) = \{chunk_1\}$ and $UafSet = \{(26, chunk_1)\}$, meaning that $chunk_1$ is dangling and dereferenced on line 26.

For the above results, the static detection tool GUEB extracts the corresponding data flow subgraph from the source code. This subgraph contains all the operation instructions that trigger the UAF vulnerabilities. In these subgraphs, all program points between the *allocate* point and the entry point of the program are marked in blue. All program points between the *free* point and the *allocate* point are marked in green. All program points between the *use* point and the *free* point are marked in orange, and all *use* points are marked in red.

Figure 4 shows a subgraph of the data flow extracted from the binary code for example (Fig. 2). This subgraph contains a path that triggers the UAF vulnerability in the example scenario 2. However, the above results are generated with a specified order of thread execution. There is uncertainty about the order of execution between threads (there are three execution scenarios in Fig. 2), so the analysis results are different for each scenario. For example, the analysis results for scenario 3 are not UAF vulnerability. Therefore, it requires an exhaustive scheduling order when using the static detection methods such as GUEB to detect multi-threaded programs, resulting in inefficient execution.

Another static detection tool, Pinpoint [42], applies points-to analysis, inter-procedural analysis and alias analysis to detect UAFs. It traces the flow of data by constructing a symbolic expression graph (SEG) and combining it with sparse value flow analysis (SVFA) [13, 29, 44] to obtain the value flow paths of possible UAF vulnerabilities. Pinpoint uses the SMT solver to determine the feasibility of the value flow paths to determine the real UAF vulnerabilities. The specific implementation process is as follows:

1. Constructing SEGs and Finding Value Flow Paths

Pinpoint proposes a new type of SVFG (sparse value flow graph), the SEG, which contains the necessary data dependencies and control dependencies (data and conditions related to the freed memory variables) in the program code and allows easy and efficient querying of "efficient path conditions" [44].

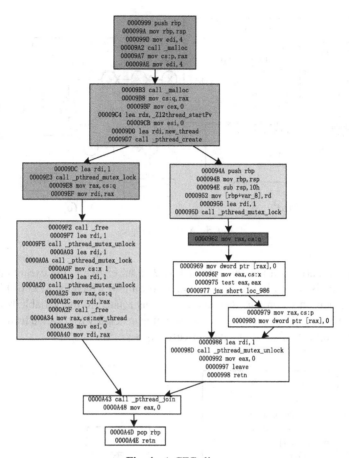

Fig. 4. A CFG slice.

As shown in Fig. 5, Fig. 5(a) presents a source code containing the UAF vulnerability, and Fig. 5(b) is the SEG of the *bar* function in Fig. 5(a). In the SEG, rectangles, circles and hexagons together form the vertex V. Rectangular vertices are denoted by $v@s$, indicating that the variable v is defined or used in the statement s; circular vertices are the set of all *Boolean* variables; hexagonal vertices are monadic or binary symbolic expressions. The SEG contains two directed edges, where the solid edge indicates a data dependency and the data on the edge indicates the condition for the data dependency to hold. The dashed edge indicates the control dependency, and the data on the edge indicate that $v@s$ holds only if these data are satisfied. If there are no data on the edge, the default is true.

To detect the UAF vulnerability, it is necessary to traverse the SEG starting from the vertex *free(c)* to obtain the complete value flow path $\langle$free(c), c, Y, returnY, L, f, print(*f)$\rangle$.

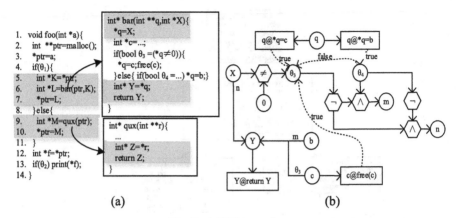

Fig. 5. A SEG example.

Taking Scenario 2 in the example of Fig. 2 as an example, the execution trace of its source code is shown in Fig. 6(a) below, according to which the SEG of Fig. 6(b) is obtained, which only contains data and conditions related to the variable being freed. To obtain the value flow path for this instance, it is necessary to traverse from *free(q)* to obtain the complete value stream path $\langle free(q), q, *q = 0 \rangle$.

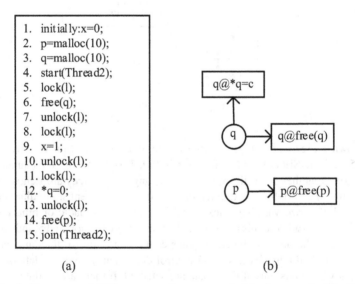

Fig. 6. The trace and SEG corresponding to scenario 2 in the example (Fig. 2).

2. Query "Efficient Path Condition" on SEG

As shown in Fig. 5(b), the valid path condition that c can be passed to Y is θ_3 because there is a data-dependent edge with a value of θ_3, whereas the valid path condition that the *returnY* statement can reach is *true* because there is no value for the data-dependent edge and no control-dependent edge exists. Therefore, given a value flow path, the basic idea of computing its "valid path condition" is to combine the data dependencies and control dependencies on that path. To do this pinpoint, it introduces two functions $DD(v@s)$ and $CD(v@s)$ to describe data dependence and control dependence, respectively. The path conditions are calculated as follows:

$$PC(\pi) = \bigwedge_{i=1\cdots n} CD(v_i@s_i) \wedge \bigwedge_{i=2\cdots n} (v_{i-1}@s_{i-1} = v_i@s_i) \wedge$$

$$\bigwedge_{i=2\cdots n} \mathcal{L}_d((v_{i-1}@s_{i-1} = v_i@s_i)) \wedge \bigwedge_{i=2\cdots n} DD(\mathcal{L}_d((v_{i-1}@s_{i-1} = v_i@s_i)))$$

$CD(v_i@s_i)$ in the above equation represents the conditions under which s_i is accessible at runtime. $v_{i-1}@s_{i-1} = v_i@s_i$ describes the transfer of values from line s_{i-1} to line s_i, and the remainder represents the feasible conditions for the flow of values from $v_{i-1}@s_{i-1}$ to $v_i@s_i$.

For example, a value flow path $\pi_2 = \langle \text{free}(c)@s_{20}, c@s_{19}, Y@s_{22}, \text{retY}@s_{23}\rangle$ can be obtained according to the local SEG shown in Fig. 5(b), which has the valid path condition:

$$PC(\pi_2) = (\theta_3@s_{18} = \text{true} \wedge x@s_{16} \neq 0 \wedge c@s_{19} = \text{free}(c)@s_{20} \wedge c@s_{19}$$
$$= Y@s_{22} \wedge Y@s_{22} = \text{retY}@s_{23}).$$

Additionally the value flow path $\pi_1 = \langle L@s_7, f@s_{12}, \text{print}(*f)@s_{13}\rangle$ is computed in the *foo* function, whose valid path condition is $PC(\pi_1) = (\theta_2@s_{13} = \text{true} \wedge \theta_1@s_4 = \text{true} \wedge L@s_7 = f@s_{12} \wedge f@s_{12} = \text{print}(*f)@s_{14})$. At this point two local valid path conditions are obtained.

The complete value flow path $\pi = \langle \text{free}(q)@s_6, q@s_3, *q = 0@s_{12}\rangle$ with respect to variable q is obtained according to the SEG shown in Fig. 6(b), and the valid path condition for this path is $PC(\pi) = (q@s_3 = \text{free}(q)@s_6 \wedge *q = 0@s_{12} = q@s_3)$.

3. Global Value Flow Analysis

The above path conditions formula can be obtained from the local valid path conditions, but if we want to detect UAF vulnerabilities, the value flow path needs to be considered globally. Therefore we need to calculate the path conditions of the global value flow path. At the same time, due to the existence of the local value flow path, its calculated path conditions may be miss its associated caller and callee constraints, and these missing constraints need to be restored before calculation of the path conditions of the global value flow path.

To determine which constraints are missing from the path condition, it is first rewritten as $PC(\pi)_R^P$, where P and R both denote the set of functions, with P representing the set of function parameters and R representing the set of return value receivers. Constraints are recovered using the following formula:

$$PC(\pi)_\varnothing^{P'} = PC(\pi)_R^P \wedge \bigwedge_{v_i@s_i \in R} \underset{(1)}{v_i@s_i = M(v_i@s_i) \wedge DD(M(v_i@s_i))_\varnothing^{Q_i}} \wedge$$

$$\bigwedge_{v_j@s_j \in Q_i} \underset{(3)}{v_j@s_j = M(v_j@s_j) \wedge DD(M(v_j@s_j))_\varnothing^{P_i}}$$

$PC(\pi)_R^P$ in the above equation indicates the local path condition obtained, (1) indicates that the return value recipient is equal to the corresponding return value, or a pair of formal and real parameters; (2) indicates the constraint on the range of return values restricted by the called function, i.e., the data dependency corresponding to the return value; (3) indicates the data dependency between real parameters.

The exact path condition calculation for the global value flow path can be performed once, and the missing constraint is restored and is calculated as follows:

$$PC(\pi_1\pi_2)_\varnothing^P = PC(\pi_1)_\varnothing^{P_1} \wedge \mathbf{PC(\pi_2)_\varnothing^{P_2}} \wedge v_n@s_n = u_1@r_1 \wedge$$

$$\bigwedge_{v_i@s_i \in P_2} v_i@s_i = M(v_i@s_i) \wedge DD(M(v_i@s_i))_\varnothing^{Q_i}$$

where $PC(\pi_1)_\varnothing^{P_1} \wedge \mathbf{PC(\pi_2)_\varnothing^{P_2}}$ denotes two local path conditions and the bolded part denotes the path condition of the called function; $u_1@r_1$ and $v_n@s_n$ are a pair of formal and real parameters or a pair of return values and their recipients, where $v_n@s_n$ belongs to π_1 as the main called function and $u_1@r_1$ belongs to π_2 as the called function.

The two local value flow paths π_1 and π_2 obtained in Fig. 5(b) lead to the global path condition of $PC(\pi_1\pi_2) = PC(\pi_1) \wedge PC(\pi_2) \wedge L@s_7 = retY@s_{23}$. Finally, the SMT solver is used to solve whether the path condition is satisfied, and if so, UAF vulnerability exists.

According to Fig. 6(b), the value flow path is complete, so the resulting path condition 4 can be solved directly using the SMT solver, and the result is "SAT", so there is a UAF vulnerability, in line with the analysis of Scenario 2.

Pinpoint based on points-to analysis, alias analysis and inter-procedural analysis has the same limitations as GUEB in that the above results are generated in scenarios where the order of execution of threads is specified, whereas multi-threaded detection can involve a number of different scenarios, each with a different order of execution and results, so using pinpoint also requires an exhaustive scheduling order, resulting in inefficient execution. Unlike GUEB, pinpoint has a more accurate inter-procedural analysis and can be used for large program detection with greater scalability.

Other static detection methods for UAF vulnerabilities include UAFChecker [26], which integrates classic static analysis techniques, alias analysis and inter-procedural analysis in the same way as pinpoint, while being based on taint analysis and symbolic execution. The Pointsto static detection tool [36] and the decompilation and data flow analysis-based approach [11] focus on applying data flow analysis to build data flow graphs to track the flow of data to detect UAF vulnerabilities.

Figure 7 shows the structure of UAFChecker, where the program source code is first taken as input and compiled into an LLVM IR using the LLVM front-end compiler. The LLVM IR is passed to the UAFChecker Core, which consists of five engines each performing a different analysis of the LLVM IR.

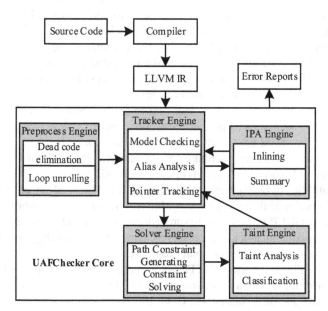

Fig. 7. Architecture of UAFChecker.

The Preprocess Engine is mainly used to eliminate dead code and expand loops, eliminating dead code improves the efficiency of static analysis, while expanding loops makes symbolic execution easier. Tracker Engine primarily applies precise alias analysis to determine the MustAlias set [24] and MayAlias set [24] of a pointer, during which a finite state machine is constructed to model UAF vulnerabilities. Freeing the memory pointed to by the pointer p in the "*start*" state will change the state to "*free*", and if a MustAlias set or MayAlias set using this pointer is detected in the free state, it is likely that a UAF vulnerability exists.

IPA Engine uses inter-procedural analysis in the same way as the pinpoint tool, inlining the callee into the caller in a bottom-up manner. However, inter-procedural analysis alone is not suitable for large programs and can result in call graphs that are too large to inline each function call and lack of precision. Some candidate UAF vulnerabilities are reported in the Tracker Engine, but not all are real, and this report is subject to false positives, so symbolic execution is applied in the Solver Engine to reduce false positives. The Solver Engine will call the SMT solver to check whether the path constraint from the release point to the point of use can be satisfied; thus, removing the vulnerabilities from the candidate vulnerabilities will never occur, reducing the false positives to a certain extent, but there will still be false positives. The last Taint Engine is designed to help programmers focus on the highest risk vulnerabilities.

UAFChecker integrates classical static analysis techniques, but alias analysis and inter-procedural analysis of large-scale programs can still introduce false positives, while using the SMT solver can only partially exclude contradictory constraints and reduce some of the false positives.

PointsTo and an approach based on decompilation and data flow analysis are similar to UAFChecker, again compiling the program source code into LLVM IR and applying data flow analysis and alias analysis on top of LLVM bitcode. The flow of pointers through the program is also tracked, including the assignment, copying, use and release of pointers. All pointer-related operations are represented in a data flow diagram, and finally the data flow diagram is searched to find the process from assignment to release to use, and if present, a UAF vulnerability is reported.

Both of these use alias analysis, but alias analysis introduces considerable false positives due to issues such as pointer renaming. Pointsto's view is to trade off false positives for misses to ensure that the program executes correctly.

The latest static detection literature "Scalable Static Detection of Use-After-Free Vulnerabilities in Binary Code" [52] proposes that UAFDetector is designed to detect UAF vulnerabilities in binary code. While GUEB is the most advanced static tool for detecting UAF vulnerabilities and has some binary code detection capabilities, its inlining technique in GUEB is used for inter-procedural analysis and multiple function calls are repeated, introducing much unnecessary system overhead, making this approach not very scalable and only suitable for small programs.

As shown in Fig. 8, the latest UAFDetector integrates two parts: pre-processing and detection, converting the binary code into an intermediate representation during the pre-processing phase and then building the CFG. Unlike GUEB, which uses the static tool IDA Pro to directly analyze the code to build the CFG, this method has a high code coverage. However, indirect jump addresses are not available, resulting in some areas of code that cannot be reached. In contrast, UAFDetector first uses a static method to obtain the basic control flow without indirect jumps, then it runs the target program, where it uses dynamic binary staking to obtain indirect jumps, and finally combines these two steps to obtain the complete CFG.

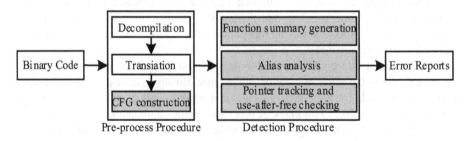

Fig. 8. Architecture of UAFDetector.

The detection phase traces pointer passing based on inter-procedural analysis and alias analysis to find the program path from release to use to identify the presence of a UAF vulnerability on the CFG. UAFDetector models an additional node, DESTROY, compared to GUEB and specifies that if there is a DESTROY node between a release node and a use node in a path, then the path is not vulnerable to UAF. Overall, UAFDetector has less system overhead and is more scalable than GUEB.

Static detection methods have low false negatives compared to dynamic detection. They do not increase the extra overhead of program execution and low system overhead because they do not require runtime information. However, static detection methods also have drawbacks. Static detection is mainly based on alias analysis, inter-procedural analysis, data flow analysis and points-to analysis. In general, static detection can lead to high false positives. Using static detection to detect vulnerabilities in multiple threads is inefficient due to the uncertainty of the scheduling order and the need to exhaust all scheduling orders.

3.2 Dynamic UAF Detection Methods

Dynamic detection methods usually take the program behavior information captured during program execution as input. AddressSanitizer [41] and KASAN (Kernel Address Sanitizer) [45] are representative tools for the dynamic detection of UAF vulnerabilities. AddressSanitizer is included in KASAN, which can only detect memory problems in user space, while KASAN is implemented in the Linux kernel, so KASAN can be seen as a kernel-space AddressSanitizer.

They use shadow memory to determine whether each byte of memory accessed in a program is safe, and after they insert shadow memory before each memory access during compilation, then they feed back detected error messages to identify and detect UAF vulnerabilities.

Shadow memory is extra memory used to mark the state of available memory. The corresponding shadow memory is checked each time, and the memory is *loaded/stored* to determine whether the memory is accessible. Converting the application address to its corresponding shadow address uses a direct map with a scale and offset. Given the memory address *Addr* of a program to be tested, the address of the shadow byte is calculated by using $(Addr >> 3) + Offset$. One byte of shadow memory stores 8 bytes of application memory; if all 8 bytes of program memory that is tested are accessible, the shadow value is 0; $n(1 <= n <= 7)$ indicates that the first n bytes are accessible; any negative value indicates that the entire 8-byte word is inaccessible. The shadow value is used to determine whether a memory operation is feasible:

$$ShadowAddr = (Addr >> 3) + Offset;$$

$$if\ (*ShadowAddr! = 0)$$

$$ReportAndCrash(Addr);$$

When an 8-byte memory access is detected, the address of the corresponding shadow byte is first calculated, and the byte is loaded. Then, AddressSanitizer checks whether the shadow value it is zero. Zero indicates that all 8 bytes of memory that is tested are accessible.

$$ShadowAddr = (Addr >> 3) + Offset;$$

$$k = *ShadowAddr;$$

$$If\,(k! = 0\&\&((Addr\&7) + AccessSize > k))$$

$$ReportAndCrash(Addr);$$

When a memory access of 1-, 2-, or 4-byte is detected, if the shadow value is a positive number k, we need to compare the last three bits of the memory address plus the access byte with k. If $< k$, the byte memory is accessible. Shadow memory is used for detection with low false positives.

AddressSanitizer was also combined with dynamic instrumentation [33, 43] technology. Dynamic instrumentation inserts probes into the program based on the original logic integrity of the procedure that is tested. The probes are essentially a code segment of information collection. The program executes the probes and throws out the characteristic data of the program running; then, we analyze the data to obtain information on the program's control flow or data flow. For the detection of UAF vulnerabilities, generally, the system needs to capture heap and pointer operations and then insert probes before and after the pointers are released and the memories are used. For example, AddressSanitizer's dynamic instrumentation replaces all *malloc* and *free* during program runtime. When the *malloc* function is called, it allocates memory A of a specified size and marks the memory around A (redzone) as *poisoned*. When the *free* function is called, the freed memory A is also marked as *poisoned*. All subsequent *read* and *write* to memory will determine whether the memory region is *poisoned*.

AddressSanitizer was used to test the program in Fig. 2, and the results are as follows:

```
==2256==ERROR: AddressSanitizer: heap-use-after-free on address 0x602000000030 a
t pc 0x56514f991ef3 bp 0x7f70ad2fdea0 sp 0x7f70ad2fde90
WRITE of size 4 at 0x602000000030 thread T2
SUMMARY: AddressSanitizer: heap-use-after-free /home/tang/Documents/UFO/uaf.cc:2
6 in Thread2(void*)
```

Fig. 9. The detection results of AddressAanitizer for the program in Fig. 2.

Figure 9 shows a *write* operation free memory on line 26 of *Thread2*, and the program is detected to have a UAF vulnerability.

Figure 10 shows another program containing a UAF vulnerability; there are two pointers p and q. The program first allocates memory to p, then assigns a value to the memory through p and releases the memory referenced by p, at which pointer p becomes a dangling pointer. After the program allocates memory to q, if $q! = p$, then the program enters the while loop. The loop the program releases the memory referenced by q at first, then it reallocates memory to q and judges the loop conditions again. $q = p$ at the end of the loop means that q is reallocated to the memory that has been released by p. Then the program assigns a value to the memory through q, and it outputs the value stored in the memory through p. This time, an operation on its free memory through pointer p results in a UAF.

```
1    #include <stdio.h>
2    #include <stdlib.h>
3    #include <unistd.h>
4    int *p;
5    int *q;
6    int main(){
7      p=(int*)malloc(sizeof(int));
8      *p=5;
9      free(p);
10     q=(int*)malloc(sizeof(int));
11     while(q!=p){
12        free(q);
13        q=(int*)malloc(sizeof(int));
14     }
15     *q=8;
16     printf("p %d\n",*p);
17     return 0;
18   }
```

Fig. 10. Program 2, which contains a UAF vulnerability.

Using AddressSanitizer to detect UAF vulnerability for the above program, the detection reports that there is no UAF vulnerability caused by the dangling pointer p.

The AddressSanitizer dynamic detection tool does not detect the UAF vulnerability on line 18. This UAF vulnerability is caused by a *read* operation on freed memory through the dangling pointer p, resulting in a false negative. The reason is that after freeing the memory space pointed to by p and q, when reallocating memory space to q, the memory space allocated is exactly the memory already freed by p. AddressSanitizer thinks that p has not been freed, so the UAF vulnerability on line 18 is not detected and the program executes correctly.

Another representative memory security detection tool, Valgrind [34], is composed of the kernel and other kernel-based debugging tools, including Memcheck, Cachegrind, Callgrind, Helgrind, Massid, and Extension. Valgrind's architecture is shown in Fig. 11 below, where memcheck is the most commonly used tool for detecting memory security issues in programs, and the UAF vulnerability discussed in this paper is detected by using memcheck.

The key to memcheck's ability to detect memory security is the creation of two global tables: Valid-Value and Valid-Address. Each byte in the process's entire address space corresponds to 8 bits, and each register in the CPU corresponds to slightly vector. The bits in the valid value table record whether the byte or register has a valid, initialized value. The bits in the valid value table record whether the byte or register has a valid and initialized value. Each byte of the address space also corresponds to a bit, and the Valid-Address table is used to record whether the address can be read or written.

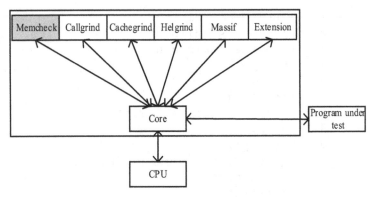

Fig. 11. Architecture of Valgrind.

When a byte in memory is read or written, the bit in the Valid-Address table corresponding to that byte is checked, and if it is shown to be invalid, memcheck reports a read/write error. When a byte in memory is loaded into the real CPU during program operation, the bit in the Valid-Value table corresponding to that byte is also loaded into the virtual CPU. When the value in that register is used to generate a memory address or to affect program output, memcheck checks the bit in the corresponding Valid-Value table, and if the corresponding memory address has been freed, memcheck will report a "read/write of freed memory block".

Using the memcheck tool in Valgrind to test the program example in Fig. 2, the following test results were obtained:

```
==15609== Thread 2:
==15609== Invalid write of size 4
==15609==    at 0x108A39: thread_start(void*) (uaf.cc:26)
==15609==    by 0x4E466DA: start_thread (pthread_create.c:463)
==15609==    by 0x517F61E: clone (clone.S:95)
==15609== Address 0x544f090 is 0 bytes inside a block of size 4 free'd
==15609==    at 0x4C32FA8: free (vg_replace_malloc.c:540)
==15609==    by 0x108A96: deallocate(void*) (uaf.cc:36)
==15609==    by 0x108B3C: main (uaf.cc:52)
==15609== Block was alloc'd at
==15609==    at 0x4C31DFB: malloc (vg_replace_malloc.c:309)
==15609==    by 0x108AB8: main (uaf.cc:41)
==15609==
==15609==
==15609== HEAP SUMMARY:
==15609==     in use at exit: 0 bytes in 0 blocks
==15609==   total heap usage: 3 allocs, 3 frees, 280 bytes allocated
==15609==
==15609== All heap blocks were freed -- no leaks are possible
==15609==
==15609== For lists of detected and suppressed errors, rerun with: -s
==15609== ERROR SUMMARY: 1 errors from 1 contexts (suppressed: 0 from 0)
```

Fig. 12. The detection results of Valgrind for the program in Fig. 2.

Figure 12 shows a *write* operation to free memory on line 26 of Thread2, and the program is detected to have a UAF vulnerability.

The example program in Fig. 10 was tested by using memcheck and the results were as follows (Fig. 13):

```
==15621==
p1 42
==15621==
==15621== HEAP SUMMARY:
==15621==     in use at exit: 4 bytes in 1 blocks
==15621==   total heap usage: 5,000,004 allocs, 5,000,003 frees, 20,001,036 bytes allocated
==15621==
==15621== LEAK SUMMARY:
==15621==    definitely lost: 0 bytes in 0 blocks
==15621==    indirectly lost: 0 bytes in 0 blocks
==15621==      possibly lost: 0 bytes in 0 blocks
==15621==    still reachable: 4 bytes in 1 blocks
==15621==         suppressed: 0 bytes in 0 blocks
==15621== Rerun with --leak-check=full to see details of leaked memory
==15621==
==15621== For lists of detected and suppressed errors, rerun with: -s
==15621== ERROR SUMMARY: 0 errors from 0 contexts (suppressed: 0 from 0)
```

Fig. 13. The detection results of Valgrind for the program in Fig. 10.

The memcheck tool in Valgrind also fails to detect a UAF vulnerability caused by a read operation on free memory via the dangling pointer p in line 18 of Fig. 10, so the dynamic detection tool Valgrind generates false negatives, which are caused by the same reallocation of memory as the AddressSanitizer.

AddressSanitizer, KASAN and Valgrind are all dynamic inspection tools, while Valgrind is simpler than the other two and therefore may not be able to resolve complex memory issues. Moreover, Valgrind is a programmatic tool and will be less efficient than the other two when inspecting memory safety. AddressSanitizer and KASAN, on the other hand, are compiled toolsets, that will be more efficient and have more accurate results. However, both are subject to false negatives in regard to detecting memory reallocation problems.

In addition to the memory detection tools mentioned above, there is a class of methods designed to avoid vulnerabilities by automatically cancelling all pointers to memory objects when they are detected as being freed. Examples include DangNull [28], FreeSentry [50], DangSan [27] and pSweeper [29].

DangNull identifies pointer passes based on taint tracking technology, automatically tracks the pointing relationship between the pointer and the memory object, and then cancels all pointers to the memory object when it is freed as a way to avoid UAF vulnerabilities. However, DangNull can only track pointers on heap objects and ignore pointers on the stack and global memory. At the same time, DangNull introduces significant memory overhead, with experiments showing that DangNull has a system overhead of up to 80%.

FreeSentry also applies taint tracking techniques to track pointers to memory objects and invalidate these pointers when memory is freed, thus mitigating the vulnerability. Unlike DangNull, FreeSentry can track all types of pointers and reduce the system overhead to 25%, but FreeSentry does not support multi-threaded program detection.

To complement these two approaches, researchers have proposed DangSan, which can detect multi-threaded programs and reduce the system overhead by half while extending to read and write to a large number of different types of pointers. However, DangSan requires locking when detecting multi-threaded programs to avoid competition from multiple threads of the application.

Based on the shortcomings of the above three methods, pSweeper has been proposed, which iteratively searches for dangling pointers in concurrent threads and uses object origin tracking (OOT) to invalidate dangling pointers. pSweeper uses an idle CPU to reduce the detection latency, so it consumes more CPU resources than the above three methods, but pSweeper will be more efficient.

The latest dynamic detection method, xTag [6], proposes a software-based pointer tagging scheme in which a 4-bit identifier is embedded in an allocated heap pointer, and this identifier is then additionally stored in shadow memory. At each memory access, the identifier embedded in the pointer is first verified to match the identifier stored in the disjoint shadow memory, and if it does not match, then the UAF vulnerability is considered to exist.

This method uses the same shadow memory to store logos as AddressSanitizer and KASAN, except that AddressSanitizer and KASAN map 8 bytes of heap memory to one byte of shadow memory, whereas xTag maps 16 bytes of heap memory to one byte. Therefore, xTag is more memory efficient than AddressSanitizer and KASAN and does not incur unreasonable overheads, making better use of the CPU's caching mechanism and reducing memory overheads.

xTag does not require close tracking of all pointers as opposed to the method of eliminating dangling pointers to avoid vulnerabilities. In the same way as DangNull, xTag also only tracks pointers on heap objects and ignores pointers on the stack and global memory, as xTag believes that most UAF vulnerabilities occur on pointers to heap memory. As DangNull is built on an older LLVM version (version 3.8) and xTag is based on a newer LLVM version, it is not possible to compare the performance of the two.

The advantages of using dynamic detection methods to detect security vulnerabilities are that there are fewer false positives and the operations are simple, without an in-depth understanding of the structure and implementation of the internal code of the program. The disadvantages are that if the code is complex and not understood, it is difficult to cover the entire code, resulting in low code coverage and false negatives. According to the above detection results, the dynamic detection tools AddressSanitizer, KASAN and Valgrind have a low false positive rate when used to detect UAF vulnerabilities but can generate false negatives due to pointer redistribution. DangNull, FreeSentry, DangSan and pSweeper, which focus on eliminating dangling pointers to avoid vulnerabilities, suffer from high system overhead, inability to detect multi-threaded programs and high resource consumption.

In summary, the static methods of detecting multi-threaded programs are inefficient due to the uncertainty in the order of execution between programs and the fact that static methods such as points-to analysis, alias analysis, inter-procedural analysis and data flow analysis can only be carried out if the scheduling order of threads is specified. The dynamic detection methods do not require the scheduling order of threads to be specified, and they execute the program to determine whether there is a memory safety problem, making them more efficient than static detection.

Static detection has a lower system overhead than dynamic detection because it does not require execution of programs and tracking of memory space usage. Dynamic detection methods based on pointer tagging, shadow memory, taint tracking techniques and dynamic symbolic execution require tracking the use of memory objects, tracking the pointing relationship between memory objects and pointers, and marking memory objects and pointers. Therefore, dynamic detection has a higher memory overhead than static detection.

In addition, static detection generates a high level of false positives due to the lack of information about the program runtime. However, because the static detection methods require a full and thorough inspection of the code, the code coverage is higher, resulting in a lower rate of false negatives. On the other hand, dynamic detection cannot cover the entire code situation by executing the code once, so some code information will be missing, resulting in false negatives. Experiments have shown that dynamic detection can also cause false negatives due to memory reallocation. However, because dynamic detection makes full use of runtime information such as memory usage and the relationship between memory and pointers, there are very few false positives.

3.3 Combination of Static and Dynamic UAF Detection Methods

The combination of static and dynamic detection methods aims to combine the advantages of both static and dynamic detection methods. First, using static detection is used to analyze as much code as possible to improve code coverage, thus more comprehensively uncovering hidden security vulnerabilities, providing more information for subsequent dynamic detection and reducing false negatives at the dynamic stage. Second, using dynamic detection is used to verify the true vulnerability and eliminate false positives at the static analysis stage.

Static detection and dynamic detection methods work together to better detect security vulnerabilities and reduce false negatives and false positives. Finding the Needle in the Heap [18] is a representative combination of mixed detection methods, where static detection is implemented using GUEB to find all possible UAF vulnerabilities and then verify the real UAF although dynamic symbolic execution.

Dynamic symbolic execution (DSE) [7–9, 20, 39, 40, 47] causes a program to follow a specific trace by analyzing the program in response to a specific input value. For each path p, a symbolic path predicate Φ_p is computed as a set of input constraints to the program. Thereby, the program is caused to follow that path during runtime, where the path predicates are conjunctions of all conditional branches c_i in that path p. The existence judgment statement is detected according to the steps. If it does not exist, the path is directly checked for oracles σ. The detection steps are as follows:

1. The program is executed from the initial input i_0 to generate the first path p_0. The path predicate Φ_{p0} is added to a working list WL with an initial value of empty;
2. Extract the path predicate $\Phi_p = c_1 \wedge c_2 \wedge \cdots \wedge c_n$ from WL to detect other branching statements. The branch conditions c_i in path predicate Φ_p are negated in turn, creating a new path predicate Φ_p';
3. The SMT solvers are used to solve the path predicate Φ_p'. If the solvers have a solution then output an input i. At this time, the trace t is obtained, and oracle σ determines whether the UAF exists for trace t. If it does, then return i. At this point, detect the UAF and stop retrieval.

4. If it does not exist or solver has no output, add Φ'_p to *WL* and resume at step 2 until *WL* is empty to stop retrieval.

A UAF vulnerability exists on execution trace t if and only if the following property Φ holds:

$$t = (..., n_{alloc}(size_{alloc}), ..., n_{free}(a_f), ..., n_{use}(a_u)) \tag{1}$$

$$a_f \text{ is a reaching definition of the } n_{alloc} \text{ return block.} \tag{2}$$

$$a_u \text{ is a reaching definition of an address in the } n_{alloc} \text{ return block.} \tag{3}$$

where n_{alloc}, n_{free}, and n_{use} denote *allocation, free* and *use* nodes, respectively, and $size_{alloc}$ denotes the allocation size.

To verify whether the above properties hold, definition Φ' denotes the negation of properties (2) and (3):

$$\Phi' = (a_f \neq S_{alloc}) \vee (a_u \notin [S_{alloc}, S_{alloc} + size_{alloc} - 1])$$

where S_{alloc} corresponds to the value returned by the allocation site n_{alloc}. S_{alloc} is used to detect if there exists of a data flow relationship between this address and the one used by n_{free}. $n_{use}.a_f \neq S_{alloc}$ indicates that the pointer used as the parameter for *free* is not a pointer allocated at n_{alloc}. $a_u \notin [S_{alloc}, S_{alloc} + size_{alloc} - 1]$ indicates that the pointer used by the *use* operation is not the reaching definition of the pointer allocated at n_{alloc}. Additionally, φ_t indicates a trace predicate for trace t.

Define oracle σ as $\varphi_t \wedge \Phi'$, if and only if on SMT $\varphi_t \wedge \Phi'$ is *UNSAT*, indicating that properties (2) and (3) are satisfied: a UAF is present.

Take the CFG slice path in Fig. 4 as an example. In this slice allocation, *free* and *use* do not involve conditional branches. The trace shown in Fig. 14 can be directly $\varphi_t \wedge \Phi'$ judged as follows:

$\varphi_t = (q_0 = S_{alloc})$, $\Phi' = (q_0 \neq S_{alloc}) \vee \neg(S_{alloc} \leq q_0 \leq S_{alloc} + 4)$, then $\varphi_t \wedge \Phi'$ is UNSAT: a UAF is present.

The dynamic detection methods based on DSE also have the problem of execution inefficiency for the detection of multi-threaded programs. Due to the uncertainty of the multi-threaded execution order, exhausting all scheduling orders will cause a path explosion, leading to increased system overhead. The latest method, UFO [25], is mainly used for the detection of multi-threaded programs. Its detection process consists of two stages: online tracing of the program trace and offline analysis of the detection. It captures the trace of the program at runtime, then uses MaxModel to obtain the largest feasible set of traces, and it encodes the intra-thread and inter-thread causality in the trace and UAF conditions into constrains. Finally, UFO using the Z3 solver in SMT to determine the feasibility of the constraints and thus verifies UAF vulnerabilities. UFO is more efficient than "Finding the Needle in the Heap", and the system overhead of UFO is determined by the number of trace events; the larger the number is, the higher the overhead.

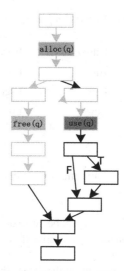

Fig. 14. The trace in Fig. 4.

In summary, static detection methods can improve code coverage by comprehensively screening program code, but the lack of runtime information makes it difficult to accurately analyze and reason about the program behavior; therefore, many false positives occur. Dynamic detection methods can accurately obtain the information of the program runtime and reduce false positives, but can also be largely false negatives due to low code coverage or memory reallocation. A combination of static and dynamic detection methods can be used and using dynamic methods to reduce false positives, using static methods to improve code coverage, but these methods usually increase system overhead. These combination methods are inefficient in detecting multi-threaded programs due to the uncertainty of the execution order and the need to exhaust the scheduling order.

4   Comparison of UAF Detection Methods and Outlook

As mentioned above, much research work has been carried out to detect UAF vulnerabilities. The comparison of various detection methods is shown in Table 2.

Among them, static detection methods can improve code coverage and reduce false negatives, do not require running programs and reduce system overhead. However static detection methods lack of runtime information, leading to a high false positives and detection inefficiency. Dynamic detection methods can provide runtime information and reduce false positives, but they have high false negatives due to low code coverage. The combination of static and dynamic detection methods uses static methods to improve code coverage and reduce false negatives caused by dynamic detection. Then, they use dynamic detection methods to replay to verify the real UAF vulnerabilities and reduce false positives. However, these methods usually introduce significant system overhead.

Table 2. The compare results.

Detection methods	False positives	False negatives	System overhead
Static detection methods	High	Low	Low
Dynamic detection methods	Low	High	Low
Combination of static and dynamic detection methods	Low	Low	High

However, the various UAF detection methods mentioned above currently have the following problems:

1. UAF vulnerability detection is inefficient because detection methods are often blind. Target-oriented UAF detection may be effective if they can target potential statements that could lead to UAF vulnerabilities;
2. Existing detection methods can accurately detect the specific lines of code where UAF vulnerabilities occur but do not complete automatic or semi-automatic repair of vulnerabilities;
3. Existing detection methods are mainly oriented toward serial programs for UAF detection, and there are execution inefficiencies and weak detection abilities when detecting parallel programs such as multi-threaded programs.

In view of the above shortcomings, our next research includes the following:

1. Improvement of detection efficiency: Target-oriented detection with specific, potentially UAF causes two statements as input to avoid the blindness of the detection method and improve the efficiency of UAF detection;
2. Vulnerability repair: Proposed methods can automatically repair UAF vulnerabilities while detecting them, improve the execution efficiency of software systems and reduce security problems;
3. Vulnerability detection for parallel programs: Proposed UAF detection methods for parallel programs to improve the efficiency of UAF vulnerability detection.

5 Conclusion

This paper provides an overview of existing UAF vulnerability detection methods and points out the root causes of UAF vulnerabilities. Then, taking several typical UAF detection methods as examples, the principles of static UAF detection methods, dynamic UAF detection methods, and a combination of static and dynamic UAF detection methods are illustrated. After analyzing their respective advantages and shortcomings, this paper makes a comparison and looks ahead for future research work. This has an important reference value for subsequent research on UAF detection.

References

1. Ainsworth, S., Jones, T. M.: MarkUs: drop-in use-after-free prevention for low-level languages. In: 2020 IEEE Symposium on Security and Privacy (SP), pp. 578–591. IEEE (2020)
2. Allen, F.E.: Interprocedural data flow analysis. In: The 6th IFIP Congress 1974, pp. 398–402. North-Holland (1974)
3. Ayewah, N., Hovemeyer, D., Morgenthaler, J., et al.: Using static analysis to find bugs. IEEE Softw. **25**(5), 22–29 (2008)
4. Balakrishnan, G., Reps, T.: Analyzing memory accesses in x86 executables. In: Duesterwald, E. (ed.) CC 2004. LNCS, vol. 2985, pp. 5–23. Springer, Heidelberg (2004). https://doi.org/10.1007/978-3-540-24723-4_2
5. Balakrishnan, G., Reps, T., Melski, D., Teitelbaum, T.: WYSINWYX: what you see is not what you execute. In: Meyer, B., Woodcock, J. (eds.) VSTTE 2005. LNCS, vol. 4171, pp. 202–213. Springer, Heidelberg (2008). https://doi.org/10.1007/978-3-540-69149-5_22
6. Bernhard, L., Rodler, M., Holz, T., & Davi, L.: Xtag: mitigating use-after-free vulnerabilities via software-based pointer tagging on intel x86-64. arXiv e-prints (2022)
7. Caballero, J., Grieco, G., Marron, M., et al.: Undangle: early detection of dangling pointers in use-after-free and double-free vulnerabilities. In: The 2012 International Symposium on Software Testing and Analysis, pp. 133–143. ACM (2012)
8. Cadar, C., Dunbar, D., Engler, D.R.: KLEE: unassisted and automatic generation of high-coverage tests for complex systems programs. In: Usenix Conference on Operating Systems Design & Implementation, pp. 209–224. USENIX Association (2009)
9. Cadar, C., Ganesh, V., Pawlowski, P. M., et al.: EXE: automatically generating inputs of death. In: ACM Conference on Computer & Communications Security, pp. 322–335. ACM (2006)
10. Cadar, C., Sen, K., et al.: Symbolic execution for software testing: three decades later. Commun. ACM **56**(2), 82–90 (2013)
11. Cesare, S.: Bugalyze.com - detecting bugs using decompilation and data flow analysis
12. Cheng, W., Zhao, Q., Yu, B., et al.: TainTrace: efficient flow tracing with dynamic binary rewriting. In: The 11th IEEE Symposium on Computers and Communications, ISCC 2006, pp. 749–754. IEEE (2006)
13. Cherem, S., Princehouse, L., Rugina, R.: Practical memory leak detection using guarded value-flow analysis. ACM SIGPLAN Not. **42**(6), 480–491 (2007)
14. Costa, M., Crowcroft, J., Castro, M., et al.: Vigilante: end-to-end containment of Internet worms. In: The 20th ACM Symposium on Operating Systems Principles 2005, pp. 133–147. ACM (2005)
15. Common vulnerabilities and exposures. https://cve.mitre.org/index.html
16. Dullien, T., Porst, S.: REIL: a platform-independent intermediate representation of disassembled code for static code analysis. CanSecWest (2009)
17. Suh, G.E., Lee, J.W., Zhang, D., Devadas, S.: Secure program execution via dynamic information flow tracking. In: The 11th International Conference on Architectural Support for Programming Languages and Operating Systems, pp. 85–96. ACM (2004)
18. Feist, J., Mounier, L., Bardin, S., et al.: Finding the needle in the heap: combining static analysis and dynamic symbolic execution to trigger use-after-free. In: The 6th Workshop on Software Security, Protection, and Reverse Engineering, pp. 1–12. ACM (2016)
19. Feist, J., Mounier, L., Potet, M.-L.: Statically detecting use after free on binary code. J. Comput. Virol. Hacking Tech. **10**(3), 211–217 (2014)
20. Godefroid, P., Levin, M.Y., Molnar, D.A.: SAGE: whitebox fuzzing for security testing. Queue **10**(3), 20 (2012)

21. Gui, B., Song, W., Xiong, H., et al.: Automated use-after-free detection and exploit mitigation: how far have we gone. IEEE Trans. Softw. Eng. (2012)
22. Hastings, R., Joyce, B.: Purify: fast detection of memory leaks and access errors. In: Proceedings of the Winter 1992 USENIX Conference, pp. 125–136. USENIX Association (1991)
23. Han, X., Wei, S., Ye, J., et al.: Detect use-after-free vulnerabilities in binaries. Qinghua Daxue Xuebao/J. Tsinghua Univ. **57**(10), 1022–1029 (2017)
24. Hind, M., Burke, M., Carini, P., Choi, J.-D.: Interprocedural pointer alias analysis. ACM Trans. Program. Lang. Syst. **21**(4), 848–894 (1999)
25. Huang, J.: UFO: predictive concurrency use-after-free detection. In: 2018 IEEE/ACM 40th International Conference on Software Engineering (ICSE), pp. 609–619. IEEE Computer Society (2018)
26. Ye, J., Zhang, C., Han, X.: Poster: UAFchecker: scalable static detection of use-after-free vulnerabilities. In: The 2014 ACM SIGSAC Conference on Computer and Communications Security, pp. 1529–1531. ACM (2014)
27. Kouwe, E.V.D., Nigade, V., Giuffrida, C.: DangSan: scalable use-after-free detection. In: 12th European Conference on Computer Systems. ACM (2017)
28. Lee, B., Song, C., Jang, Y., et al.: Preventing use-after-free with dangling pointers nullification. In: Network & Distributed System Security Symposium, Internet Society (2015)
29. Liu, D., Zhang, M., Wang, H.: A robust and efficient defense against use-after-free exploits via concurrent pointer sweeping. In: Proceedings of the 2018 ACM SIGSAC Conference on Computer and Communications Security, pp. 1635–1648. ACM (2018)
30. Liu, T., Curtsinger, C., Berger, E.D.: DoubleTake: fast and precise error detection via evidence-based dynamic analysis. In: The 38th International Conference on Software Engineering, pp. 911–922. ACM (2016)
31. Livshits, V.B., Lam, M.S.: Tracking pointers with path and context sensitivity for bug detection in c programs. ACM SIGSOFT Softw. Eng. Notes **28**(5), 317–326 (2003)
32. Nguyen, M.D., Bardin, S., Bonichon, R., et al.: Binary-level directed fuzzing for use-after-free vulnerabilities. In: The 23rd International Symposium on Research in Attacks, Intrusions and Defenses, RAID 2020 (2020)
33. Nethercote, N.: Dynamic binary analysis and instrumentation. University of Cambridge (2004)
34. Nethercote, N., Seward, J., Seward, J.: Valgrind: a framework for heavyweight dynamic binary instrumentation. ACM SIGPLAN Not. **42**(6), 89–100 (2007)
35. Newsome, J., Song D.X.: Dynamic taint analysis for automatic detection, analysis, and signature generation of exploits on commodity software. In: 12th Annual Network and Distributed System Security Symposium, NDSS 2005 (2005)
36. Goodman, P.: Pointsto: static use-after-free detector for c/c++. https://blog.trailofbits.com/2016/03/09/the-problem-with-dynamic-program-analysis/
37. Chow, J., Pfaff, B., Garfinkel, T., Christopher, K., Rosenblum, M.: Understanding data lifetime via whole system simulation (2004)
38. Sanyal, A.: Data Flow Analysis, 1st edn. CRC Press, Boca Raton (2009)
39. Sen, K.: DART: directed automated random testing. In: The 5th International Haifa Verification Conference on Hardware and Software: Verification and Testing, pp. 213–223. ACM (2009)
40. Sen, K., Marinov, D., Agha, G.: CUTE: a concolic unit testing engine for C. In: The 10th European Software Engineering Conference held jointly with 13th ACM SIGSOFT International Symposium on Foundations of Software Engineering, 2005, pp. 263–272. ACM (2005)
41. Serebryany, K., Bruening, D., Potapenko, A., et al.: AddressSanitizer: a fast address sanity checker. In: Usenix Conference on Technical Conference, p. 28. USENIX Association (2012)

42. Shi, Q., Xiao, X., Wu, R., et al.: Pinpoint: fast and precise sparse value flow analysis for million lines of code. ACM SIGPLAN Not. **53**(4), 693–706 (2018)
43. Singh, B., Soni, M.: Dynamic instrumentation: US, US20110154297 A1
44. Snelting, G., Robschink, T., Krinke, J.: Efficient path conditions in dependence graphs for software safety analysis. ACM Trans. Softw. Eng. Methodol. **15**(4), 410–457 (2006)
45. The kernel address sanitizer. https://www.kernel.org/doc/html/latest/dev-tools/kasan.html
46. Wang, X., Xue-Xin, L.I., Zhou, Z.P., et al.: Analysis of the software testing platform: S2E. Netinfo Secur. **2012**(07), 16–19 (2012)
47. Williams, N., Marre, B., Mouy, P.: On-the-fly generation of K-path tests for C functions. In: The 19th IEEE International Conference on Automated Software Engineering, pp. 290–293. IEEE (2004)
48. Xu, G., et al.: Defending use-after-free via relationship between memory and pointer. In: Gao, H., Wang, X., Iqbal, M., Yin, Y., Yin, J., Gu, N. (eds.) CollaborateCom 2020. LNICSSITE, vol. 349, pp. 583–597. Springer, Cham (2021). https://doi.org/10.1007/978-3-030-67537-0_35
49. Yamauchi, T., Ikegami, Y., Ban, Y.: Mitigating use-after-free attacks using memory-reuse-prohibited library. IEICE Tras. Inf. Syst. **E100.D**(10), 2295–2306 (2017)
50. Younan, Y.: FreeSentry: protecting against use-after-free vulnerabilities due to dangling pointers. In: Network & Distributed System Security Symposium (2015)
51. Zhen, F., Nie, S., Wang, Y., Zhi, X.: Use-after-free vulnerabilities detection scheme based on S2E. Comput. Appl. Softw. **33**(04), 273–276 (2016)
52. Zhu, K., Lu, Y., Huang, H.: Scalable static detection of use-after-free vulnerabilities in binary code. IEEE Access **8**, 78713–78725 (2020). https://doi.org/10.1109/ACCESS.2020.2990197

Coverage Optimization of Field Observation Instrument Networking Based on an Improved ABC Algorithm

Xingyue Deng[1], Jiuyuan Huo[1,2(✉)], and Ling Wu[1,2]

[1] School of Electronic and Information Engineering, Lanzhou Jiaotong University, Lanzhou 730070, People's Republic of China
huojy@mail.lzjtu.cn
[2] Lanzhou Ruizhiyuan Information Technology Co. LTD, Lanzhou 730070, People's Republic of China

Abstract. The severe conditions of cold and arid areas seriously affect the progress of data collection and analysis for field observation instruments. Therefore, this study adopted the modified artificial bee colony (ABC) algorithm to optimize the coverage of nodes and designed an energy-efficient node coverage optimization method. In the coverage optimization, the coverage rate and the number of working nodes are considered comprehensively, and the fitness value calculation is improved. The experimental results reveal that the modified ABC algorithm has better coverage optimization performance than the original ABC algorithm, genetic algorithm (GA), and particle swarm optimization (PSO) algorithm.

Keywords: Field observation instrument networking · Wireless sensor network · ABC algorithm · Coverage optimization

1 Introduction

The severe conditions of cold and arid areas seriously affect the progress of data collection and analysis for field observation instruments, which delays information acquisition and limits geological research in these areas [1]. After the sensor nodes are deployed, the formed network can be evaluated from several aspects, such as the node coverage rate and the feedback quality of the target area information provided by the network. To extend the life cycle of field observation instrument networking, an energy-efficient node coverage optimization method to better allocate various resources of the whole network needs to be studied and designed. Therefore, it is of great research value to study how to deploy nodes to improve network coverage and connection quality [2]. Coverage control technology refers to finding the best location where network nodes will be deployed on the premise of ensuring the quality of service of the network to establish a reliable network of field observation instruments to maximize the coverage of the target monitoring area and optimize the allocation of network space resources [3–9]. In the past few years, metaheuristic algorithms such as the ABC algorithm have been used to optimize the deployment of nodes to achieve this goal [10–16]. Based on the original ABC algorithm, this paper designs an energy-saving node coverage optimization method. In the coverage optimization, the coverage and the number of working nodes are considered comprehensively, and the calculation of the fitness value is improved so that the improved ABC algorithm has better coverage optimization performance than the original ABC algorithm.

2 Related Work

In practical applications, when the WSN scale is relatively large, the deployment mode of network nodes and whether the target area can meet the coverage requirements must be considered [3, 6]. For network coverage, previous researchers have studied coverage control algorithms through relevant comparative experiments, which can prolong the life of WSNs with minimum power consumption [17–19]. The desired application requirements can be achieved by designing these control algorithms to construct a reasonable network topology. In addition to coverage, when the distance between nodes is too large, the quality of communication links will be reduced, resulting in the destruction of network performance such as energy consumption, delay, and throughput [20]. Romoozi et al. [21] pointed out that more dispersed sensor nodes will improve network coverage but will consume more energy, which means that there is a conflict between network coverage and the energy consumption of nodes. This situation has led many researchers to become interested in the study of sensor node deployment. To make node deployment more efficient, most researchers choose to use computation intelligence methods, especially bioinspired-based technologies, to optimize node deployment [22].

3 Improved ABC Algorithm

3.1 Improvement of the ABC Algorithm

Optimizing the Employed Bee Search Process

The traditional ABC algorithm uses probabilities to randomly select new food sources for location updates, leading to greater randomness of particles even though it allows for a thorough particle search [23]. Therefore, it is considered to increase the optimal food source position X_{best} in the update formula, that is, the position with the largest fitness value, which reduces the randomness of the particles. The position updating formula of the improved algorithm can be expressed by Eq. (1):

$$V_{ij} = X_{ij} + \phi_{ij}(X_{ij} - X_{kj}) + X_{best} \tag{1}$$

The employed bee X_{ij} interacts with another employed bee X_{kj} at the old location, chooses a dimension to advance the distance randomly $alt = \phi_{ij}(X_{ij} - X_{kj})$ and then generates a new food source $V_{ij} = X_{ij} + alt$. Then, compare the fitness value of the new food source V_{ij} with the old food source X_{ij} and update the location of the food source. The employed bees only have one random transformation in each search process of updating food sources, and the algorithm does not consider the convergence speed. Therefore, a feedback strategy was combined to improve the employed bees' search mechanism. The employed bee X_{ij} only interacts with X_{kj} once while gathering honey, but two random transformations may occur. To some extent, the search strategy of the improved employed bee can solve the shortcomings of easily falling into a local optimum and improve the search efficiency of the ABC algorithm.

Calculation of Fitness Value

To maximize the field observation instrument network's node coverage and minimize the monitoring blind spots in the target area, it is necessary to increase the distribution density of the nodes in the field observation instrument networking. However, if the distribution density of the networking nodes is too large, it will lead to redundant nodes in the field observation instrument networking. This situation may cause data conflicts during the transmission process, and the nodes will waste more energy, ultimately shortening the survival time of the field observation instrument networking. Therefore, on the premise of ensuring network connectivity, the quantity of working nodes in the observation instrument network should be minimized to decrease the overall power consumption of the network. Therefore, the number of working nodes and network coverage should be considered in the deployment stage of network nodes. The final objective function f_i is expressed by formula (4).

$$f_1 = \frac{\sum_{i=1}^{M \times M} P(A, S_i)}{M \times M} \tag{2}$$

$$f_2 = \frac{|A| - |A^i|}{|A|} \tag{3}$$

$$f_i = \omega_1 f_1 + \omega_2 (1 - f_2), i \in \{1, 2, \ldots, SN\} \tag{4}$$

Among them, f_1 of formula (2) is network coverage; f_2 of formula (3) is the function of the node dormancy rate, A is the sum number of nodes, and A^i is the quantity of nodes in work; ω_1 and ω_2 satisfy the condition that $\omega_1 + \omega_2 = 1$.

3.2   The Improved ABC Algorithm

The working process of the modified ABC algorithm is summarized as follows:

Step 1: Initialize the parameters of the algorithm: population number SN, the maximum iterations MCN, and the maximum times the food source is collected $limit$, determine the search range of the problem to be solved, and randomly generate an original solution $X_i(i = 1, 2 \ldots SN)$.
Step 2: Calculate and evaluate the fitness value fitness$_i$ of each original solution that is generated randomly;
Step 3: Set the cycle condition and start the cycle;

Step 4: Employed bees search the neighborhood of the solution X_i and generate a suitable new food source V_i. Then, calculate its fitness value fitness$_i$ and update the location of the new food source according to the search strategy.

Step 5: Calculate the probability P_i related to the quality of the food source;

Step 6: Food sources were chosen by onlooker bees according to the probability P_i related to food source quality. A new food source V_i was searched and generated, and its fitness value was calculated. Then, the new food source's location was updated according to the strategy.

Step 7: Determine if there are food sources that should be discarded. If there is, the scout bees will randomly generate a new food source to replace the discarded food source;

Step 8: Record the best food source (solution) produced thus far;

Step 9: Determine whether the algorithm meets the termination condition. If it is, terminate the cycle and export the best food source (solution); if not, please return to Step 4 to continue the search.

4   Experimental Analysis

In this section, the coverage optimization experiment is carried out on the MATLAB R2017a platform using the original ABC algorithm and the modified ABC algorithm.

4.1   Coverage Optimization Strategy Based on the ABC Algorithm

In this experiment, the coverage optimization problem of the field observation instrument network is solved by the improved ABC algorithm. Set the target area as a square two-dimensional space of 100×100. To calculate the network coverage, 100×100 pixels with the same area and all 1 pixel in the monitoring area were divided, 50 sensor nodes were randomly placed in the monitoring area, and "·" was used to represent node positions, as shown in Fig. 1.

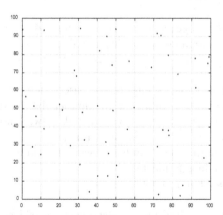

Fig. 1. Random distribution of sensor nodes.

The overall optimization steps are as follows:

Step 1: Initialize the following parameters: the area of the target monitoring area, the sensing radius, the initial quantity of bee colonies, the maximum number of iterations *MCN*, and the threshold limit.

Step 2: Randomly generate an initial solution and form an initial food source;

Step 3: calculate the initial coverage rate;

Step 4: Update the node position;

Step 5: Calculate the coverage of the new node f_1 and the dormancy rate of the node f_2, calculate its fitness value f_i, and compare it with the fitness value of the current node to reserve a better node.

Step 6: Record the node position as the local optimal solution;

Step 7: Cycle times $+1$;

Step 8: If the number of cycles exceeds the limit or the current fitness value f_i of the node is greater than the target fitness value, the best fitness value is returned, and the program ends. Otherwise, return to Step 2.

Step 9: After node deployment is complete, select cluster heads, form clusters, and transfer data in each round.

4.2 Coverage Model and Simulation Parameter Configuration

Assuming that fifty nodes are randomly deployed in a two-dimensional observation area with a size of 100×100, the sensor node's sensing radius r is set to 10, and the communication radius $R = 2r$ is set to 20. The distance from the pixel point to the node can be calculated by formula (5):

$$d(A, S_i) = \sqrt{(x_i - x)^2 + (y_i - y)^2} \tag{5}$$

According to the previous description of the Boolean perception model, the probability of event occurrence is a binary distribution, as shown in formula (1). In this paper, regional coverage is represented by formula (2). The initial environment settings for networking are summarized as follows:

- Each sensor node is heterogeneous and has a different id;
- All networking nodes are randomly deployed in fixed positions of the network;
- All nodes can calculate their remaining energy, and some nodes have the ability to replenish energy.
- All nodes know the specific location of the base station and can determine their position based on the intensity of the information sent by the base station.
- To facilitate calculation, the perception model of all network nodes adopts a Boolean perception model and has the same physical structure;
- To reasonably allocate the energy in the network and prolong the network life, all nonworking nodes are in a dormant state;
- The power of the base station is not restricted.

The network coverage optimization of static nodes considered in this paper can be seen as using an optimization algorithm to select a group from a vast number of nodes in the network, which can keep the network connected and achieve the maximum area coverage.

4.3 Simulation and Analysis

To carry out the experimental comparative analysis of algorithm coverage optimization, it is considered to deploy nodes in a redundant way in the network so that the deployed nodes can cover 100% of the target monitoring area as much as possible and then use different algorithms to select a certain number of nodes to compare the coverage of nodes selected by different algorithms. The deployment process of the simulation experiment of field observation instrument networking coverage optimization is as follows:

First, 49 nodes were deployed evenly in the target observation area to optimize their distribution, as shown in Fig. 2. Then, 50 nodes are randomly placed in the same monitoring area, as shown in Fig. 3. The optimal distribution of nodes shown in Fig. 2 and the randomly placed nodes in Fig. 3 are superimposed to form a new node distribution, as shown in Fig. 4. Take the new node distribution as the candidate nodes for network coverage optimization and use the optimization algorithm to select the 50 most suitable nodes from the candidate nodes so that they can cover the network to the greatest extent, as shown in Fig. 5.

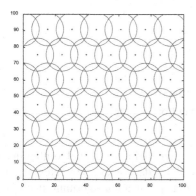

Fig. 2. Optimal distribution of nodes

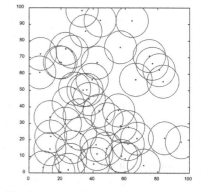

Fig. 3. Randomly deploy 50 sensor nodes

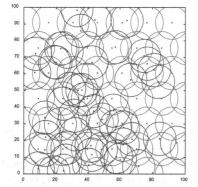

Fig. 4. Superposition of random deployment and uniform deployment

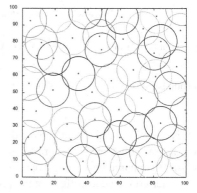

Fig. 5. Optimized node distribution

In the function optimization experiment, the improved ABC algorithm has better convergence speed and solution accuracy than the original ABC algorithm. As shown in Fig. 6, to verify the utility of the improved algorithm in the optimal coverage of field observation instrument networking, the improved algorithm and the original ABC algorithm were iterated 300 times in the process of solving the optimization coverage of the field observation instrument networking, and the test was repeated 30 times to obtain the results of the average network coverage of the two algorithms with the quantity of iterations.

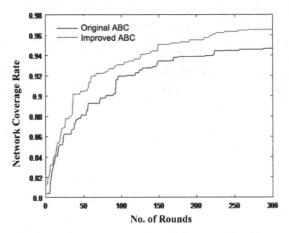

Fig. 6. Performance comparison of Coverage algorithms.

The coverage data obtained by applying the improved intelligent algorithm in coverage optimization in this study are compared with the data obtained by references [24] and [25]. The coverage results of several intelligent optimization algorithms in coverage optimization are shown in Table 1.

Table 1. Coverage comparison of different algorithms.

Intelligent algorithms	Improved ABC	ABC	GA	PSO
Coverage (%)	96.55	94.68	89.11	92.23
The number of iterations	300	300	600	457

It is clear from Fig. 6 and Table 1 that the improved ABC algorithm also has a good optimization effect in the coverage optimization of field observation instrument networking. Compared with similar optimization algorithms, the improved ABC algorithm has a faster convergence effect on the premise that nodes reach the required coverage, and the quantity of nodes distributed is reduced, which reduces the network redundancy and prolongs the network's life. Finally, one very positive conclusion can be drawn that the improved ABC algorithm in this study is better than the traditional ABC algorithm to some extent.

5 Conclusion

Field observation instruments are mostly deployed in cold and arid regions where harsh natural conditions result in the backwardness of data acquisition ability and seriously restrict the geographical study in these areas. Therefore, an improved ABC algorithm is applied to the field observation instrument network node coverage optimization simulation experiments. The analysis shows that the improved coverage optimization algorithm can improve the coverage of nodes and prolong the network life.

Acknowledgment. This work is supported by the National Nature Science Foundation of China (Grant No. 61862038), Gansu Province Science and Technology Program - Innovation Fund for Small and Medium-sized Enterprises (21CX6JA150), the Lanzhou Talent Innovation and Entrepreneurship Technology Plan Project (2021-RC-40), and the Foundation of a Hundred Youth Talents Training Program of Lanzhou Jiaotong University.

References

1. Yang, J., Huo, J., Al-Neshmi, H.M.M.: Multi-objective decision-making of cluster heads election in routing algorithm for field observation instruments network. IEEE Sens. J. **21**(22), 25796–25807 (2021)
2. Xu, L., Zhang, H., Lü, T., Shi, W., Gulliver, T.A.: Performance analysis of mobile wireless sensor network system under n-Rayleigh fading channels. Chin. J. Sens. Actuators **28**(2), 265–270 (2015)
3. Chowdhury, A., De, D.: Energy-efficient coverage optimization in wireless sensor networks based on Voronoi-Glowworm Swarm Optimization-K-means algorithm. Ad Hoc Netw. **122**(1), 102660 (2021)
4. Ling, H., Zhu, T., He, W., Luo, H., Jiang, Y.: Coverage optimization of sensors under multiple constraints using the improved PSO algorithm. Math. Probl. Eng. **2**, 1–10 (2020)
5. Li, K., Feng, Y., Chen, D., Li, S.: A global-to-local searching-based binary particle swarm optimisation algorithm and its applications in WSN coverage optimisation. Int. J. Sens. Netw. **32**(4), 197 (2020)
6. Amutha, J., Sharma, S., Nagar, J.: WSN strategies based on sensors, deployment, sensing models, coverage and energy efficiency: review, approaches and open issues. Wirel. Pers. Commun. **111**(2), 1089–1115 (2020)
7. Priyadarshi, R., Gupta, B., Anurag, A.: Wireless sensor networks deployment: a result oriented analysis. Wirel. Pers. Commun. **113**(2), 843–866 (2020)
8. Li, W., Tu, X.: Quality analysis of multi-sensor intrusion detection node deployment in homogeneous wireless sensor networks. J. Supercomput. **76**, 1331–1341 (2020)
9. Zaimen, K., Brahmia, M.-A., Dollinger, J.-F., Moalic, L., Abouaissa, A., Idoumghar, L.: Coverage maximization in WSN deployment using particle swarm optimization with Voronoi diagram. In: Bellatreche, L., Chernishev, G., Corral, A., Ouchani, S., Vain, J. (eds.) MEDI 2021. CCIS, vol. 1481, pp. 88–100. Springer, Cham (2021). https://doi.org/10.1007/978-3-030-87657-9_7

10. Ganesan, T., Rajarajeswari, P.: A novel genetic algorithm with 2D CDF 9/7 lifting discrete wavelet transform for total target coverage in WSNs deployment. Int. J. Commun. Netw. Distrib. Syst. **26**(4), 464–483 (2021)

11. Cao, M., Xiong, H.: Robust pollution source parameter identification based on the artificial bee colony algorithm using a wireless sensor network. PLoS ONE **15**(5), e0232843 (2020)

12. Sowndeswari, S., Kavitha, E.: An energy-competent enhanced memetic artificial bee colony-based optimization in WSN. In: Bindhu, V., João, M.R., Tavares, S., Ke-Lin, Du. (eds.) Proceedings of 3rd International Conference on Communication, Computing and Electronics Systems: ICCCES 2021, pp. 615–625. Springer, Singapore (2022). https://doi.org/10.1007/978-981-16-8862-1_40

13. Wei, Y., Zhou, Y., Luo, Q., Bi, J.: Using simplified slime mould algorithm for wireless sensor network coverage problem. In: Huang, D.-S., Jo, K.-H., Li, J., Gribova, V., Bevilacqua, V. (eds.) ICIC 2021. LNCS, vol. 12836, pp. 186–200. Springer, Cham (2021). https://doi.org/10.1007/978-3-030-84522-3_15

14. Elma, K.J.: Clustering and coverage using artificial bee colony (ABC) optimization in heterogeneous WSN (HWSN). J. Adv. Res. Dyn. Control Syst. **12**(3), 182–194 (2020)

15. Lu, C., Li, X., Yu, W., Zeng, Z., Li, X.: Sensor network sensing coverage optimization with improved artificial bee colony algorithm using teaching strategy. Computing **103**(7), 1439–1460 (2021)

16. Khalaf, O.I., Abdulsahib, G.M., Sabbar, B.M.: Optimization of wireless sensor network coverage using the bee algorithm. J. Inf. Sci. Eng. **36**(2), 377–386 (2020)

17. Rajpoot, P., Dwivedi, P.: MADM based optimal nodes deployment for WSN with optimal coverage and connectivity. IOP Conf. Ser. Mater. Sci. Eng. **1020**(1), 012003 (2021)

18. Anurag, A., Priyadarshi, R., Goel, A., Gupta, B.: 2-D coverage optimization in WSN using a novel variant of particle swarm optimisation. In: 2020 7th International Conference on Signal Processing and Integrated Networks (SPIN). IEEE (2020)

19. Xu, Y., Ding, O., Qu, R., Li, K.: Hybrid multiobjective evolutionary algorithms based on decomposition for wireless sensor network coverage optimization. Appl. Soft Comput. **68**(42), 268–282 (2018)

20. Younis, M., Akkaya, K.: Strategies and techniques for node placement in wireless sensor networks: a survey. Ad Hoc Netw. **6**(4), 621–655 (2008)

21. Romoozi, M., Vahidipour, M., Romoozi, M.: Genetic algorithm for energy efficient & coverage-preserved positioning in wireless sensor networks. In: 2010 International Conference on Intelligent Computing and Cognitive Informatics, ICICCI 2010, Kuala Lumpur, Malaysia, pp. 22–25 (2010)

22. Sheikh-Hosseini, M., Hashemi, S.R.S.: Connectivity and coverage constrained wireless sensor nodes deployment using steepest descent and genetic algorithms. Exp. Syst. Appl. **190**, 116164 (2021)

23. Karaboga, D.: An idea based on honey bee swarm for numerical optimization. Engineering Faculty, Computer Engineering Department, Erciyes University, Kayseri, Turkey (2005)

24. Fu, H., Han, S.: Optimal sensor node distribution based on the new quantum genetic algorithm. Chin. J. Sens. Actuators **21**(7), 1259–1263 (2008)

25. Lin, Z.-L., Feng, Y.-J., Yu, L.: Research on the strategy of wireless sensor networks coverage by the particle optimization evolutionary. Chin. J. Sens. Actuators **22**(6), 873–877 (2009)

Education Track

Predicting Student Rankings Based on the Dual-Student Performance Comparison Model

Yijie Chen[✉] and Zhengzhou Zhu

School of Software and Microelectronics, Peking University, Beijing, China
chenyijietongxue@163.com

Abstract. Currently, learning early warning mainly uses two methods, student classification and performance regression, both of which have some shortcomings. The granularity of student classification is not fine enough. The performance regression gives an absolute score value, and it cannot directly show the position of a student in the class. To overcome the above shortcomings, we will focus on a rare learning early warning method — ranking prediction. We propose a dual-student performance comparison model (DSPCM) to judge the ranking relationship between a pair of students. Then, we build the model using data including class quiz scores and online behavior times and find that these two sets of features improve the Spearman correlation coefficient for the ranking prediction by 0.2986 and 0.0713, respectively. We also compare the process proposed with the method of first using a regression model to predict scores and then ranking students. The result shows that the Spearman correlation coefficient of the former is 0.1125 higher than that of the latter. This reflects the advantage of the DSPCM in ranking prediction.

Keywords: Learning early warning · Student ranking prediction · Class quiz score · Online behavior time

1 Introduction

Learning early warning refers to establishing a model to assess students' learning status and issuing early warning signals according to the evaluation results, finally providing reasonable suggestions and interventions [1]. Teachers and education administrators can obtain the learning situations of students in time through learning early warning and improve the quality of teaching [2].

At present, learning early warning has received increasing attention from the academic and educational fields, and related research and applications are also increasing [3]. The data sources of learning early warning are becoming increasingly abundant, not only traditional classroom data such as questionnaires [4], previous grades [5], and class performance [6] but also online platform data such as online behaviors [7] and facial expressions [8]. The method of establishing learning early warning models is also becoming increasingly mature. In addition to traditional algorithms [9], various novel means [10] have also begun to be applied in this field.

However, most researchers have overlooked the question — what is the most appropriate method of learning early warning? At present, the methods of learning early warning can mainly be divided into two types. One is student classification to identify at-risk students. We believe that the main disadvantage of this method is that the warning granularity is relatively coarse, and the warning result output by the model is only one of the limited categories. The other is performance regression, which involves model design or feature selection for more accurate prediction of scores. We think that the main disadvantage is that the score is an absolute indicator. If we only focus on the score of a single student and ignore the overall situation of the whole class, it may produce an insufficiently comprehensive judgment on the student's performance.

To this end, we will research a rare learning early warning method — student ranking prediction — in this thesis. This method gives an integer indicating the student's ranking, which is much more detailed than classification; the ranking can also show the position of each student in the class, which shows a more comprehensive display of the students' performances. Rankings are also used more broadly than scores, so predicting rankings is more practical.

We will present a method for predicting student rankings in this thesis. First, we propose a deep learning model to judge the ranking relationship between two students. After training the model, we can first obtain the ranking relationship between each pair of students and then obtain the ranking of all students in the class. Of course, if a regression method is used for learning early warning, we can predict the students' scores first and then rank them according to the prediction scores. Consequently, we will conduct an experiment to compare the difference in ranking prediction effects between these two methods.

The main contributions of this thesis are as follows:

- In view of the fact that the current learning early warning mainly focuses on the two methods of student classification and performance regression, this thesis focuses on a very rare learning early warning method — predicting the rankings of students.
- This thesis designs a model for predicting student rankings, which can be transferred to various learning early warning situations.
- This thesis compares the difference between directly predicting rankings and indirectly predicting rankings according to scores.

The rest of this thesis is structured as follows: in Sect. 2, we will provide an overview of the current state of learning early warning methods; in Sect. 3, we will propose our prediction model; in Sect. 4, we will use the proposed method to predict student rankings in specific contexts and compare it with the corresponding regression model; in Sect. 5, we will discuss the experimental results further; and in the final chapter, we will summarize our research and look forward to future work.

2   Literature Review

In recent years, due to the research results in the artificial intelligence field, many new tools have been used in the learning early warning field, such as LSTM [11] and attention networks [12]. However, most of the current learning early warning methods are limited in two categories—student classification and performance regression. In addition, there are very few studies using ranking prediction as the learning early warning method.

2.1   Related Research on Student Classification

The general process of learning early warning research by classification is to first specify which type of students is risk, then train a model to classify students, and finally give early warning to students who are classified as positive examples. The training objectives of such models are mostly to improve the accuracy, recall, F1 value and other indicators of classification [13].

The most popular criterion for this method is to regard students who cannot pass the course as risk students. Tomasevic [14], Baneres [15], Chen [16], Sisovic [17], Zafra [18], Hachey [19] and many other researchers aim to identify such students and give many learning early warning models suitable for various situations. Another common criterion is to regard students who will drop out as risk students. Ramesh [20], Qiu [21] and Tan [22] have all contributed to identifying these students.

Some studies have further supplemented and improved the classification criteria: the research of Rizvi [23] and Gitinabard [24] divided performances into three grades: excellent, fair, and risky; Gray [25] divided the reason why students could not pass the course into four more specific subcategories; Mhetre [26] used the learning speed as the classification criterion and identified students who were learning too slowly. Some studies have carried out learning early warnings for other learning processes rather than a single course: Olivé [27] regarded a single assignment as the early warning object to predict whether students could submit assignments in time; Botelho [28] proposed the causes that students could not finish assignments in a timely manner, mainly by stopping learning and doing useless work, and then identified these two groups of students. There are also studies that chose a longer time span for learning early warning. For example, Agnihotri [29] and Raheela [30] judged the students' success in graduation as the indicator and used the early data of students to establish a graduation risk model.

However, most of the learning early warning models that use classification as the method can only classify students into two categories: risk students and non-risk students; even if some studies have refined the classification, the output is within only a few discrete values. Therefore, most of these learning early warning models have the shortcoming that the early warning results are not detailed enough, and it is necessary to use more quantitative methods for learning early warning.

2.2   Related Research on Performance Regression

The research on learning early warning by regression mainly includes two categories: score prediction and feature selection. The general process of the former is to train a regression model to predict student scores as accurately as possible, trying to reduce the mean square error, mean absolute error and other indicators of regression during the training process [31]. The latter mainly uses feature engineering, regression analysis and other methods to find more suitable input features for learning early warning.

Designing regression models to predict student scores has always been one of the core issues in the field of learning early warning. Sweeney [32], Almutairi [33], and Voß [34] all used different data and algorithms to establish their own models and obtained satisfying prediction results. Arsad [35] focused on the entire university period and used the grades of the basic courses of the first semester to predict the grade points of the eighth semester.

On the other hand, researchers are also looking for input features that are more effective for learning early warning. Li [36] explored the correlation between different clickstream behaviors and course grades based on the clickstream data of students on the MOOC platform. Lu [37] found some traditional and online factors that had the greatest impact on academic performance through regression analysis. Vaessen [38] summarized five help strategies from students and analyzed their impact on course performance. You [39] pointed out that the measurement of meaningful learning behavior is a more effective feature than mere behavior counts. Hung [40] also found that compared with absolute features, using relative features can build more accurate learning early warning models.

Although the regression models of learning early warning have achieved excellent results, researchers have mostly ignored the fact that only by comparing a student's score with other students in the class can the final judgments on performance be made. Therefore, in addition to predicting individual students' scores, researchers should also pay attention to the value of rankings in learning early warning.

2.3 Related Research on Ranking Prediction

At present, a small number of learning early warning studies have used ranking prediction as the method. For example, Cao [41] and Ma [42] collected relevant data from students' daily lives and predicted student rankings based on them, obtaining ideal learning early warning effects.

However, these studies mainly focus on the analysis and processing of student data but lack sufficient understanding of the practical significance of student ranking prediction and the advantages of ranking prediction compared with performance regression. This paper hopes to supplement these deficiencies.

3   Methodology

3.1   Model Overview

Because the basic operation of sorting is to compare the elements in the pairwise, the goal of our model is to predict the ranking relationship of a pair of students. We refer to these two students as student i and student j, respectively.

The input of the model must be an ordered pair of learning features of two students. If the features of the students are recorded as X_i and X_j, respectively, the feature pair input can be written as $\langle X_i, X_j \rangle$. Although the type and format of the two students' features must be exactly the same, we do not impose any restrictions on them. Researchers, teachers, and educational administrators can decide on their own input features based on specific learning situations and data sources.

The output of this model is the ranking order of the two students. We record the ranking of students as r_i and r_j, and then the output of the model is the probability $p(r_i > r_j)$. If $p(r_i > r_j) > 0.5$, then we consider that student i is ranked ahead of student j; otherwise. If $p(r_i > r_j) = p(r_j > r_i) = 0.5$, then we consider student i to be tied with student j.

3.2 Model Structure

Now, we give the specific structure of our dual-student performance comparison model (DSPCM), as shown in Fig. 1.

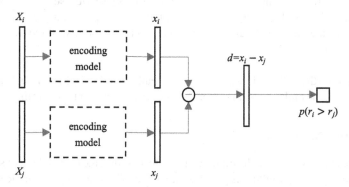

Fig. 1. The structure of the DSPCM

The features X_i and X_j will first be input into an encoding model. The role of the encoding model is to process a student's original features and finally integrate them into a vector, which we call learning feature code, as x_i and x_j are shown. We do not impose any other restrictions on the encoding model, except that the encoding models of the two students must have the same algorithm, structure, and parameters. Researchers, teachers, and education administrators can design encoding models by themselves according to specific learning situations, input characteristics, and model effects.

Then, we subtract the learning feature codes of the two students to obtain a vector that represents the gap between the learning situations of the two students, which we call the learning gap code, which is shown as d in Fig. 1.

Finally, we put d into a fully connected layer to obtain the model's output $p(r_i > r_j)$. This layer has only one unit and uses a sigmoid activation function so that its output value is a real number between 0 and 1, which conforms to the basic properties of probability. More importantly, this fully connected layer does not use the bias parameter. Without this parameter, the output of our model can satisfy the symmetry and coherence of the ranking. The loss function of our DSPCM is binary cross entropy.

3.3 Generation of the Training Set and Test Set

Since the data we initially have are the features and scores (X_i, s_i) of each student, but the input of the DSPCM is the features of a pair of students $\langle X_i, X_j \rangle$, we first need to pair the features of the students to match the input format. Therefore, the training and test sets are generated as follows:

1. All students were divided into a training group and a test group according to certain rules.
2. For each pair of students (i, j) in the training group (test group), do the following:

    (1) Add ordered pairs of their features $\langle X_i, X_j \rangle$ and $\langle X_j, X_i \rangle$ into the input list of the training set (test set).
    (2) If the scores of the two students satisfy the relationship $s_i > s_j$, then labels "1" and "0" are added into the output list of the training set (test set); if $s_i < s_j$, then labels "0" and "1" are added into the output list of the training set (test set); if $s_i = s_j$, then labels "1" and "1" are added to the output list of the training set (test set).

After training the DSPCM, we can use the ranking relationship of each pair of students to obtain the rankings of all students in the groups, similar to the sorting process.

3.4 Evaluation Indicators

Suppose there are m students in the training group (test group), the true ranking of student i is t_i, and the prediction ranking is p_i. This paper uses the following two indicators to evaluate the final ranking effect.

Spearman Correlation Coefficient ρ
The coefficient ρ of the prediction rankings and the true rankings is calculated as

$$\rho = 1 - \frac{6 \sum_{i=1}^{m} (p_i - t_i)^2}{m(m^2 - 1)} \tag{1}$$

It is a real number between -1 and 1, representing the correlation of the prediction rankings and the true rankings. The higher its value is, the better the prediction effect is.

Average Percentage Ranking Error e
It is defined as

$$e = \frac{1}{m} \sum_{i=1}^{m} \frac{|p_i - t_i|}{m} \tag{2}$$

It is a real number between 0 and 1, representing the error between the prediction rankings and the true rankings. The smaller its value is, the better the prediction effect is.

4 Experiment

4.1 Experiment Background

The data of this experiment come from the second-degree course "Software Engineering" for undergraduates majoring in software engineering at a university. The course dates ranged from September 21, 2020, to January 23, 2021, lasting 18 weeks. A total of 84 students took the course. The course adopted a mixed teaching mode. In addition to attending classes in the classroom every week, students also needed to watch the course videos on the online platform Moodle.

The first set of student features in this experiment is 13 class quiz scores. One of the quizzes is a preschool test, and the remaining 12 are unit tests.

The second set of student features in this experiment is the number of online behaviors. The Moodle platform can record 16 types of operations from students: opening the video interface, starting watching, pausing watching, continuing watching, fast-forward, rewinding, closing the video interface, minimizing the Moodle page, restoring the Moodle page, opening the full screen, closing the full screen, turning on the mute, turning off the mute, adjusting the volume, ending the video show, and checking the user's online heartbeat packet. During the class date range, the platform recorded a total of approximately 85 k student operations.

The final score of a student in this course was determined by 5 parts: class attendance, class participation, online study duration, final exam score and group project score.

4.2 Encoding Model

Handling of Class Quiz Scores

We only standardize the scores of these 13 quizzes for the time being. Standardization means that the mean of each feature in the data becomes 0 and the standard deviation becomes 1. The specific approach is to calculate the average avg and standard deviation std of each feature in the training set and then perform the following process on the feature value c of each sample in the training set and test set:

$$c' = \frac{c - avg}{std} \tag{3}$$

The result c' is the normalized feature. After normalizing the scores of each student's class quizzes, a 13-dimensional vector quiz is obtained.

Handling of Online Behavior Times

We treat the characteristics of students' online behavior times as time series data. Since the teacher's requirement was that students needed to watch videos online for at least 1 h per week and considering the sparseness of the data, we chose one week as a time step to count the times of students' various online behaviors. The statistical results are then normalized to obtain a matrix *behavior* of 18×16 dimensions for each student.

We use a long short-term memory (LSTM) neural network to process the behavior. We input each student's matrix into an LSTM with 8 units and obtain an 8-dimensional vector h_{18}, which represents the processed online behavior time features.

Encoding Model Structure

Now, we combine the processed class quiz scores and online behavior times to obtain the encoding model of the experiment. Its detailed structure is shown in Fig. 2.

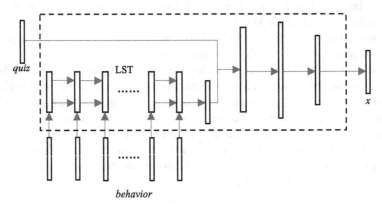

Fig. 2. The structure of the encoding model

After obtaining *quiz* and h_{18}, we concatenate them together to obtain a 21-dimensional vector. Then, we input it into a three-layer fully connected neural network to fuse the two sets of features together. The three fully connected layers all use ReLU as the activation function, and the numbers of their units are 16, 12, and 8. The output of the last layer is an 8-dimensional vector, which is the learning feature code x obtained by deeply processing all the features of a student.

4.3 The Role of Different Features on Ranking Prediction

In this study, we use two sets of student features — class quiz scores and online behavior times. We will examine their role in ranking prediction effects.

We delete the input of the *quiz* and *behavior* in the encoding model and observe the change in ranking prediction effects. Due to the reduction of input features, maintaining the original scale of the fully connected layers will lead to overfitting of the model. To this end, we reduce the number of units of these layers to 12, 8 and 6. The final prediction effects of the test set are shown in Table 1.

When two sets of features are both input, the ranking prediction result has the highest Spearman correlation coefficient and the smallest percentage ranking error, so the ranking prediction effect is the best. This shows that both class quiz scores and the number of online behaviors have a positive effect on ranking prediction.

Table 1. Comparison of ranking prediction effects of different input features

	Spearman correlation coefficient ρ	Average percentage ranking error e
quiz + behavior	0.7038	0.1596
only *behavior*	0.4052	0.2370
only *quiz*	0.6325	0.1765

Specifically, when quizzes are not input, the Spearman correlation coefficient decreases by 0.2986, and the percentage ranking error increases by 0.0774. Obviously, class quiz scores have a significant effect on ranking prediction. This is mainly because the quiz scores can largely reflect the students' mastery of knowledge.

When *behavior* is deleted, the Spearman correlation coefficient decreases by 0.0713, and the percentage ranking error increases by 0.0169. This shows that the number of online behavior times also has a certain promoting effect on ranking prediction, but the effect is not as significant as that of class quiz scores. This may be because the times of online behaviors cannot sufficiently reflect students' attitudes and the effects of online learning. Hence, we can consider collecting features such as facial expressions of students during online learning to judge the students' attitudes or set up several exercises after each video to test students' learning effects.

4.4 Comparison with Regression Models

Regression models can also be used to predict student rankings by first predicting students' scores and then using the prediction performances to rank students. Therefore, it is necessary to compare the ranking prediction effects of the DSPCM with those of the regression model.

We take the following measure to establish the corresponding regression model: except for the last layer, the input and structure of the regression model are consistent with the encoding model proposed earlier, so its penultimate layer will obtain an 8-dimensional vector. Its last layer is a fully connected layer with only 1 unit, without an activation function, and the bias parameter is included. In this way, the last layer of the regression model outputs a real number, which represents the student's prediction score. The mean square error is used as the loss function of the regression model. In this way, the structure of the regression model we established is basically the same as our DSPCM, with only one difference in the number of parameters, i.e., the bias of the last fully connected layer.

We also selected some regression algorithms from the field of machine learning for comparison. Although most of these regression algorithms have their corresponding classification algorithms, their judgment results do not satisfy the symmetry and coherence of rankings, so we do not choose them to predict student rankings. The final prediction effects of the test set are shown in Table 2.

Table 2. Comparison of ranking prediction effects of different input features

	Spearman correlation coefficient ρ	Average percentage ranking error e
DSPCM	0.7038	0.1596
Regression model	0.5913	0.1782
Linear regression	0.3146	0.2627
RandomForest	0.4931	0.2152
Adaboost	0.5194	0.2030
GBDT	0.5287	0.2036
LightGBM	0.5402	0.1921

It is obvious that compared with the corresponding regression model, our DSPCM has a better prediction effect. The correlation coefficient of the results is 0.1125 higher than that of the regression model, and the percentage error is 0.0186 lower than that of the regression model. In addition, other machine learning regression algorithms' results also have lower Spearman correlation coefficients and higher percentage ranking errors than the DSPCM, indicating that our model also outperforms these algorithms in ranking prediction. Consequently, we can conclude that if we choose ranking prediction as the learning early warning method, then compared it with first predicting students' scores and then sorting the scores to obtain the student rankings, we can directly use our DSPCM to achieve more accurate results.

To further reveal the source of the difference in effect between DSCPM and the regression model, we do the following:

(1) Calculate the percentage ranking of each student.
(2) Calculate the difference between the percentage rankings of the two students in each student pair.
(3) Based on the difference between the percentage rankings of the two students, all student pairs were grouped into groups of 10%. That is, divide the student pairs whose percentage ranking difference is 0 to 10, 10 to 20…, 90 to 100 into a group. Finally, we obtained 10 groups.
(4) Count the accuracy of DSCPM and the regression model on each group.

The result is shown in Fig. 3.

Obviously, the accuracy of both models increases as the difference in percentage rankings increases. This is mainly because as the difference between the percentage rankings of two students gradually increases, the gap between their learning features also increases, so it becomes easier to judge their ranking relationship. On the other hand, compared to the regression model, DSPCM always has better accuracy, especially on the groups where the difference in percentage rankings is 10–50. This is exactly where the difference in the ranking prediction effect between the DSPCM and the regression model comes from.

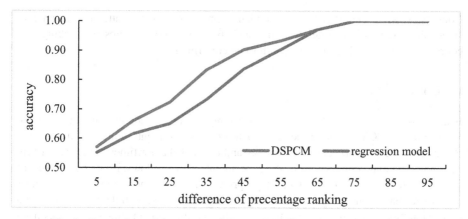

Fig. 3. Comparison of the accuracy between DSPCM and the regression model

The main reason why the DSPCM predicts ranking better than the regression model is that their prediction goals are different. The goal of the former is to predict the ranking order of two students without paying attention to the numerical value of their scores. However, the goal of the latter is to predict the value of the students' scores, without considering whether the prediction results conform to the true order of the students' rankings. Sometimes the regression model can make the size relationship of the prediction results in line with the true situation, but sometimes it cannot.

5 Discussion

Although we have emphasized the value of rankings in learning early warning in this thesis, we do not negate the importance of scores. If a regression model can rank students accurately enough from prediction scores, then it is natural to use it to predict scores and rankings at the same time. Therefore, we will discuss the suitable situations of the two learning early warning models.

A main characteristic of the DSPCM is that it only cares about the size relationship between the scores of the two students, not the specific value of the students' scores. Furthermore, it does not care about the specific gap between the two students' scores. Therefore, during training, it will mainly focus on those pairs of students for whom it is difficult to determine the ranking relationship because of similar features and small score gaps. In other words, it will mainly focus on the middle score segment students. The regression model, however, only cares about the error between the prediction scores and the real scores. Because the gap between the scores of high or low score segment students and the average score is larger, the regression model may focus more on the students in these two groups, while not paying enough attention to the middle score segment students to accurately predict the rankings for them.

From the discussion above, we can see that when the distribution of student scores is concentrated, it may be more appropriate to use our DSPCM to predict rankings.

Conversely, when the distribution of student scores is relatively scattered, we can consider using a regression model to simultaneously obtain student scores and rankings. Of course, this view needs further research to be confirmed.

6 Conclusion

In this study, we note a rare learning early warning method—predicting student rankings. We use the DSPCM to judge the relationship between students' rankings and use the two datasets of students' class quiz scores and online behavior times to predict student rankings. The main conclusions of the experiments are as follows: First, the two sets of features both have a positive effect on the prediction of student rankings, and the class quiz scores have a more significant effect. Second, compared with the method of using regression models to predict student scores first and then obtain student rankings from prediction scores, directly using the DSPCM proposed in this thesis to rank students can obtain a better prediction effect.

There are two main directions for our future work. One is to perform feature engineering to improve the ranking prediction effects, such as which learning features are more suitable for predicting rankings and how to preprocess the student features to obtain better results. The other is to build more timely learning early warning models. We will attempt to build learning early warning models using data collected in earlier stages and make the model output results more prompt, enabling teachers and students to have longer reaction times. Therefore, we hope to develop a more accurate and practical learning early warning model by ranking prediction.

References

1. Macfadyen, L.P., Dawson, S.: Mining LMS data to develop an "Early Warning System" for educators: a proof of concept. Comput. Educ. **54**(2), 588–599 (2010)
2. Siemens, G.: Learning analytics: the emergence of a discipline. Am. Behav. Sci. **57**(10), 1380–1400 (2013)
3. Papamitsiou, Z., Economides, A.A.: Learning analytics and educational data mining in practice: a systematic literature review of empirical evidence. Educ. Technol. Soc. **17**(4), 49–64 (2014)
4. Ramaswami, M., Bhaskaran, R.: A CHAID based performance prediction model in educational data mining. Int. J. Comput. Sci. Issu. **7**(1), 10–18 (2010)
5. Golding, P., McNamarah, S.: Predicting academic performance in the School of Computing & Information Technology (SCIT). In: Proceedings Frontiers in Education 35th Annual Conference, pp. 230–233 (2005)
6. Parack, S., Zahid, Z., Merchant, F.: Application of data mining in educational databases for predicting academic trends and patterns. In: 2012 IEEE International Conference on Technology Enhanced Education (ICTEE), pp. 1–4. IEEE, Amrita University, Amritapuri Campus (2012)
7. Du, X., Yang, J., Hung, J.: An integrated framework based on latent variational autoencoder for providing early warning of at-risk students. IEEE Access **8**, 10110–10122 (2020)
8. Sarah, D., et al.: Computer vision for attendance and emotion analysis in school settings. In: IEEE 9th Annual Computing and Communication Workshop and Conference, pp. 134–139. IEEE, Las Vegas (2019)

9. Marquezvera, C., Romero, C., Ventura, S.: Predicting school failure using data mining. In: International Conference on Educational Data Mining, pp. 271–276. (2011)

10. Yang, Z.K., Yang, J., Rice, K., Hung, J.L., Du, X.: Using convolutional neural network to recognize learning images for early warning of at-risk students. IEEE Trans. Learn. Technol. **13**(3), 617–630 (2020)

11. Min, W., et al.: DeepStealth: game-based learning stealth assessment with deep neural networks. IEEE Trans. Learn. Technol. **13**(2), 312–325 (2020)

12. Sun, B., Zhu, Y., Yao, Z., Xiao, R., Xiao, Y., Wei, Y.: Tagging reading comprehension materials with document extraction attention networks. IEEE Trans. Learn. Technol. **13**(3), 567–579 (2020)

13. Nawang, H., Makhtar, M., Shamsudin, S.N.W.: Classification model and analysis on students' performance. J. Fundam. Appl. Sci. **9**(6), 869–885 (2017)

14. Tomasevic, N., Gvozdenovic, N., Vranes, S.: An overview and comparison of supervised data mining techniques for student exam performance prediction. Comput. Educ. **143**, 1–18 (2020)

15. Baneres, D., Rodríguez-Gonzalez, M.E., Serra, M.: An early feedback prediction system for learners at-risk within a first-year higher education course. IEEE Trans. Learn. Technol. **12**(2), 249–263 (2019)

16. Chen, W., Brinton, C.G., Cao, D., Mason-Singh, A., Lu, C., Chiang, M.: Early detection prediction of learning outcomes in online short-courses via learning behaviors. IEEE Trans. Learn. Technol. **12**(1), 44–58 (2019)

17. Sisovic, S., Matetic, M., Bakaric, M.B.: Clustering of imbalanced moodle data for early alert of student failure. In: IEEE 14th International Symposium on Applied Machine Intelligence and Informatics (SAMI), pp. 165–170. IEEE, Herlany (2016)

18. Zafra, A., Romero, C., Ventura, S.: Multiple instance learning for classifying students in learning management systems. Expert Syst. Appl. **38**(12), 15020–15031 (2011)

19. Hachey, A.C., Wladis, C.W., Conway, K.M.: Do prior online course outcomes provide more information than G.P.A. alone in predicting subsequent online course grades and retention? An observational study at an Urban Community College. Comput. Educ. **72**, 59–67 (2014)

20. Ramesh, A., Goldwasser, D., Huang, B., Daume, H., Getoor, L.: Interpretable engagement models for MOOCs using hinge-loss markov random fields. IEEE Trans. Learn. Technol. **13**(1), 107–122 (2020)

21. Qiu, L., Liu, Y.S., Hu, Q., Liu, Y.: Student dropout prediction in massive open online courses by convolutional neural networks. Soft. Comput. **23**(20), 10287–10301 (2019)

22. Tan, M.J., Shao, P.J.: Prediction of student dropout in E-Learning program through the use of machine learning method. Int. J. Emerg. Technol. Learn. **10**, 11–17 (2015)

23. Rizvi, S., Rienties, B., Khoja, S.A.: The role of demographics in online learning; a decision tree based approach. Comput. Educ. **137**, 32–47 (2019)

24. Gitinabard, N., Xu, Y., Heckman, S., Barnes, T., Lynch, C.F.: How widely can prediction models be generalized? Performance prediction in blended courses. IEEE Trans. Learn. Technol. **12**(2), 184–197 (2019)

25. Gray, C.C., Perkins, D.: Utilizing early engagement and machine learning to predict student outcomes. Comput. Educ. **131**, 22–32 (2019)

26. Mhetre, V., Nagar, M.: Classification based data mining algorithms to predict slow, average and fast learners in educational system using WEKA. In: 2017 International Conference on Computing Methodologies and Communication (ICCMC), pp. 475–479. IEEE, Surya Engineering College (2017)

27. Olivé, D., Du, Q.H., Reynolds, M., Dougiamas, M., Wiese, D.: A quest for a one-size-fits-all neural network: early prediction of students at risk in online courses. IEEE Trans. Learn. Technol. **12**(2), 171–183 (2019)

28. Botelho, A.F., Varatharaj, A., Patikorn, T., Doherty, D., Adjei, S.A., Beck, J.E.: Developing early detectors of student attrition and wheel spinning using deep learning. IEEE Trans. Learn. Technol. **12**(2), 158–170 (2019)

29. Agnihotri, L., Ott, A.: Building a student at-risk model: an end-to-end perspective from user to data scientist. In: Proceedings of the 7th International Conference on Educational Data Mining (EDM), pp. 209–212 (2014)

30. Raheela, A., Agathe, M., Mahmood, P.: Predicting student academic performance at degree level: a case study. Int. J. Intell. Syst. Appl. **7**, 49–61 (2014)

31. Pardo, A., Jovanovic, J., Mirriahi, N., Dawson, S., Gaevi, D.: Generating actionable predictive models of academic performance. In: International Conference on Learning Analytics and Knowledge, pp. 474–478. ACM, Edinburgh (2016)

32. Sweeney, M., Rangwala, H., Lester, J., Johri, A.: Next-term student performance prediction: a recommender systems approach. J. Educ. Data Mining **8**, 22–51 (2016)

33. Almutairi, F.M., Sidiropoulos, N.D., Karypis, G.: Context-aware recommendation-based learning analytics using tensor and coupled matrix factorization. IEEE J. Select. Top. Signal Process. **11**(5), 729–741 (2017)

34. Voß, L., Schatten, C., Mazziotti, C., Schmidtthieme, L.: A transfer learning approach for applying matrix factorization to small ITS datasets. Int. Educ. Data Mining Soc. **8**, 372–375 (2015)

35. Arsad, P.M., Buniyamin, N., Manan, J.A.: A neural network students' performance prediction model (NNSPPM). In: 2013 IEEE International Conference on Smart Instrumentation, Measurement and Applications (ICSIMA), pp. 1–5. IEEE, Kuala Lumpur (2013)

36. Li, S., Wang, S., Du, J., Li, M.: The characteristics of attention flow of MOOC students and their prediction of academic persistence. In: 2019 Eighth International Conference of Educational Innovation through Technology, pp. 91–98. IEEE, Biloxi (2019)

37. Lu, O., Huang, A.Y.Q., Huang, J.C.H., Lin, A.J.Q., Ogata, H., Yang, S.: Applying learning analytics for the early prediction of students' academic performance in blended learning. Educ. Technol. Soc. **21**(2), 220–232 (2018)

38. Vaessen, B.E., Prins, F.J., Jeuring, J.: University students' achievement goals and help-seeking strategies in an intelligent tutoring system. Comput. Educ. **72**, 196–208 (2014)

39. You, J.W.: Identifying significant indicators using LMS data to predict course achievement in online learning. Int. High. Educ. **29**, 23–30 (2016)

40. Hung, J.L., Shelton, B.E., Yang, J., Du, X.: Improving predictive modeling for at-risk student identification: a multistage approach. IEEE Trans. Learn. Technol. **12**(2), 148–157 (2019)

41. Cao, Y., et al.: Orderness predicts academic performance: behavioral analysis on campus lifestyle. J. R. Soc. Interface **15**(146), 20180210 (2018)

42. Ma, Y., Zong, J., Cui, C., Zhang, C., Yang, Q., Yin, Y.: Dual path convolutional neural network for student performance prediction. In: Cheng, R., Mamoulis, N., Sun, Y., Huang, X. (eds.) WISE 2020. LNCS, vol. 11881, pp. 133–146. Springer, Cham (2019). https://doi.org/10.1007/978-3-030-34223-4_9

M-ISFCM: A Semisupervised Method for Anomaly Detection of MOOC Learning Behavior

Shichao Zhou[1], Liefeng Cao[1], Ruizhe Zhang[1], and Guozi Sun[1,2(✉)]

[1] School of Computer Science, Nanjing University of Posts and Telecommunications, Nanjing 210023, China
sun@njupt.edu.cn
[2] Key Laboratory of Urban Land Resources Monitoring and Simulation, MNR, Shenzhen 518000, China

Abstract. Massive online courses (MOOCs) are becoming increasingly vital in the modern era, yet tools to track and detect MOOC learners' progress are inadequate. In reality, labeled MOOC data are difficult to acquire, whereas unlabeled data make up the majority of the data, and these massive unlabeled data are difficult to analyze, resulting in data waste. This paper tackles this issue by presenting a MOOC learning behavior anomaly detection model (M-ISFCM) for the supervision and inspection of MOOC learners' learning that combines semisupervised fuzzy C-mean clustering (SFCM) and an isolated forest algorithm. To optimize MOOC data usage, the model leverages unlabeled and labeled MOOC data as prior assumptions. The MOOC detection runtimes are enhanced by integrating the outliers of the isolated forest approach in SFCM. The results show that the model has a higher precision rate, recall rate, and AUC than the traditional anomaly models in MOOC data. Therefore, the model is effective for recognizing anomalous MOOC learning behaviors.

Keywords: MOOC · Self-supervised · Anomaly detection · Data analysis

1 Induction

Because online education is vital in this modern era, it is crucial to monitor and research students' online learning habits. Most online learning today lacks essential oversight, which is merely reliant on the results to evaluate students' learning status, such as by evaluating students' learning through the final scores of online examinations. However, this type of outcome-based evaluation has failed to provide process-oriented monitoring to help students and teachers better grasp their learning and teaching status.

The majority of current research on MOOC learning (Massive Open Online Course) is based on supervised learning, yet MOOC data are frequently unlabeled data in the real production process; therefore, supervised learning algorithm research is irrelevant. As a result, we present a strategy for detecting MOOC learning behavior anomalies that is semisupervised (M-ISFCM).

© The Author(s), under exclusive license to Springer Nature Singapore Pte Ltd. 2022
Y. Wang et al. (Eds.): ICPCSEE 2022, CCIS 1629, pp. 323–336, 2022.
https://doi.org/10.1007/978-981-19-5209-8_22

The following are some of the benefits of our algorithm:

(1) To solve the issue of massive MOOC data and a lack of labels, a semisu-pervised clustering anomaly detection approach is utilized to train the model with few labeled samples of data and massive samples of unlabeled data as known information, ensuring the training model's performance while lowering labor costs.

(2) The isolated forest algorithm is added to the semisupervised clustering model for MOOC data to detect the outliers of learners, and the isolated forest model's outliers of learning behaviors are combined with the maximum similarity of learners to obtain the credibility of MOOC unlabeled data, which provides relatively scientific and reasonable prior information for semisupervised clustering training on MOOC data.

(3) Several sets of learning behavior anomaly detection experiments using various models and methods are designed for MOOC datasets, with the results demonstrat-ing that the M-ISFCM proposed in the paper outperforms other unsupervised and semisupervised anomaly detection algorithm models.

2 Related Work

Anomaly detection is a technique for recognizing abnormal patterns that do not cor-respond to predicted behavior. It is a branch of machine learning. Network intrusion detection, industrial machinery defect detection, fraud detection, and log anomaly detec-tion are a few of the sectors where anomaly detection is now frequently employed. The following are the primary approaches that have been used: due to the heterogeneity of incoming traffic and incomplete hardware, neural network-based methods [1], using a hybrid data processing model of gray wolf optimization (GWO) and convolutional neural network (CNN), are only relevant to network anomaly detection, and anomaly identifi-cation may be difficult; the distance-based technique [2], which determinesliers based on ROD, overcomes the limits of existing approaches and improves anomaly detection; the log-based approach [3] may detect numerous anomalies and rate the severity of these abnormal occurrences.

According to the availability of data labels, anomaly detection in artificial intelli-gence may be divided into three categories: supervised learning, unsupervised learning, and semisupervised learning. In most cases, a limited number of labeled samples may be obtained in practical circumstances. Unsupervised learning methods are prone to omitting important information, but supervised learning methods require manual label-ing, which is time consuming and labor intensive. Semisupervised anomaly detection approaches, which are more relevant to practical applications, bridge the gap between unsupervised and supervised methods and have grown in popularity in recent years. The literature [4] introduces a stable and adaptive discriminant analysis rule to address the problem of class membership being unreliable for certain training units (label noise); the literature [5] utilizes supervised and unsupervised support vector machine algorithms while introducing expert feedback for anomaly detection, and a large number of studies have proposed many anomaly detection models for various applications [6–8]. There is also relevant work in the literature [9] that classifies and compares approaches.

The major focus of studies on MOOC online education is on dropout prediction, course difficulty prediction, and graduation difficulty prediction [10]. Its goal is to better identify educational anomalies and encourage educational progress. The literature [11] concentrates on the topic of online test cheating and recommends the use of machine learning to detect exam cheating situations. This strategy has been shown to be quite effective in identifying cheating.

2.1 Semisupervised Fuzzy C-means Clustering

Semisupervised fuzzy c-mean clustering is a derivative of fuzzy c-mean clustering (FCM), which is an objective function-based clustering algorithm that determines the class affiliation of sample points by optimizing the objective function to obtain the affiliation of the class center corresponding to each sample point to classify the sample data. By integrating labeled sample information into the objective function of FCM, Pedrycz [12] proposed a semisupervised FCM method (SFCM). To improve the effectiveness of fuzzy clustering, SFCM employs a small quantity of labeled data as a priori knowledge to guide the clustering process.

The FCM is simple to use and can solve a wide range of problems; however, it is heavily reliant on initialization clustering centers and readily slips into local optimum solutions, which is troublesome for detecting anomalous MOOC learning patterns. By processing a limited quantity of labeled MOOC learning behavior data as a priori knowledge, we employ the theory of SFCM to increase the performance of MOOC learning behavior anomaly detection in this thesis.

2.2 Unsupervised Isolated Forest

The isolation forest is an unsupervised anomaly detection approach based on integrated learning that was formed by Prof. Zhihua Zhou's team [13] in 2008. Unlike other anomaly detection methods, which employ statistical, distance, and density quantifiers to represent the sparsity of data samples in comparison to other samples, the isolation forest explicitly depicts the sparsity of data for anomalous data features. By randomly isolating the data space, the isolation forest method creates an isolation tree (*iTree*), which subsequently isolates the samples. The anomalous samples are segregated by a modest number of random feature segmentations since they are typically sparsely distributed and far from the high-density group.

In actuality, labeled MOOC learning behavior data are difficult to obtain, and unlabeled data make up the majority of the data. To address this issue, this thesis employs the isolated forest method in conjunction with fuzzy C-mean clustering to process unlabeled data, increase MOOC learning behavior data utilization, and enhance MOOC learning behavior anomaly detection efficacy.

3   M-ISFCM: SFCM Combined with Isolated Forest for MOOC Learning Behavior Anomaly Detection Model

The semisupervised MOOC learning behavior anomaly detection model (M-ISFCM) is described in detail in the section. Section 3.1 explains the definition of anomalies in learning samples and describes the MOOC datasets; Sect. 3.2 describes how to process the MOOC datasets; Sect. 3.3 introduces feature extraction of MOOC learning behavior features; and Sect. 3.4 specifies the model structure for using SCFM conjunction with isolated forest anomaly detection for MOOC data.

3.1 Definition and Explanation of MOOC Data

MOOC Learning Anomaly Sample Definition. The M-ISFCM model takes the learner's learning behavior X as the detection sample inside the divided learning time range T_c and classifies the learner as an abnormal sample if an aberrant learning behavior is discovered within divided learning time T_c, otherwise as a normal sample.

Acquisition and Description of MOOC Datasets. The data in this thesis originate from the MoocData platform, which is based on the MOOC website XuetangX, which is operated by a member of Tsinghua University's Knowledge Engineering MOOC team and offers scholars online educational data and contests. The data accurately capture learners' behaviors on the XuetangX platform over a period of time, allowing it to determine who is a normal learner and who is an aberrant learner.

The data are split into two sections. The learner's tracking log file, which is the clickstream log file generated by the learner's interaction with the MOOC platform, is one main portion of the data.

The learner's trace log file is saved as a JSON file with the following format:

$$[[course\_id, \{session\_id : [enevent1, time1], [event2, time2], \ldots\}]]$$

It is difficult to analyze the learner's learning behavior data directly since it is buried in layers of JSON files; thus, JSON parsing of the tracking log files is needed. Furthermore, because the JSON file is too large, it must be processed via stream reading, which is performed as shown in the following figure.

3.2 Preparation of MOOC Data

Because of the original MOOC dataset's size and complexity, as well as the inclusion of missing values, it is preprocessed before feature extraction, including missing value processing and normalization.

Handling of Missing User Values. The learning behavior data generated and collected during the interaction between learners and the MOOC platform are incomplete in real-world scenarios, and there are missing examples. Missing values can cause the system to lose much relevant information and make it difficult to discover abnormalities in MOOC learning patterns; thus, they must be addressed. The missing values are addressed in this study in two ways: first, samples with a high proportion of missing MOOC learning behavior information are removed; second, samples with less missing MOOC learning behavior information are approximated with the KNN method.

Learning Behavior Data Normalization. Different types of data characteristics and distribution intervals of data feature values can be found in the extracted MOOC datasets. To eliminate the influence of dimensionality between features, the data are often normalized to improve the comparability between data features. Data normalization assists in maintaining accuracy and speeding up processing. The following is how the learner's learning behavior data are standardized in this paper:

$$x' = \frac{x - x_{\max}}{x_{\max} - x_{\min}} \tag{1}$$

where x is the original value of the normalized point in a series, $x_{\max}$ and $x_{\min}$ are the sequence's largest and smallest values, respectively, and x' is the value after the original value x has been normalized.

3.3 Feature Extraction of MOOC Learning Behavior

This paper employed behavioral characteristic measurements for MOOC learners with the goal of more completely and profoundly reflecting on learning. Course content learning behavior, course discussion interaction behavior, course resource access behavior, and course quiz assessment behavior are the four primary aspects. The following table lists the specific behavioral trait indicators (Table 1):

Table 1. Description of MOOC learner behavior characteristic indicators.

Indicator category	Feature description	Indicator category	Feature description
Course learning behavior	Number of video views	Quiz and assessment behavior	Number of homework acquired
	Length of video views		Number of homework submissions
	Density of video views		Number of homework inspections
	Number of units involved		Number of errors in questions
	Number of visits to courseware		Number of correct questions
Discussion of interaction behavior	Number of forum visits	Resource access behavior	Number of downloads of courseware
	Number of discussions		Number of visits to the announcement

3.4 M-ISFCM for MOOC Learning Behavior Anomaly Detection

This section explains how to discover anomalies in learning behavior using a combination of semisupervised clustering and unsupervised isolated forest.

We present a semisupervised clustering method for MOOC learning behavior anomaly identification in this research, which integrates isolated forest anomaly detection to improve the model's performance. The structure of the model is shown in the following figure (Fig. 1).

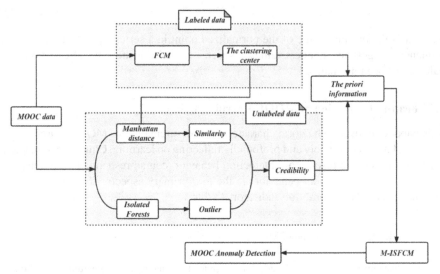

Fig. 1. Overall flowchart for M-ISFCM.

The procedure is broken down into three steps: first, the solution of cluster centers: FCM is performed on the labeled data to calculate the cluster centers of labeled learners in MOOC data; second, the solution of credibility: the isolated forest is performed on some unlabeled learning behavior data, and the credibility is calculated by combining the similarity degree of users and cluster centers; third, semisupervised clustering is performed: multiple iterations are performed to obtain the affiliation matrix of MOOC learners and take their maximum credibility as the category of users, that is, to detect whether learners are anomalous or not.

Calculation of Clustering Centers with Labeled MOOC Learners. The learner sets in this thesis are supposed to have several categories of aberrant learners and one category of normal learners. Therefore, cluster analysis is performed to generate the cluster centers for each learner type using the labeled data.

Let set the labeled learners $X = \{x_1, x_2, \cdots, x_n\}$, each with f behavioral features $x_i = (x_{i1}, x_{i2}, \cdots, x_{if})$, and define the number of categories of X as c, $2 \leq c \leq n$, the matrix consisting of C clustering centers as $V = (v_1, v_2, \cdots, v_c)^T$, $V \in R^{c \times f}$, and let $U = (u_{ik})_{n \times c}$ be the fuzzy classification matrix and u_{ik} be the affiliation value.

To calculate the optimal case of fuzzy classification, J.C. Bezdek [14] introduced the proposed objective function J_m. The formula is as follows.

$$J_m(U, x_k, v_i) = \sum_{k=1}^{n} \sum_{i=1}^{c} (u_{ik})^m d_{ik}^2(x_k, v_i) \ s.t. \begin{cases} \sum_{i=1}^{c} u_{ik} = 1, \forall k \\ u_{ik} \in [0, 1], \forall i, k \\ \sum_{k=1}^{n} u_{ik} > 0, \forall i \end{cases} \tag{2}$$

where $d_{ik}(x_k, v_i) = \|x_k - v_i\|^2$ is expressed as the euclidean distance between the learning behavior feature x_k and the i-th class clustering center v_i; m is the fuzzy index greater than 1, and with the increase of m, the higher the fuzzy degree of classification, the value is generally taken as 2.

The Algorithm Flow of FCM

Input.   Sample set X and required parameters for FCM.

Output. Clustering center $V(v_1, v_2, \cdots, v_c)$ and affiliation matrix $U(U = (u_{ik})_{n \times c})$.

Step 1.   Given categories c, the parameter m, the value of the allowable error ξ.

Step 2.   initialize the cluster centers $v_i(k), i = 1, 2, \cdots, c$ and make cycles $k = 1$.

Step 3.   Affiliation formula U.

$$u_{ik}(p) = \left(\sum_{p=1}^{c} \frac{d_{ik}(x_k, v_i)}{d_{pk}(x_k, v_p)} \right)^{\frac{2}{m-1}} \Big)^{-1}, \ i = 1, 2, \cdots, c \ ; \ k = 1, 2, \cdots, n \tag{3}$$

Step 4.   Modify all clustering centers $v_i(k), i = 1, 2, \cdots, c$;

$$v_i = \frac{\sum_{p=1}^{n} (u_{ip})^m \cdot x_k}{\sum_{p=1}^{n} (u_{ip})^m} \ , \ i = 1, 2, \cdots, c \tag{4}$$

Step 5.   Calculation error $e = \sum_{i=1}^{c} \|v_i(p+1) - v_i(p)\|^2$;
    If $e < \xi$, end; Else $p = p+1$, to Step3;

Step 6.   Sample Categorization.
    If $d_{ik}(x_k, v_i) < d_{pk}(x_k, v_p)$, $p = 1, 2, \cdots, c$, $p \neq i$;

x_k belongs to the i-th category.

Because FCM is overly sensitive to the beginning center, this work uses statistical principles to compute the first clustering center for MOOC learning and provides a threshold [15] α as a constraint to locate alternative feasible domains beyond the identified sample viable domains.

As a result, the iterative clustering method can solve for the clustering centers of MOOC learners in many viable domains, eliminating the possibility of local convergence to extremely tiny due to random beginning centers while also simplifying the calculation.

Calculation of the Credibility of MOOC Data. Based on rigorous considerations, it is assumed that learners in the same category in the MOOC datasets have similar learning behavior characteristics; that is, the closer a learner's feature value is to the center, the more likely the learner belongs to that type.

The notion of credibility is introduced in this work. Credibility refers to the chance that a MOOC participant belongs to a specific learner group. As a result, the credibility for the unlabeled MOOC datasets is generated to offer somewhat scientifically a priori information for the semisupervised clustering algorithm that follows. The credibility is made up of two primary components: one is an isolated forest model that calculates MOOC learning behavior outliers, and the other is a procedure that Calculates the similarity value of MOOC Learners' data closeness. The specific process is as follows.

Calculation of Outliers. The formula for isolated forest outliers $s(x, n)$ is as follows:

$$s(x, n) = 2^{\frac{-E(h(x))}{b(x,n)}} \tag{5}$$

where $E(h(x))$ is the desired path length of the sample, $b(x, n)$ is the average path length of the tree, and n is the number of samples.

$$b(x, n) = 2H(n - 1) - \frac{2(n - 1)}{n} \tag{6}$$

where $H(n - 1)$ is the harmonic series and γ is the Euler constant.

$$H(n - 1) = \ln(n - 1) + \gamma \tag{7}$$

The closer the outlier is to 1, the more likely the MOOC learner is abnormal; the closer the outlier is to 0, the more likely the learner is normal; the outlier is approximately 0.5, it is difficult to tell whether the learner is abnormal or normal.

As a result, utilizing the outlier value alone to reliably quantify the existence of aberrant learners is problematic; hence, this paper combines the closeness of the unlabeled learner data in conjunction with the determined clustering centers (similarity value). The following is how the similarity values are expressed.

Calculation of Similarity Values. The similarity value represents the closeness of unlabeled MOOC learner data to known clustering centers (learner types). Taking MOOC learner x as an example, the closeness between x and each cluster center is determined, and the cluster center of x is chosen from the class of cluster centers with the highest similarity value.

The Manhattan distance [16] is used to calculate the closeness between two learners data based on the idea of simplifying the computation. The following is the formula.

$$D(x_k, c_i) = |x_k - c_i| \tag{8}$$

where x_k is the k-th learner data; c_i is the i-th clustering center; $D(x_k, c_i)$ is the Manhattan distance; and $d(x, c)$ is the similarity value.

$$d(x, c) = \frac{1}{1 + D(x_k, c_i)} \tag{9}$$

The closer the similarity value is to 1, the more likely it is that the learner belongs to the group. In this study, a notion of credibility is presented to integrate outlier information and similarity values. The specifics are as follows.

Calculation of Credibility. Credibility defines the likelihood size of the user category to which MOOC learners belong, which is determined by both outliers and similarity. Outliers, on the other hand, represent only anomalous learning behavior data, whereas similarity values describe all learning behavior data; hence, credibility is calculated on a case-by-case basis.

Only the similarity is utilized to determine the credibility if the cluster center that a learner is near is the typical user class, which implies that the learner is close to the normal learner. The following is the formula.

$$P(x, c) = d(x, c) \tag{10}$$

If a learner is close to a clustering center that is an abnormal user class, it signifies that the learner is close to an anomalous learner, and the credibility must be calculated by combining the anomaly and similarity values. The following is the formula.

$$P(x, c) = \omega \cdot s(x, n) + (1 - \omega) \cdot d(x, c) \tag{11}$$

where $\omega \in (0, 1)$ denotes the weight coefficient, which is used to balance the outliers of unlabeled MOOC learner data and the similarity of labeled MOOC learner data.

M-ISFCM for MOOC Learning Behavior. M-ISFCM uses both the computed credibility of partially unlabeled data and the clustering centers of labeled data as a priori information, thus addressing the problem of poor outcomes in SFCM owing to insufficient a priori information. The goal function is depicted in the diagram below.

$$J_n(U, x_k, v_i) = \sum_{k=1}^{n} \sum_{i=1}^{c} (u_{ik})^m d_{ik}^2(x_k, v_i) + \varepsilon \sum_{k=1}^{n} \sum_{i=1}^{c} (u_{ik} - f_{ik} b_k)^m d_{ik}^2(x_k, v_i) \tag{12}$$

where ε is the balance factor, which is proportional to the ratio of the total number of MOOC learners n to the number of labeled learners and is used to balance the supervised and unsupervised optimization mechanisms; f_{ik} represents the a priori information, which is expressed as the credibility that x_k belongs to c_i; b_k is the introduced binary vector, where b_k takes 1 to indicate it belongs to labeled learners; and b_k takes 0 to indicate that it belongs to unlabeled learners.

The Algorithm Flow of M-ISFCM

Input.   Sample set X, the labeled sample set Y, and required parameters.

Output. Clustering center $V(v_1, v_2, \cdots, v_c)$ and affiliation matrix $U(U = (u_{ik})_{n \times c})$.

Step 1.   Cluster Y and get the cluster centers $(v_1, v_2, \cdots, v_c)$.

Step 2.   Initialize cluster centers with $(v_1, v_2, \cdots, v_c)$.

Step 3.   Affiliation matrix U.

$$u_{ik}(P) = \frac{1}{1+\varepsilon}\left(\frac{1+\varepsilon(1-b_k\sum_{i=1}^{c}f_{ik})}{\sum_{p=1}^{c}(\frac{d_{ik}(x_k,v_i)}{d_{pk}(x_k,v_p)})^{\frac{2}{m-1}}}+\varepsilon f_{ik}b_k\right) \tag{13}$$

Step 4.   Calculate the objective function J_n.
   If (J_n is less than the threshold), then output; Else, to step 5.

$$J_n(U,x_k,v_i) = \sum_{k=1}^{n}\sum_{i=1}^{c}(u_{ik})^m d_{ik}^{\ 2}(x_k,v_i) + \varepsilon\sum_{k=1}^{n}\sum_{i=1}^{c}(u_{ik}-f_{ik}b_k)^m d_{ik}^{\ 2}(x_k,v_i) \tag{14}$$

Step 5.   Calculated new clustering centers $(v_1, v_2, \cdots v_c)$ and back to step2.

$$v_i = \frac{\sum_{p=1}^{n}(u_{ip})^m \cdot x_k}{\sum_{p=1}^{n}(u_{ip})^m}, \quad i = 1,2,\cdots,c \tag{15}$$

---

4   Experimental Results and Analysis of MOOC Data

4.1   Experimental Setup

Experiments were conducted with Python 3.8 as the programming language, restricted the MOOC learning scope to $T_C = 7$ days, and divided the MOOC datasets into small chunks with small batch training to analyze and study the learning behavior data from the learners of the XuetangX website from the MoocData platform presented above.

4.2   Criteria for the Assessment of Abnormal Learning Behaviors

To evaluate the results of the M-ISFCM model, these experiments use the precision rate (*pre*), recall rate (*recall*), and *AUC* as the metrics for learning behavior anomaly detection.

   Pre describes the probability that an abnormal learner sample is detected as true, and in the MOOC learning behavior abnormality detection scenario, the model has a high precision rate to prevent excessive false positives, which is calculated as follows:

$$Pre = \frac{TP}{TP + FP} \tag{16}$$

Recall describes the proportion of anomalous learner samples detected. In the MOOC learning behavior anomaly detection scenario, the model is expected to hold a high-level recall to avoid excessive omissions, which is calculated as follows:

$$Recall = \frac{TP}{TP + FN} \tag{17}$$

AUC describes the area under the characteristic curve ROC of MOOC data, and since the ROC curve cannot explicitly depict the classification performance of the *M-ISFCM* model, AUC is selected as one of the evaluation metrics.

In the preceding formula, *TP* represents the number of learners who are truly abnormal among those judged abnormal. *FP* represents the number of learners who are classified as abnormal by the model but whose actual situation is normal. *FN* represents the number of learners deemed abnormal but whose actual situation is abnormal. *TN* represents the number of learners who are truly normal among those who are judged to be normal.

4.3 Experimental Comparison of MOOC Data

Effect of Equilibrium Factor ε in the M-ISFCM Model on Experimental Results.
The parameter ε is the semisupervised clustering equilibrium factor, which is used to balance the degree of guidance of labeled data on semisupervised clustering. The influence of parameter ε on the detection of abnormal MOOC learning behavior is depicted in Fig. 3 below. As we can see, as ε increases, the labeled data guide the semisupervised clustering deeper, and Pre, Recall, and AUC show a rising trend, demonstrating that the effect of MOOC abnormal learner detection is on the increase.

It can be observed that when $\varepsilon \approx 0$, the worst effect of MOOC abnormal learning behavior detection is achieved when the credibility value is not taken into account at all, indicating that unlabeled data are useful for M-ISFCM. At $\varepsilon \approx 1$, the credibility value weight of the M-ISFCM increases, resulting in a decrease in the effectiveness of the M-ISFCM. According to the graph, the best effectiveness of the M-ISFCM is at $\varepsilon \approx 0.8$, which is therefore chosen as the optimal value of the parameter (Fig. 2 and Tables 2, 3).

Effect of the Weight Coefficient ω in Credibility P on the Experimental Results.
The parameter ω is the weight coefficient for calculating the credibility value *P*, which is used to determine the weight of unlabeled data outliers and the similarity of labeled data clustering centers. Figure 4 above depicts the effect of the parameter on the detection of anomalies in MOOC learning behavior.

As the parameter ω is gradually increased in the graph, the weight of unlabeled data anomalies decreases, and the fold plots of *Pre*, *Recall*, and *AUC* roughly show concave arcs. The best effect of the M-ISFCM is achieved at $\omega \approx 0.7$, so this value is chosen as the optimal value.

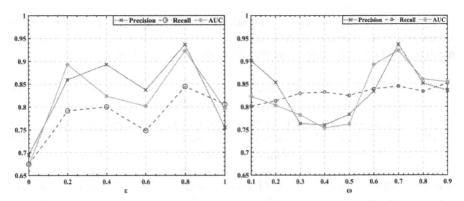

Fig. 2. Impact of ε on detection effect.   **Fig. 3.** Impact of ω on detection effect.

Table 2. Data of the ε impact factor.

x	0	0.2	0.4	0.6	0.8	1.0
Precision	0.6952	0.8593	0.8931	0.8372	0.9367	0.7544
Recall	0.6744	0.7915	0.8001	0.7482	0.8445	0.8054
AUC	0.6722	0.8933	0.8237	0.8017	0.9231	0.7962

Table 3. Data of the ω impact factor.

x	0.1	0.2	0.3	0.4	0.5	0.6	0.7	0.8	0.9
Precision	0.9014	0.8534	0.7632	0.7599	0.7833	0.8333	0.9367	0.8511	0.8354
Recall	0.8011	0.8131	0.8293	0.8321	0.8244	0.8389	0.8445	0.8333	0.8511
AUC	0.8233	0.8033	0.7821	0.7533	0.7621	0.8921	0.9231	0.86	0.8543

Comparisons of Anomaly Detection Algorithms. To validate the effectiveness of the M-ISFCM in this paper, we performed cross-sectional comparisons on the same datasets with the unsupervised isolated forest algorithm, unsupervised fuzzy C-mean clustering, and semisupervised fuzzy C-mean clustering, using accuracy, recall, and AUC as performance metrics.

According to Fig. 5, the accuracy rate of the M-ISFCM algorithm reached 93.7%, the recall rate reached 84.5%, and the AUC reached 92.3% . When compared to the simple unsupervised algorithm isolated forest algorithm and fuzzy C-mean clustering, the accuracy rate improved by 17% and 14.9%, the recall rate improved by 13.2% and 12.4%, and the AUC improved by 16.8% and 12.4%, respectively, which shows that M-ISFCM outperforms other anomaly detection algorithms in MOOC learning behavior anomaly detection (Table 4).

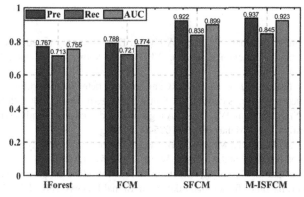

Fig. 4. Effectiveness of the M-ISFCM compared to other detection algorithms.

Table 4. Comparison of runtimes.

	M-ISFCM	FCM	SFCM	IF
Runtime/s	576	876	1223	153

The above table compares the runtimes of the M-ISFCM and other anomaly detection algorithms. Although IF runs the fastest, it has low model performance. The table shows that the improved M-ISFCM runs in less time than the SFCM before the improvement, demonstrating that the improved M-ISFCM effectively reduces the algorithm's complexity.

5   Conclusion

In this paper, we propose an M-ISFCM model for MOOC learning behavior. The model detects anomalies in MOOC learning behaviors by combining the SFCM model with the isolated forest algorithm.

The model is used to discover anomalies in a large quantity of unlabeled MOOC learning behavior data, and it presents a novel strategy that combines semisupervised and unsupervised learning behavior anomaly identification. Experiments on MOOC datasets show that the model can detect MOOC learning behavior anomalies more effectively, and the next work plans to study the MOOC learning behavior data generated in real time, analyze the situation of the model in the actual production process, and further optimize the model.

Acknowledgments. The authors would like to thank the anonymous reviewers for their elaborate reviews and feedback. This work was supported by the National Natural Science Foundation of China (No. 61906099), the Provincial Undergraduate Training Program for Innovation and Entrepreneurship (No. SYB2021019), and the Open Fund of Key Laboratory of Urban Land Resources Monitoring and Simulation, Ministry of Natural Resources (No. KF-2019–04-065).

References

1. Garg, S., Kaur, K., Kumar, N., et al.: A Hybrid Deep Learning-based model for anomaly detection in cloud datacenter networks. IEEE Trans. Netw. Serv. Manag. **16**(3), 924–935 (2019)
2. Ji, C., Zou, X.N., Liu, S.J., et al.: ADARC: an anomaly detection algorithm based on relative outlier distance and biseries correlation. Softw.-Pract. Exp. **50**(11), 2065–2081 (2020)
3. Duan, X.Y., Ying, S., Yuan, W.L., et al.: A log anomaly detection method based on Q-learning algorithm. Inf. Process. Manag. **58**(3), 102540 (2021)
4. Cappozzo, A., Greselin, F., Murphy, T.B., et al.: Anomaly and Novelty detection for robust semisupervised learning. Stat. Comput. **30**(5), 1545–1571 (2020)
5. Lesouple, J., Baudoin, C., Spigal, M., et al.: How to introduce expert feedback in one-class support vector machines for anomaly detection. Signal Process. **188**, 108197 (2021)
6. Akcay, S., Atapour-Abarghouei, A., Breckon, T.P.: GA nomaly semi-supervised anomaly. Lect. Notes Artif. Intell. **11363**, 622–637 (2019)
7. Yang, L., Chen, J.J., Wang, Z., et al.: Semisupervised log-based anomaly detection via probabilistic label estimation. In: 2021 IEEE/ACM 43rd International Conference on Software Engineering (ICSE 2021), pp. 1448–1460. Electrical Network (2021)
8. Domaika, F., Baradaaji, A., EI Traboulsi, Y.: Soft label and discriminant embedding estimation for semi-supervised classification. In: 2020 25th International Conference on Pattern Recognition (ICPR), pp. 7250–7257 (2021)
9. Martino, G., Gruenhagen, A., Branlard, J., et al.: Comparative evaluation of semi-supervised anomaly detection algorithms on high-integrity digital systems. In: 2021 24th Euromicro Conference on Digital System Design (DSD 2021), Palermo, Italy, pp. 123–130 (2021)
10. Teng, G., Bai, X., Tian, X., et al.: Educational anomaly analytics: features, methods, and challenges. Front Big Data **4**, 811840 (2021)
11. Kamalov, F., Sulieman, H., Calonge, D.S.: Machine learning based approach to exam cheating detection. Plos One 16(8), e0254340 (2021)
12. Bouchachia, A., Pedrycz, W.: A semisupervised clustering algorithm for data exploration. Lect. Notes Artif. Intell. **2715**, 328–337 (2003)
13. Liu, F.T., Ting, K.M., Zhou, Z.H.: Isolation forest. In: ICDM 2008: Eighth IEEE International Conference on Data Mining, Proceedings, p. 413 (2008)
14. Havens, T.C., Bezdek, J.C., Leckie, C., et al.: Fuzzy c-means algorithms for very large data. IEEE Trans. Fuzzy Syst. **20**(6), 1130–1146 (2012)
15. Kwon, S.H.: Threshold selection based on cluster analysis. Pattern Recogn. Lett. **25**(9), 1045–1050 (2004)
16. Mukhopadhaya, S., Kumar, A., Stein, A.: FCM approach of similarity and dissimilarity measures with alpha-cut for handling mixed pixels. Remote Sens. **10**(11), 1707 (2018)
17. Hai, D.T., Son, L.H., Le, V.T.: Novel fuzzy clustering scheme for 3D wireless sensor networks. Appl. Soft Comput. **54**, 141–149 (2017)

Classroom Teaching Effect Monitoring and Evaluation System with Deep Integration of Artificial Intelligence

Aili Wang[1], Yingluo Song[1], Haibin Wu[1(✉)], Shanshan Ding[1], Linlin Yang[1], Yunhong Yang[1], and Yuji Iwahori[2]

[1] Heilongjiang Province Key Laboratory of Laser Spectroscopy Technology and Application, Harbin University of Science and Technology, Harbin 150080, China
woo@hrbust.edu.cn

[2] Department of Computer Science, Chubu University, Kasugai 487-8501, Japan

Abstract. With the progress and development of the times, education informatization has become the primary content to promote the progress of education. However, with the rapid advancement of technology, many problems have arisen in the integration of online education and information technology. From a practical point of view, advanced information technology should be fully integrated with school education activities to promote education reform and development, and students will definitely obtain a better learning experience. This study aimed to introduce a classroom teaching effect monitoring and evaluation system that incorporates artificial intelligence. In view of the background conditions for the development of this system, reforming measures and specific implementation methods for the integration of information technology and online classroom education are proposed.

Keywords: Classroom teaching · Artificial intelligence · Integrated practice

1 Introduction

To conduct more in-depth research on the integration of information technology and classroom teaching, analyze the current situation and propose improvement measures, this paper first elaborates on the current development of information technology and online education.

In 2000, the American Educational Technology CEO Forum defined the integration of information technology and curriculum and expounded the theory and implementation of the integration. Subsequently, some short video course resources emerged on the Internet, which were favored by the public due to their refined characteristics. In 2008, two Canadian professors developed the first MOOC course in the true sense, called "Connectivity and Connected Knowledge", which was widely sought by online learning. Linyun Zhang found that MOOCs have more advantages than traditional online education and studied new teaching models, which provides certain theoretical conditions for teaching reform [1]. Xinlei Wang proposed that information technology is

Y. Wang et al. (Eds.): ICPCSEE 2022, CCIS 1629, pp. 337–350, 2022.
https://doi.org/10.1007/978-981-19-5209-8_23

changing the teaching environment and teaching patterns and provides a beneficial platform for personalized teaching and learning [2]. Kekang He proposed that the integration of information technology and courses should create a new information teaching environment, which can realize a variety of new activities, such as scenario creation, independent learning, resource sharing, and collaborative communication [3]. Cranton Patricia and Torrisi Steele Geraldine discussed the potential for fostering transformative learning in an online environment. They mainly discussed the value of learning theory reform, including the evolution of teaching reform theory and so on [4]. Lemay David John, Bazelais Paul and Doleck Tenzin took the form of a survey. The main content is the idea of students transitioning the learning process to online learning. According to their research, it can be found that online teaching also needs teachers in addition to the technical and pedagogical aspects, and educators are always on the emotional dimension of social online learning [5].

In view of the above related discussion, it can be understood that there are still obvious shortcomings between current classroom education and the integration of information technology, and the current related research does not involve the content of intelligent classroom effect detection based on deep learning. Therefore, this article will complete the discussion of the detection and evaluation system of classroom teaching effects integrated with artificial intelligence.

2 Related Background

2.1 The Lack of Integration of Information Technology into Teaching Information Among Teachers

According to the relevant materials consulted and the investigation of the current education environment, it can be found that there are many problems in the integration of current online classroom education and information technology. Due to certain barriers between the development of information technology and teaching, teachers have insufficient understanding of modern teaching technology, and many teachers do not understand the value of intelligent fusion technology. At present, the application of multimedia technology in colleges and universities has become a common phenomenon, and multimedia technology has also been optimized, which belongs to the primary application stage of modern information technology in education [6]. In actual education and teaching, teachers only use electronic teaching plans to replace the original textbook teaching plans, computers and projections to replace blackboards and other methods to use multimedia technology, and this relatively single technology application accounts for a large proportion of college education and teaching.

2.2 Lower Student Participation in Online Teaching

In addition to providing priority conditions for more flexible time and space, online teaching can also enhance students' interest in learning and greatly improve teaching effectiveness. When teachers carry out teaching plans, teachers need to invest much energy to implement online education courses, but there is often a lack of supervision of students in the actual process. At the same time, students themselves lack autonomy and participation. At present, many colleges and universities have developed their own online courses, but they are limited to some specific subjects, many online course resources have not been connected to each other, and most of them are restricted to the teachers and students of the school.

2.3 Inefficient Use of Information Technology Among Teachers

The main application characteristics of modern information technology used in education and teaching are networking, multimedia and intelligence. The use of modern information technology to assist classroom teaching can improve students' interest in learning and at the same time can strengthen students' understanding and mastery of teaching knowledge, thereby enhancing the learning effect. However, in practical applications, such as science and engineering experimental courses, some teachers rely too much on information technology to demonstrate the effects of experiments, which will lead to unsatisfactory learning effects for students [7]. In addition, the level of information technology of teachers is uneven. When individual teachers use information technology to assist teaching, they directly use ready-made resources obtained through online downloads rather than design and production by themselves, and some teachers are not yet able to use commonly used information equipment proficiently.

2.4 Insufficient Development and Application of VR Technology

Virtual reality (VR) is an immersive interactive environment based on computable information. It is mainly based on modern technology to achieve real and indistinguishable vision, environment and other feelings and a virtual environment. Users can feel the state they are in and can also take a series of virtual actions. Virtual reality technology can break through the dual limitations of space and time. Teachers can use this technology to teach students, which can not only effectively improve the teaching effect but also strengthen students' learning enthusiasm to a certain extent. At present, much educational efficiency requires the improvement of this technology in the actual teaching process.

3 Reform Measures

This research mainly uses artificial intelligence fusion technology to conduct background research and analysis on the online classroom and at the same time proposes specific optimization measures. It can not only monitor the learning status of the classroom but also obtain the actual student learning behavior.

3.1 Face Recognition System Based on Deep Learning

Modern face recognition is based on deep learning technology and convolutional neural networks, which extract face eigenvalues from input images to calculate and analyze facial expressions. Face recognition technology builds a data collection sample set by analyzing the current students' concentration and completing the classroom concentration analysis through training calculation. Through the distribution results of students' concentration in the classroom, the teacher's classroom learning situation is analyzed, and the education and teaching reform of the machine learning course is carried out. Deep learning is the current basic research based on deep-level artificial intelligence network technology, and it is also one of the teaching courses of machine learning courses. In the machine learning class, this face recognition system algorithm is explained to realize the combination of information technology and classroom teaching and enhance mutual integration.

Face recognition technology recognizes the input face image or video stream according to the facial features of the person. The first is to analyze the image to determine whether there is a face in the image. If there is a face in it, it will take the next step to obtain other information such as the location of each face, perform feature extraction based on the above information, and compare it with the current face database to identify the specific identity information. When establishing a data collection group sample and using the principles of statistics to complete the analysis of the concentration of students in the classroom, the analysis of the teaching process and the concentration can be correlated to form the result of the concentration distribution of the classroom process [8]. This technology can not only improve students' concentration in class but also allow students to learn further about machine learning courses.

3.2 Online Course Performance Prediction Model Based on Behavior Analysis

A very effective research direction has been established in the development process of the modern information technology field, which is the prediction model based on the analysis of online behavior learning, and then scientifically analyzes the relevant factors that affect performance. The model uses knowledge of data mining, neural networks and machine learning, which is in line with the teaching content of the "machine learning" course. This model can analyze and find the learning behavior index data closely related to online course performance to realize performance prediction, which provides a good reference for the implementation, development, and reform of "machine learning" online courses.

First, the establishment of a performance prediction model requires a systematic analysis of learning behavior and performance prediction strategies and a mechanism for online platform data processing, performance prediction algorithm design, performance prediction and algorithm optimization. Second, the model uses the behavior data collected by mining technology to analyze the behavior data in combination with the operating characteristics of online users, obtains the data of 10 behavior indicators closely related to the performance and then stores it in the database. Finally, the model uses the back-end database of the "Machine Learning" course of Harbin University of Science and Technology on the MOOC platform as the experimental data basis and analyzes the learning behavior of students based on the characteristics of the course implementation. This model determines the level of behavior indicators, extracts and converts behavior data, and uses neural networks to predict online course performance. This model uses modern information technology to give students fair and equitable course scores, which is in line with the teaching needs of "machine learning" courses.

3.3   Based on the VR Virtual Deep Operation Drill Learning Platform

Modern information technology includes big data technology, cloud computing, blockchain, and 5G communication technology. VR technology is an advanced technology that has emerged with the rapid progress of modern data technology and is applied in the practical operation of "machine learning" courses. VR technology is convenient for solving problems in deep learning practice teaching, guiding teaching activities in a standardized and orderly manner, and driving rapid student machine learning practice literacy [9].

Based on the smart laboratory of VR virtual reality technology, students can maintain a lasting interest in learning through computer programming, Python system installation or smart robots arranged by the school to maintain a lasting interest in learning and continue to acquire machine learning knowledge. With the deepening of learning, students can use information technology to complete many professional learning tasks to meet teaching needs. Through the establishment of a VR intelligent laboratory, the school creates a real simulation situation and uses the organic interaction between learning content and practical teaching so that students can stimulate deep thinking and feelings while observing and operating and promote technological innovation. In addition, teachers can collect computer hardware data from the Internet to realize 3D modeling of computer hardware based on 3D technology and create interactive practice scenes. On this basis, the teaching model, sound, text and other elements are integrated into it, and these designs are vividly presented in front of students through the form of web pages. In this way, it is helpful for students to observe the neural network structure from multiple angles, grasp the latest developments in machine learning, and strengthen the students' exploration ability. While enriching the computer knowledge reserve.

Different from other courses, "machine learning" practical teaching has distinct practical characteristics and is in close contact with actual practical applications, which has far-reaching significance to speed up the reform of intelligent teaching. Teachers need to demonstrate the learning content for students, but due to many factors, they cannot be fully presented in front of students, which greatly affects the practical teaching effect of "machine learning". The establishment of a virtual experimental practice operation platform based on VR technology, where students perform operations and exercises, has a positive effect on the cultivation of students' creativity and moldability.

3.4   EEG Attention Monitoring Technology to Monitor Classroom Teaching

As an intelligent information technology, EEG attention detection technology has an extremely important position. The current collection and application of EEG signals by this technology has shifted from clinical research to various neighborhoods. For example, in the aspect of intelligent integrated teaching, EEG attention monitoring technology can be used to analyze and record the brain signals of students to strengthen their attention, thereby helping teachers improve teaching effects. Especially in the "machine learning" classroom, EEG attention monitoring technology analysis can realize the combination of algorithms and equipment to create a smart classroom.

Under the current teaching mode, due to the lack of understanding and communication between teachers and students and the limitation of intelligent teaching, students cannot find problems and feedback to teachers in a timely manner. It is also difficult for teachers to obtain the detailed learning status of students from relevant intelligent teaching systems, and the implementation effect of teaching programs cannot be accurately obtained. Therefore, EEG attention monitoring technology is introduced into education and teaching, combined with artificial intelligence technology and machine learning education and teaching to solve the problem. Under the traditional teaching mode, the method for teachers to know students' attention is single, and the results are not accurate. The EEG wearable system uses the principle of EEG attention monitoring technology to sense the brain wave frequency spectrum, EEG signal quality, and original brain waves of the tested student through EEG sensing equipment and collects the blink frequency, relaxation and seriousness. The resulting parameters will then be collected and sent out by wireless sensor devices. These preprocessed data can be used to reflect the concentration of the student.

With the advent of the Internet era and the continuous innovation and development of technology, intelligent technologies such as language processing, face recognition, and virtual technology have been widely used and developed. Smart technology devices have long been common in daily life, but they are always changing people's lifestyles. As communication entertainment software, WeChat has been widely used in people's lives. Therefore, the WeChat platform can be used to release the data after processing and analysis. When students query their attention concentration data on the platform, they can also know the change in their attention degree through data comparison and chart display to realize self-improvement. Teachers applying EEG attention monitoring technology to assist students in learning can not only change traditional learning methods and ensure the rationality and effectiveness of teaching programs but also help teachers accurately select teaching priorities and difficulties and correctly distinguish differences between students and put forward solutions.

4   Specific Method

After discussing the current status of online education and technology integration and the measures that need to be reformed and implemented, this research proposes specific methods to solve the above problems. The implementation process is as follows.

4.1   Machine Learning Classroom Teaching Based on the Course Library on the Cloud

Teachers should actively hold conferences on the intelligent integration of teaching information technology, etc., which can not only promote the development of "machine learning" course teaching but also explore new forms of learning on the cloud. In addition to the first time, it is also necessary to pay attention to the existing advanced experience of other institutions in time to explore the curriculum integration method suitable for the characteristics of the students of the school.

During the teaching process, the "machine learning" course can be selected as the experimental content. Teachers use information acquisition technology to select resources suitable for students from the Internet according to the standard of the course syllabus. At the same time, teachers can use information storage technology to choose a suitable platform to build a "cloud-based curriculum library" structure and set the management authority of teachers and student groups in the platform so that all students can access it, as shown in Fig. 1.

Teachers need to strengthen the learning of advanced information technology. After learning, teachers should combine the learning content with the algorithms used in machine learning, convey them to students in a timely manner, and pay attention to their feedback and suggestions. In addition, teachers can use information storage technology to choose a suitable platform to build a "course library on the cloud" structure.

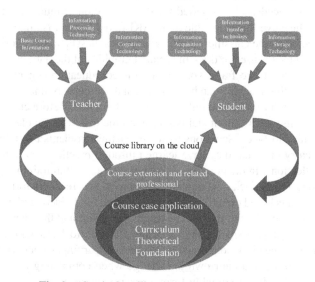

Fig. 1. "Curriculum library on the cloud" structure

4.2 Hybrid Teaching Mode Based on a Deep Neural Network

Blended teaching is a teaching form that combines traditional and modern learning methods in the current teaching environment. This method can not only improve students' learning enthusiasm but also speed up the teaching process. To a certain extent, it can make teachers play a leading role in the guidance, initiation and follow-up process. In the mixed teaching mode, compared with traditional teaching, online teaching has a single process and data, but the complexity will be significantly improved, and a large amount of teaching data can be generated.

Blended teaching mainly involves four aspects: learning theory, learning environment, learning methods, and the integration of learning resources. Under the guidance of blended learning theory, hybrid teaching with Internet + college education as the background reasonably constructs a student-based learning strategy, and this model supports online, offline, virtual and diversified learning environments and can integrate online with face-to-face learning. This model includes multiple types of high-quality learning resources online and offline, which can promote the overall development of students. The following takes the course teaching of "Machine Learning" as an example to design a mixed teaching model based on deep neural network teaching student portraits.

In the teaching of artificial intelligence, the classification problem in machine learning is a more popular and comprehensible problem. The Naive Bayes Algorithm in the classification problem is clearer for students with a certain mathematical foundation, and it is easier for students to understand and accept the teaching as a case. This algorithm uses the classic dataset called iris in machine learning and uses the Python list operation to extract the last two dimensions in the feature space of the iris dataset to form two-dimensional data, which can construct a two-dimensional plane image. In addition, the numpy library needs to be imported for the calculation of multidimensional arrays and matrices. The naive Bayes module in the SKLEARN library is imported, and GaussianNB is selected as the classifier of the naive Bayes classification algorithm. Finally, the matplotlib library is used to draw the related graphics of the dataset.

According to the basic framework of the mixed education organization process, naive Bayes algorithm teaching can be designed in three steps: online instruction before class, offline research in class and online practice after class. Before class, students can build a self-learning algorithm model based on online resources; in class, teachers and students have face-to-face discussions to solve existing problems together; after class, students will carry out what they have learned through practical tests, summaries and consolidate reflections. In the practice of naive Bayes algorithm training, the main task of online guidance before class is how to cause problems. Teachers formulate preclass guidance documents based on the teaching content and send video guidance documents to students in advance, listing the plan of teaching content and the questions that need to be answered in class, which can encourage students to learn with questions. As shallow learning in the classroom transfers to preclass learning, students understand the difficulties of learning and their own knowledge when completing preclass tasks and use social tools (WeChat) to quickly form a discussion group to discuss each other's gains and questions. In the classroom, the main task of offline research is how to collect and solve problems. Teachers use traditional classrooms to promote in-depth interaction between students. On the one hand, teachers understand and solve students' difficulties based on students' preclass guidance. On the other hand, students' collaborative research can provide more problem-solving methods and increase the depth of learning. The main research task of online practice after class is how to deepen the problem. Teachers help students evaluate what they have learned by assigning homework and conducting after-school tests [10]. Teachers use the evaluation function of the learning platform to provide feedback to students to help students adjust their learning status at any time and improve learning efficiency. This method allows students to intuitively and quickly harvest the results of machine learning and stimulates students' interest to seek causes, avoiding the situation that students are daunted by the incomprehensibility of artificial intelligence courses [11]. Because of this, using the above cases to participate in the actual teaching of students not only accelerates the development of artificial intelligence teaching but also improves the quality of students' teaching [12].

This teaching mode is based on the operation of the mass data of mixed teaching. During the operation, this mode extracts the characteristic values related to the student's learning effect, reasonably introduces some other data to supplement the missing data, and finally constructs a student's learning portrait based on the deep neural network [13]. According to relevant research findings, students' classroom learning, after-school learning and learning attitudes are also factors that affect academic performance [14]. The behavioral portrait method, which is based on the actual situation of students based on various dimensional data, can more accurately understand and analyze the characteristics of students and can effectively improve the learning assistance for students and can also improve students' learning to a certain extent. This method can greatly improve the learning efficiency of students. AI's means to serve teaching can improve teachers' and students' cognition of current teaching effects and provide more rigorous technical support for mixed teaching [15].

5   Experimental Methods

5.1   Experimental Setup

Throughout this section, sets of experiments are provided for obtaining test data collection, data annotation techniques, annotator-observed attention level assessments, and correspondence with student behavior. The conditions of the experiments are described below.

1. The brightness of the test environment (computer lab room) was fixed.
2. The size of the classroom was set during the study with 6 undergraduate students as subjects, consisting of 1 female and 5 male students.
3. The camera used was a visionseed sensor attached directly to a laptop on top of the teacher's desk. The camera was positioned at the front of the classroom at a height of 1.6 m and an angle of 35°. The camera was able to clearly capture the students and the entire lab.
4. Video recording and image capture were performed during one and a half hour computer lab session. Six separate recordings were made to collect data.

5.2   Dataset Collection and Preparation

A series of recordings was made for the two classes based on the set environment, Attentive and Not Attentive scenario. Two videos were recorded wherein the students made the Attentive and Not attentive behavior that is visible from student faces. A webcam was used to generate a series of images that were used to generate the dataset [16]. The type of data was the format of the images. Once the video was successfully recorded, the recording was separated frame by frame into pictures. This resulted in over 14,000 frames, composed of attentive and nonattentive images of the students. With these images, it was possible to create a dataset by selecting good quality frames in which the student's attentiveness and not attentiveness were evident. These are prepared for later data processing. The methods used to predict student behavior are shown below [17].

Once the dataset is collected, it is prepared for annotation and segmentation [18], training and testing to generate models for face recognition and to predict student behavior based on the set facial features. Figure 2 shows the framework of the system for collecting the dataset set [19].

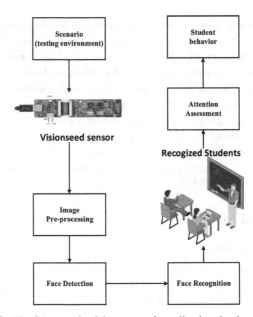

Fig. 2. The framework of the system for collecting the dataset set

5.3 Annotation and Splitting of Dataset

After building a dataset consisting of several images, these images were annotated to mark whether each student's face was attentive or not. Based on the format of the trainer in CNN [20]. The files were labeled for this reason. This procedure helps to create boxes that enclose the students' faces for subsequent facial recognition and labeling in their

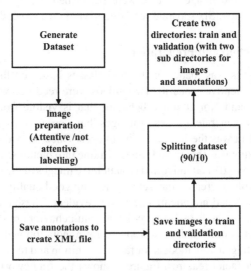

Fig. 3. Annotation and splitting of the dataset

respective classrooms. Each file will contain the object class, x, y, width and height. x, y, width and height are floating values and contain the relative width and height of the image. Figure 3 shows the processing of annotation and splitting of the dataset [21].

5.4 Model Training

A total of 4087 images were used, 90% of which were used for training and 10% for dataset testing, as shown in Fig. 4. The ImageAI package was used to train the dataset. It supports the YOLOv4 [22] algorithm to detect objects, students' faces and eyes. Tencent YouTu AI labs, an online transcript provided by Tencent YouTu, used to speed up the training using visionseed sensors. The Python platform was also used in the training. The visionseed package enables the laptop to run the algorithm as it is based on its library [23].

Fig. 4. Trained examples

Training and evaluation used the YOLOv4 algorithm to generate models for student face recognition and student behavior. After an optimal dataset was developed, the model was generated using a deep learning algorithm. YOLOv4 was used for dataset processing [24]. This method takes the sample and generates a pattern that can be used in fast-paced images. This ensures that the software can be used in high frame rate videos and yet has good accuracy.

After a long period of training, several models are made each time a loss of preparation falls. This loss is the percentage of bad predictions made by the trained model. A low loss means that the model is more reliable. The final model is then used to identify interfering images. However, a lower loss model does not always mean that it is the most accurate and precise [25]. This is mainly due to overfitting. A model may be overfitted during training, which means it is biased and may lead to misclassification.

6 Experimental Results and Analysis

The test focused on a live stream of the sample video. To validate the procedure, two situations were developed: an attentive environment and an inattentive environment, in which most students would show attentive behavior to the attentive situation, and the same way would be extended to the inattentive environment. The test results are shown in Figs. 5 and 6.

Fig. 5. Not Attentive: sample test live recognition of student behavior in the laboratory classroom

Fig. 6. Attentive: sample live recognition of student behavior in the laboratory classroom

From the sample live detection, students face was recognized. Each head of the student and eye movement was also detected in the defined scenario. Assessment was based on the set class: high = Attentive and low = Not Attentive. Below is the summary of the test results of student behavior assessment (Table 1).

Table 1. Summary of test results of student behavior assessment

Subject	Class	Assessment	mAP (%)
Student01	High	Attentive	93.64
Student02	Low	Not Attentive	87.77
Student03	High	Attentive	95.68
Student04	High	Attentive	96.92
Student05	High	Attentive	94.37
Student06	High	Attentive	95.24

7 Conclusions

The "Machine Learning" course has the advantages of cutting-edge times, high technical level, and strong practical ability. This paper integrates information technology with higher education, realizes the innovation of teaching ideas and concepts, and discusses the deep integration of new technologies such as the Internet, big data, artificial intelligence, and virtual simulation with machine learning teaching.

References

1. Zhang, L.: Comparison of MOOC and traditional online education in the context of online and offline mixed teaching. J. Office Informatization **26**(19), 6–8 (2021)
2. Wang, X.: Research on educational informationization and teaching reform in colleges and universities. J. Cult. Ind. **13**(06), 149–150 (2021)
3. He, K.: On the "Chinese Wisdom" of educational information. Teach. J. **23**(09), 18–23 (2019)
4. Patricia, C., Geraldine, T.S.: transformative learning in an online environment. Int. J. Adult Educ. Technol. **12**(4), 1–11 (2021)
5. John, L.D., Paul, B., Tenzin, D.: Transition to online learning during the COVID-19 pandemic. Comput. Hum. Behav. Rep. **4**(30), 100–130 (2021)
6. Wu, G., Ni, J.: Research on the development of MOOC projects for teachers' teaching ability improvement under the background of "Internet + education". Sci. Educ. Article Collects **542**(26), 13–15 (2021)
7. Sun, Z.: Focus on personalized and precise learning in smart classrooms. China J. Multimedia Netw. Teach. **6**(07), 110–111(2021)
8. Chen, L.: Research on the strategies of deep integration of modern information technology and college education and teaching. China Manag. Informationization **24**(12), 233–234 (2021)
9. Dineshwar, R., Oren, M., Mantoo. D., Khemraj, K.: A study of university students' attitude towards integration of information technology in higher education in Mauritius. J. High. Educ. Q. **75**(02), 348–363 (2020)
10. Liu, J.: Research and thoughts on modern information technology promoting the reform of university curriculum teaching mode. Educ. Teach. Forum **01**(04), 64–67 (2021)
11. Liu, T., Hu, W., Liu, F., Li, Y.: Sentiment analysis for MOOC course reviews. In: Zeng, J., Qin, P., Jing, W., Song, X., Lu, Z. (eds.) ICPCSEE 2021. CCIS, vol. 1452, pp. 78–87. Springer, Singapore (2021). https://doi.org/10.1007/978-981-16-5943-0_7

12. Mao, Y., et al.: Learning behavior-aware cognitive diagnosis for online education systems. In: Zeng, J., Qin, P., Jing, W., Song, X., Lu, Z. (eds.) ICPCSEE 2021. CCIS, vol. 1452, pp. 385–398. Springer, Singapore (2021). https://doi.org/10.1007/978-981-16-5943-0_31

13. Zhao, X., Chen, C., Li, Y.: Implementation of online teaching behavior analysis system. In: Zeng, J., Qin, P., Jing, W., Song, X., Lu, Z. (eds.) ICPCSEE 2021. CCIS, vol. 1452, pp. 399–409. Springer, Singapore (2021). https://doi.org/10.1007/978-981-16-5943-0_32

14. Thongkoo, K., Panjaburee, P., Daungcharone, K.: Integrating inquiry learning and knowledge management into a flipped classroom to improve students' web programming performance in higher education. Knowl. Manage. E-Learn. 11(3), 304–324 (2019)

15. Alhazzani, N.: MOOC's impact on higher education. Soc. Sci. Humanit. Open 2(1), 1–6 (2020)

16. Lu, F.: Research on the development and application of English MOOC system based on neural network. Autom. Instrum. 7, 226–228 (2019)

17. Wang, Y., Liang, Z., Cheng, X.: Fast target tracking based on improved deep sort and YOLOv3 fusion algorithm. In: Zeng, J., Qin, P., Jing, W., Song, X., Lu, Z. (eds.) Data Science. ICPCSEE 2021. Communications in Computer and Information Science, vol. 1451, pp. 360–369. Springer, Singapore (2021). https://doi.org/10.1007/978-981-16-5940-9_27

18. Wu, Y., Li, Z., Li, Y., Liu, Y.: Teaching reform and research of data structure course based on BOPPPS model and rain classroom. In: Zeng, J., Qin, P., Jing, W., Song, X., Lu, Z. (eds.) ICPCSEE 2021. CCIS, vol. 1452, pp. 410–418. Springer, Singapore (2021). https://doi.org/10.1007/978-981-16-5943-0_33

19. Long, J.: Research on the application of micro-courses in higher vocational teaching reform. Int. J. Educ. Econ. 5(2), 44–45 (2022)

20. Li, H., Li, A., Zhu, X.: A brief analysis of the comprehensive application of MOOC teaching mode and virtual reality technology in engineering education. Int. J. High. Educ. Teach. Theory 3(2), 14–15 (2022)

21. Yuan, T.: Research on MOOC teaching mode in higher education based on deep learning. Comput. Intell. Neurosci. 1, 8031602 (2022)

22. Zhang, X.: Computer basic teaching reform in applied undergraduate colleges based on "MOOC". In: Proceedings of 3rd International Conference on Education, Economics and Management Research (ICEEMR 2019) (Advances in Social Science, Education and Humanities Research), vol. 385, pp. 25–28 (2019)

23. Mao, W.: Video analysis of intelligent teaching based on machine learning and virtual reality technology. Neural Comput. Appl. 1, 1–12 (2021)

24. Bihui, C.: Design and implementation of mathematical MOOC based on BP Neural Network Algorithm. In: 2021 IEEE International Conference on Artificial Intelligence and Computer Applications (ICAICA), pp. 20–22 (2021)

25. Lu, F.: Research on the development and application of English MOOC system based on neural network. Autom. Instrum. 7, 226–228 (2019)

Audit Scheme of University Scientific Research Funds Based on Consortium Blockchain

Pengfan Yue[1,2] and Heng Pan[1,2(✉)]

[1] Zhongyuan University of Technology, Zhengzhou 450007, Henan, China
panheng@zut.edu.cn
[2] Henan International Joint Laboratory of Blockchain Data Sharing, Zhengzhou 450007, Henan, China

Abstract. In the current process of university scientific research funding audits, the auditors do not obtain audit evidence in a timely manner, resulting in a lag in audit time; at the same time, there is the possibility that the audit evidence may be tampered with, and the sensitive expenditure cannot be effectively protected. Based on this, this paper proposes an audit scheme for university scientific research funds based on a consortium chain. The scheme saves relevant audit evidence through blockchain and IPFS, which changes the storage method of traditional audit evidence and effectively ensures the integrity, auditability of audit evidence and verifiability, and facilitates real-time auditing. We use Pedersen Commitment and Zero-Knowledge Range Proof to ensure the auditability of sensitive data; use smart contracts to automate auditing to further ensure that auditors can find relevant audit issues in a timely manner; and separate different departments and construct an audit tree through multi-channel technology to ensure the integrity and reliability of data and improve audit efficiency and accuracy. Finally, functional comparison analysis and security analysis show that the scheme in this paper has certain feasibility and robustness.

Keywords: Blockchain · Real-time audit · Pedersen Commitment · Zero-knowledge range proof · Audit tree

1 Introduction

The auditing of scientific research funds refers to the internal control, budget management, revenue and expenditure management, balance management, and capital utilization efficiency of various types of scientific research funds of the school in accordance with the national, provincial, municipal and school regulations on the management of scientific research funds and project assignment monitoring and evaluation. A scientific research funding audit is conducted to scientific, rational and legal compliance audits on a series of budget forms, invoices, fund balances, etc.

There are two main problems in the traditional audit of scientific research funds in colleges and universities. On the one hand, because the original documents of scientific research expenditures are not open and transparent enough, generally only the corresponding business personnel and financial personnel keep these documents. There are certain difficulties when auditors obtain relevant audit evidence. If the audit object is unwilling to actively provide and open the data of the department, it will lead to a lag in the audit time, and the audit cannot be conducted in real time. Then a series of serious problems such as temporary fraudulent accounting will occur, which will bring problems to the audit findings. There are great challenges; on the other hand, some sensitive expenditures are directly presented on the documents, and all parties can view them directly on the documents, which cannot meet the needs of privacy protection.

The openness, transparency, non-tampering, non-repudiation and traceability of blockchain technology provide a powerful solution for the auditing of scientific research funds in universities. First, each node is open and transparent, and the relevant audit evidence (such as project budgets, project detailed accounts, expenditure details, and expenditure vouchers, etc.) of expenditures is uploaded to the chain, and nodes on each chain can see it. On the one hand, it is the same type. The business processes of different batches of business can be compared with each other, which can play the role of mutual restriction and supervision; and prevent collusion among employees; on the other hand, auditors can directly understand the business process and obtain the required original documents. Second, the data on the chain cannot be tampered with and cannot be denied. The audit evidence related to the expenditure cannot be modified once it is uploaded to the chain. If there is a problem, it can only be reuploaded, which will keep the modification records. During the audit process, you can view the corresponding original documents and modify records to determine some problems or fraudulent behaviors in the use of funds, and those who have problems can be held accountable and punished.

However, the characteristics of blockchain also bring some problems. On the one hand, malicious nodes can see the specific amount of an item on the chain, and will conduct financial reimbursement fraud (overreporting, false reporting, etc.) based on this amount to seek profit. On the other hand, the privacy of some expenditures cannot be effectively protected, and encrypted expenditures cannot well support multi-party circulation and auditing in the business process. Therefore, for this kind of sensitive spending, this paper adopts Pedersen Commitment to ensure the privacy of sensitive spending; and cooperates with zero-knowledge scope proof technology to ensure the auditability of sensitive spending.

Based on this, we propose an audit scheme of university scientific research funds based on a consortium chain. The main work is as follows:

1. Combining blockchain technology with auditing of university research funds, using IPFS to store the clear text of relevant audit evidence, and using blockchain to store the address returned by IPFS, it is open and transparent while ensuring that the original documents cannot be tampered with and non-repudiation, which is conducive to real-time audits. Improve the audit time lag problem in traditional audits.
2. Automated auditing is realized through smart contracts. One is the smart contract for integrity verification, which judges whether the relevant audit evidence has been tampered by calling the smart contract; the other is the smart contract for financial auditing, which verifies the zero-knowledge scope proof and commitment value. To ensure that all parties are honest; the third is the smart contract for time limit judgment, which regularly judges the time limit of the unfinished project to determine

whether the project is within the time limit to ensure that the audit office can timely detect the unfinished project in time.

3. The project audit tree is designed using the LBT tree structure, and the faculty audit tree and university audit tree are designed using the Merkle tree structure. The system automatically constructs the project when the project is completed, which increases the cost of fraud. The integrity of the relevant data can be quickly verified through the path, and can effectively judge whether the relevant information stored in the database has been tampered with, and improve the efficiency and accuracy of auditing.

4. Ensure privacy and auditability of sensitive spending with Pedersen Commitment and Zero-Knowledge Range Proof. Among them, the obscurity of Pedersen Commitment is used to protect the privacy of sensitive expenditures; the binding and homomorphism of Pedersen Commitment are used to ensure the auditability of sensitive expenditures; and the zero-knowledge range proof can be used to test the effective range of commitment values.

2 Related Work

At present, there are few studies on the auditing of university scientific research funds based on blockchain, but auditing work involves data integrity verification, amount auditing, user traceability, confidentiality and auditability of sensitive data, and other related technical studies are relatively mature.

In terms of the integrity verification of audit evidence, that is, data integrity verification, reference [5] designed a flexible cloud data audit scheme based on blockchain. To eliminate the dependence on third-party auditors, a decentralized audit framework was proposed, thus improving the stability, security and performance of the whole scheme. Cloud service providers can automatically generate audit credentials to reduce the burden of cloud service providers. A Merkel hash tree is also used to improve validation performance. Reference [6] proposed an audit model based on blockchain to solve the problems of a single point of failure, performance bottleneck and disclosure of user privacy in cloud data integrity audits. The Merkle tree MA-MHT was constructed by composing a verification transaction block itemBlock* of partitioned data. The root node verifies that the CSP stores user-managed files completely and without error. Reference [8] proposed a data integrity verification mechanism based on a large branching path tree (LBT), which simplifies the dynamic update process and achieves full dynamic update by constructing a simple certification tree. Reference [9] designed a block chain-based data integrity verification scheme for cloud storage. By generating a data integrity proof tree (DIPT) and auxiliary validation information tree (AVIT), smart contracts are used to replace traditional third-party auditing institutions to verify whether CSP's response is legitimate, storing data integrity certificates in blocks to establish accountability integrity verification. The above solutions are stored in the block chain block the hash value of data, or metadata, cipher, etc., through the block chain tamper-resistant features for authentication, but each node in the audit can be obtained from the chain to tamper with the relevant documents. This article uses the block chain store IPFS address to enable each node through the address to get to the relevant documents, At the same time, it also relieves the storage pressure of blockchain.

In terms of amount audit and traceability, most of them are for the audit of on-chain transactions. Reference [12] uses Elgamal encryption, digital signature and improved hidden address technologies to achieve the privacy protection of transaction content, transaction sender identity and transaction receiver identity. By introducing the two roles of auditor and supervisor, it can audit the amount of a single transaction on the blockchain, judge whether the amount conforms to the limit of a single transaction; and track the identities of both parties through tracking keys when problems are found. Reference [13] puts forward a confidential transaction scheme ACT that can be audited by the total amount of transactions. It uses signatures to authenticate the identity of the source of audit requests to ensure that only the auditor has the right to audit. Paillier homomorphic encryption is introduced in the audit to protect the privacy of a single transaction and the amount of a single user, and a zero-knowledge proof is adopted to effectively detect dishonest users' falsification of data while ensuring the privacy of the transaction; to ensure the accuracy of audit data. The scheme in the above literature only realizes the audit of a single transaction, which is not enough to meet the efficiency requirements in the audit scenario of university scientific research funds. Therefore, the scheme in this paper adopts the method of constructing an audit tree to verify various audit evidence quickly.

Regarding the confidentiality and auditability of sensitive data, the reference [15] proposes FAPC (Fully Auditable privacy-preserving Cryptocurrency), which is mainly composed of three schemes: Traceable and linkable ring signature schemes (TLRS), Proof of Traceable Range (TRP) and Long-term Address tracking schemes (TSLA). In FAPC, auditors can track their identity, the amount of transactions and the corresponding long-term address; and maintain anonymity and confidentiality from others. TLRS and TRP are simple and modular in construction, using only standard ring signatures as components, without any additional one-time signatures or zero-knowledge proofs. TSLA makes use of standard ring signature and TSLA encryption to achieve long term address traceability of transactions. Reference [16] states the definition of scope proof, research results, classification of algorithms, security and efficiency of each algorithm, and research direction of future research. For example, Monero uses an invisible address and ring signature to hide transaction parties and transaction amounts and submits transaction amounts to hide. Scope proof algorithms are used to prove that the submitted values are within a given range, but third-party institutions cannot monitor and manage transactions, which do not meet regulatory requirements because criminals may conduct illegal transactions through them. The application of zkLedger [17] is the first auditable blockchain system, which can not only protect user privacy; but also enable auditors to implement effective audit supervision. In previous blockchain systems, some information about transactions either needed to be disclosed, or it needed to be hidden from auditors to effectively audit them. As a result, blockchain still faces significant challenges in the financial sector. ZkLedger's proposal provides a new direction to solve these problems. By using a Schnedger-type noninteractive zero-knowledge proof, it provides a new proof scheme for zero-knowledge audits. Unlike ZK-Snarks, this technology does not require a trusted setup and only relies on extensive encryption assumptions, enabling auditing without disclosing information to auditors. This system consists primarily of actors (banks, auditors), transactions, and accounts. In view of the privacy problem of sensitive data on blockchain, the reference [19] proposes a decentralized and location-aware architecture to solve the data integrity and privacy protection problems in blockchain-based traffic management systems, which integrates a blockchain network and a noninteractive zero-knowledge range proof (ZKRP) protocol. The system innovatively integrates zero-knowledge range proof into a gateway mechanism to verify vehicles connected between adjacent blockchain networks without revealing any sensitive information.

3   Pre-knowledge

3.1   Merkle Tree

A Merkle tree is a tree in a data structure. It can be a binary tree or a multi-fork tree. It has all the characteristics of a tree structure. Merkle trees were first proposed by Ralph Merkle in 1980 [1] and were widely used in file systems and P2P systems before the advent of blockchain systems (Fig. 1).

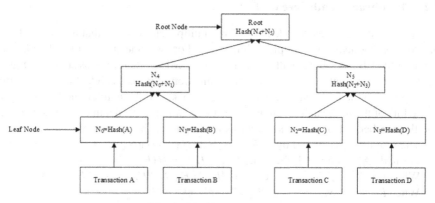

Fig. 1. Merkle tree structure

Merkle tree was originally used to establish a public directory of digital signature certificates, which can ensure that the data blocks transmitted in the peer-to-peer network are complete and have not been tampered with. In the Bitcoin network, each block contains the hash value of the transaction information; this hash value is not directly connected to the transaction sequence and then calculates their hash, but is generated through a Merkle tree. The Merkle tree generation algorithm will hash each transaction once, and then hash the calculated hash value two by two until the root value of the tree is calculated, and this root value contains all the transaction information. In this way, the space occupied by the wallet can be greatly saved. When verification is needed, it is only necessary to find a hash path from the leaf node of the transaction information to the root node, instead of downloading all the data of the blockchain.

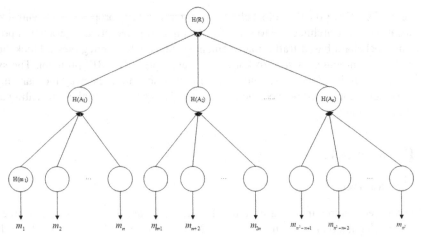

Fig. 2. LBT structure

3.2 Multi-branch Path Tree LBT

Compared with MHT, each node in LBT has multiple child nodes, that is, each node has only one parent node but multiple sibling nodes. For the same number of leaf nodes, the depth of the LBT is much smaller than that of the MHT. Therefore, as shown in Fig. 2, LBT is introduced into the data integrity verification scheme. Each leaf node in the tree corresponds to the hash value $H(m_i)$ of a data block, and the non-leaf node is the form of the link of the hash value of all its child nodes. In Fig. 2 the depth of the tree is $l = 2$, and the number of leaf nodes is n^2, corresponding to the hash value of n^2 data blocks.

The root node is denoted by R: $R = H(H_{A_1} \| H_{A_2} \| \cdots \| H_{A_n})$;

For node A, when $i = 1$, the value H_{A_1} is $H_{A_1} = H(H_{m_1} \| H_{m_2} \| \cdots \| H_{m_n})$;

When $i \neq 1 \wedge i \neq n$, the value H_{A_i} is $H_{A_i} = H(H_{m_i} \| H_{m_{i+1}} \| \cdots \| H_{m_{i+n-1}})$;

When $i = n$, the value H_{A_n} is $H_{A_n} = H(H_{m_{n^2-n+1}} \| H_{m_{n^2-n+2}} \| \cdots \| H_{m_{n^2}})$;

3.3 Pedersen Commitment

Pedersen Commitment is a type of commitment in cryptography, involving both the committer and the verifier. The process is divided into two phases. The first phase is the commitment phase. The committer selects a message v and sends it to the receiver in the form of ciphertext, which means that it will not change v. The second stage is the opening stage. The promiser publishes the message v and the blinding factor (equivalent to the secret key), and the receiver uses this to verify whether it is consistent with the message received in the promise stage. The promise scheme has two basic properties: hiding and binding. Concealment means that the promise value will not reveal any information about the message v; binding means that any malicious promiser cannot open the promise as a message other than v and verify it, that is, the receiver can be sure that v corresponds to the promise news.

Taking the elliptic curve as an example, Pedersen's core formula is expressed as follows:

$$C = r * G + v * H \tag{1}$$

In the above formula, C is the generated commitment value, G and H are the generation points on a specific elliptic curve, r stands for Blinding factor, and v stands for original information. Since G and H are generative points on a specific elliptic curve, $r * G$ and $v * H$ can be regarded as public keys on the corresponding curve (r and v can also be regarded as private keys).

The process of commitment generation and disclosure is shown in Fig. 3:

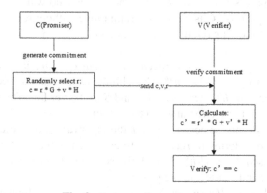

Fig. 3. Pedersen Commitment

Due to the introduction of the random blind factor r, different commitments c can be generated for the same v. Even if sensitive privacy data v remain unchanged, the final commitment c will change with the change of r, thus providing the concealment of information theory security.

The homomorphism of Pedersen Commitment means that if $comm_1$ and $comm_2$ are commitments to v_1 and v_2 using blind factors r_1 and r_2 respectively; $comm(v_1 + v_2)$ is commitments to $v_1 + v_2$ using blind factors $r_1 + r_2$:

$$
\begin{aligned}
comm(v_1 + v_2) &= (r_1 + r_2) * G + (v_1 + v_2) * H \\
&= (r_1 * G + v_1 * H) + (r_2 * G + v_2 * H) \\
&= comm(v_1) + comm(v_2)
\end{aligned} \tag{2}
$$

The homomorphism of Pedersen Commitment and the Zero-Knowledge Proof have a wide range of applications in the blockchain. Currently, it is mainly in the form of a hidden ledger to provide flexible private data ciphertext on-chain certificate and transaction ciphertext numerical correlation third party verification. In the specific scheme design, after the relevant business parties complete the business interaction under the chain, the corresponding value changes are expressed as Pedersen commitments, and then the corresponding commitment data are uploaded to the chain. In this process, there is no need to disclose any private data in plaintext. After the chain, it is difficult for an unrelated third party to deduce the plaintext of private data in the form of ciphertext promised by Pedersen, but it can verify the binding relationship between commitments and verify the legitimacy of business interaction.

3.4 Zero-Knowledge Range Proof

The zero-knowledge range proof (ZKRP) allows one to prove that a secret integer belongs to an interval. For example, a person can use the ZKRP scheme to prove that they are over 18 without disclosing their age, so the range can be defined as all integers between 18 and 200. In the expense audit, ZKRP can be used to prove that the amount of expenditure is non-negative or within a certain range when actual expenditure is incurred.

Taking Bulletproofs [3] as an example, the prover who wants to prove that secret $v \in [0, 2^n - 1]$ in promise C is within a certain range without revealing the value of v needs to prove the following relationship: $(g, h \in G, C, n; v, \gamma \in \mathbb{Z}_p) : V = g^v h^v \wedge v \in [0, 2^n - 1]$.

Prove(): Certifier must first construct a_L and a_R, meet $< a_L, 2n >= v, a_R = a_L - 1^n$, and construct promised $A = h^\alpha g^{a_L} h^{a_R}$. Then, blind factors s_L and s_R are selected, and the construct promised $S = h^\beta g^{s_L} h^{s_R}$. Send A and S to the verifier. The verifier randomly selects y, z and sends these two values to the prover.

The prover randomly selects τ_1, τ_2 for the eigenvalues of the coefficients of the first and second order terms in $t(x)$. The structure of the $t(x)$ primary and secondary coefficients promised $T_i = g^{t_i} h^{\tau_i}, i = \{1, 2\}$; send T_1, T_2 to the verifier. The verifier randomly selects x and sends it to the prover.

The prover calculates $l(x)$ and $r(x)$ according to the x given by the verifier; $t(x) =< l(x), r(x) >$; Calculate $\tau_x = \tau_2 \cdot x^2 + \tau_1 \cdot x + z^2 \cdot \gamma$, where γ is random; Calculate $\mu = \alpha + \rho \cdot x$; generate a promise C about v: $C = g^v h^\gamma$; Send $\tau_x, \mu, t(x), l(x), r(x), C$ to the verifier.

Vertify(): validators verify $t(x)$: determine $g^{t(x)} h^{\tau_x}$ is equal to $C^{z^2} \cdot g^{\delta(y,z)} \cdot T_1^x \cdot T_2^{x^2}$; according to a_L, a_R, s_L, s_R generated promises A, S, generate the promise of $l(x)$ and $r(x)$: $P = A \cdot S^x \cdot g^{-z} \cdot (h')^{a \cdot y^n + z^2 \cdot 2^n}$, where $h' = (h_1, h_2^{y^{-1}}, h_3^{y^{-2}}, ..., h_n^{y^{-n+1}})$. Determine whether the generated commitment P is equal to $h^\mu \cdot g^l \cdot (h')^r$; determine whether $t(x)$ is equal to $l(x), r(x)$.

If all of the above judgments are correct, the verifier will admit that the verifier's v is in the range $[0, 2^n - 1]$.

4 Audit Scheme of University Scientific Research Funds Based on Consortium Blockchain

4.1 Framework

As shown in Fig. 4 above, the scheme in this paper is based on the assumption that the Audit Office is a trusted entity in the whole process; and realizes real-time audit and rapid verification in the blockchain environment. The generation and storage of audit evidence is realized by the Finance Office. Considering that some document information may contain large-memory data such as pictures, IPFS is used for storage, and the blockchain

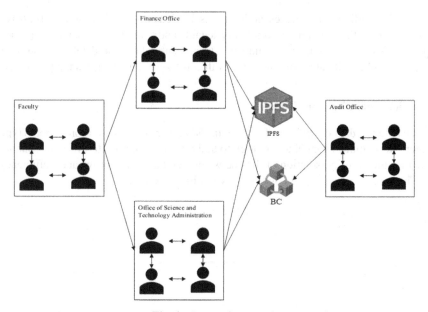

Fig. 4. System framework

is used to store the addresses returned by IPFS and other related information. The Audit Office completes the audit work through real-time audits and entrusted audits. Combined with smart contracts, audit work is more complete and automated.

The entity participants involved in the scheme include Interplanetary File System (IPFS), Blockchain (BC), Faculty (F), Finance Office (FO), Office of Science and Technology Administration (OSTA) and Audit Office (AO). The roles are described as follows:

IPFS: Responsible for storing documents generated by business processes and responding to node users' file download requests.

BC: The staff of each department join in different channels through the client. The nodes of the finance office and Audit Office join all channels to obtain information. The union link node is used to store the address returned by IPFS. The CA is responsible for generating global public system parameters and issuing digital certificates to system users.

F: It consists of project leaders. Producer of audit evidence (budget forms, expenditure documents, final accounts, etc.).

FO: On the one hand, the audit of the budget; on the other hand, the expenses of the project approved by the Department of Science and Technology should be reimbursed and accounted for. After verifying the commitment value of sensitive expenditures, the proof of zero knowledge range should be generated for up-chain storage. Another aspect is to verify the final statement prepared by the project leader.

OSTA: On the one hand, the budget of the scientific research project group is examined; on the other hand, the expenditure registered by the project leader is audited; and on the other hand is the examination and approval of budget adjustments.

AO: The audit of scientific research funds is divided into real-time audit and entrusted audit, including the authenticity, rationality, legality and compliance of scientific research expenditures; verification of the audit tree to prevent data in the database from being tampered with, verification of committed value and zero-knowledge range proofs, etc.

4.2 Blockchain Network Topology

To separate the data of different faculties in the university, this solution adopts Hyperledger Fabric multi-channel technology to add different faculties to different channels; to realize the security isolation of data between faculties and improve audit efficiency.

The specific network topology is shown in Fig. 5 below:

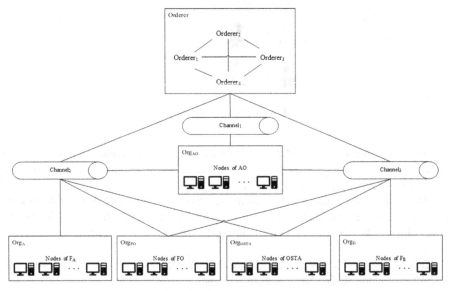

Fig. 5. Network topology

Suppose there are two faculties A and B in the university, and faculty A is represented by F_A and faculty B is represented by F_B, then F_A belongs to Org_A and F_B belongs to Org_B, then F_A and F_B nodes are added to $Channel_2$ and $Channel_3$ respectively. Each node represents the leader of a research project and joins the Fabric multi-channel network through the client. The nodes of the FO and OSTA are added to $Channel_2$ and $Channel_3$, and the nodes of the AO are separately added to $Channel_1$. Among them, $Channel_1$ is used to store the tree root value of the audit tree, and $Channel_2$ and $Channel_3$ are used to store the IPFS storage address of audit evidence, commitment value, zero-knowledge range proofs, etc.

4.3 Scheme

The scheme in this paper is mainly divided into two types of expenditure audits: audit of
the expenditure of general funds and audit of sensitive expenditures. The overall process
is shown in Fig. 6 below.

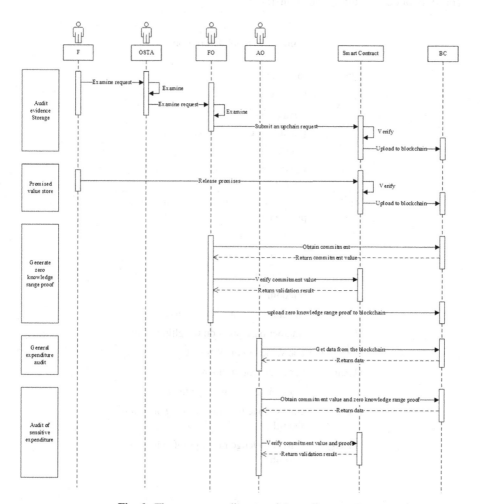

Fig. 6. Time sequence diagram of the audit process

The scheme is mainly divided into two parts: audit evidence storage and audit,
involving general expenditure and sensitive expenditure. The difference between the two
is that when audit evidence is stored, in addition to uploading relevant audit evidence to
IPFS and storing it on the chain, the audit evidence of the sensitive expenditure needs to
additionally store the commitment value and the zero-knowledge scope proof; during the
audit, for the audit of the general expenditure, the audit office obtains the relevant audit
evidence on the chain in real time for auditing. For the audit of sensitive expenditures,

the AO first verifies the zero-knowledge scope proof, opens the commitment after the verification is passed, verifies the commitment value, and then conducts further audit work after the verification is passed.

Symbol Description

The symbol description is shown in Table 1:

Table 1. Symbol description.

Symbol	Meaning
P_k	Public key
S_k	Private key
$Enc()$	Use public keys to encrypt related information
$Dec()$	Decrypt encrypted information using a private key
$Sig()$	The information is signed with its own private key
SC	Smart contract
TID	Block transaction number
v_i	Amount paid
r_i	Random number
G	Generators on an elliptic curve
H	Another base point on the elliptic curve
c_i	The value of commitment
$Open()$	Verify the commitment
π	Proof of zero knowledge range
$Prove()$	Zero knowledge range proof generation algorithm
$Verify()$	Zero knowledge range proof validation algorithm

Detailed Scheme Flow

Based on the network topology structure in Fig. 5, the specific business process of the scheme in this paper is as follows:

(1)   Storage of audit evidence

① The storage of project budget is shown in Fig. 7 below. When applying for the project, the project leader will draw up the source budget, expenditure budget and related explanatory materials, which will be submitted to the Science and Technology Division and Finance Division for examination Sig_1 and Sig_2 represent the approved Project. The system will combine this information into PB_1(Project Budget, PB) and upload it to IPFS for preservation. The storage address returned by IPFS $addr_{PB_1}$ will be uploaded to the channel where the project leader is located, and the nodes in the channel will verify and reach a consensus and store it in the distributed blockchain ledger.

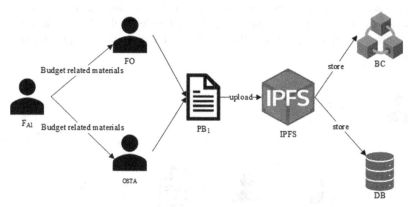

Fig. 7. Project budget storage process

Even if the files stored in PFS have only minor changes, their storage addresses will be very different. Second, the IPFS addresses stored on the blockchain are also immutable, which fully guarantees the non-tampering and non-repudiation of relevant documents. At the same time, the local database will also store $addr_{PB_1}$ and its corresponding block transaction number TID_{PB_1}, which is convenient for retrieval.

② When the project needed budget adjustment, as shown in Fig. 8 below, the project leader reported the budget adjustment to the OSTA for examination and approval. Reupload to the IPFS after approval and obtain a new address $addr^*_{PB_1}$. Update the relevant data information on the chain and in the database.

Fig. 8. Project budget adjustment process

③ The audit evidence storage process for general expenditures is shown in Fig. 9 below. When the actual expenditures are generated, the project leader fills in RV_1(reimbursement vouchers, RV), and submits it to OSTA for review in turn. After the FO has passed the review and made corresponding business processing, the information will be combined into EAE_1 (electronic audit evidence, EAE), and uploaded to IPFS for storage. The address $addr_{EVE_1}$ returned by IPFS will also be stored in the corresponding channel on the chain. After the nodes in the channel verify and reach a consensus, they are stored in the distributed blockchain ledger. All nodes in the channel can download the electronic audit evidence actually stored in IPFS through $addr_{EVE_1}$ stored on the chain.

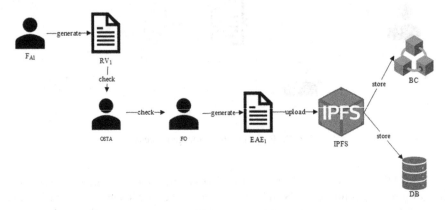

Fig. 9. General expenditure audit evidence storage process

④ Audit evidence storage for sensitive expenditures differs from general expenditures because when the project leader sends the reimbursement voucher RV_2, the commitment value c_1 of this expenditure v_1 will be calculated according to the random value r_1: $c_1 = r_1 * G + v_1 * H$. c_1 is released within the chain channels, using the public key of the FO to encrypt v_1 and r_1: $Enc(P_{k(FO)}, v_1, r_1)$, then send to FO. The FO after receiving the information, using its own private key to decrypt the information: $Dec(S_{k(FO)}, Enc(P_{k(FO)}, v_1, r_1))$, through the $Open(c_1, r_1, v_1)$ verify the commitment; calculated value $c_1' = r_1 * G + v_1 * H$, judge c_1' and chain store c_1 are equal, if equal, $Open(c_1, r_1, v_1) \rightarrow true$. The project leader honestly published the commitment value on the chain, synthesized the information into electronic audit evidence EAE_2 and uploaded it to IPFS for storage. The address $addr_{EVE_2}$ returned by IPFS will also be stored in the corresponding channel on the chain.

⑤ When the project is concluded, the project leader prepares the final statement. The FO calculates the sum of sensitive expenditure of the project: $sum = \sum_{i=1}^{n} v_i = v_1 + v_2 + \cdots + v_n$, and the sum of the corresponding random promised number $r\_sum = \sum_{i=1}^{n} r_i = r_1 + r_2 + \cdots + r_n$, and calls the algorithm $Prove()$ to generate a range proof π

that *sum* does not exceed 120% of the budget to verify the following level:

$$(g, h \in \mathbb{Z}_p, C, n; sum, r\_sum) : C = r\_sum * g + sum * h \wedge sum \in [0, 2^n - 1] \quad (3)$$

Form the EFS_2 (electronic final statement, EFS) of the project. At the same time, the FO will make a summary of all the scientific research expenditures of F, form an EDL_2 (electronic detailed ledger, EDL), and upload EFS_2 and EDL_2 to IPFS for storage. The returned address is stored in the corresponding channel.

At the same time, the system will automatically follow the project. The courtyard (department) classification of the intelligent contract calls for the conclusion of the project audit tree construction project audit, auditing departments tree and the tree, preventing data in the database in the follow-up audit from being malicious tampering, convenient audit for fast verification at the same time improving the efficiency of the audit and fraud cost, See Sect. 4.3.4 Audit tree construction for details.

(2) Audit

The audit in this scheme mainly includes three parts: real-time audit, entrusted audit and smart contract time limit judgment.

① Real-time audit (Fig. 10):

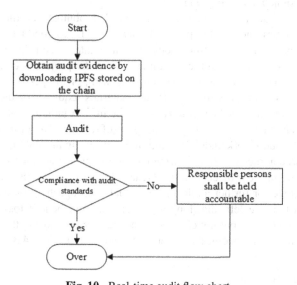

Fig. 10. Real-time audit flow chart

The AO can obtain the IPFS storage address of the latest audit evidence ($addr_{EVE_1}$, $addr_{EFS_1}$, $addr_{EDL_2}$, etc.) stored on the chain in real time from the blockchain. Carry out real-time auditing, find and deal with existing problems in time, reduce economic losses, and improve the problem of audit time lag in traditional auditing. The audit of EFS and EDL involves the integrity verification of the audit tree. The smart contract *SC_VerifyDataIntegrity* needs to be called to verify the integrity of the IPFS address stored in the database. If the verification is passed, the corresponding audit evidence will be downloaded for further audit work.

Different from a general expenditure audit, in audit of EFS_2 and EDL_2 for sensitive spending, the auditor verifies the zero-knowledge evidence and commitment value of sensitive expenditure by invoking smart contract SC_Audit. $C' = r\_sum * G + sum * H$ is calculated according to the sum and r_sum sent by the Finance Office, and then the commitment value on the chain is calculated:

$$
\begin{aligned}
C &= \sum_{i=1}^{n} c_i \\
&= c_1 + c_2 + \cdots + c_n \\
&= (r_1 * G + v_1 * H) + (r_2 * G + v_2 * H) + \cdots + (r_n * G + v_n * H) \\
&= (r_1 + r_2 + \cdots + r_n) * G + (v_1 + v_2 + \cdots + v_n) * H \\
&= r\_sum * G + sum * H
\end{aligned}
\tag{4}
$$

Judge whether C' and C are equal. If they are equal, it will prove that the total expenditure of the project is sum, and then conduct further audit. Otherwise, it indicates that the project leader and finance department did not honestly release the commitment value and hold them accountable.

② Commissioned audit (Fig. 11):

In addition to auditing through the real-time data obtained on the chain, the AO can also be commissioned by the project leader or OSTA to conduct final accounting audits. In this case, the Audit Office does not immediately audit the final accounts of project-related scientific research expenditures when the project is concluded. In this process, the off-chain database may be tampered with, or the person in charge of the audited project does not provide the relevant materials required for the audit in time. Therefore, this paper adopts the method of constructing the audit tree when the project is concluded. Ensure that FO can verify the integrity of relevant data in subsequent audits, and complete the audit work under the condition of ensuring the authenticity of the data.

③ The smart contract $SC\_TimeAudit$ will regularly judge the time limit of unfinished projects, and submit the projects that have reached the time limit but not yet completed to the Audit Office for further audit. In traditional auditing, the problems of scientific research funds that are left idle, long-term pending, and benefit loss caused by the completion of the project but not closing the account or not closing the project in time can be improved. For the specific smart contract algorithm, see Sect. 4.3.4 Smart Contract Design.

Audit Tree Construction

To prevent data in the database from being tampered before the audit, the audit tree structure can be constructed to quickly verify the integrity of audit evidence through the root value, which improves the audit efficiency and increases the cost of fraud for malicious users. The specific construction method is as follows.

(1) Construction of project audit tree: Construct a financial audit tree for EAE of completed scientific research projects, and store root value in $Channel_1$. The details are shown in Fig. 12 below:

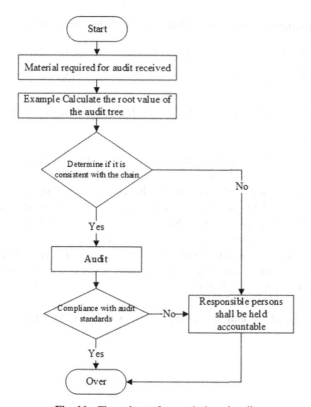

Fig. 11. Flow chart of commissioned audit

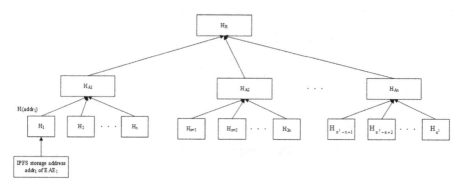

Fig. 12. Project audit tree structure

Since many expenses will be incurred in the process of a project, if the structure of Merkle tree is adopted, many calculations will be generated. Therefore, the construction of the project audit tree is based on the large branching tree (LBT) structure proposed in reference [8]. First, take the IPFS storage address $addr_1$ of EAE_1 as a leaf node: $H_1 = H(addr_1)$; Divided into procurement expenditure, equipment maintenance expenditure and so on, classification and combination and obtain the combination of hash value:

$$H_{A_1} = H(H_1 \| H_2 \| \cdots H_n) = H(H_{addr_1} \| H_{addr_2} \| \cdots \| H_{addr_n})$$
$$H_{A_1} = H(H_1 \| H_2 \| \cdots H_n) = H(H_{addr_1} \| H_{addr_2} \| \cdots \| H_{addr_n})$$

$$\cdots$$

$$H_{A_n} = H(H_{n^2-n+1} \| H_{n^2-n+2} \| \cdots H_{n^2}) = H(H_{addr_{n^2-n+1}} \| H_{addr_{n^2-n+2}} \| \cdots \| H_{addr_{n^2}})$$

$$(5)$$

Hash the combined hash value to obtain the tree root value $H_R = H(H_{A_1} \| H_{A_2} \| \cdots \| H_{A_n})$, which represents the expenditures for this project. When the audit office wants to verify the integrity of the audit evidence stored in the database, it only needs to take the hash value of the IPFS storage address of the electronic audit evidence corresponding to the expenditure, combined with other hash values in this category and the hash value of other categories. Calculate the tree root value HR^*, and compare it with the one stored in the blockchain to complete the verification. Therefore, on the premise of ensuring authenticity and integrity, the actual expenditures are audited for rationality and legal compliance.

(2) The structure of the audit tree of the faculty is shown in Fig. 13:

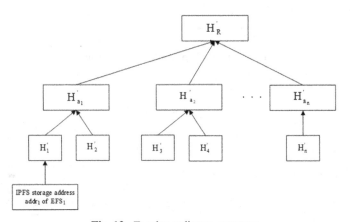

Fig. 13. Faculty audit tree structure

The construction of the audit tree of the faculty takes the IPFS storage address of the electronic final account EFS of a completed scientific research project as the leaf node $H_1' = H(addr_{el_1})$. According to the structure of the Merkle tree, the audit tree is constructed by two-by-two hashes, and the root value of the tree is obtained:

$$H_{R'} = H(H_{A'} \| H_{a_n'})$$
$$= H(H(H_{a_1'} \| H_{a_2'}) \| \cdots \| H_{a_n'})$$
$$= H(H(H_{1'} \| H_{2'}) \| H(H_{3'} \| H_{4'}) \| \cdots \| H_{a_n'})$$
$$= H(H(H_{addr_{el1}'} \| H_{addr_{el2}'}) \| H(H_{addr_{el3}'} \| H_{addr_{el4}'}) \| \cdots \| H_{addr_{elan}'})$$

$$(6)$$

Summary of scientific research expenditures of the faculty. When the AO wants to verify the integrity and authenticity of the electronic accounts of a project stored in the database, it only needs to take the hash value of the electronic accounts of the project expenditure and combine the path information to quickly check its integrity and authenticity and improve the audit efficiency.

(3) The details of the structure of the audit tree in faculty and universities are shown in Fig. 14 below:

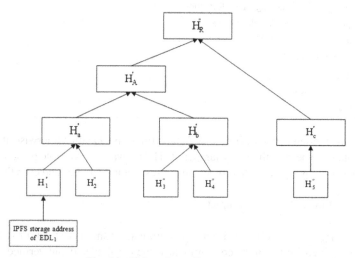

Fig. 14. University audit tree structure

The construction of the university audit tree takes the IPFS storage address of the detailed general ledger EDL of the faculty as the leaf node. The construction method is consistent with that of the faculty audit tree, and the root value HR'' represents the summary of the detailed expenditure of the scientific research funds of the university. When auditing, the AO only needs to verify the root value according to the faculty to quickly verify the integrity and authenticity of the general ledger of the detailed expenditure of scientific research funds of the university and improve the audit efficiency.

Smart Contract Design

Smart contracts mainly include integrity verification smart contract $SC\_VerifyDataIntegrity$, expenditure audit smart contract SC_Audit and audits for time limit smart contract $SC\_TimeAudit$. The specific algorithm design of smart contract is as follows.

(1) Integrity verification smart contract

$SC\_VerifyDataIntegrity(addr, Path, HR) \rightarrow b$ is the integrity verification smart contract. The node IPFS storage address, path information and root value are taken as input to verify the integrity of the address and determine whether it is maliciously tampered, with to ensure that the AO audits the data without tampering.

Algorithm 1. *SC_VerifyDataIntegrity*

Input : $addr_1, Path, H_R$

Output : *true or false*

1. $H_1 = HASH(addr_1)$ // Compute the hash value
2. *get Path*   // Obtaining path information
3. *get* H_R *   // Calculate the root value
4. IF $H_R* = H_R$ THEN
5.   *return true*
6. **ELSE**
7.   *return false*
8. **END IF**

If they are the same, that is, $HR^* = HR$, perform the next audit. Otherwise, it indicates that the data have been maliciously tampered. The relevant responsible persons of F and FO can be traced by signature to protect the interests of universities and even the country.

(2) Expenditure audit smart contract

$SC\_Audit(\pi, sum, r, C) \rightarrow b$ is an expenditure audit smart contract. First, the range proof π is verified, and then the commitment value is calculated and verified.

Algorithm 2. *SC_Audit*

Input : π, sum, r_sum, C

Output : *true or false*

1. *Verify*(π)   // Verify zero knowledge range proof
2. IF *true* THEN
3. $C' = sum*G + r\_sum*H$ // Calculate commitment values on audit evidence
4.   IF $C = C'$ THEN // Determine whether the commitment values are equal
5.   *return true*
6.   **ELSE**
7.   *return false*
8.   **END IF**
9. **ELSE**
10. *return false*
11. **END IF**

If the calculated commitment value is consistent with that stored on the chain, returning true means that the Finance Department is honest and the audit evidence corresponding to the expenditure is credible. Otherwise return false and report to finance office for accountability.

(3) Audit for the time limit smart contracts

$SC\_TimeAudit(L_1, L_1) \rightarrow b$ is the audit for the time limit smart contracts. The time limit L_1 and duration L_2 are taken as input. If within the time limit, output 1; otherwise, output 0 and submit relevant information to the auditor. At regular intervals, the deadline of uncompleted scientific research projects shall be automatically checked, and relevant information shall be submitted to the audit office for timely discovery of problems.

Algorithm 3. *SC_TimeAudit*

Input : L_1, L_2

Output : *true or false*
1.   IF $L_1 \geq L_2$ THEN
2.   *return true*
3.   ELSE
4.   *return false*
5.   END IF

If $L_1 \geq L_2$, return true for normal; otherwise, return false and submit the relevant information to the auditor to ensure that the auditor can discover the unfinished project in a timely manner.

5 Scheme Analysis

5.1 Functional Comparison of Schemes

Table 2. Functional contrast.

Functional characteristics	[6]	[9]	[12]	[18]	[28]	[30]	Article
Data integrity	√	√	×	√	×	×	√
Auditability	√	×	√	×	√	√	√
Real-time	√	√	√	√	√	√	√
Homogeneity	×	×	×	×	√	√	√
Zero-Knowledge	×	×	×	×	√	×	√

A functional comparison was made between the scheme in this paper and the scheme in the reference [6, 9, 12, 18, 28, 30], and the results are shown in Table 2. The application environment of [6, 9] is the cloud server, [12] is the audit scheme in the anonymous environment, and [28, 30] uses the idea of co-encryption to audit the expenditure ciphertext or commitment value. By comparison, it was found that [12, 28, 30] could not achieve data integrity verification. Reference [9] and reference [18] do not meet the auditability; Reference [6, 9, 12, 18] cannot audit encrypted data and verify ciphertext. Reference [6, 9, 12, 18, 28, 30] can obtain the required data in real time through blockchain to meet the real-time requirements; Reference [28] uses a zero-knowledge proof to verify the validity of homomorphic ciphertext. Compared with the schemes mentioned above, this scheme has certain advantages in functionality.

5.2 Security Analysis

Data Security

(1) Store the address returned by IPFS via blockchain. First, files stored on IPFS can return vastly different addresses for even a single punctuation change. Second, blockchain data are stored in blocks, each block has a block head, the block head stores a hash value of all data in the block through hash algorithm integration, and each block stores the hash value of the previous block to form a block chain structure. If a malicious user attempts to tamper with one of the data, it will inevitably lead to change the hash value of the blocks. To ensure that the distortion of information is cut node identification, a malicious user must be in the block node as the starting point of tampering, and recalculating the hash value of all blocks behind it, however, to admit all the nodes after tampering blocks, is basically impossible.

(2) Construct an audit tree to ensure that the on-chain retrieval information stored in the database will not be tampered with while fast verification. According to the nature of Merkle trees, any changes in the underlying data will be passed to the parent node, all the way to the root. Therefore, only by verifying the consistency of the root values of the audit tree; the integrity of relevant document information can be quickly judged and audit efficiency be improved.

(3) By Pedersen Commitment and zero-knowledge range proof, can make the finance office to verify this commitment value and generate scope of zero knowledge proof, audit can be stored by chain and compare the available big commitment value chain, to ensure that the bursar's office of honest business, and other nodes cannot be learned from the chain of sensitive spending specific data, Ensure the security of sensitive expenditures that require confidentiality.

Openness and Transparency

The nodes in the channel can obtain the data on the chain at any time without authorization, which ensures the openness and transparency of the data. On the one hand, business nodes in the same channel can obtain documents generated by different business processes and compare them, thus playing the role of mutual restraint and supervision. On the other hand, auditors can obtain the data in the channel in real time, which greatly improves the time lag in the traditional audit.

6   Conclusion

In this paper, we propose an audit scheme for university scientific research funds based on a consortium chain, which can conduct real-time, fast and efficient auditing of audit evidence related to the fund expenditures. Blockchain technology and IPFS technology are used to ensure that audit evidence information is open and transparent, non-tampering, non-repudiation and traceable. Automated auditing through smart contracts effectively reduces problems such as untimely and omission in traditional auditing; Pedersen commitments and zero-knowledge scope proofs are used to ensure the confidentiality and auditability of sensitive expenditures. The scheme is analyzed from the two aspects of function comparison and security.

Although this paper excavates the connection between blockchain and university research funding audits, but due to the relationship between time and knowledge level, the system still has the following defects: a) The scheme in this paper only supports static data to complete the construction of the audit tree; and cannot be constructed dynamically based on real-time data. Therefore, it is the direction of future efforts to support the construction of audit trees with dynamic data. b) The solution in this paper can only judge the amount through smart contracts; and cannot perform a series of analyses on the amount of data. Therefore, it is also a direction to be improved in the future to make more detailed judgments on the amount through machine learning training models. c) The scheme of this paper only considers the audit of the internal research funds of universities, but in practice other universities, manufacturers, etc. will be involved, and the design of relevant external confirmations needs to be improved.

Acknowledgement. This paper was supported by Major Science and Technology Project of Henan Province (project number 20130021030).

References

1. Merkle, R.C.: A digital signature based on a conventional encryption function. In: Pomerance, C. (ed.) CRYPTO 1987. LNCS, vol. 293, pp. 369–378. Springer, Heidelberg (1988). https://doi.org/10.1007/3-540-48184-2_32
2. Shao, Q., Jin, C., Zhang, Z., Qian, W., Zhou, A.: Blockchain: architecture and research progress. Chin. J. Comput. **41**(05), 969–988 (2018)
3. Bünz, B., Bootle, J., Boneh, D., Poelstra, A., Wuille, P., Maxwell, G.: Bulletproofs: short proofs for confidential transactions and more. In: 2018 IEEE Symposium on Security and Privacy (SP), pp. 315–334 (2018)
4. Wang, Q., Wang, C., Li, J., Ren, K., Lou, W.: Enabling public verifiability and data dynamics for storage security in cloud computing. In: Backes, M., Ning, P. (eds.) ESORICS 2009. LNCS, vol. 5789, pp. 355–370. Springer, Heidelberg (2009). https://doi.org/10.1007/978-3-642-04444-1_22
5. Fan, K., Li, F., Yu, H., Yang, Z.: A blockchain-based flexible data auditing scheme for the cloud service. Chin. J. Electron. **30**(06), 1159–1166 (2021)
6. Zhou, J., Jin, Y., He, H., Li, P.: Research on cloud data audit scheme based on blockchain. Appl. Res. Comput. **37**(06), 1799–1803 (2020)

7. Lu, N., Zhang, Y., Shi, W., Kumari, S., Choo, K.R.: A secure and scalable data integrity auditing scheme based on hyperledger fabric. Comput. Secur. **92**(C), 101741.1–101741.16 (2020)

8. Li, Y., Yao, G., Lei, L., Zhang, X., Yang, K.: LBT-based on cloud data integrity verification scheme. J. Tsinghua Univ. (Sci. Technol.) **56**(05), 504–510 (2016)

9. Liu, F., Zhao, J.: Cloud storage data integrity verification scheme based on blockchain. J. Appl. Sci. **39**(01), 164–173 (2021)

10. Weng, J., Weng, J., Zhang, J., Li, M., Zhang, Y., Luo, W.: DeepChain: auditable and privacy-preserving deep learning with blockchain-based incentive. IEEE Trans. Dependable Secure Comput. **18**(5), 74–85 (2019)

11. Putz, B., Menges, F., Pernul, G.: A secure and auditable logging infrastructure based on a permissioned blockchain. Comput. Secur. **87**(C), 101602.1–101602.10 (2019)

12. Zhao, X., Li, Y.: Auditable and traceable blockchain anonymous transaction scheme. J. Appl. Sci. **39**(01), 29–41 (2021)

13. Jiang, Y., Li, Y., Zhu, Y.: ACT: auditable confidential transaction scheme. J. Comput. Res. Dev. **57**(10), 2232–2240 (2020)

14. Huang, L., Zhang, G., Yu, S., Fu, A., Yearwood, J.: SeShare: secure cloud data sharing based on blockchain and public auditing. Concurr. Comput. Pract. Exp. **31**, e4395 (2019)

15. Li, W., Wang, Y., Chen, L., Lai, X., Zhang, X., Xin, J.: Fully auditable privacy-preserving cryptocurrency against malicious auditors. Cryptology ePrint Archive, Report 925 (2019)

16. Deng, C., et al.: A survey on range proof and its applications on blockchain. In: 2019 International Conference on Cyber-Enabled Distributed Computing and Knowledge Discovery (CyberC), China, pp. 1–8 (2019)

17. Narula, N., Vasquez, W., Virza, M.: ZkLedger: privacy-preserving auditing for distributed ledgers. In: Proceedings of the 15th USENIX Conference on Networked Systems Design and Implementation (NSDI 2018), pp. 65–80. USENIX Association, USA (2018)

18. Tan, H., et al.: Archival data protection and sharing method based on blockchain. J. Softw. **30**(09), 2620–2635 (2019)

19. Li, W., Guo, H., Nejad, M., Shen, C.-C.: Privacy-preserving traffic management: a blockchain and zero-knowledge proof inspired approach. IEEE Access **8**, 181733–181743 (2020)

20. Hang, L., Ullah, I., Kim, D.H.: A secure fish farm platform based on blockchain for agriculture data integrity. Comput. Electron. Agric. **2**(5), 170–175 (2020)

21. Zhang, L., Liu, Z., Xie, G., Xue, X.: Secure data sharing model based on smart contract with integrated credit evaluation. Acta Automatica Sinica **47**(03), 594–608 (2021)

22. Mrinalni Vaknishadh, M.: Enabling public auditability and data dynamics for storage security in cloud computing. Int. J. Innov. Res. Dev. **1**(5), 46–57 (2012)

23. Hirano, T., et al.: Data validation and verification using blockchain in a clinical trial for breast cancer: regulatory sandbox. J. Med. Internet Res. **22**(6), 24–29 (2020)

24. Sim, S., Jeong, Y.: Multi-blockchain-based IoT data processing techniques to ensure the integrity of IoT data in AIoT edge computing environments. Sensors **21**(10), 3515 (2021)

25. Ding, Y., Xiang, H., Luo, D., Zou, X., Liang, H.: Scheme for electronic certificate storage by combining Fabric technology. J. Xidian Univ. **47**(05), 113–121+158 (2020)

26. Gao, S., Zheng, D., Guo, R., Jing, C., Hu, C.: An anti-quantum e-voting protocol in blockchain with audit function. IEEE Access **7**, 115304–115316 (2019)

27. Yin, X., He, J., Guo, Y., Han, D., Li, K., Castiglione, A.: An efficient two-factor authentication scheme based on the Merkle tree. Sensors **20**(20), 5735 (2020)

28. Zhou, X., Liu, Y.: Supervisable transaction privacy-preservation scheme for air travel consumption. Appl. Res. Comput. **1**(04), 1–7 (2022)

29. Wang, Q., Wang, C., Ren, K., Lou, W., Li, J.: Enabling public auditability and data dynamics for storage security in cloud computing. IEEE Trans. Parallel Distrib. Syst. Publ. IEEE Comput. Soc. **22**(5), 25–31 (2011)

30. Li, B., Zhang, W., Wang, J., Zhao, W., Wang, H.: Sealed-bid auction scheme based on blockchain. J. Comput. Appl. **41**(04), 999–1004 (2021)
31. Li, J., Wu, J., Jiang, G., Srikanthan, T.: Blockchain-based public auditing for big data in cloud storage. Inf. Process. Manage. **57**(6), 102382 (2020)
32. Sheldon, M.D.: A primer for information technology general control considerations on a private and permissioned blockchain audit. Curr. Issues Audit. **13**(1), A15–A29 (2019)
33. Wang, Q., Wang, C., Li, J., Ren, K., Lou, W.: Enabling efficient verification of dynamic data possession and batch updating in cloud storage. KSII Trans. Internet Inf. Syst. **12**(6), 15–23 (2018)
34. Gao, H., Li, L., Lin, H., Li, J., Deng, D., Li, S.: Research and application progress of blockchain in area of data integrity protection. J. Comput. Appl. **41**(03), 745–755 (2021)
35. Yang, X., Pei, X., Wang, M., Li, T., Wang, C.: Multi-replica and multi-cloud data public audit scheme based on blockchain. IEEE Access **8**, 246–261 (2020)
36. Pranto, T.H., Noman, A.A., Mahmud, A., Haque, A.B.: Blockchain and smart contract for IoT enabled smart agriculture. PeerJ. Comput. Sci. **7**, 73–82 (2021)
37. Farooq, M.S., Khan, M., Abid, A.: A framework to make charity collection transparent and auditable using blockchain technology. Comput. Electr. Eng. **83**(C), 137–143 (2020)

Performance Evaluation for College Curriculum Teaching Reform Using Artificial Neural Network

Jia Li[1](✉) and Siyang Zhi[2]

[1] UCSI Graduate Business School, UCSI University, Taman Connaught, Malaysia
`1002161154@ucsiuniversity.edu.my`
[2] China Publishing Group Digital Media Co., Ltd., Beijing, China

Abstract. To address the problems of poor performance evaluation and performance management of college curriculum reform, the performance evaluation method of college curriculum reform using artificial neural networks is proposed. First, the performance evaluation index system of college curriculum reform using artificial neural network technology is constructed. Second, the performance evaluation algorithm of college curriculum reform is improved, and the performance evaluation process of college curriculum reform is simplified. The experiment proves that the performance evaluation method of college curriculum reform using artificial neural networks has higher practicality than the traditional method and fully meets the research requirements.

Keywords: Artificial neural network · College curriculum · Reform in education · Performance evaluation

1 Introduction

Teaching performance evaluation is an important part of teaching management and has the function of improving teaching performance. However, there are problems in the existing teaching evaluation work of most domestic colleges and universities, which cannot play its due function and even cause many negative effects, which has become an obstacle to the improvement of teaching levels in colleges and universities [1]. The core of teaching evaluation activities in colleges and universities is the construction of an evaluation index system, and most of the problems in the process of teaching evaluation are related to whether the evaluation index is reasonable or not. Therefore, it is of great practical significance to develop and improve the existing evaluation index system. Higher education should firmly establish the central position of talent training in the work of colleges and universities, improve the quality of talent training and increase teaching investment. We visited some colleges and universities and found that there are many problems in the design of the existing teaching evaluation system in colleges and universities, even the deviation of goal orientation, which has become a serious obstacle for colleges and universities to improve the level of running a school. Specifically, the existing teaching evaluation activities show the following characteristics: from the perspective of evaluation objects, they pay attention to the individual evaluation of teachers and ignore the evaluation of disciplines or teaching units. Most schools' internal teaching evaluation objects are limited to individual teachers, and their subordinate teaching institutions are basically not evaluated and assessed [2]. This leads to the lack of teamwork spirit in many teachers' teaching activities, wastes resources and is not conducive to the improvement of the overall teaching level of the school. Emphasize the evaluation of the classroom effect and ignore the evaluation and assessment of other links in the teaching process. The classroom effect is one of the core elements of teaching performance, but there are many difficulties in the accurate evaluation of classroom effect. Therefore, limited to the evaluation of classroom effect, it is easy to lead to distortion and misjudgment of teaching performance evaluation. The evaluation subject is single, and some schools tend to pay too much attention to students' evaluation of teaching. Students' opinions can reflect teachers' teaching situations to a certain extent [3]. However, it cannot be denied that due to the lack of knowledge accumulation of students, there are limitations in the ability to judge the teaching situation of teachers. Therefore, it is difficult for a single student evaluation to fully reflect teachers' teaching level. It weakens the role of teachers in guiding students, weakens the leading position of teachers in teaching, and has an adverse impact on the realization of teaching objectives. There is a tendency to overuse the evaluation results of teaching. Some schools directly use the teaching evaluation results in the annual evaluation and professional title evaluation of schools, which has caused some negative effects. The evaluation of professional titles in colleges and universities involves the fundamental interests of the majority of teachers, which leads to the excessive attention of young teachers to the evaluation results and the alienation of the relationship between teachers and students. The vicious competition among teachers also affects the cooperation among teachers, which is not conducive to the construction of teaching teams and restricts the improvement of the overall teaching level of the school. Due to the simplification of the evaluation subject, paying too much attention to students' evaluation of teaching has brought some negative consequences [4]. Alienation of the teacher-student relationship. Because many young teachers are afraid of the impact of students' evaluation and scoring on themselves, they neglect to manage the negative phenomena of students in the classroom and ignore them, which contributes to the poor habits of students. It inhibits the teaching reform and leads to teachers' excessive pursuit of interest and neglect of science in class. Students' evaluations of teachers have limitations. Students' evaluation focuses on the classroom effect and attaches importance to classroom interest. According to our survey, some students do not adapt to new teaching methods, such as research-based teaching, from not adapting to conflict. The reason for the maladjustment lies in the influence of the traditional teaching view, which believes that the more teachers speak, the better, the more specific, and the more detailed. There is also resistance to the innovation of teaching methods [5]. Teachers talk more, students see less, students spend less time, and learning is easier. Therefore, when teachers hope to let students read, think and study more by themselves by speaking less, students are unwilling. If teachers assign more homework, students will be unhappy; if you give him another challenging question, he will feel that the teacher is difficult for them. Third, taking students' evaluation of teaching as the only standard for teachers' reward and punishment worsens the teacher relationship. To obtain good evaluation results, teachers' cooperative relationships have evolved into competitive relationships and even vicious competition, which affects team cooperation [6]. To take the lead in the competition, they guard against and block each other in the use of teaching plans and courseware, which is not conducive to the improvement of the overall teaching level and the optimal allocation of resources. With the continuous improvement of people's understanding of the curriculum, teaching, as a means of cultivating people with all-round development, plays an increasingly important role in the process of talent training. Teaching performance evaluation is an important way for colleges and universities to comprehensively improve teaching quality and strengthen teaching management. It is not only the logical end point of guiding teaching activities but also the behavioral starting point of new teaching activities. The evaluation methods and the construction of an index system of teaching performance evaluation in colleges and universities will directly affect the smooth development of teaching work in colleges and universities [7]. Therefore, systematic research on teaching performance evaluation in colleges and universities has important practical significance. Through the research review of domestic college teaching performance evaluation articles, this paper summarizes the consensus of academic research, the existing shortcomings and the prospect of follow-up research to grasp the research context of college teaching performance evaluation.

2 Performance Evaluation of Curriculum Teaching Reform in Colleges and Universities

2.1 Performance Evaluation Indicators of Curriculum Teaching Reform in Colleges and Universities

Education is a special cause of human resource development. Educational activities do not directly generate profits. The direct economic benefits of educational activities are mainly reflected in the utilization rate of resources [8]. The social benefits of educational activities are the impact of educational activities on individuals, society and as a rational organism. Performance evaluation is mostly applied to the evaluation of teachers' teaching performance in teaching. At this stage, a considerable number of areas take the students' academic year (or semester) and the total score (or average score) of entrance examination results as the only standard to evaluate teachers' teaching performance. At the same time, this score is used as the only basis for teachers' evaluation, promotion and post evaluation so that teachers pay attention to teaching regardless of everything. Taking the academic year or semester examination results as the performance of teachers' teaching work cannot objectively reflect the level of teaching quality [9]. Examination alone cannot truly reflect the quality of students' learning, nor can it objectively, fairly and comprehensively reflect the level of teachers' teaching performance. The comprehensive evaluation of teachers' teaching performance should be the evaluation of the results of the whole process of teaching. In reality, it is biased to simply define performance as result, output or process [10]. Performance as a result and process has its own advantages and disadvantages. We can understand this definition as that it depends not only on what you do but also on how you do it. Excellent performance depends not only on the result of doing something but also on the behavior or quality of doing it. It is illustrated in Table 1.

Table 1. Comparison of the advantages and disadvantages of the performance definition of curriculum reform in colleges and universities

Compare	Advantage	Shortcoming
Results oriented	It advocates attaching importance to output, which is easy to form a "result oriented" culture and atmosphere in the organization; Strong sense of achievement of organization members; It is convenient to evaluate quantifiable results	It is not easy to find inappropriate behavior before the results are formed; In case of external factors beyond the control of the responsible person, the evaluation fails; Unable to obtain information in the process of personal activities and provide guidance and help in time; Easy to lead to short-term benefits; High risk
Process oriented	Be able to obtain the personal activity information of members at any time and adjust inappropriate behaviors in time, which is helpful to guide and help the members of the organization; Effectively avoid risks and irreparable results; Considerable long-term benefits	It is not conducive to the development of innovative members, and sometimes stifle their enthusiasm; Overemphasis on working methods and steps sometimes leads to neglect of actual work results

Under the guidance of the new curriculum concept, the new curriculum advocates independent, cooperative and inquiry learning. Therefore, everyone's ability to cooperate should be included in the evaluation. Therefore, the relationship performance should also be included in the investigation. The term "effect" is equivalent to performance when some people define performance and think that performance is effect. With the continuation of performance research, only the combination of "result" and "process" to define performance is the most comprehensive and universally applicable, which is also the most recognized definition. Then, the efficiency that can be reflected in real time in the process must be considered [11]. The term "performance" itself originated and developed in enterprises. Enterprises pursue benefits, so there is no doubt that benefits are a part of performance. The understanding of educational performance is the comprehensive reflection of educational effect, educational efficiency, educational benefit and relationship performance under the guidance of certain educational objectives. Based on this, combined with the principle of neural networks, the performance evaluation method of teaching reform is studied. Artificial neural networks are information processing systems that imitate the structure and function of biological brains [12]. It is a highly complex nonlinear dynamic system. Perceptron belongs to supervised learning. Its basic idea is to gradually input the samples into the network and adjust the weight matrix in the network according to the difference between the output result and the ideal output. Figure 1 is a model diagram of a multioutput perceptron.

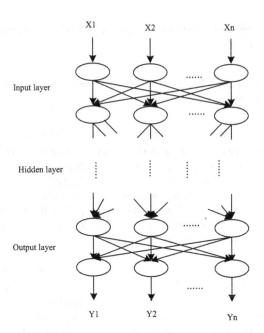

Fig. 1. Structural model of a multioutput perceptron (artificial neural network)

Although the structure and function of each neuron constituting the neural network are relatively simple, the behavior of the system composed of a large number of networks is rich and colorful. At present, people still have a superficial understanding of the neural network structure and operation mechanism of the brain and even the principle of a single neural cell. However, based on the distributed storage, parallel processing and adaptive learning of biological neural systems, an artificial neural network with a certain level of intelligence has been constructed. It has the commonness of general nonlinear systems, such as high-dimensional neural networks, extensive continuity, adaptability and self-organization among neurons [13]. The ultimate goal of error correction learning is to minimize an objective function based on the error signal, and the actual output of each output unit in the network is the closest to the expected output in a statistical sense. Once the objective function form is selected, the error correction learning sentence becomes a typical optimization problem. The most commonly used objective function is the mean square error criterion.

$$J = E\left(\frac{1}{2k}\right) / \sum_k e_k^2(n - 1) \tag{1}$$

where e is the error signal, E is the expected operator, k is the kth neuron of the input, and n is the objective function of the mean square error. The specific method can be the steepest gradient descent method [14]. When using J as the objective function directly, it is necessary to know the statistical characteristics of the whole process, which is usually difficult. To solve this difficulty, the instantaneous value at time n is generally used $\varepsilon(n)$

$$\varepsilon(n) = EJ \sum_k e_k^2(n)/2 \tag{2}$$

Competitive learning means that the output units in the network compete with each other, and finally, there is only one strongest activator. The most commonly used competitive learning rules are as follows:

$$\Delta w_{kg} = \begin{cases} \varepsilon(n) - \eta \\ 0 \end{cases} \tag{3}$$

Performance competition evaluation should be scientific and reasonable, and performance evaluation work η should be implemented. We should thoroughly implement the scientific outlook on development, fully implement the party's education policy, take serving and promoting the scientific development of compulsory education as the goal, improve the quality of teachers as the core, promote teachers' performance as the guidance, and strive to build a teacher performance evaluation system that is in line with the law of education, teaching and teachers' growth, with clear guidance, scientific standards and a perfect system [15]. We encourage teachers to contribute wisdom and strength to running education to the satisfaction of the people. The teacher performance evaluation index is the foundation and basis of teacher performance evaluation, whether it evaluates a teacher with deep experience in higher vocational colleges or a young teacher. Whether it is the evaluation of teachers' teaching or teachers' scientific research, we should first consider establishing a set of evaluation standards to determine from which aspects to evaluate the work level of college teachers [16].

2.2 Performance Evaluation Algorithm of Classroom Teaching Reform in Colleges and Universities

With the combination of quantitative and qualitative methods, too much quantitative evaluation will have many negative effects. The judgment and analysis of teachers' teaching performance is very complex. Improper performance evaluation will lead to negative effects, frustrate teachers' enthusiasm and lead to unnecessary contradictions [17]. Therefore, the combination of qualitative and quantitative factors should be emphasized in the design of evaluation indicators. Level 1 Evaluation Index: including teaching attitude, teaching ability, teaching contribution and classroom effect. Secondary indicators include code of conduct, concept of time, teaching tasks, team contribution, basic ability, student evaluation, peer evaluation, etc. Determination of index weight [18]. We designed a study comparing the importance between the primary indicators and the secondary indicators under the primary indicators and distributed it to the professionals engaged in teaching supervision and teaching management in colleges and universities. Multiple judgment matrices are formed, and the weight of each index is calculated by the analytic hierarchy process, as shown in Table 2.

The process of discovering various models, profiles, and derived values from known data sets. Using an artificial intelligence network needs to pay attention to that it is a process, not just a simple application of tools and technologies [19]. The process of data

Table 2. Evaluation indicators of teachers' teaching performance

Primary index	Secondary index	Leave early
Teaching attitude	Time concept; Code of conduct	Answer and make calls in class; clothing
Teaching ability	Basic ability; Professional ability	Qualification certification; Scientific research assessment
Teaching contribution	Sharing of teaching resources; Team communication; Teaching methods; Team participation; Student evaluation; Peer evaluation	Teaching plan; Guide new teachers; Teaching and research achievements; Observe teaching activities; Open class
Classroom effect	Observation points; late	Satisfaction; Fuzzy comprehensive evaluation

mining is generally divided into four steps: data selection, data transformation, mining data and interpreting results. The specific processes and steps are shown in Fig. 2.

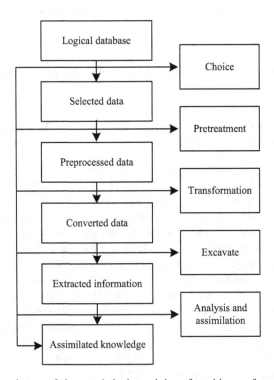

Fig. 2. Process and steps of characteristic data mining of teaching performance indicators

Based on characteristic data, the key to constructing an artificial neural network is to reasonably select the number of hidden layers and the number of neurons in the input layer, hidden layer and output layer. To reduce the large scale of the network, the number of hidden layers is not large, and one layer is generally set [20]. Therefore, the number of nodes in the input layer (m) of the teacher performance evaluation model is the number of evaluation indicators, and the number of nodes in the output layer ($n = 1$) is the performance evaluation index. The number of hidden layer neurons is small, the network accuracy is not enough, the curve fitting is poor, and the training time is long. Generally, the more neurons in the hidden layer of the artificial neural network, the higher the learning accuracy, but the ability of the network to apply to the input without learning will be worse. Therefore, if the number of neurons in the hidden layer is too high, the training time will increase, and the uncoordinated fitting and accuracy will deteriorate. The number of hidden layer nodes $L = \sqrt{m + n} + a$ or $\frac{m+n}{2}$ is the function between the hidden neural layers, and e^{-x} is sigmoid, which is the activation constant between the hidden neural layers. The mathematical expression of the sigmoid activation function is

$$f(x) = \frac{1}{\Delta w_{kg} + Le^{-x}} \tag{4}$$

The excitation function of the output layer is a pure linear function. The learning rate is an important factor to be considered when establishing the model. In the gradient descent algorithm used in the artificial neural network, if the learning rate is too small, the convergence speed is very slow. If it is too large, the iterative solution will oscillate violently. The selection range of the learning rate is generally between 0.01–0.8. After the experts score the d-th teacher's qualitative index, the expert opinions of each index should be gathered by the arithmetic average method. Namely,

$$q = f(x) \sum_{s=1}^{T} V_{rs} - r/Tc \tag{5}$$

where V_{rs} is the expert scoring value of the r-th index of the s-th teacher, R is the serial number of the evaluation index, c is the scoring value of the s-th expert on the r-th index of the d-th teacher, r is the expert serial number, and R is the label of an index. Table 3 presents the text of the "informatization teaching performance evaluation gauge" in the form of a qualitative description, including each level of standards and various indicators.

A brief description of the four levels of each indicator. Set of principal factors

$$U = (u_1, u_2, u_3, u_4) \tag{6}$$

where u_1 is the artificial neural network platform; u_2 is the practical teaching mode; u_3 is the teacher factor; and u_4 is the student factor. Set of subfactors

$$u_1 = (u_{11}, u_{12}, u_{13}) \tag{7}$$

$$u_2 = (u_{21}, u_{22}, u_{23}) \tag{8}$$

$$u_3 = (u_{31}, u_{32}, u_{33}) \tag{9}$$

Table 3. Performance evaluation gauge of teachers in the lesson preparation stage

Primary index	Secondary index	Grade I	Grade II	Grade III	Grade 4
teaching effectiveness	instructional design	Design teaching according to students' characteristics; Preparation for the integration of this course; Planned and organized instructional design	Fully prepare for the integration of this course; Planned and organized instructional design	Integrate instructional design according to requirements	According to the requirements of teaching design, the lack of consideration of curriculum integration
Teaching benefit	Resource preparation	The skilled use of hardware used in class, the selection of resources and teaching purposes are fully adapted to the age characteristics of students	Skillfully operate the software used in class, fully adapt the selection of resources to the teaching purpose, and adapt to the age characteristics of students	Proficient in the hardware used in class, and the selection of resources is basically consistent with the purpose, but it is lack of originality	Proficient in the hardware used in class, and the selection of resources can not effectively adapt to the purpose
Teaching efficiency	Time allocation plan	Reasonable allocation of class time, detailed plan, and pay attention to reserving time for students to think and discuss	Have a plan for class time allocation, and be able to reserve some time to communicate with students	There is a general plan for the allocation of class time, but there is no consideration of emergencies in class	No plan

$$u_4 = (u_{41}, u_{42}, u_{43}) \tag{10}$$

The evaluation gauge has four levels of indicators, of which economic and social benefits belong to the evaluation indicators of teaching benefits. There are Ruo level indicators under each level of indicators. Each indicator item is divided into four levels. The scores from level 1 to level 4 are 4, 3, 2, and 1. Finally, all the scores are added to obtain the total score. Through the reference of relevant literature and interviews with experts, this paper preliminarily establishes the dimension of teaching performance evaluation in colleges and universities, which mainly includes four aspects. The preparation work before the teaching work is mainly completed by teachers, which is not only an important premise but also a key link to successfully complete the teaching objectives. It is inevitable that there will be some interference with the expected teaching process. Therefore, it is necessary for teachers and students to carry out necessary communication to enable the teaching work to continue. The teaching process also includes the implementation of teachers' teaching preparation. It is not only the transmission of teaching knowledge and experience but also a very important link in the evaluation index. The third is the effect of students' learning. The performance level of teaching work can directly reflect the improvement of students' sports ability, and the main body of teaching work is students. All teaching work should be student-centered. The fourth is teaching management. To evaluate teaching performance, the content of teaching management is indispensable because teaching management runs through the process of teaching work. Moreover, the scientific formulation and steady implementation of the teaching management work plan also plays an obvious role in the improvement of teaching performance.

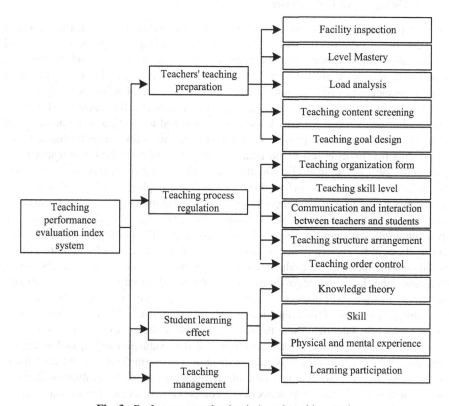

Fig. 3. Performance evaluation index of teaching work

Based on this, the teaching performance evaluation index model is constructed, as shown in Fig. 3.

In the teaching work of colleges and universities, the inspection of the safety of teaching facilities by teachers and other relevant staff before and during class, as well as the reasonable layout of equipment, can effectively protect the physical safety of students, and there is no concern about the learning of students' skills and theoretical knowledge, which is further reflected in the learning effect, improve the learning effect of students, and then improve the work performance of teaching. It can be seen from the survey that in the inspection of site facilities, the safety inspection score of the site is the highest, with a score of 0.682. Therefore, we can see the importance of site safety inspection, and we should pay attention to this evaluation result when evaluating teaching performance. In addition, the score of the index of a sufficient number of required teaching facilities is 0.423, which is the lowest under this secondary index. It can be seen that whether the number of teaching facilities is sufficient is a relatively weak interference factor affecting the improvement of teaching performance.

2.3 Implementation of Performance Evaluation of Curriculum Teaching Reform in Colleges and Universities

Teacher performance appraisal is a teacher appraisal system to deepen the reform of the educational personnel system, promote the smooth implementation of the school performance salary system, strengthen the construction of teachers and promote the scientific development of education. The result of performance appraisal is the main basis of performance salary distribution. To implement the reform of performance-based pay distribution in compulsory education schools, we must establish a teacher performance appraisal system in line with the laws of education and teaching and the professional characteristics of teachers, provide institutional guarantees for the performance-based pay distribution to better reflect the actual performance and contribution of teachers and give better play to the incentive function, and put forward requirements for grasping the basic requirements of performance appraisal and exploring effective methods of performance appraisal. Comprehensive evaluation is a complex process that is generally divided into four stages. The first is the stage of determining the evaluation project and evaluation index system. On the one hand, according to the problems to be evaluated, clarify the purpose and requirements of evaluation, define the scope of evaluation objects and determine the items of evaluation objects. On the other hand, we comprehensively understand the characteristics of the evaluation object and determine the evaluation index system by using a combination of qualitative and quantitative methods. The second is the construction stage of the evaluation model algorithm. According to the requirements of the evaluation date, determine the type attribution of the comprehensive evaluation, clarify the source of the measurement value of the evaluation index, and select the data processing methods such as standardization (i.e., normalization), the calculation of the weight of the evaluation index and the calculation of the evaluation index of the evaluation project in the process of the comprehensive evaluation to construct the corresponding evaluation model algorithm. The third is the basic data investigation stage. According to the evaluation model algorithm, evaluation items and evaluation index system, this paper puts forward the contents and methods of the investigation, designs the corresponding questionnaire, and carries out the collection of statistical data or typical investigation or expert consultation. The fourth is the basic data processing stage. According to the constructed evaluation model algorithm, the basic data are processed to obtain the evaluation index weight and evaluation project evaluation index. Fifth, the analysis and decision-making stage. Analyze the results of basic data processing and determine the reliability of the results. The corresponding technical route process of comprehensive evaluation is shown in Fig. 4.

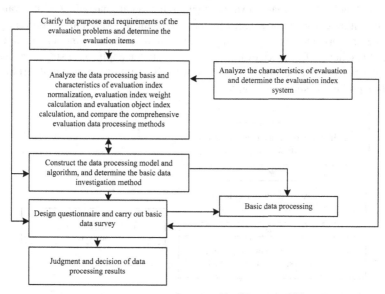

Fig. 4. Corresponding technical route process of comprehensive teaching evaluation in colleges and universities

The establishment of an evaluation index system should be analyzed according to specific problems. Generally, the selection of evaluation indexes mainly follows the following principles: scientific principle: the establishment of a comprehensive evaluation index system must be based on science and objectively reflect the level of evaluation objects. Concise system principle: the design of the index system should be concise and accurate and must fully reflect the basic characteristics of all aspects of the evaluated object. The indexes are independent and interrelated to form an organic whole. The minimum number of indexes should be used to describe the characteristics of the system. Dynamic principle: everything in society is constantly changing, and the design of the index system should also be forward-looking, which can not only reflect the current development level but also consider the impact of future competition. Representativeness principle: systematically study the evaluation objects in theory, grasp the universality of the evaluation objects, design according to the common characteristics, and build a representative index system. The formulation of evaluation indexes and evaluation standards should be objective and practical, covering a large amount of information.

Data availability principle: if the indicators included in the index system are quantitative, we should be able to find real and reliable data. If they are qualitative, we should strive to have hierarchical evaluation criteria and find appropriate personnel for objective evaluation. The basic process of teacher performance evaluation includes four stages. The basic process of teacher performance evaluation is relatively complex. The first is the initial stage, which is to understand the purpose, requirements and research process of the research; the second is the index system construction stage, which analyzes the research methods such as index screening and weight and establishes the index system suitable for teacher performance evaluation in higher vocational colleges: the evaluation model algorithm construction stage, which analyzes the algorithm model currently used for teacher performance evaluation and establishes the evaluation model and algorithm system of teacher performance evaluation. Finally, in the application stage, we evaluate the teacher performance evaluation and test the reliability of the model algorithm. The evaluation process is shown in Fig. 5.

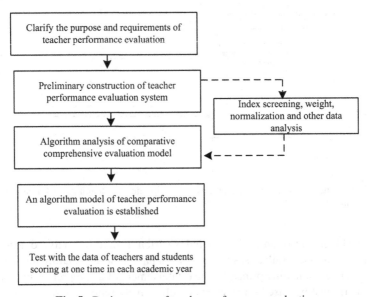

Fig. 5. Basic process of teacher performance evaluation

Judging whether teachers match their posts and whether teachers' teaching attitude, teaching ability and teaching performance meet their job requirements must be based on the evaluation results of teachers' teaching performance over a long period of time, which is more conducive to the healthy development of academic teams and the reasonable optimization of management teams. Teaching performance evaluation is a long-term, dynamic, continuous and systematic process. Therefore, the results of teaching performance evaluation can also provide managers with long-term, dynamic, continuous and systematic information about teachers' teaching behavior and teaching results to effectively help managers analyze teachers' post setting and job appointments. The results of long-term good teaching performance evaluation show that the matching degree between teachers and their posts is high, and teachers' teaching performance is good. We can consider maintaining teachers' posts and improving teachers' professional titles and positions. The poor teaching performance evaluation results obtained many times show that the teachers' personal quality does not match their posts, or the teachers' working attitude is negative and their working ability is low. After further analyzing the reasons, the teachers can be adjusted, demoted, and eliminated, and other corresponding treatments can be made.

3   Results and Discussion

To reflect the research results objectively and accurately and to verify the accuracy of each index weight and the effectiveness of each index item of the classroom teaching performance evaluation system, I have determined an observation principle: take the classroom teaching performance evaluation gauge as the basis of observation; develop a more concise and clear observation scale and use it in observation; in the process of observation, each lesson is judged and diagnosed according to the "observation scale". On the other hand, the deficiencies of "classroom teaching - teaching performance evaluation gauge" and "observation scale" are found in the observation, and they are improved on this basis. The same information-based lesson was scored with the students according to the classroom observation table. This candidate selected the students who have an educational technology background and are now graduate students in a college. According to the evaluation criteria, there are no emotional color or subjective factors, so the score is rational. Compare the satisfaction scoring of traditional performance evaluation and this performance evaluation, record the experience of scoring, the accuracy of each index weight and the effectiveness of each index item. There is little difference between the evaluation results and the expected value, so we can believe that the classroom teaching performance evaluation gauge in this study is effective. If the score difference is large, it should be analyzed and adjusted. Based on this, the teaching performance evaluation values of different subjects are compared in Table 4.

Table 4. Comparison of teaching performance evaluation results

Subject	Expected value	Traditional method	Paper method
Physics	82	42	85
Chemistry	84	42	86
History	84	43	88
English	85	53	87
Mathematics	83	56	87
Language	82	53	88
Composition	82	60	85
Geography	85	52	89

Based on the analysis of the above table, the performance evaluation results of this method are basically consistent with the expected values, which is obviously better than the traditional methods. After further training of the artificial neural network, we find that the network can be used to evaluate the performance of teachers. The generalization value of the teacher performance assessment of the five teachers in the above courses is taken as the judgment basis of evaluation error. The lower the generalization value is, the lower the evaluation error is. Based on this, the generalization error value of teacher performance assessment of the traditional method and this method is statistically compared. See Tables 5 and 6.

Table 5. Generalization error value of teacher performance appraisal under this method

Subject	Date value								
Physics	0.3293	0.2985	0.2985	0.2356	0.2568	0.2593	0.2286	0.3578	0.2235
Chemistry	0.2828	0.4885	0.3698	0.2985	0.3185	0.2568	0.2258	0.3265	0.2898
History	0.2583	0.2293	0.3827	0.2369	0.3956	0.3674	0.3585	0.3985	0.2568
English	0.3368	0.3828	0.3538	0.3785	0.2856	0.3856	0.3522	0.3565	0.3958
Mathematics	0.5083	0.2366	0.2956	0.3308	0.2298	0.2568	0.2228	0.4368	0.3385
Language	0.3368	0.3368	0.2828	0.2865	0.2298	0.3368	0.2368	0.2522	0.3827
Composition	0.2828	0.3827	0.3385	0.2828	0.3827	0.2385	0.2522	0.2789	0.2898
Geography	0.3268	0.3298	0.4368	0.4368	0.3865	0.2298	0.3898	0.2685	0.3859

Table 6. Generalized value of teacher performance assessment under traditional methods

Subject	Date value								
Physics	0.8075	0.8536	0.8825	0.8985	0.7265	0.9826	0.568	0.8598	0.8365
Chemistry	0.8868	0.9325	0.8658	0.7685	0.8358	0.4985	0.9325	0.6985	0.7265
History	0.8368	0.7582	0.8289	0.7868	0.8536	0.9652	0.7652	0.9856	0.7885
English	0.7198	0.9515	0.7652	0.8065	0.8668	0.9685	0.8685	0.8652	0.8698
Mathematics	0.7509	0.7258	0.8658	0.8598	0.8652	0.7998	0.7865	0.9325	0.8538
Language	0.9325	0.7768	0.9325	0.9325	0.8536	0.8536	0.8698	0.9325	0.9325
Composition	0.7265	0.9325	0.8289	0.7768	0.8289	0.9325	0.8698	0.8536	0.9325
Geography	0.9325	0.8698	0.7265	0.8698	0.9325	0.7768	0.9325	0.8289	0.8536

The data verification in the above table shows that the generalization value of teaching effect and teaching performance under the guidance of traditional methods is low, while under the guidance of teaching methods, the generalization value is high, which can better improve the accuracy of performance evaluation. After analyzing the generalization value of each lesson, the difference in the total score of performance evaluation was compared according to the index level. Through comparison, the reliability of each index item is further analyzed. As shown in Fig. 6.

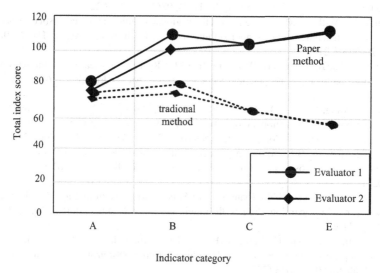

Fig. 6. Comparison statistics of performance evaluation indicators

Based on the above comparative test results, compared with the traditional methods, this method has better evaluation accuracy and less error in the process of practical application. Therefore, it is feasible to use ANNs to evaluate teachers' performance.

4   Conclusions

At present, in the field of practical teaching, the case study of using ANNs to promote practical teaching has achieved some results, but there is still a lack of in-depth practical exploration. In the field of artificial neural network research, there is a lack of theoretical research and teaching cases for artificial neural networks to support practical teaching in China. There is little research on the practical teaching model and its effect evaluation based on ANNs. This paper studies the design of a performance evaluation index system for practical teaching from the perspective of ANNs. Through the analysis of four factors affecting the performance of practical teaching based on the perspective of artificial neural networks, including artificial neural network platforms, practical teaching modes, teachers and students, the relevant evaluation index system is established. Explore how to evaluate the teaching effect of the new practical teaching mode of introducing artificial neural network technology.

References

1. Liu, F.: Language database construction method based on big data and deep learning. Alexandria Eng. J. **61**(12), 9437–9446 (2022)

2. Park, J., Kim, J., Lee, S., Choi, J.K.: Machine learning based photovoltaic energy prediction scheme by augmentation of on-site IoT data. Fut. Gener. Comput. Syst. **134**, 1–12 (2022)

3. Baashar, Y., et al.: Evaluation of postgraduate academic performance using artificial intelligence models. Alexandria Eng. J. **61**(12), 9867–9878 (2022)

4. Wipfli, H., Withers, M.: Engaging youth in global health and social justice: a decade of experience teaching a high school summer course. Glob. Health Action **15**(1), 1987045–1987045 (2022)

5. Wang, C., Li, B., Cheng, B., Yang, J., Zhou, L.: Research on learning initiative Based on behavior quantization and potential value clustering. Alexandria Eng. J. **61**(7), 5621–5627 (2022)

6. Wu, D., Wang, S., Liu, Q., Abualigah, L., Jia, H., Razmjooy, N.: An improved teaching-learning-based optimization algorithm with reinforcement learning strategy for solving optimization problems. Comput. Intell. Neurosci. **2022**, 1535957–1535957 (2022)

7. Gill, H.S., Khehra, B.S.: Apple image segmentation using teacher learner based optimization based minimum cross entropy thresholding. Multimedia Tools Appl. **81**(8), 11005–11026 (2022)

8. Lu, W., Vivekananda, G.N., Shanthini, A.: Supervision system of English online teaching based on machine learning. Prog. Artif. Intell., 1–12 (2022). https://doi.org/10.1007/s13748-021-00274-y

9. Mashwani, W.K., Shah, H., Kaur, M., Bakar, M.A., Miftahuddin, M.: Large-scale bound constrained optimization based on hybrid teaching learning optimization algorithm. Alexandria Eng. J. **60**(6), 6013–6033 (2021)

10. Sokoli, D., Širca, N.T., Koren, A.: Quality of teaching in Kosovo's higher education institutions: viewpoints of institutional leaders and lecturers1. Hum. Syst. Manage. **40**(5), 685–700 (2021)

11. Ma, H., Yang, S., Feng, D., Jiao, L., Zhang, L.: Progressive mimic learning: a new perspective to train lightweight CNN models. Neurocomputing **456**, 220–231 (2021)

12. Yang, N.-C., Liu, S.-W.: Multi-objective teaching–learning-based optimization with pareto front for optimal design of passive power filters. Energies **14**(19), 6408 (2021)

13. Pratama, M., Za'in, C., Lughofer, E., Pardede, E., Rahayu, D.A.P.: Scalable teacher forcing network for semisupervised large scale data streams. Inf. Sci. **576**, 407–431 (2021)

14. Mathur, G., Chauhan, S.A.: Teacher evaluation of institutional performance: managing cultural knowledge infrastructure in knowledge organizations. Int. J. Knowl. Manage. **17**(4), 93–108 (2021)

15. Hua, L., Liu, G.: Development of basketball tactics basic cooperation teaching system based on CNN and BP neural network. Comput. Intell. Neurosci. **2021**, 1–11 (2021). https://doi.org/10.1155/2021/9497388

16. Tsai, F.H., Hsiao, H.S., Yu, K.C., Lin, K.Y.: Development and effectiveness evaluation of a STEM-based game-design project for preservice primary teacher education. Int. J. Technol. Des. Educ. **3**, 1–22 (2021)

17. Tian, Y., Zhang, L., Sun, J., Yin, G., Dong, Y.: Consistency regularization teacher–student semisupervised learning method for target recognition in SAR images. Vis. Comput., 1–14 (2021). https://doi.org/10.1007/s00371-021-02287-z

18. Zhang, B., Velmayil, V., Sivakumar, V.: A deep learning model for innovative evaluation of ideological and political learning. Prog. Artif. Intell., 1–13 (2021). https://doi.org/10.1007/s13748-021-00253-3

19. Tamai, T., Okamoto, K., Iuchi, K., Kawada, K.: Development of teaching material to design a vehicle on data science in junior high school technology education. IEEJ Trans. Electr. Electron. Eng. **16**(10), 1407–1413 (2021)

20. Dietrich, J., Greiner, F., Weber-Liel, D., Berweger, B., Kämpfe, N., Kracke, B.: Does an individualized learning design improve university student online learning? A randomized field experiment. Comput. Hum. Behav. **122**, 106819 (2021)

Document-Level Sentiment Analysis of Course Review Based on BG-Caps

Jing Wu[✉], Tianyi Liu, and Wei Hu

College of Computer Science and Technology, Wuhan University of Science and Technology,
Wuhan, China
{wujingecs,Lty,huwei}@wust.edu.cn

Abstract. With the development of the Internet in various fields, the combination of education and the Internet is close. Many users choose courses they are interested in to study on the MOOC platform and leave text reviews with emotional colors. However, the traditional word vector representation method extracts text information in a static way, which ignores text location information. The convolutional neural network cannot fully utilize the semantic features and correlation information, so the results of text sentiment analysis are inaccurate. To solve the above problems, this paper proposes a sentiment analysis method based on BG-Caps MOOC text review. The ALBERT pretraining model was used to obtain the dynamic feature of the text. Combined with the BiGRU and capsule network model, the features were trained to obtain deep semantic features. We evaluated our mode on the MOOC review dataset. The results show that the proposed method achieved effective improvement in accuracy.

Keywords: Emotional analysis · Capsule network · MOOC · ALBERT

1 Introduction

MOOC (Massive Open Online Courses) is a large open online course. Since 2012, large-scale online open courses have been rapidly expanded, and the teaching process of traditional education has been innovated in the expansion process. The MOOC platform provides innovative breakthroughs in many aspects, such as learning methods, knowledge dissemination scope and time and place constraints of offline courses. The MOOC platform [1] retains learning records in the teaching process, including course reviews and learning duration. As part of educational big data, course reviews can provide reasonable and objective course evaluation references for educators, learners and platform managers of sentiment analysis of natural language processing and the field of educational course teaching construction. Its evaluation reference has important theoretical value and practical application value.

Although online learning platforms have been widely studied and sought by learners in recent years, there are still some problems [2]. For example, the quality of the platform courses, the number of courses, the needs of learners, etc., and some problems existing in the platform itself will affect the quality of the MOOC platform. Therefore, research

Y. Wang et al. (Eds.): ICPCSEE 2022, CCIS 1629, pp. 394–405, 2022.
https://doi.org/10.1007/978-981-19-5209-8_26

on MOOC course evaluation is of great significance. Due to the large amount of data in MOOC course reviews, we use artificial methods to count and analyse MOOC course reviews and extract the content information of interest. This is not only inefficient but also causes many errors due to human subjectivity.

Therefore, we urgently need a method that can efficiently obtain the required information and fully display the analysis results. How to effectively use sentiment analysis technology to extract and extract the content and information we truly want from massive MOOC course reviews [3] and obtain the sentiment orientation of these reviews through sentiment analysis technology has turned into the research difficulty of natural language processing. It is also a research challenge of education at home and abroad, which has led to the in-depth study of many scholars.

Therefore, to address the above challenges, this paper proposes a text analysis model. The BG-Caps model for sentiment analysis of MOOC text reviews is as follows:

(1) The ALBERT pretraining model is used to obtain the text dynamic feature representation.
(2) We designed a sentiment analysis model BG-Caps, which was composed of the BiGRU network, attention mechanism module and capsule network emotion classification layer. BiGRU carried out feature extraction and assigned attention weight to important emotional features combined with an attention mechanism. Then, the feature vector was transformed by a capsule network to capture the information that can characterize the emotional intensity of the text. The attribute features are extracted, and feature reorganization is carried out to effectively classify different types of texts.
(3) The softmax function was used to classify semantic features, and the sentiment polarity of comments was obtained. Our method was evaluated on the MOOC review dataset and compared with other review sentiment analysis methods. The results indicate that our review sentiment analysis model achieved improvement in accuracy.

2 Related Work

Document-level sentiment analysis refers to the classification based on documents. This analysis classifies the whole document as a whole and judges whether the document expresses positive, neutral or negative emotions. At present, the mainstream research methods of document-level sentiment analysis include sentiment dictionary [4], machine learning [5] and deep learning methods [6].

Yang [7] considered that the method based on an emotional dictionary has the problem of emotional deviation and added threshold parameters to the calculation formula of emotional analysis, which effectively reduced the polar deviation rate. They used the emotional dictionary SentiWordNet and Amazon four products as the experimental dataset, and the results showed that the performance of the emotional analysis method based on the emotional dictionary was significantly improved. Vilares [8] used the Twitter corpus of emotion label code conversion, which largely solved the difficulty of multilingual emotion classification on Twitter. Xu [9] constructed an extended sentiment dictionary for classifying emotions. The accuracy was improved by combining it with a naive Bayesian classifier.

Wang [10] proposed a method of SVM for emotion classification and compared it with Doc2vec, indicating the effectiveness of this method. Huang [11] proposed a topic sentiment method with a multiple feature model and finally realized the synchronous detection of emotions and topics in microblogs.

Machine learning is based on sentiment analysis as a text classification problem, usually using supervised learning algorithms for sentiment classification [12]. Although this method does not require predefined semantic rules, it requires manual annotation of features. Therefore, the method is not only time consuming and labor intensive, but the quality of these features also has an impact on the results. At present, deep learning methods have advantages that other analysis methods do not have. For example, computing and learning ability. Moreover, the deep learning algorithm can capture the grammatical and semantic features of the text without a large number of feature engineering [13].

Mikolov [14] proposed a distributed word vector representation with a neural network, namely, CBOW and skip gram, which has become one of the most commonly used technologies for deep learning to process natural language tasks. A recurrent neural network is a model based on time series. Liu [15] used the temporal logic of sentence vectors as the upper input of RNNs to effectively solve the problem of low accuracy. Chen [16] proposed multichannel convolutional neural networks for sentiment classification so that the model can learn all aspects of the characteristics. Trofimovich [17] used GRU to classify Russian microblog comments, and the experimental effect was better than that of a convolutional neural network. Zhang [18] used a bidirectional long-term and short-term memory network combined with an attention model to encode microblog text and its emoticons, which has better performance than the known model in multiple tasks. Many scholars propose combining deep learning structures for sentiment analysis. Yan [19] proposed that an attention mechanism should be applied to a convolutional neural network and BIGRU at the same time, and then the feature fusion obtained by the two networks was applied to sentiment analysis. Zhang [20] proposed introducing an emotion integral into a CNN to obtain local features of text and obtain global features of text through BiGRU and an attention mechanism. Then, the two features were fused and applied to short text emotion analysis. Yuan [21] proposed first extracting text features with a multichannel convolutional neural network and then sending them to BiGRU. Finally, combined with the attention mechanism, the contextual emotional features of the text are obtained.

Inspired by all of the above work, we proposed a new text classification model composed of ALBERT and a capsule network and introduced the attention mechanism for text classification. Evaluated on the basis of the obtained dataset, compared with the general sentiment analysis model, we obtained better sentiment analysis results and indicated the validity of the proposed model.

3   Method

We consider the role of the text representation model in judging the emotional polarity of text and propose an emotional analysis model BG-Caps.

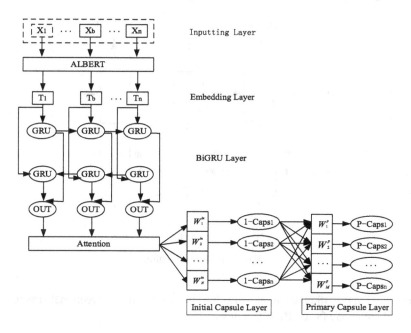

Fig. 1. BG-Caps network structure.

In this model, ALBERT is used for the dynamic representation of text features. BiGRU performs feature extraction, and the attention mechanism is used to assign attention weight to important emotional features. Then, the capsule network is used to feature vectorize, and the information that can characterize the emotional intensity of the text is captured. The attribute features are extracted, and feature reorganization is carried out to effectively classify different types of texts.

The structure consists of the BiGRU network, attention mechanism module and capsule network emotion classification layer. As shown in Fig. 1.

3.1 BiGRU Module

BiGRU (bidirectional gated recurrent unit) can effectively integrate the association of text context features and has a good classification effect on emotional texts. BiGRU also avoids the gradient disappearance problem in the traditional RNN and has a stronger memory function. TO capture the useful information features of text, the BG-Caps model vectorizes scalar information, making the representation of text features richer. To make the feature description more accurate, it needs to be adjusted on this basis.

The GRU module in our method is shown in Fig. 2.

In the proposed model, each word vector in the text sequence corresponds to a BiGRU unit, and the number of BiGRU units needs to be equal to the number of words in each text in the dataset. Therefore, for the MOOC review dataset, the number of BiGRU units was set to 50. At the same time, the output vector dimension of each GRU unit was set

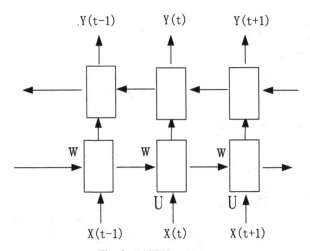

Fig. 2. BiGRU structure.

to the word vector dimension, namely, 768. Because of the bidirectional structure, the output dimension of each BiGRU unit was 1536.

3.2 Attention Mechanism Module

BiGRU is able to obtain contextual features of text sentence sequences. However, in the actual language context, the semantic information of each sentence word has different contributions to the emotional expression of the discourse. Each text contains a large amount of useless word information, which will lead to information redundancy, so it is difficult to obtain useful information from sentence sequences simply by using BiGRU.

Therefore, this paper proposes using a multiple self-attention mechanism to further obtain important information after the BiGRU network captures context features. The multiattention mechanism can better assign attention weights to important emotional features to achieve a more accurate understanding of sequence feature semantics. The self-attention mechanism mainly includes three kinds of elements: Query, Key and Value, in which the query is a sequence of key-value mappings. Considering self-attention does not fully obtain the meaningful features of the text from multiple perspectives. Therefore, we used the multiple self-attention mechanism, which uses the output of the upper BiGRU as the input of multiple self-attention, and calculates feature contexts of different dimensions in the vector space. After repeated calculations, the results are spliced to obtain a comprehensive feature representation.

The multiple self-attention mechanism is as follows:

$$multi(h_t) = concat(score_1(h_1), score_2(h_2), ..., score_h(h_t))W^O \qquad (1)$$

where h_t represents the hidden layer output of the BiGRU, score i represents the output of attention, and h is the number of repetitions. The formula for the score i is as follows:

$$score_i(h_t) = attention\left(h_t W_i^Q, h_t W_i^K, h_t W_i^V\right) \qquad (2)$$

where W_i^Q, W_i^K, W_i^V and W^O are parameter matrices for mapping input h_t to vector spaces of different dimensions.

The attention function is the point product operation in the self-attention mechanism. The formula is:

$$attention(Q, K, V) = softmax(\frac{QK^T}{\sqrt{d}})V \tag{3}$$

3.3 Capsule Network

Initial Capsule Layer. We used the transformation matrix w to produce the prediction g and used weight sharing. For each capsule layer unit, the connection strength is increased or decreased by dynamic routing to express the characteristics in the text, which is more effective than the maximum pool strategy in cnn. Without losing meaningful spatial information, it detects whether there are similar or identical features in different positions of the text.

Main Capsule Layer. The main capsule layer used the feature vector u as the input and uses the dynamic routing algorithm to recombine these inputs between features. In this way, important information that can better characterize the text features is extracted.

Dynamic Routing Algorithm. For each capsule layer unit, the connection strength is increased or decreased by dynamic routing to express the characteristics in the text, which is more effective than the maximum pool strategy in CNN.

Dynamic routing builds a nonlinear mapping to ensure that each prediction is sent to the appropriate unit in the subsequent capsule layer:

$$\left\{ \hat{g}_{j|i} \in R^d \right\}_{i=1,2,3...H,j=1,2,...,N} \rightarrow \left\{ u_j \in R^d \right\}_{j=1}^N \tag{4}$$

H represents the number of GRU units. For the given predictor $g_{j|i}$, the connection strength $c_{j|i}$ at each iteration is updated as follows:

$$c_{j|i} \leftarrow softmax(b_{j|i}) \tag{5}$$

where $b_{(j|i)}$ is the coupling coefficient, and the sum of weight $s_{(j|i)}$ is obtained by accumulating all the prediction quantities $g_{(j|i)}$:

$$s_{j|i} \leftarrow \sum_j c_{j|i}\hat{g}_{j|i} \tag{6}$$

The feature vector u_j of the initial capsule layer and its probability p_j^l are obtained by using the squash nonlinear activation function to compress the weight sum. The specific representation is:

$$u_j \leftarrow squash(s_{j|i}), p_j^l \leftarrow |u_j| \tag{7}$$

After each feature vector u_j of the initial capsule layer is generated, each coupling coefficient $b(j|i)$ is updated:

$$b_{j|i} \leftarrow \langle \hat{g}_{j|i}, u_j \rangle \tag{8}$$

The squash function can reflect the salient degree of vector feature u_j in a bounded range. The squash function is divided into two parts: the first is used to normalize p_j^l, and the second ensures that $s_{(j|i)}$ and u_j are in the same direction. In this paper, to better achieve the amplification effect when the u_j module length tends to 0, the constant parameters were adjusted to 0.5 in the first item. This adjustment makes the experimental results show a more accurate classification effect. The specific calculation method is as follows:

$$squash(s_{j|i}) = \frac{\left\| s_{j|i} \right\|^2}{0.5 + \left\| s_{j|i} \right\|^2} \frac{s_{j|i}}{\left\| s_{j|i} \right\|} \tag{9}$$

Output Layer. The softmax function is used to calculate the input and then classify the text.

4 Results

4.1 Experimental Details

We perform experiments on the homemade MOOC course review dataset and focus on verifying the methods proposed in this paper for sentiment analysis of student review texts. We determine the advantages and disadvantages of this method through experimental analysis. To evaluate the validity of the ALBERT-BiGRU-Caps method for comment text, the model is constructed based on PyTorch, and different learning rates are used for different layers. The ALBERT layer parameters should not be too high because of the pretraining, so the learning rate should not be too high to avoid overfitting, so it was set to 2e−5, and its learning rate was 1e−3. We used the NVIDIA RTX 1660.

4.2 Metrics

To better evaluate the model, we used three evaluation metrics: precision, recall and F1 value. The calculation formula is as follows:

$$precision = \frac{TP}{TP + FP} \tag{10}$$

$$recall = \frac{TP}{TP + FN} \tag{11}$$

$$F1 = 2 * \frac{precision * recall}{precision + recall} \tag{12}$$

where TP is true positive, FN is false negative, and TN is true negative. Precision denotes the proportion of correct predictions in all instances with a model predictive value of 1. Recall reflects the proportion of correct predictions in all instances with a real value of 1. The F1 combines the results of precision and recall. The model has good performance when the F1 value is 1 and poor performance when the F1 value is 0.

4.3 Experimental Results

To demonstrate the validity of our method in review text, we compared word embedding performance and model performance.

Experimental Comparison of Word Vector Performance. To demonstrate the effectiveness of different types of word vectors in emotion classification tasks, we conduct experiments on the basic model framework. The albert module used to generate word encoding in the albert-bigru-caps model is replaced by pytorch's own word embedding layer, word2vec word embedding and glove word embedding. The bigru-caps deep semantic feature extraction layer and output layer are retained. Table 1 shows the classification accuracy obtained by the model using various word vectors.

Table 1. Table of experimental results on word vector performance.

Word vector type	Accuracy	Recall	F1
Random word embedding	81.3	82.1	81.7
Word2Vec	84.4	84.7	84.5
GloVe	89.2	88.8	89.0
BERT	91.9	91.4	91.2
ALBERT	91.3	90.6	90.9

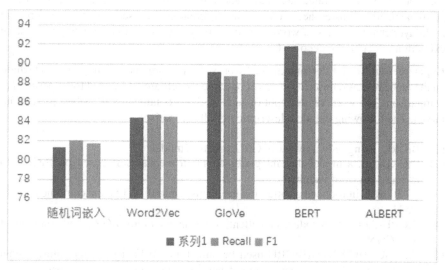

Fig. 3. Experimental results of word vector

The results were visualized as shown in Fig. 3.

It is obvious that the model achieves the worst classification performance using embedded random words. The random word embedding is obtained through the word embedding layer in the deep learning framework. The context word sequence marked by integers is randomly transformed into the real number vector of the specified dimension through the embedding layer. Random words embedded in the initial stage do not have any semantic information, resulting in insufficient representation ability.

When Word2Vec and GloVe pretraining models are used to generate word embedding, the classification performance is significantly improved. Both traditional word embeddings have implicit semantic information. The performance of GloVe is slightly higher than that of Word2Vec. The Word2Vec algorithm mainly uses adjacent context to train word vectors and lacks global dependency support. The GloVe algorithm uses global information to train word vectors based on the co-occurrence relationship of words, so its semantic representation is more abundant and complete. The use of the ALBERT pretraining model greatly improves the classification performance, and the accuracy was improved by 10%, 6.9% and 2.1%.

From Table 1, the BERT is slightly better than the ALBERT, but the gap is not large. However, it is worth noting that the size of ALBERT is only 65 MB, while the model size of BERT is 390 MB. The former model size is only one-sixth of the latter, which is small in performance.

We believe that this result is because the word vector obtained based on ALBERT can solve the problem of polysemy. ALBERT can also generate different dense vectors for the same word according to different semantic environments. The analysis model can extract more abundant semantic features according to these vectors, highlighting the multilayer characteristics of words. Most Chinese words contain complex features, including grammatical semantics. Because Word2vec and GloVe have simple structures, the word vectors generated are static, and the distribution of word vectors will not be dynamically adjusted according to the different context semantics. Therefore, the multilayer characteristics of words cannot be fully expressed, thus affecting the feature extraction effect on the deep text.

Experimental Model Performance. To prove the performance of our proposed method on the mooc review dataset, a variety of baseline models are introduced for experimental comparison. The word vector input of each model is obtained by the albert pretraining model. The following detailed description of all comparison models:

(1) BiGRU, using only a two-way GRU structure model.
(2) BiGRU-Att, a model that integrates attention mechanisms based on BiGRU to increase keyword weights.
(3) AT-BL & C introduced an attention mechanism into BiLSTM and combined it with CNN.
(4) BiGRU-AttCNN introduces an attention mechanism into BiGRU and combines it with CNN.
(5) The MC-AttCNN-AttBiGRU used the attention mechanism in the multichannel CNN and BiGRU at the same time and then fused the obtained features for emotional analysis.
(6) BiGRU-CapsNet, using a generic capsule network model with BiGRU.

The experimental results are shown in Table 2 and Fig. 4. The experimental results indicate that the proposed method is effective and superior.

Table 2. Results.

Method	Accuracy	Recall	F1
BiGRU	78.7	78.2	76.4
BiGRU-Att	81.3	82.1	81.7
AT-BL&C	84.4	84.7	84.5
BiGRU-AttCNN	88.1	87.7	87.9
MC-AttCNN- AttBiGRU	89.2	88.8	89.0
BiGRU-CapsNet	89.5	90.4	89.9
ALBERT-BiGRU-Caps (ours)	91.3	90.6	90.9

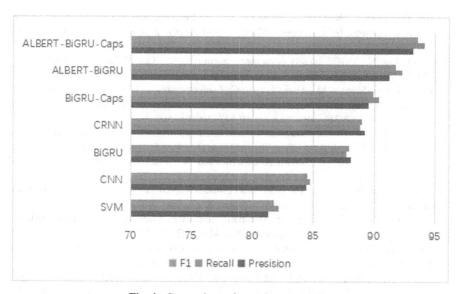

Fig. 4. Comparison of experimental results

It can be found that the capsule network with BiGRU can better extract feature vectors, the ALBERT model is used to pretrain the text, and the text feature representation can be better obtained. The ALBERT-BiGRU-Caps method was used to analyse the MOOC review text, which truly shows the application function of the model.

5 Conclusion

The large amount of educational big data brought by large-scale online open courses has important significance. It has application value for emotional analysis of natural language processing and teaching construction of education courses. Therefore, the analysis of the emotional colors of MOOC course reviews is important for Internet education.

To analyse the overall emotional tendency of text review, this paper designed the ALBERT-BiGRU-Caps model to analyse the MOOC review text and truly shows the application function of the model. It can be seen from the experimental results that the capsule network based on BiGRU can make full use of local feature information and context semantic association and better extract deep semantic features. At the same time, the ALBERT model was used to pretrain the text, and the dynamic feature representation of the text can be better obtained. Finally, the softmax function was used to classify the semantic features. Because the ALBERT model still has many parameters and the training time is too long, we may compress the ALBERT model and reduce the complexity of the model to improve the training efficiency in future work.

References

1. An, Y.-H., Pan, L., Kan, M.-Y., Dong, Q., Yan, F.: Resource mention extraction for MOOC discussion forums. IEEE Access 7, 87887–87900 (2019). https://doi.org/10.1109/ACCESS.2019.2924250
2. Li, Q., Liu, N.: MOOC quality assurance system research. Open Educ. Res. 21(05), 66–73 (2015)
3. Elena Alonso-Mencía, M., Alario-Hoyos, C., Estévez-Ayres, I., Kloos, C.D.: Analysing self-regulated learning strategies of MOOC learners through self-reported data. Australas. J. Educ. Technol., 56–70 (2021). https://doi.org/10.14742/ajet.6150
4. Zhou, Z.: Chinese short comment sentiment classification based on sentiment dictionary and supervised learning. J. Zhangzhou Normal Univ. (Nat. Sci. Ed.) 4, 23–28 (2013)
5. Jing, L., Li, M., He, T.: Sentiment classification of online reviews combining extended dictionary and self-supervised learning. Comput. Sci. 47 (2), 78–82 (2020)
6. Pan, X., Zhao, P., Zhao, Q.: The sentiment classification model of e-commerce reviews based on BLSTM and attention mechanism. Comput. Digit. Eng. 9, 2227–2232 (2019)
7. Yang, X., Zhang, Z., Wang, L.: Automatic construction and optimization of sentiment dictionary based on Word2Vec. Comput. Sci. 44 (1), 42–47 (2017)
8. Vilares, D., Alonso, M.A., Gomez-Rodríguez, C.: Supervised sentiment analysis in multilingual environments. Inf. Process. Manag. 53(3), 595–607 (2017)
9. Guixian, X., Ziheng, Y., Yao, H., Li, F., Yueting Meng, X.: Chinese text sentiment analysis based on extended sentiment dictionary. IEEE Access 7, 43749–43762 (2019). https://doi.org/10.1109/ACCESS.2019.2907772
10. Wang, Y.X., Zheng, X., Hu, X.: Short text sentiment classification of high dimensional hybrid feature based on SVM. Comput. Technol. Dev. 28(2), 88–93 (2018)
11. Huang, F.-L.: Mining topic sentiment in microblog based on multifeature fusion. J. Comput. 40 (4), 872–888 (2017)
12. Peng, Y., Wan, C., Jiang, T., Liu, D., Liu, X., Liao, G.: Product features and sentiment words extraction based on semantic constraint LDA. J. Softw. 28(03), 676–693 (2017)
13. Hu, T., Dan, Y., Hu, J., Li, X., Li, S.: News named entity recognition and sentiment classification based on attention-based bidirectional long short-term memory neural network and conditional random field. J. Comput. Appl. 40(7), 1879–1883 (2020)
14. Mikolov, T., Sutskever, I., Chen, K.: Distributed representations of words and phrases and their compositionality. In: Advances in Neural Information Processing Systems, NIPS (2013)
15. Jinshuo, L., Zhi, Z.: A food safety information sentiment classification model based on joint deep neural network. Comput. Sci. 43(12), 277–280 (2016)
16. Chen, K., Liang, B., Wende, K.: Chinese microblog sentiment analysis based on multichannel convolutional neural network. Comput. Res. Dev. 55(5), 945–957 (2018)

17. Trofimovich, J.: Comparison of neural network arChiteCtures for sentiment analysis of Russian tweets. In: Computational Linguistics and Intellectual Technologies: Proceedings of the International Conference Dialogue (2016)
18. Zhang, Y., Zheng, J., Huang, G.: Microblogging sentiment analysis method based on dual attention model. J. Tsinghua Univ. (Nat. Sci. Ed.) **58**(02), 122–130 (2018)
19. Cheng, Y., Yao, L., Xiang, G., Zhang, G., Tang, T., Zhong, L.: Text sentiment orientation analysis based on multi-channel CNN and bidirectional GRU with attention mechanism. IEEE Access **8**, 134964–134975 (2020). https://doi.org/10.1109/ACCESS.2020.3005823
20. Zhang, T., Liu, X., Gao, Y.: Emotional analysis of attention mechanism based on convolutional neural network and two-way gated loop unit network. Sci. Technol. Eng. **21**(01), 269–274 (2021)
21. Yuan, H., Zhang, X., Niu, W.: Text sentiment analysis based on multichannel convolution and bidirectional GRU model integrating attention mechanism. J. Chin. Inf. **33**(10), 109–118 (2019)

Research on Effective Teaching Based on the BOPPPS Model Under the Background of Engineering Education Professional Certification

Fang Yin[✉], Chengyan Li, Dongpu Sun, and Libo Zhou

School of Computer Science and Technology, Harbin University of Science and Technology, Harbin 150080, China
yinfang@hrbust.edu.cn

Abstract. According to the general requirement of the engineering education professional certification and aiming at the problem of how to develop undergraduates with the ability to solve complex engineering problems by teaching research and curriculum construction, a new effective teaching model is built by analyzing the relationship between BOPPPS and effective teaching. Our contributions are as follows: we combine professional training objectives with curriculum characteristics, focus on the importance of participating in the main position of students in learning, and emphasize that students' participatory interaction and feedback and diversified assessment methods for evaluation are adopted to focus on the development and improvement of the comprehensive quality of students and classroom teaching quality and teaching effect to ensure the effectiveness of teaching. The practice in the Oracle database course proves that the teaching mode proposed in this paper can effectively improve the teaching effect, and it plays a positive role in strengthening students' practical abilities to solve practical engineering problems.

Keywords: Engineering education professional certification · BOPPPS model · Effective teaching · Diversified assessment

As an international general engineering education quality assurance system, colleges and universities all try to promote the professional certification of engineering education, which is an important foundation to achieve international mutual recognition of engineering education and engineer qualification. Its main purpose is to make students' achievements outcome-based and ensure the quality of teaching. It requires confirming that the major can continuously achieve its self-defined educational objectives and that its graduates have the core competencies required by the major, maintain educational quality through the certification mechanism and pursue continuous improvement.

Its core philosophy is result oriented, student-centered and continuous improvement, and it is mainly reflected in seven aspects, among which student is the core, and the resource allocation, teaching design and teaching implementation in the process of school running are guided by the learning outcomes and achievement of graduates [1]. The basic standard of effective teaching is the progress and development of students, which

Y. Wang et al. (Eds.): ICPCSEE 2022, CCIS 1629, pp. 406–411, 2022.
https://doi.org/10.1007/978-981-19-5209-8_27

is consistent with the core concept of engineering education professional certification, which regards students as the leading role and takes achievements as the final result. Under the student-centered teaching concept, it is imperative to establish a teaching model that is conducive to teachers' teaching improvement and can promote students' effective learning. It focuses on learning objectives, students' participatory learning and access to teaching feedback [2, 3].

1   The Connotation and Essence of BOPPPS

The BOPPPS model based on constructivism was proposed by Canadian ISW, and it is a new teaching model different from the traditional one [4]. It focuses on learning objectives, students' participatory learning and timely access to teaching feedback. A patterned teaching process can be developed through it to improve teachers' teaching ability to make students accept better and significantly promote classroom teaching effectively [5].

BOPPPS is the combination of six steps. In the whole teaching process, learning is completed according to the process of introduction - autonomous learning - participatory classroom learning - after class review - summary. Among them, bridge-in is to lead out the learning content of this section by cases or problems, introduce topics and mobilize students' learning interest to clarify their learning motivations; objective is to make students clear the requirements and level that should be achieved after the class, that is, to clarify the teaching objectives; preassessment is to test students before class to master the knowledge and ability they have gained before class, to arrange the following teaching content according to the specific situation; participatory-learning is a student-centered teaching activity in the class, and mobilize students' learning enthusiasm to fully participate in teaching activities; postassessment is the test after class to investigate students' mastery of knowledge and evaluate the teaching effect; summary is that teacher and students summarize the learning situation and existing problems of class teaching content together, adjust or extend the application of teaching content, and improve the teaching effect.

This teaching method introduces topics with questions so that students have clear learning objectives and easily burst their initiative and enthusiasm to cultivate Innovation Ability [6]. On the premise that students have clear learning objectives, they can learn relevant knowledge independently before class, find problems encountered in the learning process, and ensure that they meet the required level of knowledge and ability.

2   BOPPPS and Effective Teaching

The BOPPPS teaching mode focuses on students' development needs, deepens students in learning, and obtains feedback on effective study or problems from students. It is known as an effective teaching model because of its efficiency, effectiveness and benefit [7].

Effect refers to the effect of class teaching, that is, the achievement degree of teaching objectives by class teaching activities. For teaching activities, the key point is learning rather than teaching. The key point of teaching research should also be how to enable students to learn and obtain knowledge, which is the only standard to measure the teaching effect. Efficiency reflects the ratio between output and input in the teaching process. The goal we pursue is not only to complete the teaching tasks in class and achieve the teaching objectives but also to achieve the students' learning effect and meet the standards in class. Benefit refers to the degree of conformity between the teaching effect or goal and personal or social needs. It mainly measures whether the knowledge learned by students can be applied. Therefore, benefit should be the basic premise for teaching.

Measure the BOPPPS teaching mode from the above three aspects: introduce the teaching content of this class through cases or questions to mobilize students' learning enthusiasm and ensure the learning effect; clearly set the learning objectives of this class so that teachers and students can complete teaching and learning activities driven by objectives, and the effect and efficiency can both be guaranteed. Students complete autonomous learning independently before class and evaluate their mastery of autonomous learning through a pretest to ensure the efficiency and effect of class teaching. In the middle of the class, through participatory learning, students can practice the knowledge gained in this class in the way of group cooperation to complete relevant tasks to ensure the knowledge learning effect, truly apply what they have learned, and ensure the benefits of teaching activities. Verify the learning effect by posttest, and finally, teachers and students summarize, reflect on the problems existing in the teaching process, find the problems in the students' learning process together to adjust and make up in the follow-up teaching through the feedback closed-loop teaching mode to ensure the effect, efficiency and benefit of learning activities [8].

In essence, effective teaching refers to making the expected best effect of students' learning by a series of teaching activities, that is, the specific progress or development of students after the teacher has been teaching for a period of time. Therefore, the progress and development of students become the standard to judge and measure whether teaching is effective or not.

In summary, the BOPPPS teaching mode and effective teaching have the same ultimate goal, which will achieve a better teaching effect than the traditional teaching mode. The research on effective teaching based on BOPPPS in this paper is of practical significance.

3 Diversified Evaluation System and Its Application

The basic idea of BOPPPS and engineering education professional certification is to emphasize the embodiment of students' dominant position in the teaching process to make students truly masters of the class and emphasize the importance of participatory learning. To ensure the learning effect of participatory learning in the class, autonomous learning before and after class plays a very important role here. However, there are not too many constraints, supervision and management means in BOPPPS to ensure the effectiveness of teaching activities, which is not conducive to the development of students and is also contrary to the content of the current effective teaching theory "paying attention to measurement and quantification, paying attention to all students". Therefore, it is a beneficial supplement to effective teaching research under the BOPPPS to conduct a set of diversified evaluation systems to assess students' comprehensive application ability, supervise and evaluate the teaching process and cultivate students' practical ability and system abiliby [9, 10].

In this system, various teaching content and activities designed in the teaching process are all evaluated. For example, the students' mastery of basic theoretical knowledge is assessed through the participation of class activities, the completion of after-school homework and phased assessment. The students' practical ability to apply basic theoretical knowledge to solve specific engineering problems is assessed through experiments and practical tasks. Through the final examination, students' mastery of basic theoretical knowledge and their ability to comprehensively apply basic theoretical knowledge to analyze and design specific problems are assessed.

It was applied in the Oracle Database course, which is a professional course of computer specialty with high requirements for practical ability. Its teaching objective is to integrate professional knowledge, ability and quality into the teaching process; to stimulate students' interest in learning; to cultivate students' problem-solving skills and self-learning ability; to improve students' ability to analyze, design and solve practical problems; and to be committed to the cultivation of engineering talent. Based on the character of the course, effective teaching and BOPPPS are integrated for application in Oracle Database teaching. Students can further have a good and effective attitude toward using scientific thinking and practice methods to analyze and solve practical problems and will have the ability to actively learn, ask, think and solve problems.

The diversified assessment system applied in the Oracle Database is designed into five parts: usual learning performance, homework, experiment, comprehensive task and final exam, whose compositions are shown in Fig. 1, where the final exam accounts for 50%, the practical task accounts for 20% and the other parts account for 10% each. The evaluation basis of each link is described as shown in Fig. 1.

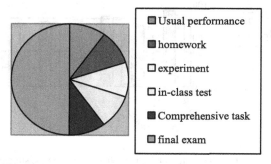

Fig. 1. Composition of the diversified assessment system

The usual performance is evaluated by whether or not they are active in class activities and the degree of basic theories grasp. The experiment is evaluated by the quality of operation and report, whose proportions are 70% and 30%, respectively. The practical task is evaluated by the system design and implementation, the proportion of which are 60% and 40%, respectively, as shown in Table 1.

Table 1. Score distribution of compositions of the diversified assessment system

Items	Usual performance	In-class test	Homework	Experiment		Practical task		Final exam	
Details				Oper-ation	Report	Design	Implementation	Theory	Tools
(%)	10	10	10	7	3	6	4	20	30

Grade 2018 is selected to show the score rate of each item of the diversified assessment system, as shown in Fig. 2. In this figure, the score rate, but not the absolute score, is used because the score rate is a uniform standard to compare each aspect. For example, it is easy to see that theory in the final exam and design of practical tasks is worse than other aspects, so it is necessary to strength these two abilities. To prove that the evaluation system is effective in promoting students' learning enthusiasm and improving the learning effect, the score rates of the two grades are compared in Fig. 3.

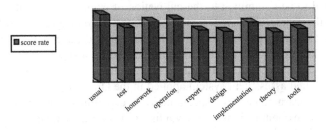

Fig. 2. Score rate of each item in Table 1 of Grade 2018

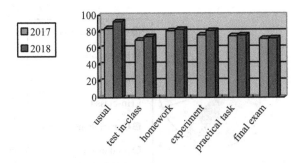

Fig. 3. Comparison of the score rate in the diversified assessment system of the two grades

4   Conclusion

Following the idea of engineering education certification and effective teaching, students are regarded as the lead role, achievements are taken as the final result, and improvements must be made continuously. The study and practice of the methods proposed in this paper are implemented on the Oracle database to emphasize the central leading goal of students in the class, guide students to study independently, adopt diversified teaching and assessment methods guided by teaching objectives and graduation requirements and strengthen students' ability to solve complex engineering problems and achieve good teaching results. The method proposed in this paper may be applied to other computer courses with high practical requirements. In the future, the methods of class interaction and practical activities can be continuously explored to better students' practical ability.

Acknowledgments. This work was financially supported by the Heilongjiang Higher Education Teaching Reform Project (SJGY20200309).

References

1. Wang, X., Wang, Yan., Chen, J.: Research on effective teaching Strategies of "Robot Technology" course under the background of professional certifications. Educ. Mod. **6**(53), 114–116+133 (2019)
2. Zhang, S., Lian, X., Yu, X.: Flipped classroom teaching design under the framework of BOPPPS model. Comput. Educ. **01**, 18–22 (2017)
3. Wu, Z., Lan, T., Wang, R.: Golden Course construction of software engineering basic course based on blended learning. In: ICPCSEE 2021, pp. 375–384. Springer, New York LLC (2021). https://doi.org/10.1007/978-981-16-5943-0_30
4. Zhang, Y.: Innovative application of LMYBK and BOPPPS effective teaching mode in the reform of teaching reform of ideological and political course in Colleges. J. Xinjiang Vocat. Univ. **29**(03), 72–77 (2021)
5. Wu, Y., Li, Z., Li, Y., Liu, Y.: Teaching reform and research of data structure course based on BOPPPS model and rain classroom. In: Zeng, J., Qin, P., Jing, W., Song, X., Lu, Z. (eds.) ICPCSEE 2021. CCIS, vol. 1452, pp. 410–418. Springer, Singapore (2021). https://doi.org/10.1007/978-981-16-5943-0_33
6. Xi, Y., Chen, X., Li, Y.: Exploration and Practice for the cultivation mode of college students' innovation ability. In: Zeng, J., Qin, P., Jing, W., Song, X., Lu, Z. (eds.) ICPCSEE 2021. CCIS, vol. 1452, pp. 456–464. Springer, Singapore (2021). https://doi.org/10.1007/978-981-16-5943-0_37
7. Zhang, J., Zhu, L.: On the effective classroom teaching design based on BOPPPS model. Teach, Learn. Method **37**(11), 25–28 (2016) (2021)
8. Zhang, Y., Cao, H.: Online education resource evaluation systems based on MOOCs. In: Zhou, Q., Miao, Q., Wang, H., Xie, W., Wang, Y., Zeguang, L. (eds.) ICPCSEE 2018, Part II. CCIS, vol. 902, pp. 605–615. Springer, Singapore (2018). https://doi.org/10.1007/978-981-13-2206-8_51
9. Zhang, W., Wang, R., Tang, Y., Yuan, E., Wu, Y., Wang, Z.: Research on teaching evaluation of courses based on computer system ability training. In: Zeng, J., Qin, P., Jing, W., Song, X., Lu, Z. (eds.) ICPCSEE 2021. CCIS, vol. 1452, pp. 434–442. Springer, Singapore (2021). https://doi.org/10.1007/978-981-16-5943-0_35
10. Huang, Q., Ye, F., Chen, Y., Xu, P.: A Cloud-based evaluation system for science-and-engineering students. In: Zhou, Q., Miao, Q., Wang, H., Xie, W., Wang, Y., Lu, Z. (eds.) ICPCSEE 2018. CCIS, vol. 902, pp. 530–538. Springer, Singapore (2018). https://doi.org/10.1007/978-981-13-2206-8_43

A Study of Vocational College Students' Anxiety in English Vocabulary Learning

Han Wu[✉]

Sanya Aviation and Tourism College, Sanya 572000, Hainan, China
516095898@qq.com

Abstract. With the development of the Hainan Free Trade Port, a large number of talented people who are proficient in English will be demanded. In Hainan Province, there are many vocational colleges that play a very important role in providing talent for future FTP construction. To improve these students' English ability, vocabulary learning is the key factor. English vocabulary learning involves many factors, and emotion is just one of them. When learning English vocabulary, students who have a positive attitude will achieve more than students who are negative. Anxiety, as a negative physiological factor, has a great impact on vocational students' English vocabulary learning. In this study, a questionnaire was used to test the anxiety state of vocational college students' English vocabulary learning, and SPSS 22.0 was used to analyze the collected data based on the questionnaire. From this study, vocational college students' vocabulary learning anxiety state can be known, and some useful suggestions are given to relieve the anxiety of students' vocabulary learning based on the results of the survey.

Keywords: Vocational college students · Vocabulary learning anxiety · Alleviation

1 Introduction

To further promote the rapid development of the Hainan Free Trade Port, in early 2019, the Hainan provincial government issued the *Action Plan to Comprehensively Improve the Foreign Language Proficiency of Hainan Citizens,* popularizing foreign languages for the whole people, cultivating the awareness of openness and improving the foreign language cultural literacy and cross-cultural communication ability of citizens. As an international language, English will play a pivotal role in the construction of Free Trade Port in the future. According to the plan, schools in Hainan, as one of the key areas, should set goals and plans for foreign language learning at different levels and in different industries to facilitate foreign language learning and communication in the whole society. In view of the current situation that Hainan Island has fewer undergraduate colleges and more vocational colleges, vocational colleges on the island will be given the function of training more specialized talent with international vision for the construction of Free Trade Port in the future. Therefore, it is urgent to improve the English level of students in vocational colleges.

Y. Wang et al. (Eds.): ICPCSEE 2022, CCIS 1629, pp. 412–426, 2022.
https://doi.org/10.1007/978-981-19-5209-8_28

English learning includes some basic skills, such as listening, speaking, reading and writing, but the basis of learning these skills is vocabulary. Wilkins, the English linguist, has ever said, without grammar people express very few things, and without words people can express nothing. Therefore, vocabulary learning is of great importance in English learning. However, the current English vocabulary of vocational college students is very small. They don't know how to learn English vocabulary in a proper method or strategy, so their enthusiasm for learning vocabulary is not high, which seriously hinders the improvement of their English level. In the process of English vocabulary learning, there are many factors affecting vocational college students' vocabulary acquisition. According to the affective-filter hypothesis proposed by Krashen (1982), learners' learning anxiety can cause language absorption disorders. Meanwhile, from humanistic theory, we know that learners with learning anxiety will not show obvious discomfort if they can relieve these anxieties properly. Otherwise, the learners will produce psychological obstacles due to failing to deal with the anxiety. Therefore, English vocabulary learning anxiety is one of the most important factors affecting students' vocabulary acquisition.

In recent years, English language learning anxiety has become a hot topic in English teaching research. Horwitz (2001) proposed that anxiety can reduce students' listening comprehension ability, word memory ability and verbal expression ability and lower their scores in language courses and standard tests. Hao Mei and Hao Ruoping (2001) conducted a study on the relationship among English score, achievement motivation and the anxiety state. It was found that students' English scores were generally affected by the degree of anxiety. Yang Jin (2000) also believed in his study that appropriate anxiety was beneficial to learning, but anxiety levels that were too high or too low hindered learning. Based on the results of previous studies, this study chose 284 students from Sanya Aviation and Tourism College as the objects of study to conduct a questionnaire survey to determine their current level of anxiety in English vocabulary learning and then put forward some corresponding methods to relieve this anxiety to improve the level and ability of college students' vocabulary acquisition.

2 The Relationship Between the English Vocabulary Level of Vocational College Students and Related Tests

2.1 English Vocabulary Proficiency Test for Vocational College Students

To clarify the current vocabulary level of vocational college students, this study took 83 first-grade students who majored in Mechanical and Electrical Equipment Maintenance from Sanya Aviation and Tourism College as the test objects to conduct a word test on the vocabulary that they have learned in the course of College English. In the test, students should choose the right spelling of a word according to its Chinese meaning or choose the right Chinese meaning according to its spelling, and the result of the test is shown in Fig. 1.

According to the test results, only 22% of the students passed the vocabulary test, while 78% of them failed. Moreover, half of the students scored less than 40 on the vocabulary test. The vocabulary of this test is all the basic core words in the College English Course, but the test results show that the students' vocabulary acquisition ability is too weak. Although they are basic words, their mastery level is not satisfactory.

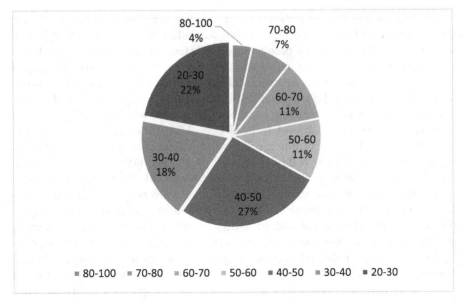

Fig. 1. The results of the vocabulary quiz

2.2 The Correlation Between English Vocabulary Level and the Score of Level A Test and Final Examination for Vocational College Students

Level A Test is the abbreviation for Practical English Test for Colleges Level A, which is a very important test for vocational college students. Students who can obtain the certificate of the test will have an enhanced chance for their future employment. Therefore, it is necessary to analyze the correlation between English vocabulary level and the score of the Level A Test and final examination.

In the following, SPSS correlation analysis was conducted on students' vocabulary test scores, final exam scores in the first semester and Level A Test scores to further determine there relationship, and the results can be seen in Table 1.

According to the results from Table 1, the Pearson correlation coefficient between students' vocabulary test scores and Level A test scores is 0.563, which indicates that there is a moderate correlation between them. This means that the higher students' vocabulary test scores are, the higher students' Level A test scores are. The Pearson correlation coefficient between students' vocabulary test scores and final exam scores is 0.644, indicating a strong correlation between the two. This means that the higher students' vocabulary test scores are, the higher students' final exam scores are. The Pearson correlation coefficient between students' final exam scores and Level A test scores was 0.762, indicating a strong correlation between the two. Therefore, it can be determined that there is a strong positive correlation among students' vocabulary test scores, Level A test scores and final exam scores, and it can be seen from Table 1 that the significance value P is $0.000 < 0.05$, indicating that there is a significant difference among the three.

Table 1. Correlation analysis

		Level A test scores	Vocabulary test scores	Final test scores
Level A test scores	Pearson correlation	1	.563**	.762**
	Sig. (2-tailed)		.000	.000
	N	42	42	42
Vocabulary test scores	Pearson correlation	.563**	1	.644**
	Sig. (2-tailed)	.000		.000
	N	42	55	55
Final test scores	Pearson correlation	.762**	.644**	1
	Sig. (2-tailed)	.000	.000	
	N	42	55	55

** Correlation was significant at 0.01 level (two-tailed)

3 A Survey of English Vocabulary Learning Anxiety of Vocational College Students

3.1 Research Method

Questionnaire survey.

3.2 Research Questions

English vocabulary learning anxiety level of vocational college students from different dimensions;

Differences in English vocabulary learning anxiety among students of different genders, majors and regions.

3.3 Research Objectives and Tools

The targets of the research are freshmen of non-English majors in Sanya Aviation and Tourism College. The research tool is a questionnaire.

The questionnaire was distributed through the website www.wjx.cn, and a total of 284 questionnaires were collected. The questionnaire referred to the Foreign Language Classroom Anxiety Scale (FLCAS) designed by Horwitz et al. in 1986 and the English Vocabulary Learning Anxiety Scale (CEVLAS) designed by Zhang Feng in 2015. Apart from the personal information about the subjects, such as gender, location of their home and major, the questions in the survey mainly focus on general vocabulary anxiety as well as vocabulary learning anxiety in listening, reading, writing, oral English, pronunciation, spelling, test and strategies. Each question adopts a 5-point Likert-type scale.

Five alternatives represent the different degrees to which the subjects respond to each question in the questionnaire (1 = strongly disagree, 2 = basically disagree, 3 = sometimes agree, 4 = basically agree, 5 = strongly agree). For each of the subjects, his or her English vocabulary learning anxiety score was produced by calculating his or her ratings of the 33 items. All the data from the questionnaires were entered into SPSS. 22. 0 to perform a statistical analysis to clarify the level of students' vocabulary learning anxiety from different dimensions and differences in vocabulary learning anxiety among students of different genders, majors and regions.

3.4   Analysis of Research Results

Testing of Reliability and Validity. The overall reliability and validity of the questionnaire were retested to determine whether the testing efforts of the questionnaire on the anxiety level of English vocabulary learning of vocational college students were credible and effective. The results are shown in Tables 2 and 3. According to Table 2, the Cronbach's alpha coefficient value of this questionnaire is 0.964. When the value of this coefficient exceeds 0.80, the reliability of the questionnaire is quite good. According to Table 3, the KMO value of this questionnaire is 0.948. According to the KMO metric standard, when its value is above 0.9, it indicates that a questionnaire is very suitable for factor analysis, and there is a strong correlation among the variables contained in each factor. The Bartlett test mainly tests whether each variable is independent. According to the results of the analysis based on SPSS 22.0, the value of Sig. in Bartlett's test is $0.00 < 0.05$ (P value < 0.05), indicating that each variable is correlated and that factor analysis is effective.

Table 2. Reliability statistics

Cronbach'Alpha	N of items
.964	33

Table 3. KMO and Bartlett test

Kaiser-Meyer-Olkin measure of sampling adequacy		.948
Bartlett's Test of Sphericity	About chi-square	8192.646
	df	528
	Sig.	.000

Exploratory Factor Analysis. Dimension reduction was carried out on the questionnaire to determine the factors in the questionnaire and identify the dimensions of English vocabulary learning anxiety of vocational college students. The results of the questionnaire analysis are shown in Table 4.

Table 4. Rotating element matrix

	Component				
	1	2	3	4	5
29. My biggest problem with writing is that words are poor and monotonous	.813	.218	.264	.122	.159
33. My vocabulary has affected my English test scores and English proficiency	.812	.256	.258		.140
30. I often cannot remember the words I have learned in the exam	.775	.277	.292	.117	
32. I always worry that my vocabulary will not meet the test requirements	.753	.312	.297	.137	.112
26. When I write an English composition, I often have difficulty in using the vocabulary I have learned to express my meaning properly	.747	.459	.208	.109	
27. When I write in English, I am always anxious about my vocabulary	.685	.440	.287	.144	.165
25. I am very confused about the new vocabulary collocation in English articles, which affects my reading speed	.543	.539	.138	.203	.152
36. I spend a lot of time on vocabulary study, but it did not work well	.491	.275		.431	.240
19. When I read or communicate in English, I feel embarrassed when people point out that certain words are not pronounced correctly	.198	.771	.249	.128	.166
16. I was worried that I was using a bad vocabulary in front of foreigners	.357	.691	.246		.166
20. For words that sound similar, I need to be able to tell them apart in specific context	.326	.688	.201	.163	
18. I have difficulty pronouncing words and often mispronounce them	.254	.681	.252		.148
15. I am afraid to communicate with others in English because of my limited vocabulary	.356	.676	.299		.247
13. I get nervous if I do not understand every word my English teacher says		.631	.351	.296	.292
24. When I read, I feel nervous and fidgety whenever new words appear or I forget the meaning of words I have learned	.393	.600	.180	.197	.297
17. When I speak English in public, I try to use simple and sure words for fear of making mistakes	.486	.564	.291	.154	
14. When speaking English, it is often difficult to think of a word that conveys the meaning	.428	.555	.436	.107	

(continued)

Table 4. (*continued*)

	Component				
	1	2	3	4	5
22. I'm worried that vocabulary is hindering my English reading	.502	.502	.426		
10. In English listening tests or listening exercises, not being able to understand every word worries me a lot	.279	.349	.763	.113	.152
11. In English listening tests or listening exercises, not being able to understand every word worries me a lot	.289	.334	.761	.155	.177
9. I'm worried that vocabulary is hindering the improvement of my English listening	.348	.339	.731		.103
12. When I was listening to English, I felt nervous and could not remember the meaning of the words I heard	.260	.468	.593	.180	.270
6. I worry more about my vocabulary than any other aspect of English learning	.463	.237	.587		
4. I always feel anxious about not being able to use words	.336	.407	.424		.372
35. I often discuss the methods and experience of vocabulary learning with my classmates or teachers	.213			.853	.215
31. I like activities like word tests and word contests			.175	.793	
23. In the English reading test, I can calmly face many unfamiliar words and get high marks in reading		.214		.742	−.148
21. My Pronunciation of English words is very authentic, which increases my confidence in speaking English		.209		.739	−.140
34. When I remember words, I use associative pronunciation, repeated transcription and other ways to help memory	.461		.185	.658	
28. When I write in English, I never get confused about what part of speech a word is, and can tell whether it is a noun, verb or adjective	.343	.237		.519	.331
8. I get bored and nervous when I hear to dictate words	.174	.424	.272		.659
7. I feel afraid and headache when I see English words. Vocabulary makes me afraid of Learning English	.245	.337	.367		.629
5. I like learning and memorizing words		−.116		.526	−.599

Extraction method: main component analysis.
Axis method: Maximum variation method with Kaiser normalization.[a]
[a] Convergent cycle in 7 iterations.

According to Table 4, the whole questionnaire can be divided into five factors, and each factor shows the English vocabulary learning anxiety of vocational college students from different dimensions. Factor 1, including questions 29, 33, 30, 32, 26, 27, 25, and 36, shows vocabulary learning anxiety in writing and total vocabulary. Factor 2, including questions 19, 16, 20, 18, 15, 13, 24, 17, 14, and 22, shows vocabulary learning anxiety in oral English and reading. Factor 3, including questions 10, 11, 9, 12, 6, and 4, shows vocabulary learning anxiety in listening. Factor 4, including questions 35, 31, 23, 21, 34, 28, and 5, shows vocabulary learning anxiety in vocabulary learning strategies and the influence of positive attitudes on vocabulary learning anxiety. Factor 5, including question 7 and 8, shows the influence of negative attitudes on vocabulary learning anxiety. All the questions in the questionnaire involved in each factor have a large factor loading, indicating that the questions in each factor better reflect the English vocabulary learning anxiety level of vocational college students from different dimensions. For example, the maximum load factor of the 8 questions involved in factor 1 is 0.813, and the minimum value is 0.491, both exceeding the minimum requirement of 0.4, indicating that each question in factor 1 belongs to the dimension of factor 1. By measuring these questions and calculating their scores, we can determine the English vocabulary learning anxiety level of vocational college students in writing and total vocabulary. The above situation also applies to other factors.

The Level of English Vocabulary Learning Anxiety from Different Dimensions. SPSS 22.0 was used for data analysis, and the statistical analysis results of each factor are described in the following tables. In Table 5, the mean of factor 1 is 25.7606, and then 25.7606 divides 8 by 3.213, so the mean of each question in factor 1 has been determined. According to Oxford and Burry-Stock's interpretation of the Likert five-point scale, the mean of each question is higher than 3.5, indicating a state of high anxiety, the mean of each question is equal to or lower than 2.4, indicating a state of low anxiety, and the mean from 2.5 to 3.4 is considered a moderate anxiety state. Therefore, students' vocabulary learning anxiety in writing and total vocabulary shows a moderate anxiety state.

Table 5. Descriptive statistics

	N	Minimum	Maximum	Mean	Standard deviation
Factor 1	284	8.00	40.00	25.7606	8.66285
Valid N (listwise)	284				

In Table 6, the mean of each question in factor 2 is 29.704, which is then averaged to each item, and the calculated average value of each question is 2.97. According to the interpretation of the reference results, students' vocabulary learning anxiety in oral English and reading also shows a moderate anxiety state, and the value of anxiety for factor 2 is lower than that for factor 1, indicating that students are more anxious about their vocabulary in writing and their total vocabulary.

In Table 7, the mean of each question in factor 3 is 18.743, which is then averaged to each item, and the calculated average value of each question is 3.079. According to the interpretation of the reference results, students' vocabulary learning anxiety in listening also shows a moderate anxiety state.

In Table 8, the mean of each question in factor 4 is 20.1232, which is then averaged to each item, and the calculated average value of each question is 2.876. In this factor,

Table 6. Descriptive statistics

	N	Minimum	Maximum	Mean	Standard deviation
Factor 2	284	10.00	50.00	29.7042	10.47547
Valid N (listwise)	284				

Table 7. Descriptive statistics

	N	Minimum	Maximum	Mean	Standard deviation
Factor 3	284	6.00	30.00	18.7430	6.47644
Valid N (listwise)	284				

students who have a positive attitude toward word learning will have lower vocabulary learning anxiety, that is, the higher the mean is, the lower vocabulary learning anxiety is. According to the value of the mean 2.876, it is known that half of the students do not have a positive attitude or proper strategy for word learning, so the words learning anxiety in this factor also show a moderate state.

Table 8. Descriptive statistics

	N	Minimum	Maximum	Mean	Standard deviation
Factor 4	284	7.00	35.00	20.1232	6.07977
Valid N (listwise)	284				

In Table 9, the mean of each question in factor 5 is 5.5423, which is then averaged to each item, and the calculated average value of each question is 2.771. In this factor, students who have a negative attitude toward word learning will have higher vocabulary learning anxiety, that is, the higher the mean is, the higher vocabulary learning anxiety is. According to the results, half of the students held a negative attitude toward word learning, so the word learning anxiety in this dimension also showed a moderate state.

Table 9. Descriptive statistics

	N	Minimum	Maximum	Mean	Standard deviation
Factor 5	284	2.00	10.00	5.5423	2.44840
Valid N (listwise)	284				

Differences in English Vocabulary Learning Anxiety Among Students of Different Genders, Majors and Regions. SPSS 22 was used to conduct independent sample T tests and one-way ANOVA on the research data to determine whether there were significant differences in English vocabulary learning anxiety among students of different genders, regions and majors.

In Tables 10 and 11, subjects were divided into two groups by sex for independent sample T tests. The results showed that the sig. The value of the Levene test of variance equation was $0.177 > 0.05$, indicating the homogeneity of variance. At this time, we need to look at the value of two-tailed significance, and it is $0.748 > 0.05$. This value still does not meet the requirement of significance level $P < 0.05$, so it can be considered that there is no difference between the two groups. In terms of this study, there was no significant difference between male and female students in English vocabulary learning anxiety. The English foundation of private vocational college students is uneven when they enter school. Some of them have received a complete high school education, but some adult education students or secondary vocational students have not received formal high school education, resulting in the phenomenon of severe polarization in English learning. Students with a good foundation in English are confident in English learning, but this group of students in higher vocational colleges accounted for a very small proportion. Most vocational college students are poor in English, and some of them had not even learned English before. These students are indifferent to English learning, especially to boring English vocabulary learning, regardless of gender. Although from the biological point of view, students of different genders have certain differences in sensitivity and comprehension of English vocabulary learning, the difference is almost minimal among higher vocational students.

Table 10. Group statistics

	Sex	N	Mean	Standard deviation	Standard error mean
Vocabulary Learning Anxiety	Male	201	100.2239	29.85674	2.10593
	Female	83	99.0241	25.20936	2.76709

Table 11. Independent-samples T test

	Levene's test forequality of variances		T test for whether the mean is equal						
	F	Sig.	T	df	Sig. (2-tailed)	Mean difference	Standard error	95% Confidence interval	
								Lower bound	Upper bound
Vocabulary learning anxiety	1.835	.177	.322	282	.748	1.19978	3.72937	−6.14116	8.54073
			.345	179.774	.730	1.19978	3.47731	−5.66182	8.06139

In Tables 12 and 13, the study subjects were classified into two groups within and outside Hainan Province for independent sample T tests. The results showed that the Sig. value of the Levene test of variance equation was $0.119 > 0.05$, indicating the homogeneity of variance. At this time, we need to look at the two-tailed Sig. That was $0.380 > 0.05$. This value still does not meet the requirement of $P < 0.05$, so it can be considered that there is no difference between the two groups. As far as this study is concerned, there is no significant difference in vocabulary learning anxiety between vocational college students in Hainan Province and students outside Hainan Province. Due to the lack of teaching resources, people's insufficient attention to education, and the negative transfer of the Hainan dialect to local students in Hainan, the English foundation of local vocational college students in Hainan is generally weak. These students have the desire to study, but they are affected by a variety of factors and feel powerless in English learning, especially vocabulary learning. Students from outside Hainan Island generally have a better English foundation than those from inside Hainan Island, but these students have not developed good learning habits and attitudes for many years. Therefore, many people hold indifferent attitudes towards English vocabulary learning, so there is no obvious difference in English vocabulary learning anxiety between students inside and outside Hainan Island.

Table 12. Group statistics

	City	N	Mean	Standard deviation	Standard error mean
Vocabulary learning anxiety	Hainan	145	98.4138	30.72065	2.55121
	Outside Hainan	139	101.3957	26.09198	2.21309

Table 13. Independent-samples T test

	Levene's test for equality of variances		T test for whether the mean is equal					95% Confidence interval	
	F	Sig.	T	df	Sig. (2-tailed)	Mean difference	Standard deviation	Lower bound	Upper bound
Vocabulary learning anxiety	2.441	.119	−.880	282	.380	−2.98189	3.38896	−9.65275	3.68897
			−.883	277.998	.378	−2.98189	3.37734	−9.63031	3.66653

In Tables 14 and 15, the study subjects were divided into 7 groups according to their majors for a one-way variance test. The results showed that the Sig. The value of the Levene test of variance equation was $0.145 > 0.05$, indicating that the variance was homogeneous. At this time, the Sig. The value of one-way ANOVA was $0.050 = 0.05$.

This value still does not meet the requirement of significance level $P < 0.05$, so it can be considered that there is no difference or no significant difference in vocabulary learning anxiety for students from different majors. As far as this study is concerned, there is no significant difference in vocabulary learning anxiety among students of different majors in vocational colleges. In recent years, higher vocational colleges have generally faced challenges in recruiting students. Most of the students with a poor foundation of cultural courses are recruited into schools, and these students are generally distributed in various majors. These students generally have low initiative and desire for knowledge in English learning, especially for English vocabulary learning, so there is no significant difference in their English vocabulary learning anxiety.

Table 14. Homogeneity test of variances

Vocabulary learning anxiety

Levene statistic	df1	df2	Sig.
1.561	7	274	.147

Table 15. ANOVA

Vocabulary learning anxiety

	Quadratic sum	df	Mean square	F	Sig.
Between-groups	11370.750	7	1624.393	2.031	.050
Ingroup	219101.619	274	799.641		
Total	230472.369	281			

4 Suggestions for Alleviating the English Vocabulary Learning Anxiety of Vocational College Students

4.1 From the Aspect of Students

Students are the major learners. To alleviate the anxiety of students' English vocabulary learning, we should start from the students' side first. According to the data of this study, students have a moderate degree of anxiety in all aspects of English vocabulary learning. It can be seen that the English level of vocational college students is poor. Some students have not even received a complete high school education when they enter college. Over the years, they have often been discriminated against by teachers, parents and students due to their poor English learning results. They have long been disgusted with English learning, especially the learning and memory of English vocabulary. Therefore, the key for such students is not how much vocabulary they need to memorize, but more importantly, it is necessary to stimulate their current interest in English vocabulary learning and cultivate their autonomous learning ability. At present, although some students are forced to learn English vocabulary due to the pressure of future employment, they cannot be successful because of a lack of proper learning skills or strategies, and it is difficult for them to stick to it for a long time. In view of these realities, we should help students cultivate correct learning concepts and attitudes as soon as possible when they are in first grade, help them build confidence in vocabulary learning and pay more attention to students' psychology. The teacher needs to understand students' difficulties and help students make a proper vocabulary learning plan. In classroom teaching, teachers should simplify and visualize complex and boring vocabulary knowledge and help students use correct learning methods and learning strategies for vocabulary learning. In students' learning process, the cooperative learning method and metacognitive strategy are used by students to cultivate their autonomous learning ability. When students use their own learning methods and strategies to improve their English vocabulary learning, their anxiety level of vocabulary learning will be gradually alleviated.

4.2   From the Aspect of Teachers

As the guide and supervisor of students' learning, teachers play an extremely important role. Students in vocational colleges are different from students in ordinary colleges and universities. Most students can be defined as students with learning difficulties. Therefore, when teachers carry out daily teaching activities, in view of the students' moderate English vocabulary learning anxiety in listening, speaking, reading, writing and examination, teachers should reconstruct the teaching contents according to the actual situation of students, make the teaching content simple and practical, and enrich the teaching methods. In English classes, teachers should pay more attention to the evaluation of students' vocabulary learning process and monitor and manage students' vocabulary learning after class so that students can gradually master learning methods and strategies and their enthusiasm for vocabulary learning will also be stimulated. In the listening class, some materials with the proper vocabulary are chosen to help students listen to the materials from easy to difficult so that students can experience the fun of learning and the sense of achievement. Gradually, students' vocabulary learning anxiety in listening will be alleviated. For oral English, it is also from simple to difficult. Before oral practice, show the required vocabulary to the students, explain them in advance, and encourage the students to express their ideas from sentences to paragraphs and then to essays. For some shy students, online man-machine dialog or video and audio recording can be used to let students practice and express their vocabulary to alleviate the anxiety of students' lack of vocabulary in oral English. Let students learn and memorize words in reading and visualize abstract words to deepen students' impression of words. When students master a certain number of words, their reading anxiety will also be relieved. In writing and examination, let students write short essays by using the words they are interested in. The topics are unlimited from simple to difficult, and students will gradually develop the habit of writing to alleviate the anxiety of their writing. Teachers should also make full use of modern teaching means and methods, networks, English learning software, XMind and other methods to let students learn and remember words in multiple ways.

4.3 From the Aspect of Schools

Students study and live on campus. A good campus environment gives students a good learning atmosphere. Vocabulary learning and memory is a boring thing. Allowing students to learn English vocabulary in a comfortable mood can improve learning efficiency. At the same time, the school should do a good job in ensuring a smooth network connection so that students can effectively use the mobile Internet for learning, and it is also convenient for teachers to supervise and manage students' vocabulary learning after class. Schools should also hold more competitions or activities related to vocabulary learning to promote students' learning enthusiasm. When students experience the good atmosphere of whole vocabulary learning, their anxiety about vocabulary learning will be relieved.

5 Conclusion

Vocational college students' vocabulary learning anxiety is reflected in all aspects of their English learning. A solid vocabulary foundation can ensure that they learn English smoothly and achieve good results in the exam. Through a questionnaire, this study clarified the current anxiety state of vocational college students' vocabulary learning and proposed corresponding suggestions to alleviate this anxiety. With the construction of the Hainan Free Trade Port and the continuous development of international exchanges, students' English level has a direct impact on their future employment and career development. Therefore, English vocabulary learning as the core of English study must draw increasing attention in the future.

References

Zhang, F.: English vocabulary learning anxiety among Non-English majors students. Xihua University (2015)

Jin, Y.: The relationship between English students' anxiety and listening comprehension. Foreign Langu. Res. (1), 54–57 (2000)

Jiang, D., Li, H.: Cooperative learning and relieving oral English anxiety. J. Univ. Shanghai Sci. Technol. (Soc. Sci. Ed.) **35**(1), 36–40 (2013)

Yang, Y.: Qualitative Research Methods in Applied Linguistics. Commercial Press, Beijing (2014)

Du, X.: Retesting the validity of the foreign Language Classroom Anxiety Scale (FLCAS). In: Proceedings of 2nd International Conference on Humanities Education and Social Sciences, pp. 627–632. Advances in Social Science, Education and Humanities Research, Xi'an, Shannxi (2019)

Horwitz, E., Horwitz, M., Cope, J.: Foreign language class room anxiety. Mod. Lang. J. **70**, 125–132 (1986)

Ren, D., Xian, D.: Study on vocational college students' communicative competence of intercultural communication. In: Zeng, J., Qin, P., Jing, W., Song, X., Lu, Z. (eds.) ICPCSEE 2021. CCIS, vol. 1452, pp. 443–455. Springer, Singapore (2021). https://doi.org/10.1007/978-981-16-5943-0_36

Liu, L., Zhang, Y., Li, M.: An empirical study on teachers' informationized teaching ability in higher vocational colleges. In: Zeng, J., Qin, P., Jing, W., Song, X., Lu, Z. (eds.) ICPCSEE 2021. CCIS, vol. 1452, pp. 419–433. Springer, Singapore (2021). https://doi.org/10.1007/978-981-16-5943-0_34

Zheng, Q., Xu, Y.: The influence of multi-modal presentation on English vocabulary learning anxiety. J. Xi'an Int. Stud. Univ. **28**(02), 49–53 (2020)

Ma, H.: A study of the effects of multimodal teaching on primary school students' English vocabulary scores and classroom anxiety. Shaanxi Normal University (2019)

Du, Q., Jia, L.: SPSS Statistical Analysis from Introduction to Mastery. Posts and Telecom Press, Beijing (2009)

Li, S., Gao, Y.: An empirical study on the effectiveness of mobile technology-assisted foreign language teaching in English vocabulary acquisition. Foreign Lang. World **4**, 73–81 (2016)

Yu, W., Shao, K., Xiang, Y.: The relationship between Eq, foreign language learning anxiety and English learning performance. Mod. Foreign Lang. **5**, 656–666 (2015)

Horwitz, E.: Foreign and second language anxiety. Lang. Teach. (2), 154–167 (2010)

Kralova, Z., Skorvaova, E., Tirpakova, A., Markechova, D.: Reducing student teachers' foreign language pronunciation anxiety through psycho-social training. System **65**(1), 49–60 (2017)

Chen, H.: Strategies for Non-English major Chinese students to learn English vocabulary. Foreign Lang. Educ. (6), 46–51(2001)

Ma, G.: Teaching and Researching English Vocabulary. Foreign Language Teaching and Research Press, Beijing (2016)

Li, C.: Factor analysis of anxiety in college English class based on R language. Mod. English **22**, 76–79 (2020)

Zhang, X.: Foreign language anxiety and foreign language performance: a meta-analysis. Mod. Lang. **4**, 763–781 (2019)

Liang, Z., Gao, Y., Xie, B., He, W.: Peer assessment anxiety in English writing and its influence on peer assessment. Technol. Enhanced Foreign Lang. (03), 41–46+67+7 (2020)

Forecast and Analysis of Passenger Flow at Sanya Airport Based on Gray System Theory

Yuanhui Li[✉] and Haiyun Han[✉]

Sanya Aviation and Tourism College, Sanya 572000, Hainan, China
576735855@qq.com, 103238991@qq.com

Abstract. The forecast of airport passenger throughput can provide a scientific basis for airport construction and management and has important reference value. A Gray model GM(1,1) is established to predict the passenger flow of Sanya Phoenix International Airport in 2018 and 2019 by collecting yearly and monthly passenger flow data from 2012 to 2017. The results indicate that the predicted values are in good agreement with the actual values and that the relative errors are very close, which means that both the monthly forecast and the annual forecast can well reflect the actual situation of the airport passenger flow.

Keywords: Passenger flow · GM(1,1) · Predict

1 Introduction

Gray prediction theory is an uncertainty system prediction theory proposed by Chinese scholar Deng Julong, which extracts valuable information by generating and modeling the known sequence to realize the accurate and effective grasp of the system operation and evolution laws. GM (1,1) is one of the most important models. Professor Deng Julong constructed the GM (1,1) model based on the data background and the incomplete information condition of the development law. Aiming at the uncertain system, he made full use of part of the known information to mine and explore the system law [1, 2].

Sanya is a popular tourist destination in China, and tourists mainly arrive by airplane. The number of passengers and airport passenger flow are affected by many factors, so the information has the characteristics of partial knowability and partial unknowability [3, 4]. An uncertain system with some known information and some unknown information is called a gray system [5]. Therefore, the airport passenger flow can be regarded as a gray system with obvious dynamic characteristics, uncertainty and ambiguity in the relationship between elements [6].

Most of the traditional prediction methods of airport passenger throughput use the regression analysis method, which requires a large amount of original data, a typical data distribution, and heavy computing, and the prediction result has a large error. The gray prediction adopts the methods of direct accumulation of raw data, weighted accumulation of moving average, etc., so that the generated sequence shows a certain regularity, the

Y. Wang et al. (Eds.): ICPCSEE 2022, CCIS 1629, pp. 427–434, 2022.
https://doi.org/10.1007/978-981-19-5209-8_29

typical curve is used to approximate its corresponding curve, and the approximated curve is used as a model to predict the system. Gray prediction does not require a large amount or typical distribution of original data, but only a time series is needed [7]. This forecast adopts the GM(1,1) model, and the forecast years are 2018 and 2019. Considering that the civil aviation industry will be seriously affected by the COVID-19 epidemic in 2020, 2021, and 2022, the data after 2020 will not be predicted for the time being [8, 9].

2   The Modeling Steps of the Gray Prediction GM(1,1)

2.1   Step 1 Judging the Feasibility of Modeling

For known raw series $x^{(0)} = \left(x^{(0)}(1), x^{(0)}(2), \cdots, x^{(0)}(n)\right)$, calculate its stepwise ratio

$$\lambda(k) = \frac{x^{(0)}(k-1)}{x^{(0)}(k)}, (k = 2, 3, \cdots, n) \tag{1}$$

If all the stepwise ratios $\lambda(k)$ fall within the tolerable coverage $\Theta = \left(e^{-\frac{2}{n+1}}, e^{\frac{2}{n+1}}\right)$, the series $x^{(0)}$ can be used as the data of the model GM(1,1) for gray prediction.

2.2   Step 2 Generate Accumulation Series $x^{(1)}$

Accumulate the raw series $x^{(0)} = \left(x^{(0)}(1), x^{(0)}(2), \cdots, x^{(0)}(n)\right)$ once, and obtain

$$x^{(1)} = \left(x^{(1)}(1), x^{(1)}(2), \cdots, x^{(1)}(n)\right) \tag{2}$$

where $x^{(1)}(k) = \sum_{t=1}^{k} x^{(0)}(t), k = 1, 2, \cdots, n$.

After the accumulation operation, the randomness and volatility of the original data are obviously weakened, and they are transformed into an increasing sequence with strong regularity, which is prepared for building predictive models in the form of differential equations, making the application process easier [10].

2.3   Step 3 Build a GM(1,1) Model

Establishment of a differential equation model

$$\frac{dx^{(1)}(t)}{dt} + ax^{(1)}(t) = b \tag{3}$$

for the generated series $x^{(1)}$, where a and b are unknown parameters. The model is a first-order differential equation with one variable, denoted as GM(1,1).

Denoted

$$u = [a, b]^T, Y = \left[x^{(0)}(2), x^{(0)}(3), \cdots, x^{(0)}(n)\right]^T,$$

$$B = \begin{bmatrix} -\frac{1}{2}\left(x^{(1)}(1) + x^{(1)}(2)\right) & 1 \\ \vdots & \vdots \\ -\frac{1}{2}\left(x^{(1)}(n-1) + x^{(1)}(n)\right) & 1 \end{bmatrix} \tag{4}$$

Then by the least square method, obtain the estimated value of

$$\hat{u} = \left[\hat{a}, \hat{b}\right]^T = \left(B^T B\right)^{-1} B^T Y. \tag{5}$$

that makes reach the minimum value u.

Solve the equation to obtain

$$\hat{x}^{(1)}(t) = \left(x^{(0)}(1) - \frac{\hat{b}}{\hat{a}}\right)e^{-\hat{a}t} + \frac{\hat{b}}{\hat{a}}, \tag{6}$$

Available predictions

$$\hat{x}^{(1)}(k+1) = \left(x^{(0)}(1) - \frac{\hat{b}}{\hat{a}}\right)e^{-\hat{a}k} + \frac{\hat{b}}{\hat{a}}, k = 0, 1, 2, \cdots. \tag{7}$$

2.4 Step 4 Verify the Model

Commonly used methods include residual tests, posteriori error tests, ratio deviation tests and correlation tests. In this paper, the residual test is used.

$$\delta(k) = \frac{\left|x^{(0)}(k) - \hat{x}^{(0)}(k)\right|}{x^{(0)}(k)}, k = 1, 2, \cdots, n. \tag{8}$$

Here $\hat{x}^{(0)}(k) = x^{(0)}(k)$. If $\delta(k) < 0.2$, the model accuracy can be considered to meet the general requirements; if $\delta(k) < 0.1$, the model accuracy can be considered to meet the higher requirements [11].

3 Prediction of Passenger Flow at Sanya Airport Based on the GM(1,1) Model

According to the tourism report of the Sanya Tourism, Culture, Radio, Television and Sports Bureau website (http://lwj.sanya.gov.cn/), the passenger flow of Sanya Airport from 2012 to 2019 is shown in Table 1.

Table 1. The monthly passenger flow of Sanya Airport from 2012 to 2019 (10,000 passengers)

Month	2012	2013	2014	2015	2016	2017	2018	2019
1	139.12	138.45	162.96	170.05	177.00	194.46	194.76	205.27
2	128.65	151.21	165.34	173.31	179.19	187.38	196.54	216.38
3	115.53	132.48	144.81	168.27	159.79	183.13	190.46	194.50
4	86.17	90.44	103.59	121.18	130.04	150.76	153.33	150.20
5	67.98	81.37	97.03	112.10	120.02	142.50	146.44	139.17
6	60.39	73.56	87.30	101.28	112.47	135.84	140.44	135.27
7	75.45	86.92	104.98	115.62	125.87	142.88	148.10	151.91
8	77.53	93.28	113.99	123.96	137.79	153.11	158.67	155.99
9	66.43	82.24	97.96	106.35	125.10	135.77	135.52	136.82
10	82.25	96.85	117.73	126.95	138.22	152.59	156.36	156.48
11	107.86	120.47	141.84	139.78	153.48	171.24	181.97	178.17
12	126.96	139.41	156.70	160.35	177.14	189.33	201.31	196.21
Total	1134.34	1286.68	1494.24	1619.19	1736.11	1938.99	2003.90	2016.37

Next, the annual passenger flow of Sanya Airport from 2012 to 2017 is taken as the training set, and the corresponding data from 2018 to 2019 are taken as the test set.

Considering that gray system theory emphasizes the inherent regularity, a model for each month is constructed. The specific idea is: to use the actual data from January 2012 to January 2017 to predict the passenger flow corresponding to January in 2018 and 2019; and then predict the passenger flow in February, ... and so on until December. Finally, the points are used. The results of the monthly forecasts are summed to produce a forecast for annual receptions.

3.1 Forecast by Yearly Data

From Table 1, the passenger flow of Sanya Airport from 2012 to 2017 can be recorded as the raw series.

$$x(0) = (8438.73, 9398.53, 9959.17, 10950.03, 11160.94).$$

According to formula (1), the range of the order ratio is $(0.8611, 0.9327)$, all of which fall within the tolerable coverage $\Theta = \left(e^{-\frac{2}{n+1}}, e^{\frac{2}{n+1}} \right) = (0.7515, 1.3307)$ which can be modeled as GM(1,1). Calculated from formula (5) $a = -0.0954$, and $b = 1151.2275$. The GM(1,1) model

$$\hat{x}(k + 1) = 13203.8487e^{-0.0954k} - 12069.5087, k = 0, 1, 2, \cdots, 7.$$

from Eq. (7).

The extrapolation predicts the annual reception numbers of Sanya Airport from 2018 to 2019, and the prediction accuracy reaches excellent and close to excellent levels, respectively (Table 2).

Table 2. Comparison of GM(1,1) Gray predicted values and actual values from 2012 to 2019 (by year)

Year	Actual value	Predicted value	Residual	Relative error
2012	1134.34	1134.34	0.00	0.00%
2013	1286.68	1321.44	−34.76	2.70%
2014	1494.24	1453.69	40.54	2.71%
2015	1619.19	1599.18	20.01	1.24%
2016	1736.11	1759.23	−23.12	1.33%
2017	1938.99	1935.29	3.70	0.19%
2018	2003.90	2128.98	−125.08	6.24%
2019	2016.37	2234.46	−218.09	10.82%

3.2 Forecast by Monthly Data

Taking the passenger flow of the corresponding months in Sanya Airport from 2012 to 2017 as the original sequence, 12 models were constructed. According to formula (1), the stepwise ratio series of the number of receptionists in 12 months is calculated (see Table 3), and the stepwise ratios all fall within the tolerable coverage (0.7515, 1.3307), which meets the stepwise ratio test requirements.

Table 3. The passenger flow ratio of the corresponding months in Sanya Airport from 2012 to 2017

Month	Stepwise ratio range	Month	Stepwise ratio range
1	(0.8496, 1.0048)	7	(0.8280, 0.9186)
2	(0.8508, 0.9672)	8	(0.8183, 0.9196)
3	(0.8606, 1.0531)	9	(0.8078, 0.9214)
4	(0.8548, 0.9528)	10	(0.8226, 0.9274)
5	(0.8354, 0.9340)	11	(0.8493, 1.0147)
6	(0.8210, 0.9005)	12	(0.8897, 0.9772)

Predict the passenger flow from January to December of each year separately, and then sum up the predicted value of the annual passenger flow. The specific data are shown in Table 4.

Table 4. Sanya Airport 2012–2019 Month by month forecast (10,000 passengers)

Month	2012	2013	2014	2015	2016	2017	2018	2019
1	139.12	144.53	155.64	167.60	180.48	194.35	209.28	213.68
2	128.65	154.61	162.52	170.83	179.57	188.75	198.40	207.09
3	115.53	135.45	145.73	156.79	168.69	181.50	195.27	206.12
4	86.17	91.58	103.62	117.23	132.63	150.06	169.77	177.11
5	67.98	83.45	95.17	108.53	123.77	141.15	160.97	170.03
6	60.39	74.09	85.91	99.62	115.51	133.94	155.31	165.62
7	75.45	90.38	101.34	113.64	127.42	142.88	160.22	165.18
8	77.53	97.65	109.47	122.71	137.55	154.18	172.83	181.11
9	66.43	84.42	95.37	107.74	121.71	137.50	155.34	155.53
10	82.25	101.74	112.80	125.05	138.63	153.68	170.38	176.75
11	107.86	123.48	133.52	144.38	156.13	168.82	182.56	196.81
12	126.96	141.38	152.09	163.63	176.03	189.38	203.74	217.11
Total	1134.32	1322.77	1453.18	1597.74	1758.13	1936.20	2134.07	2232.13

It can be seen from Table 5 that the relative error test from 2012 to 2017 is less than 3%, so the model is better. The extrapolation predicts the annual reception numbers of Phoenix Airport from 2018 to 2019, and the prediction accuracy reaches excellent and close to excellent levels, respectively.

Table 5. Comparison of GM(1,1) Gray forecast value (by month) and actual value from 2012 to 2019

Year	Actual value	Predicted value	Residual	Relative error
2012	1134.34	1134.32	0.02	0.00%
2013	1286.68	1322.77	−36.09	2.80%
2014	1494.24	1453.18	41.06	2.75%
2015	1619.19	1597.74	21.45	1.32%
2016	1736.11	1758.13	−22.02	1.27%
2017	1938.99	1936.20	2.79	0.14%
2018	2003.90	2134.07	−130.17	6.50%
2019	2016.37	2232.13	−215.76	10.70%

Figure 1 shows the comparison of forecast results and actual values of annual passenger flow of Sanya Airport from 2012 to 2019 by year and month. It can be seen that both forecasting methods fit the actual data well.

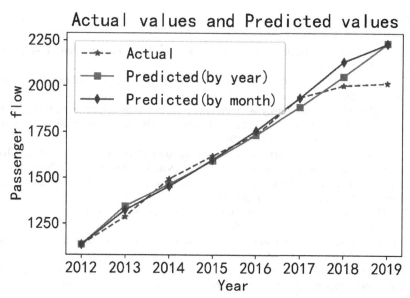

Fig. 1. Comparison of predicted and actual passenger flows at Sanya Airport from 2012 to 2019

4   Conclusions

(1) Considering the dynamic change process of the passenger flow scale in Sanya Airport as a gray system, it is feasible to construct a model and predict the passenger flow by using gray system theory [12]. In this paper, only 6 years of data are used to construct the model, and the prediction results are basically consistent with the actual values, which means that the precision of the model is rather high, so it can provide a scientific basis for the decision-making of the airport management department [13].

(2) During the development of any gray system, as time goes by, some random disturbances or driving factors will continuously enter the system, which will affect the development of the system one after another. Therefore, the GM(1,1) model is only suitable for short- and medium-term forecasting [14].

In view of the limited length of the article, how to improve the service level and supply capacity of the airport by predicting the passenger flow of Sanya Airport is not considered in this paper. The ultimate purpose of passenger flow forecasting is to balance the demand and supply, and greatly improve the tourism service level of Sanya. How to achieve this can be discussed in depth in future research [15].

Acknowledgments. This project was supported by the Education Department of Hainan Province (project number: Hnky2022ZD-25).

References

1. Liu, S.: Several basic models of GM(1,1) and their applicable bound. J. Syst. Eng. Electron. **36**(03), 501–508 (2014)
2. Xu, N.: Advances in Gray GM(1,1) forecasting model and its extension. Math. Pract. Theory **51**(13), 52–59 (2021)
3. Li, Y., Han, H.: Empirical analysis of tourism revenues in Sanya. In: Qin, P., Wang, H., Sun, G., Lu, Z. (eds.) ICPCSEE 2020. CCIS, vol. 1258, pp. 458–466. Springer, Singapore (2020). https://doi.org/10.1007/978-981-15-7984-4_34
4. Zhang, J.: Study on influential factors of Shaanxi tourism revenue in recent 10 years based on Gray relational analysis. Baoji Univ. Arts Sci. (Nat. Sci.) **40**(04), 67–72 (2020)
5. Deng, J.: The Primary Methods of Gray System Theory. Press of Huazhong University of Science and Technology, Wuhan (2005)
6. Liu, H.: Prediction and analysis of tourist reception numbers in Guangzhou based on Gray system theory. Stat. Decis. **317**(17), 64–66 (2010)
7. Yao, Y.: Application of Gray model GM(1,2) in airport passenger throughput prediction. J. Civil Aviat. Flight Univ. China **17**(4), 12–16 (2006)
8. Zhou, X.: Analysis of influencing factors of flight operation based on Gray correlation model—with Pudong airport during the epidemic as an example. J. Civil Aviat. **6**(02), 5–7+85 (2022)
9. Yang, L.: Forecast method of airport take-off and landing volume in post-epidemic period. J. North China Univ. Sci. Technol. (Nat. Sci. Ed.) **44**(01), 47–53 (2022)
10. Qiu, H.: Research on passenger throughput forecast of Yuncheng Airport based on Gray theory. J. Yuncheng Univ. **33**(04), 33–36 (2015)
11. Si, S.: Python Mathematical Experiment and Modeling. Science Press, Beijing (2020)
12. Liu, X.: Tourist flow forecast of Sanya city based on XGBoost and GM model. Sci. Technol. Econ. **32**(06), 46–50 (2019)
13. Wang, C.: Application of dynamic improved Gray model in airport throughput prediction. Comput. Simul. **36**(12), 74–77+83 (2019)
14. Xie, M.: Short-term traffic flow prediction based on GM(1, N) power model optimized by rough set algorithm. Math. Pract. Theory **51**(09), 241–249 (2021)
15. Tian, M.: Study on tourism development and economic growth based on Gray relational analysis: take Shandong Province as an example. Chongqing Univ. Technol. (Nat. Sci.) **33**(02), 208–215 (2019)

Regulatory Technology in Finance

Decentralized Counterparty Matches and Automatic Settlement of Interest Rate Swap Through Blockchain's Smart Contracts

Jian Zhao[1], Jinjing Liu[1], Hua Wang[2(✉)], Lin Lin[1], and Shaohua Xiang[1]

[1] College of Big Data and Internet, Shenzhen Technology University, Shenzhen 518118, Guangdong, China
[2] FinTech Research Center, Business School, Shenzhen Technology University, Shenzhen 518118, Guangdong, China
wanghua@sztu.edu.cn

Abstract. Interest Rate Swap (IRS) is the most vivid application of the principle of comparative advantage in the financial field. The effectiveness of interest rate swap in managing interest rate risk has been widely recognized. However, the traditional interest rate swap transaction is complicated. Meanwhile, there usually exist market risks and credit risks. To alleviate risks and cost, and improve liquidity of interest rate swap, this paper proposes a smart contract based matching platform for interest rate swap of real fiat currency. Smart contracts play a key role for sharing data among participants, which can not be forged or tampered with. In our design, an efficient peer-to-peer counterparty matching method on the chain is proposed. The whole trading process of interest rate swap is carried out on the blockchain, which has higher security. A prototype based on smart contracts running on Ethereum is implemented and validates our design.

Keywords: Ethereum · Smart contract · Interest rate swap · DeFi

1 Introduction

Interest Rate Swaps (IRS) [1, 2] refer to the exchange of fixed amount of interest and floating amount of interest on a notional principal with a mature period. Long swap position pays the fixed amount of interest and receives the floating amount of interest. As a counterparty, short swap position pays the floating amount of interest and receives the fixed amount of interest. The exchange only involves the interest under different rate, without the exchange of real principal.

Interest rate swaps are usually used to hedge interest rate risk in financial markets. When a company issues a floating interest rate bond, it can purchase long swap positions from a swap bank, which lets the company receive the floating amount of interest from the swap bank, and pay the fixed amount of interest to the swap bank. On one hand, when it provides an effective way of risk transfer, it also brings new risks, which are mainly market risks and credit risks. Market risks can be avoided by hedging, but credit risks are more difficult to avoid. On the other hand, interest rate swaps in the traditional financial system currently have some drawbacks, including requiring high fees for using traditional financial facilities, lack of the necessary transparency, being subject to counterparty risk, lack of assets and regulatory scrutiny due to geographic boundaries, and excessive and unnecessary counterparty risk for individuals.

As an emerging technology, blockchain has the potential to solve the problems faced by interest rate swap. Blockchain technology [3] has the characteristics of decentralization, tamper-proof, anonymity and traceability, which provides an unmediated and trusted Internet environment. Under the trusted network guaranteed by blockchain, the credit risk problem in interest rate swap can be well solved. Furthermore, smart contracts running on a blockchain platform allow developers to do more things in the blockchain network. Smart contracts are programming codes executed on the blockchain in essence. A smart contract can be simply understood as a special transaction contract that is driven by events. Smart contracts are well suited for applications where requirements for trust, security and persistence are high. E.g., Ethereum is a smart contract blockchain platform, which supports Turing scripting programming language running on Ethereum Virtual Machine (EVM). This allows developers to create and distribute-arbitrary decentralized applications on the platform. Each node participating in the network runs EVM as part of the block validation protocol. All nodes in the network perform the same calculations and store the same values. Contract execution is repeated multiple times across all nodes.

However, to prevent deliberate attacks or abuses on the Ethereum network, the Ethereum protocol requires a fee for every computational step of a transaction or contract call. This fee is calculated in units of gas. The block gas limit is the maximum amount of gas allowed in a single block. This is used to determine how many transactions can be packaged in a single block. When running on the Ethereum network, the first thing to consider is Ethereum's performance issues. The gas limitations make large-scale loops and recursions impossible to implement on Ethereum all DeFi projects are trying to avoid this problem. 0X protocol adopted the way of relay chain. Counterparty matching was executed on the relay chain. The emergence of Automatic Market Making (AMM) represented by Uniswap and Compound changes the original peer-to-peer transaction mode, in which loop traversal problem can be avoided.

Considering the application of blockchain technology in the financial service of interest rate swap, this paper proposes an interest rate swap model based on Ethereum for real fiat currency. A point-to-point counterparty matching scheme suitable for interest rate swaps is proposed, in which the problem of cyclic traversing matching is solved. Our contributions are as follows:

- We design a smart contract based platform for counterparty matching and automatic settlement for interest rate swaps of real fiat currency. A method suitable for on-chain counterparty matching is proposed, which does not need perform loop traversal process for matching. A new smart contract is created for one matched interest rate swap transaction. The cash flow is calculated and stored in the smart contract. The amount of exchanged interest is settled through the smart contract.
- We implement a prototype based on Ethereum to validate our design. The prototype shows the effectiveness of our design.

The remainder of this paper is organized as follows: Sect. 2 describes related work on applying blockchain technology to implement Interest Rate Swaps. Section 3 presents our design of using smart contracts to realize Interest Rate Swaps for fiat currency. An efficient automatic matching algorithm is proposed for on-chain counterparty matching. A prototype is implemented to validate the feasibility of our design. Section 4 concludes our work.

2 Related Work

Generally, current applications of blockchain technology in interest rate swaps can be divided into two categories. One is using smart contracts to implement interest rate swap for digital currency such as ERC20 Token based on the Ethereum platform [4], which requires the assistance of a decentralized bank like the DeFi platform Compound. The Compound platform provides the reliable floating interest rate information on the chain. Interest rate swaps become part of the DeFi system, whose vision is to make all assets into a borderless, open financial system on a global scale [5, 6]. However, the majority of current DeFi products focus only on replicating traditional financial products in a decentralized architecture. For example, all kinds of derivatives protocols amplify bets, and all kinds of margin trading leverage appear on Ethereum. These products based on crypto native tokens can not solve current problems of DeFi, such as small size of assets, nor can they have an important impact on the existing economy. For DeFi to make a breakthrough, it needs to connect physical assets and real fiat currency to the chain [7].

The other category is implementing interest rate swap for real fiat currency based on a consortium blockchain platform. Each node of the consortium blockchain platform is one entity organization. For assets in the physical world, they emphasize the digital contract process rather than digital assets. Compared with the public chain, the consortium blockchain platform only allow authorized nodes to participate in the blockchain network. The number of nodes in the consortium blockchain platform are limited. Therefore, the transaction processing speed of a consortium blockchain platform is faster, and meanwhile it can protect transaction privacy quite well. However, a consortium blockchain platform is not completely decentralized.

2.1 Interest Rate Swaps Based on Ethereum for Crypto Tokens

The application model of "Ethereum + Finance" is named DeFi (Decentralized Finance) [8, 9], which is one popular kind of blockchain applications with the fastest development in 2019. The goal of DeFi is to disrupt the traditional financial service system and promote a new era of digital economy. Traditional financial service infrastructure is beset by problems. The main source of these problems is the existence of centralization, monolithic institutions, etc. DeFi aims to enable every person to obtain equal financial services [10, 11]. It has the following characteristics:

- Fairness: Everyone has equal access to all financial services;
- Programmable: Smart contracts are pre-programmed and executed in the same way for everyone;

- Transparency: The internal workings of smart contracts are fully visible on the distributed ledger, thus introducing transparency into the system;
- Credibility: The whole system can be trusted by everyone because of the above characteristics.

At present, DeFi has been well developed in the mainstream forms of financial service, including lending, stable-coin [12], decentralized exchange, payments, derivatives and prediction markets, etc. Interest rates are much more volatile in crypto markets than in traditional financial markets. Compound [13] is an algorithmic money market protocol based on Ethereum blockchain, which is an application implemented by smart contracts that allows users to lend and borrow tokens. Users can invest their digital assets in the Compound system to earn interest over time. Compound lends digital assets invested by users to borrowers, earning net interest margin. The interest rate in the Compound market fluctuates all the time as different market events cause significant changes in supply and demand. Hence, the service of interest rate swaps based on the Compound platform is provided for hedging floating rate risk. Below are two interest rate swap services for crypto assets listed in the Compound system.

Cherry Swap [14] is an autonomous open source platform for Compound financial markets to trade interest rate swaps. Using the functions of Maker's stable coin DAI and Compound cDAI, Cherry Swap can be used to hedge floating rate risk. On Cherry Swap, users may predict interest rate trends and then invest DAI into Compound market to hold future interest rate positions. A long position is a bet that interest rates will rise, and a short position is a bet that rates will fall. Any interest accrued on funds locked in the asset pool is aggregated and returned to the user at the mature time. Users who predict the correct interest rate trend will gain more than they can from investing directly in the Compound market. Cherry Swap's matching rules are not traditional point-to-point. The counterparty of users is the smart contract of Cherry Swap. This matching method does not need to loop through matching counterparties, which is similar to the model of automatic market maker. As long as the liquidity in the contract is available, users can directly trade with the contract. The user can choose to go long or short a contract.

SWAP.RATE [15] is also based on Ethereum platform, and it also needs the Compound market to complete the interest rate swap. On the Compound platform, users can borrow or lend crypto-currencies at a floating interest rate, and the interest rate on the Compound market is variable. The floating interest rate is given by the Compound market, and the fixed interest rate is given by the user. After a contract is signed, SWAP.RATE freezes a certain amount of margin (usually one-tenth of the principal) for the duration of the contract to ensure that both parties pay back on time. SWAP.RATE is a product based on the Opium [16] protocol. Users can initiate or accept an interest rate swap transaction and send signed Meta transactions to an Opium protocol off-chain repeater. The repeater is responsible for matching orders and running 12 order books based on different expiration dates. When a match is completed, an interest rate swap transaction will be settled on the Ethereum blockchain as a Opium protocol smart contract, with users pledging a deposit (10% of the principal). All orders to SWAP.RATE are meta-transactions that match each other in the off-chain order book and are settled on the chain. At the settlement date, if the actual floating rate is lower than the agreed fixed rate, customers can get the margin back and receive additional interest. If the effective floating rate is higher than the agreed fixed rate, the deposit will be refunded to the user after deducting the interest owed.

2.2 Interest Rate Swaps Based on Consortium Chains

Based on Corda [17] platform, R3 [18] alliance is the world's top blockchain alliance. It was jointly established by R3 in 2014 with Barclays Bank, Goldman Sachs, J.P. Morgan and other 9 institutions. At present, it is composed of more than 300 financial service institutions, technology enterprises, and regulatory institutions. The Alliance is actively documenting, managing and enforcing financial agreements with its peers to create an unimpeded business world. Its Corda platform has expanded from financial services to applications in healthcare, shipping, insurance and other industries.

Barclays, blockchain alliance R3 and ISDA (International Swaps and Derivatives Association) have been developing based on Corda (introduced by R3 as a distributed ledger platform, part of its reference blockchain), a prototype of distributed ledger books and modification of the traditional interest rate swap, to build a bank of interest rate swaps between centralized trading platforms. The R3 Blockchain Alliance is an effort to reinvent blockchain-based digital contract processes rather than digital assets.

-NoValue-

3 Design of an Ethereum-Based Interest Rate Swap Platform for Fiat Currency

3.1 Primary Roles of the Platform

The participants of the system include two primary roles, namely traders and official verifiers. The functions of them are described as follows:

1. Traders: Traders join the system to submit their needs for long or short swap positions. The system will match a corresponding counterparty to generate an interest rate swap contract and complete the whole swap process;
2. Official verifiers: Because off-chain fiat is involved, it is necessary to verify the off-chain information of the trader to ensure the security and authenticity of the interest rate swap. For traders, different credit ratings correspond to different lending conditions, which result in different fixed and floating interest rates. Traders need to submit real information from the physical world to the system. An official verifier is responsible for verifying the trader's identity and the interest rate provided. Once the interest rate is determined, the smart contract executes the process. Fixed rates are set at the time of the swap, while floating rates require an official verifier to enter them into a rate information contract before each delivery date, e.g., using London InterBank Offered Rate (LIBOR) as a reference for the floating rate.

Traders submit identity information and related interest rate information to an official verifier, who reviews traders' credit rating. After passing the audit, the verifier will endorse the traders' Ethereum account to establish the relationship between the trader's interest rate information and his Ethereum account. Then, traders can submit their positions to the smart contract, which will be matched with a counterparty to complete the generation of interest rate swap contract.

3.2 Automatic Counterparty Matching Based on Position Index

An interest rate swap is categorized by both notional principal and the maturity period. Based on this feature, we design a multiple pool-based matching pattern as shown in Fig. 1. The value of notional principal and maturity period determines the type of the pool. One smart contract will be created correspondingly to maintain each type of pool to record the transactions for long and short swap positions. All positions in the transaction pool of the specified type have the same notional principal and maturity period.

There are two sub-pools in each transaction pool, namely short position pool and long position pool. The details of the position information are (PositionID: uint, Address: address, Amount: uint, SwapTime: uint, Margin: uint, Status: uint). The PositionID serves as the unique ID for the submitted position information in each sub-pool. In our on-chain counterparty matching scheme, PositionID also serves as the matching index value. PositionID is incremented as a new transaction position is submitted to the smart contract. Address is the blockchain account address of the user who submits the transaction position. Amount is the notional principal. SwapTime is the maturity period. Margin is the interest exchange amount up to now. Status means whether the long or short position is matched successfully or not.

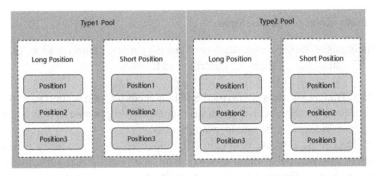

Fig. 1. Transaction pools for different types of interest rate swap positions.

Users submit long or short swap position transactions to the corresponding type of pool, then the smart contract will automatically match the couterparties in the two transaction sub-pools, which narrows the range of matching and improves the efficiency.

Each transaction sub-pool maintains a variable named ReserveId, which identifies the largest PositionID that has been matched to the counterparty in the current sub-pool, and a variable PositionAmount, which identifies the total number of positions in the sub-pool, with an initial value of 0. Each new match starts with the next position from ReserveId. If a new match is met, the match process is completed and the value of ReserveId is set to the PositionID of the newly matched position. The introduction of

ReserveId under the design of different types of transaction pools according to notional principle and maturity period greatly improve the efficiency for transaction matching. Because the matched position in a sub-pool is recorded in ReserveId, the next match can start directly from the next position of ReserveId, which avoids the loop iteration for finding a matching transaction. The specific process of matching counterparties is shown in Fig. 2, which takes short swap positions matching long swap positions as an example:

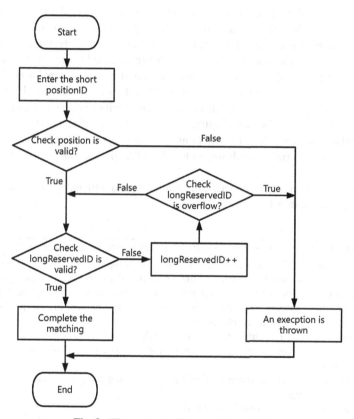

Fig. 2. The process of transaction matching

3.3 Interest Rate Swap Smart Contract for Interest Exchange Amount Calculation and Settlement

After a success match is completed, the platform will generate a corresponding smart contract for the matched interest rate swap transactions. The smart contract for a transaction pool performs as a factory smart contract to generate the interest rate swap contract. Each interest rate swap contract will have a unique Ethereum address. The position information of both parties will be recorded in the interest rate swap contract, including the Ethereum addresses of both parties, notional principal, maturity period, fixed interest rate and floating interest rate. Fixed interest rate is constant in the whole process of interest rate swap, while floating interest rate is changing, usually set as a rate related to LIBOR, which is transparent and visible to all users. The changing floating interest rate is input to the smart contract by the official verifier. Finally the smart contract will calculate the interest exchange amount of both parties and record the amount in Margin variable. The interest margin will be settled to each party when the swap's maturity period ends.

The margin of the interest rate swap contract is delivered in the form of Ethereum stablecoin [12] on the delivery date. To calculate interest margin, the interest rate swap contract will apply the nominal principal, delivery date, fixed interest rate, the reference floating interest rate and other parameters which are recorded in the contract at the time of swap maturity. Specifically, we use a separate smart contract named RateContract to record the floating rate information, which is (RateID: uint, Rate: uint, UpdateT ime: uint). The floating rate is confirmed by the official verifier and stored in the form of mapping data structure. The RateContract will record the floating rate for each semiannual period and set a rate id for the floating rate starting from id = 0. The interest rate swap will record a variable newRate, up to which the interest margin has already been calculated.

The details of specific interest margin calculation and settlement on the maturity date are shown in Algorithm 1.

Algorithm 1 Calculation and Settlement Based on Smart Contract

---

Require: $newRateID$: the latest floating interest rate ID; $processedID$: the floating interest rate ID which has been calculated for the interest margin.
Ensure: $longMargin$: interest margin summation up to now for a long swap position; $shortMargin$: interest margin summation up to now for a short swap position.
1: Get latest rate id from RateContract:
2: $newRateID = RateContract.currentRateId()$
3: Set the initial value for margin of one period of short swap position:
4: $shortProfit = 0$
5: Set the initial value for margin of one period of long swap position:
6: $longProfit = 0$
7: **for** $processedID <= newRateID$ **do**
8:     Sum up the margin for short swap positions not calculated yet:
9: $shortProfit+ = principle * (fixedRate - floatRate)$
10:     Sum up the margin for long swap positions not calculated yet:
11: $longProfit+ = principle * (floatRate - fixedRate)$
12:     Increase the $processedID$ by 1: $processedID + +$
13: **end for**
14: Update the margin of short swap position
15: $shortMargin+ = shortProfit$
16: Update the margin of long swap position
17: $longMargin+ = longProfit$

The RateContract will record the rate id for the floating interest rate. The processedID means the rate id up to which the margin has already been calculated. Unlike the traditional interest rate swap, which involves the transfer of cash flow on each settlement period, the settlement period only calculates the interest margin and record that in the smart contract. The corresponding stable coin of the fiat currency is cleared to the transaction counterparty at the time of swap maturity. When the final settlement date has reached, the trader calls the settlement function of the interest rate swap contract to transfer the corresponding stable coin locked in the interest rate swap contract to the participating party. The amount transferred equals to the value of the shortMagin or longMargin field. Finally, the state of the interest rate swap contract is set to the completed state, and the whole interest rate swap ends.

3.4   Implementation of System Prototype

We present our prototype implementation in this section, which validates our design. We use Ganache to provide an Ethereum blockchain environment. We can check the related blockchain information in Ganache. We deploy our designed smart contracts in Ganache.

Users submit their position information on the chain, including positionID, the type of the position, the amount of the principal, the maturity time of the swap, the Ethereum account address and the position status, as shown in Fig. 3.

Users match to a counterparty to generate an interest rate swap contract, as shown in Fig. 4, including the ID of the short and long positions, the amount of principal, the maturity time of the swap, the address of the interest rate swap contract, and the amount of income for both parties.

Our design succeeds in fulfilling the transaction of long and short swap positions.

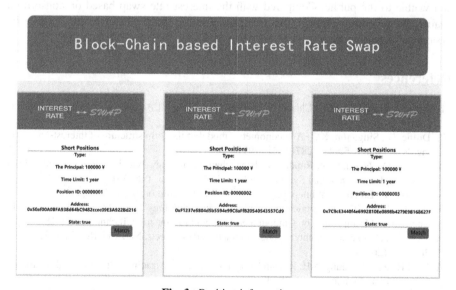

Fig. 3. Position information

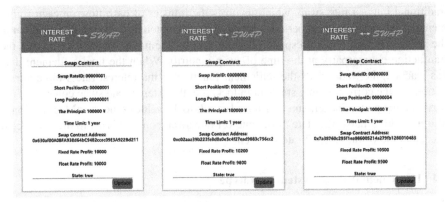

Fig. 4. Interest rate swap contract information

4   Conclusion

The Ethereum based interest rate swap platform for fiat currency developed in this paper takes advantage of the decentralized and tamper-proof characteristics of blockchain technology to solve the problems of insufficient liquidity, centralized data management, reputation risk and other problems existing in interest rate swaps. This paper proposes a peer-to-peer matching method based on Ethereum platform, which can fulfill the whole process from the transaction information submission by the user to the closure of the interest rate swap on the chain. Except the floating interest rate information needs to be provided by the verifier off chain, the interest rate information on the chain is transparent and visible to the public. Compared with the interest rate swap based on consortium chains, the decentralization degree is higher, and the threshold for traders to join the system is relatively low.

References

1. Angeris, G., K Ao, H.T., Chiang, R., Noyes, C., Chitra, T.: An analysis of uniswap markets (2019)
2. Duffie, D., Singleton, K.J.: An econometric model of the term structure of interest-rate swap yields. J. Financ. **52**(4), 1287–1321 (1997)
3. Vujicic, D., Jagodic, D., Randic, S.: Blockchain technology, bitcoin, and ethereum: a brief overview. In: International Symposium Infotehjahorina, pp. 1–6 (2018)
4. Buterin, V.: Ethereum Whitepaper (2015). https://ethereum.org/zh/whitepaper/
5. Wu, B., Duan, T.: The application of blockchain technology in financial markets. In: 2018 International Symposium on Computer Science and Engineering Technology (2018)
6. Zhao, W.: Blockchain technology: development and prospects. Natl. Sci. Rev. Eng. Edn. **6**(2), 369–373 (2019)
7. Li, Z.H., X.X.Y., Liang, Y.P.: Research on interest rate swap trading platform based on alliance chain. Wuhan Financ. 43–48 (2020)
8. Zetzsche, D., Arner, D., Buckley, R.: Decentralized finance (DEFI) (2020)
9. Yan, C.A., Cb, B.: Blockchain disruption and decentralized finance: the rise of decentralized business models. J. Bus. Ventur. Insights 13

10. Lee, J.Y.: A decentralized token economy: how blockchain and cryptocurrency can revolutionize business. Bus. Horiz. **62**(6), 773–784 (2019)
11. Dorofeyev, M., Ksov, M., Ponkratov, V., Masterov, A., Karaev, A., Vasyunina, M.: Trends and prospects for the development of blockchain and cryptocurrencies in the digital economy. Eur. Res. Stud. J. XXI (2018)
12. Lyons, R.K., Viswanath-Natraj, G.: What keeps stablecoins stable? In: NBER Working Papers (2020)
13. Robert Leshner, G.H.: Compound Whitepaper (2019). https://compound.finance/documens/Compound.Whitepaper.pdf
14. Cherry Swap, T.: Cherry Swap (2019). https://www.cherryswap.net/#/swap
15. Opium Team, T.: SWAP.RATE (2019). https://swaprate.finance/
16. Opium Team, T.: Opium (2019). https://opium.network/
17. R3 Corda Team, T.: R3 Corda (2018). https://www.corda.net/
18. Rizzo, P.: How does barclays build a smart contract template using r3 blockchain technology. Big Data Era **1**, 56–59 (2017)

Imbalanced Problem in Initial Coin Offering Fraud Detection

Yifan Zheng[1]([⊠]) and Maoning Wang[2]

[1] School of Information, Central University of Finance and Economics, Beijing 100081, China
zhengyf_cufe@163.com
[2] Engineering Research Center of State Financial Security, Ministry of Education,
Central University of Finance and Economics, Beijing 102206, China

Abstract. ICOs, the initial coin offerings, are a common way to raise funds for blockchain projects. Fraudulent ICO projects not only cause financial losses to investors but also cause a loss of confidence in the blockchain capital market. Whitepapers are usually the most important information source, so it is feasible to identify fraudulent ICO programs by analyzing whitepapers. However, the fraud samples are difficult to collect, and the classes are imbalanced. In this study, we attempt to solve this problem by extracting linguistic features from the ICO whitepaper and using a variety of cutting-edge machine learning and deep learning algorithms to train the prediction model and attempt to resample, modify the weight and modify the loss function for imbalanced samples. Our optimal method achieves an AUC of 0.94 and an accuracy of 82%, which is better than other traditional standard methods, and the results provide important implications for ICO fraud detection.

Keywords: Initial coin offering · Fraud detection · Data imbalance

1 Introduction

ICO, the initial coin offering, is now a common way to raise funds for cryptocurrency/blockchain projects (Fisch 2019). Early participants can receive the cryptocurrency originally generated as return, since tokens have market value and can be exchanged for fiat coins (Heines et al. 2021; El-Masri et al. 2019). The advent of Bitcoin in 2008 brought digital currency into people's eyes; then, in 2014, the release of Ethereum ignited enthusiasm in the coin world. 430 ICOs having raised a total of $4.6 billion by the end of November 2017, considering there were only two ICOs in 2013, it is a big step forward (Diemers 2022). In 2019, the price of Bitcoin rose from a low of $3,155 in late 2018 to $64,863 in mid-April 2021. Meanwhile, ICO metamorphosed into various new modes, such as IEO, STO, IBO, DEFI, etc. (Fan et al. 2020).

M. Wang—This work is supported by the National Natural Science Foundation of China under Grant No. 61907042 and Beijing Natural Science Foundation under Grant No. 4194090.

Y. Wang et al. (Eds.): ICPCSEE 2022, CCIS 1629, pp. 448–464, 2022.
https://doi.org/10.1007/978-981-19-5209-8_31

However, as the ICO market boomed, so did the risk of fraud (Liebau and Schueffel 2019). It was found that some ICO initiators drive up the value of the crowdfunded cryptocurrency and then quickly "dump" the coins for huge profit, and some never accomplish what they promised (Bian et al. 2018). The Securities and Exchange Commission conducted an emergency asset freeze to halt fast-moving ICO fraud that raised up to $15 million from thousands of investors by falsely promising a 13-fold profit in less than a month (SEC. 2017). On September 4, 2017, seven government agencies in China jointly issued *the Notice Regarding Prevention of Risks of Token Offering and Financing*. The Notice banned all ICOs in China and ordered any organizations or individuals who had previously completed ICOs to make arrangements such as return of token assets to investors to protect investor rights. "Investors should bear their own investment risks and invest prudently", it added. Therefore, to reduce the investment risk of investors and protect the property security of people, it is necessary to discover and identify the fraud possibility of ICO projects in a timely manner.

Reading the whitepaper of the project, which is a detailed exposition of the business mode of the project, is usually the primary source of information about the ICO and one of the important ways for investors to learn about an ICO project (Xuan et al. 2020; Dürr et al. 2020). Careful reading of whitepapers as an important step in identifying fraudulent ICO projects because "companies that have a flashy website may reveal they lack a fundamentally sound concept. However, a company with a website containing spelling errors may have a whitepaper that indicates a rock-solid concept and a carefully conceived implementation plan" (Reff 2022). However, due to the autonomy and non-mandatory disclosure of whitepapers, they often contain complex technical details and in the strongly personal style of the publishing team. The content is also complicated and not standardized enough. As a result, manually distinguishing fraudulent ICOs from legitimate ICOs is a challenging, time-consuming, and error-prone task. At this time, natural language processing technology and machine learning algorithms can provide a more efficient method for the identification of fraudulent ICO whitepapers (Toma and Cerchiello 2020; Dio and Tam 2019), that is, they can automatically convert unstructured data such as text into features, perform classification tasks on these features, determine the fraud tendency of ICO projects, and provide an important information reference for human decision-making.

While it is difficult to identify fraud cases with certainty because of the short time since the ICOs took place and the rare incidence of final judicial decisions, it is also not easy to determine that the project has not committed fraud yet is a completely reliable one, which makes it necessary to set strict screening rules to screen out positive and negative samples (Hornuf et al. 2021). After collecting and data precleaning with a conservative method to ensure reliability, we selected 120 nonfraudulent samples and 31 fraudulent samples from 203 ICO projects investigated, with a ratio of 4:1. Therefore, fraud detection of ICOs based on whitepapers is a small sample class imbalance problem.

The problem of data imbalance has attracted wide attention in the field of machine learning because data imbalance is considered to be one of the main reasons for the declining performance of machine learning algorithms in classification tasks: most mature classification algorithms are trained on the assumption that the proportion of classes is almost equal. In this context, this paper proposes the following research questions:

How can ICO fraud detection be conducted in a small sample and unbalanced context?

To answer the above question, we apply an appropriate NLP method to extract features from the text of the ICO whitepaper based on the characteristics of the language and then use a variety of cutting-edge machine learning and deep learning algorithms to train the prediction model. We focus on the imbalance of samples and try a wide variety of imbalanced data processing methods, including resampling, modifying the weight and modifying the loss function. Our experiment yields several interesting results. First, the synthetic minority oversampling technique (SMOTE) is not perfect for the current situation. Second, simply changing the weight can achieve good classification results. Third, considering the inherent characteristics of the data and models studied in this paper, it is unlikely that more benefits will be obtained by modifying the loss function. Finally, our experiment shows that the XGBoost model with a minority class weight of 7 has the best classification result, reaching an AUC of 0.93, which is higher than any other cases, and can effectively detect ICO fraud projects under the background of a small and imbalanced dataset.

2 Background

Fraud detection in the financial field is a frequent research topic (Vrij 2015; Wang et al. 2019; Carcillo et al. 2021). The practical significance of this problem has led many researchers in academia and industry to study this problem. Especially in recent years, with the rapid development of text analysis and machine learning technology, fraud detection research has entered a new stage (Dong et al. 2014; Dong et al. 2016). Specific to the research topic of this paper, the main purpose of fraud detection is to use a data analysis model to identify fraud or anomaly patterns embedded in language features extracted from ICO whitepapers (Liu et al. 2021). The general process of fraud detection based on text analysis can be summarized as follows: text acquisition, text preprocessing, word segmentation, deleting stop words, feature selection, and mining with algorithms (Karimov and Wojcik 2021). Naïve Bayes, decision tree, RNN and other algorithms are often used in the data mining stage.

Furthermore, in the stage of data mining, extreme gradient boosting (XGBoost) is a new algorithm based on boosting in recent years that has a high computing speed and good effect. Compared with GBDT, which only uses the information of the first derivative, XGBoost uses the second derivative to fit the loss function with higher speed and efficiency. Since this algorithm was proposed, XGBoost has been used in a wide variety of classification problems and has performed well in dichotomous scenarios.

However, XGBoost is not ideal for handling imbalanced datasets such as credit card risk prediction, network intrusion detection, and medical detection, so it must be combined with imbalanced data processing methods. The following table summarizes the progress of typical approaches to solve data imbalances in the literature (Table 1):

Table 1. Typical approaches to solve the data imbalance problem

Study	Method	Key contribution
Chawla et al. (2002)	Resample	Put forward the oversampling technique SMOTE, which based on k-nearest sample points of each sample point and selecting several adjacent points randomly, multiplies the distance by a [0,1] range threshold for interpolation, to achieve the purpose of synthesizing data. SMOTE reduces the risk of model overfitting compared to random oversampling
He et al. (2008)		Proposed ADASYN, which assigns different weights to different minority class samples, which depends on the number of majority classes around the sample. The larger the number, the larger the weight, thus generating more samples
Saner et al. (2019)		Compared the effect of stochastic oversampling, SMOTE, ADASYN combined with XGBoost classification model, and determined the oversampling strategy according to the ratio of positive and negative sample cross entropy loss
Bauder and Khoshgoftaar (2017)		Undersampling and oversampling methods were used to conduct medical insurance fraud detection experiments with unsupervised learning, supervised learning and hybrid machine learning methods

(*continued*)

Table 1. (*continued*)

Study	Method	Key contribution
Chen et al. (2017)	Adjust the weights toward positive and negative samples	Weighted-XGBoost algorithm is proposed to assign different weights to each sample by designing weight functions, so that the algorithm pays attention to high-weight samples
Li et al. (2018)		A Cost-sensitive and Hybrid attribute measure Multi-Decision Tree (CHMDT) method is proposed for binary classification of unbalanced datasets. The two classes are penalized imbalanced to improve classification accuracy
Tao et al. (2019)		Proposed a new cost-sensitive ensemble method for support vector machine (SVM) based on adaptive cost weight. The cost-sensitive SVM is used as the base classifier and an improved cost-sensitive boosting scheme is used, which is beneficial to slight deviation of final classification boundary from a few classes

Regarding the fraud detection of ICO projects studied in this paper, as it is a relatively new topic, relevant literature is very limited, and there is no research result on the endogenous imbalance in the process of text data collection. Therefore, for us, we divide the imbalanced data process method into three categories, which are **changing weight**, **resampling** and **changing loss function**, and then conduct experiments to research and test the validity of the above methods for the ICO project dataset we build. We focus on the processing experiment of ICO fraud from the perspective of **resampling** and **function modification** and use **SMOTE-XGBoost** and f**ocal loss-XGBoost** methods to conduct multiple comparative experimental studies with SVM, Bayes, logistic regression, decision tree, etc.

In summary, this study is the first ICO fraud detection experimental study aiming at the problem of imbalanced samples, and the results provide important implications for ICO fraud detection and ICO platform regulation.

3 Data

In the first step, we selected ICOs with good scores on the ICO evaluation websites, such as ICOBench.com, as nonfraud samples and screened ICOs in the scam category on Deadcoins.com as fraud samples. In this step, 203 samples were initially collected, including 132 possibly nonfraud samples and 71 possibly fraud samples.

Second, we verified the pending nonfraud ICO scores on multiple other platforms to ensure that it has good reviews on all websites and is in a good and normal state. For fraudulent ICOs, we searched search engines to see if there were lawsuits against the project. To be more precise, nonfraud ICOs are selected based on the following criteria: a) Score 4.0/5 or above on ICOBench.com and 8.0/10 on icomarks.com b) Raised $1 million or more, as it at least proves a successful project (Fahlenbrach and Frattaroli 2020) c) Active on social websites. In contrast, fraud ICOs on Deadcoins.com need to meet the following additional conditions: a) no longer update and interact on the social platform, b) have been delisted from the listing platform (Karimov and Wojcik 2021), and c) lawsuit against them found. After this step, we carefully discarded a considerable number of fraudulent samples and a small number of nonfraudulent samples to ensure a high degree of confidence in the original data. The final number of nonfraudulent samples was 120, and the number of fraudulent samples was 31, with a launch time range from 2014 Ethereum to 2021 Metacom.

After downloading the relevant PDFs of those whitepapers from the Internet, the Pdfplumber package of Python was used to extract the text, and preprocessing operations such as word segmentation, part-of-speech tagging and lemmatization were carried out.

4 Data Analysis

The research in this paper is divided into three steps, namely, feature extraction, outlier screening, and the use of a variety of imbalance problem solutions with experimental comparison.

4.1 Feature Extraction

In the context of fraud detection, Zhou et al. proposed the language feature extraction theory based on an experiment that studied the effectiveness of automated Linguistics Based Cues (LBC) (Zhou et al. 2004). It consists of nine categories of language behavioral features, which are *Quantity, Complexity, Uncertainty, Nonimmediacy, Expressivity, Diversity, Informality, Specificity* and *Affect*, and still achieves a decent level of performance that is comparable to those of the state-of-the-art models according to the literature (Zhang et al. 2016). Given the high intelligence and error-correcting capabilities of modern text editing tools, as well as the variety and innovation of ICO project names, in this paper, we decided not to consider *informality*, as it is likely to be meaningless or even misleading to count the number of misspelled words. Therefore, we adapt the 20 most representative features from 8 categories except *Informality*, as shown in Table 2.

Table 2. Linguistic features

Feature ID	Feature	Formula
1	WordQuantity	number of words
2	VerbQuantity	number of verbs
3	NounQuantity	number of nouns
4	AdjectiveQuantity	number of adjectives
5	AdverbQuantity	number of adverbs
6	PronounQuantity	number of pronouns
7	DiversityRatio	$\dfrac{\text{number of unique words}}{\text{number of words}}$
8	WordLength	$\dfrac{\text{number of letters}}{\text{number of words}}$
9	MarkRatio	$\dfrac{\text{number of punctuation marks}}{\text{number of sentences}}$
10	PerceptionRatio	$\dfrac{\text{number of perception words}}{\text{number of words}}$
11	Emotiveness	$\dfrac{\text{number of adjectives and adverbs}}{\text{number of verbs and nouns}}$
12	PositiveRatio	$\dfrac{\text{number of positive words}}{\text{number of words}}$
13	NegativeRatio	$\dfrac{\text{number of negative words}}{\text{number of words}}$
14	AffectRatio	$\dfrac{\text{number of positive and negative words}}{\text{number of words}}$
15	PleasureRatio	$\dfrac{\text{number of positive words}}{\text{number of words}}$
16	ModalVerbRatio	$\dfrac{\text{number of modal verbs}}{\text{number of words}}$
17	UncertaintyRatio	$\dfrac{\text{number of words connoted with uncertainty}}{\text{number of words}}$
18	GroupRatio	$\dfrac{\text{number of words connected to the group}}{\text{number of words}}$
19	IndividualRatio	$\dfrac{\text{number of words connected to individuals}}{\text{number of words}}$
20	SelfRatio	$\dfrac{\text{number of words connected to first person speaker}}{\text{number of words}}$

In this paper, the part-of-speech tagging function in the Python NLTK package, General Inquirer and self-built dictionaries are used to label and count words and then calculate language features. The number of sentences is similar to the number of periods, exclamation marks and question marks, which is not a completely correct statistical method, but experiments have shown that the result of this calculation method is not significantly different from the actual value and has little effect on the model. In addition, the number of perception words should be counted as sensory verbs rather than sensory state words (Dong et al. 2016).

4.2   Outlier Screening

Due to the natural irregularity of whitepaper data, outliers often appear in the dataset. For example, some whitepaper texts are extremely long, resulting in abnormally high eigenvalues under the Quantity category. The existence of these samples will greatly affect the classification effect of the model, so the iForest method is used to screen outliers in positive and negative samples separately. Before screening, the number of nonfraudulent and fraudulent samples was 120 and 31, respectively. The distribution of features before and after screening is shown in Figs. 1 and 2.

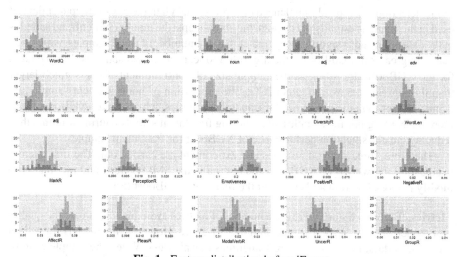

Fig. 1. Feature distribution before iForest

After screening outliers, the number of nonfraudulent samples is 107, and the number of fraudulent samples is 29. Intuitively, before iForest outlier removal, the data are particularly concentrated due to the existence of abnormal samples in the dataset. After the removal of abnormal samples, the data distribution is more even. In addition, there are significant differences in the distribution of some features between fraudulent and nonfraudulent samples. To make use of these differences to truly achieve classification, we need to train classification models.

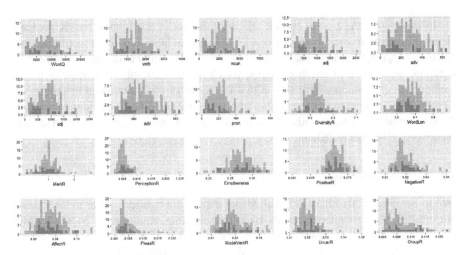

Fig. 2. Feature distribution after iForest

4.3 Imbalanced Data Processing

Fraud detection is a typical imbalanced data problem. When the ICO whitepaper dataset used in this paper is established through the above steps, the number of nonfraud samples in the data collected is significantly greater than the number of fraud samples, which will lead to a greater probability for the algorithm to learn nonfraud samples. In particular, for the XGBoost algorithm, as the sample size of nonfraudulent ICOs is much larger than the sample size of fraudulent ICOs, the proportion of nonfraudulent sample loss is much larger than the proportion of fraudulent samples, which makes XGBoost focus more on learning normal samples and ignoring fraudulent samples during training, thus reducing the classification performance of XGBoost for fraudulent samples.

Therefore, three steps of analysis and experiment are carried out. The first attempt is adjusting the weight of XGBoost, the second is the SMOTE test, and the final is to modify the loss function and carry out the focal-loss XGBoost test.

Metrics. Before we start, there are a number of metrics available to evaluate the quality of a model, such as accuracy, precision and F1-score. However, for models trained on imbalanced data, these metrics are sometimes meaningless. When the sample ratio of training data is 4:1, such as the dataset in this paper, that is, positive samples (classification target sample, fraud samples in this paper) only account for 20%, a classifier can achieve 80% accuracy as long as all samples are predicted as negative examples. Obviously, this model will not provide us with any valuable information.

At this point, we consider the ROC curve that can remain constant when the distribution of positive and negative samples in the dataset changes. The abscissa of the ROC curve is the false positive rate (FPR), and the ordinate is the true positive rate (TPR). For one classifier, we can obtain a TPR and FPR point pair according to its performance on the test set. In this way, the classifier can be mapped to a point on the ROC plane. By adjusting the threshold value used in the classification of this classifier, we can obtain a curve that passes through (0, 0) and (1, 1), which is the ROC curve of this classifier. The area under the ROC curve is AUC, when:

0.5 < AUC < 1, the model is better than random guess
AUC = 0.5, the model is the same as a random guess, and the model has no predictive value.
AUC < 0.5, the model is worse than random guess

Given a randomly chosen observation x belonging to the fraud class and a randomly chosen observation x′ belonging to the nonfraud class, the AUC is the probability that the evaluated classification algorithm will assign a higher score to x than to x′ and is independent of the proportion of positive and negative samples, so the influence of imbalance is excluded. Therefore, this paper takes AUC as the primary evaluation metric to evaluate the effect of the trained model. To compare with the results of other papers, accuracy will also be used as an additional evaluation metric.

$$Accuracy = \frac{True\ Positive + False\ Positive}{Total} \tag{1}$$

The results of different models on the original dataset are as follows (Table 3):

Table 3. AUC and accuracy of different models

AUC	LR	Decision Tree	SVM	Naïve Bayes	RNN	XGBoost
Train set	0.80	1.00	0.79	0.64	0.75	1.00
Test set	0.79	0.76	0.73	0.55	0.67	0.87
Accuracy	LR	Decision Tree	SVM	Naïve Bayes	RNN	XGBoost
Train set	83.16%	100.00%	85.26%	76.84%	76.88%	100.00%
Test set	80.49%	70.73%	82.80%	82.93%	72.22%	73.17%

We can see that XGBoost has more advantages in AUC than the other classifiers above. Next, for specific imbalanced data processing methods, we try in order:

Modified Weight. The idea of solving imbalance by weighting positive and negative differently is actually that, in the algorithm implementation process, different weights are assigned to categories with different sizes of samples in classification, and higher weights are assigned to minority samples, while smaller weights are assigned to majority samples, making the algorithm more inclined to learn the minority and then continue training the model.

XGBoost provides a variety of ways to change the weight of positive and negative samples, including adjusting the weight parameter or assigning the weight in DMatrix, which has the same effect as simply duplicating the positive sample, i.e., random over-sampling. Normally, the default weight of a negative sample is set as 1, and we optimize the scale by adjusting the weight of the positive sample.

The effect of models using different weights of the fraud class is shown below (Figs. 3 and 4, Table 4):

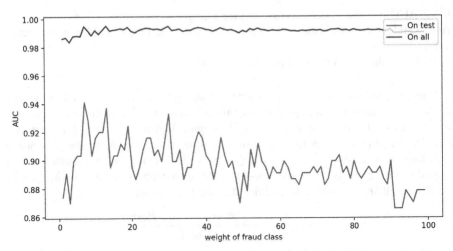

Fig. 3. AUC of models with different weights of the fraud class

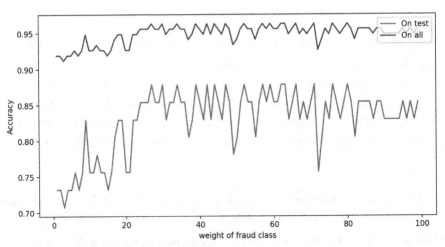

Fig. 4. Accuracy of models with different weights of fraud class

Table 4. AUC and accuracy of models with different weights of fraud class

Weight of fraud class	1	...	5	6	7	8	9	10	11	12	13
AUC on test	0.87	...	0.90	0.90	**0.94**	0.92	0.90	0.91	0.92	0.93	0.89
Accuracy on test	0.73	...	0.73	0.75	**0.82**	0.75	0.75	0.78	0.75	0.75	0.73

The effect of the model starts to improve as the weight of the fraud class increases; when the weight of the minority class is equal to 7, the model achieves the highest AUC of 0.94. With the continuous increase in minority weight, the effect of the model gradually decreases, showing a trend of increasing first and then decreasing. The accuracy value has been in an upward trend and gradually stabilized and fluctuated in a small range in the later period. The AUC and accuracy value reach local maxima when the weight of fraud samples is 7 at almost the same time, which is also the best choice when the AUC metric is given priority.

SMOTE. Resampling datasets can change the proportion of samples and solve the problem of sample imbalance at the data level, including undersampling and oversampling. More specifically, undersampling balances the dataset by reducing the majority of classes. This method retains all minority class samples and randomly selects the same number of samples as minority class samples from the majority class, that is, discarding redundant data to form a balanced dataset, which is typically represented by random undersampling. Therefore, this approach is appropriate only when the amount of data is large enough, which is definitely not the case in this paper. Under the background of a small dataset, it is more suitable to use the oversampling method, which generates more samples. However, some oversampling algorithms simply duplicate a few samples, which easily leads to overfitting of models. Therefore, it is necessary to conduct experiments on specific data.

SMOTE is an oversampling technique that synthesizes minority class samples in the characteristic space. The algorithm steps are as follows:

1. For each sample x in the minority class, the Euclidean distance is used as the standard to calculate the distance from sample x to all samples in the sample set S_{min} of a minority class, and its K-nearest neighbors are obtained.
2. Set a sampling ratio according to the sample imbalance ratio to determine the sampling ratio N. For each minority sample x, randomly select N samples $\{x_1, x_2, \ldots x_i \ldots, x_n\}$.
3. For each randomly selected neighbor xi, construct a new sample for interpolation according to the following formula:

$$x_{new} = x + rand(0, 1) \times |x - x_i| \tag{2}$$

The percentage of fraudulent and nonfraudulent samples before and after using SMOTE is shown in the following figure (Fig. 5):

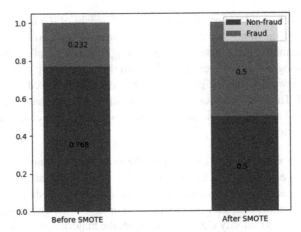

Fig. 5. Percentage of fraudulent and nonfraudulent samples

The experimental results are as follows (Table 5):

Table 5. AUC and accuracy of different models

AUC		LR	Decision Tree	SVM	Naïve Bayes	RNN	XGBoost
Without SMOTE	Train set	0.80	1.00	0.79	0.64	0.75	**1.00**
	Test set	0.79	0.76	0.73	0.55	0.67	**0.87**
SMOTE	Train set	0.81	1.00	0.80	0.56	0.72	**1.00**
	Test set	0.75	0.68	0.70	0.55	0.57	**0.88**
Accuracy		LR	Decision Tree	SVM	Naïve Bayes	RNN	XGBoost
Without SMOTE	Train set	83.16%	100.00%	85.26%	76.84%	76.88%	**100.00%**
	Test set	80.49%	70.73%	82.80%	82.93%	72.22%	**73.17%**
SMOTE	Train set	76.03%	100.00%	71.92%	56.16%	68.83%	**100.00%**
	Test set	60.98%	74.05%	65.49%	73.17%	62.45%	**74.61%**

On the dataset in this paper, SMOTE alone does not bring significant improvement to the effect of the models and even has a negative effect in the Naïve Bayes, LR algorithms, in which both AUC and accuracy decreased. The reason for this phenomenon is that SMOTE uses interpolation to generate positive samples, and it works best when the relationship between the characteristic value and label is monotonic, which does not hold in the context of this article. For example, too large or too small of SelfRatio may both indicate the probability of ICO fraud, and this correspondence is not monotonic. In such a complex characteristic space, SMOTE creates many suspicious points during the generation of minority class samples, which covers the boundary between fraudulent samples and nonfraudulent samples, thus reducing the classification effect of the model. Considering this situation, we argue that the combination of the above techniques of changing the weight of positive and negative samples and oversampling cannot continue to improve the effect of the model.

Modified Loss Function. Aside from adjusting the weights of different classes during model training, the loss function of the algorithm can also be modified to make the model more sensitive to the minority class. The ultimate goal of training the machine learning model is to minimize the loss function. By increasing the loss of misjudgment of minority samples in the loss function, the model can better identify minority samples. A common approach to modify loss functions is to penalize misclassified minority samples using a cost-sensitive loss function. In the process of training, adding a penalty term is a common means of regularization. in the context of imbalanced data, we adopt the same idea, assign a penalty coefficient to a few incorrectly classified samples and add it to the loss function. In this way, the model will naturally become more sensitive to the minority class samples during training.

In this part of the experiment, we train the XGBoost model with different loss functions and obtain multiple model training results, among which the binary logistic loss function performs best. The focal loss function produced even worse results due to the overlap of positive and negative samples. In the fifth iteration, the model has already fitted to a good effect, but the effect of the model gradually deteriorates in the subsequent iterations due to excessive concentration on difficult samples. Therefore, we argue that the combination of the above techniques of adjusting the weight of positive and negative samples and modifying the loss function cannot continue to improve the effect of the model (Figs. 6, 7, and 8).

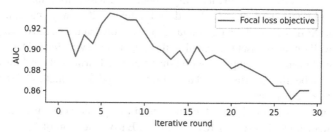

Fig. 6. AUC of XGBoost with focal loss

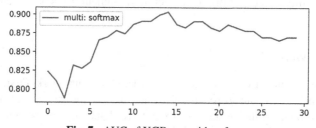

Fig. 7. AUC of XGBoost with softmax

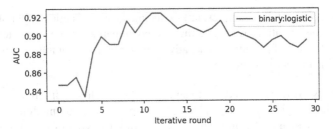

Fig. 8. AUC of XGBoost with logistic regression

In summary, we can see that after experiments of adjusting the weight of XGBoost positive and negative samples and modifying the loss function, the best-performing model is the XGBoost model with the weight change method, with the highest AUC of 0.94 and an accuracy rate of 82%. Compared with the 80% accuracy achieved by (Dürr et al. 2020) using the original XGBoost algorithm for 222 samples and the 72% accuracy achieved by (Bian et al. 2018) using the whitepaper sample alone, the model will clearly perform better after dealing with the imbalance problem, which means that we have obtained an efficient scheme that can detect ICO fraud in a small and imbalanced dataset.

5   Conclusion

5.1   Contribution

In this paper, we build a machine learning model for classification that can distinguish fraudulent and nonfraudulent ICOs based on the text of ICO whitepapers. The results show that the XGBoost model based on language features and a negative sample weight of 7 works well on the dataset presented in this paper, achieving 94% AUC, higher than other similar algorithms. Compared with other machine learning algorithms (SVM, LR, Bayes) that can be used for fraud ICO detection, the model and method designed in this paper have more advantages.

From a practical point of view, the findings of this paper can be used to support the investment decisions of potential ICO investors. Early detection of fraudulent tendencies in projects can help investors reduce investment risk and thus avoid capital loss. In addition, ICO exchange institutions can also use this model to improve the professionalism and credibility of the platform by detecting fraudulent ICOs before launching ICO projects.

5.2   Further Work

In subsequent research, two issues can be considered in future work. First, feature engineering can be carried out to try to solve the multicollinearity among features, and features based on quantitative statistics except WordQuantity can be converted to proportional features to balance the excessive influence of differences in whitepaper text length on the model. Second, interpretability analysis of the training results of the model is also the next research direction, and machine learning interpretation methods can be adopted in the future to conduct a deeper analysis and exploration of the model's judgment basis.

References

Bauder, R.A., Khoshgoftaar, T.M.: Medicare fraud detection using machine learning methods. In: 2017 16th IEEE International Conference on Machine Learning and Applications (ICMLA) (2017)

Bian, S., et al.: Icorating: a deep-learning system for scam ico identification. arXiv preprint arXiv: 1803.03670 (2018)

Carcillo, F., Le Borgne, Y.-A., Caelen, O., Kessaci, Y., Oblé, F., Bontempi, G.: Combining unsupervised and supervised learning in credit card fraud detection. Inf. Sci. **557**, 317–331 (2021)

Chawla, N.V., Bowyer, K.W., Hall, L.O., Kegelmeyer, W.P.: Smote: synthetic minority over-sampling technique. J. Artif. Intell. Res. **16**, 321–357 (2002)

Chen, W., Fu, K., Zuo, J., Zheng, X., Ren, W.: Radar emitter classification for large data set based on weighted-Xgboost. IET Radar Sonar? Navigation **11**(8), 1203–1207 (2017)

Diemers, D.: A strategic perspective: Global and Switzerland. PwC strategy& and Crypto Valley Association. https://cryptovalley.swiss/wp-content/uploads/20171221_PwC-S-CVA-ICO-Report_December_final.pdf. Accessed 9 May 2022

Dio, D.D., Tam, N.T.: On leveraging deep learning models to predict the success of ICOs. ResearchGate. unpublished preprint (2019). https://doi.org/10.13140/RG.2.2.27268.99201

Dong, W., Liao, S., Fang, B., Cheng, X., Zhu, C., Fan, W.: The detection of fraudulent financial statements: an integrated language model approach. In: 18th Pacific Asia Conference on Information Systems, PACIS 2014. Pacific Asia Conference on Information Systems (2014)

Dong, W., Liao, S., Liang, L.: Financial statement fraud detection using text mining: a systemic functional linguistics theory perspective. In: 20th Pacific Asia Conference on Information Systems (PACIS 2016). Pacific Asia Conference on Information Systems (2016)

Dürr, A., Griebel, M., Welsch, G., Thiesse, F.: Predicting Fraudulent Initial Coin Offerings Using Information Extracted from Whitepapers, ECIS (2020)

El-Masri, M., Al-Yafi, K., Samir Sherif, K.: The Digital Transformation of Fintech: Disruptions and Value Paths (2019)

Fan, W., Lin, Y., Fan, W.: Betting on the Horse, the Jockey or the Tips? Evidence from Blockchain-Based Fundraising Via Initial Coin Offerings (2020)

Fahlenbrach, R., Frattaroli, M.: ICO investors. Fin. Mark. Portfolio Mgmt. **35**(1), 1–59 (2020). https://doi.org/10.1007/s11408-020-00366-0

Fisch, C.: Initial coin offerings (ICOs) to finance new ventures. J. Bus. Ventur. **34**(1), 1–22 (2019)

Hauch, V., Blandon-Gitlin, I., Masip, J., Sporer, S.L.: Are computers effective lie detectors? A meta-analysis of linguistic cues to deception. Pers. Soc. Psychol. Rev. **19**(4), 307–342 (2015)

He, H., Yang, B., Garcia, E. A., Li, S.: Adasyn: adaptive synthetic sampling approach for imbalanced learning. In: Neural Networks, 2008. IJCNN 2008. (IEEE World Congress on Computational Intelligence). IEEE International Joint Conference on (2008)

Heines, R., Dick, C., Pohle, C., Jung, R.: The tokenization of everything: toward a framework for understanding the potentials of tokenized assets. In: PACIS 2021 Proceedings (2021)

Hornuf, L., Kück, T., Schwienbacher, A.: Initial coin offerings, information disclosure, fraud. Small Bus. Econ. **58**(4), 1741–1759 (2021)

Humpherys, S.L., Moffitt, K.C., Burns, M.B., Burgoon, J.K., Felix, W.F.: Identification of fraudulent financial statements using linguistic credibility analysis. Decis. Supp. Syst. **50**(3), 585–594 (2019). Kim, K., Lee, S.-Y. T., Assar, S.: Coin market behavior using social sentiment Markov chains (2011)

Karimov, B., Wojcik, P.: Identification of scams in initial coin offerings with machine learning. Front. Artif. Intell. **4**, 718450 (2021)

Li, F., Zhang, X., Zhang, X., Du, C., Xu, Y., Tian, Y.C.: Cost-sensitive and hybrid-attribute measure multi-decision tree over imbalanced data sets. Inf. Sci. S0020025517304784 (2018)

Liebau, D., Schueffel, P.: Cryptocurrencies & initial coin offerings: are they scams – an empirical study. J. Br. Blockch. Assoc. **2**(1), 1–7 (2019). https://doi.org/10.31585/jbba-2-1-(5)2019

Lin, T.Y., Goyal, P., Girshick, R., He, K., Dollár, P.: Focal loss for dense object detection. IEEE Trans. Pattern Anal. Mach. Intell **99**, 2999–3007 (2017)

Liu, Y., Sheng, J., Wang, W.: Technology and Cryptocurrency Valuation: Evidence from Machine Learning (2021). https://ssrn.com/abstract=3577208

Reff, N.: How to Identify Cryptocurrency and ICO Scams (2022). https://www.investopedia.com/tech/how-identify-cryptocurrency-and-ico-scams/. Accessed 9 May 2022

Saner, C.B., Kesici, M., Yaslan, Y., Genc, V.: Improving the performance of transient stability prediction using resampling methods. In: 2019 11th International Conference on Electrical and Electronics Engineering (ELECO) (2019)

Tao, X., et al.: Self-adaptive cost weights-based support vector machine cost-sensitive ensemble for imbalanced data classification. Inf. Sci. **487**, 31–56 (2019)

Toma, A.M., Cerchiello, P.: Initial coin offerings: risk or opportunity? Front. Artif. Intell. **3**, 18 (2020). https://doi.org/10.3389/frai.2020.00018

U.S. Securities and Exchange Commission. SEC Emergency Action Halts ICO Scam (2017). https://www.sec.gov/news/press-release/2017-219

Vrij, A.: A cognitive approach to lie detection. In: Granhag, P.A., Vrij, A., Verschuere, B. (Eds.): Detecting Deception: Current Challenges and Cognitive Approaches, pp. 205–229. Wiley-Blackwell (2015)

Wang, D., et al.: A semi-supervised graph attentive network for financial fraud detection. In: 2019 IEEE International Conference on Data Mining (ICDM), pp. 598–607 (2019)

Xuan, M., Zhu, X., Zhao, J.L.: Impact of social media on fundraising success in initial coin offering (Ico): an empirical investigation. In: 24th Pacific Asia Conference on Information Systems (PACIS 2020). Association for Information Systems (2020)

Zhang, D., Zhou, L., Kehoe, J.L., Kilic, I.Y.: What online reviewer behaviors really matter? Effects of verbal and nonverbal behaviors on detection of fake online reviews. J. Manag. Inf. Syst. **33**(2), 456–481 (2016)

Zhou, L., Burgoon, J.K., Nunamaker, J.F., Twitchell, D.: Automating linguistics-based cues for detecting deception in text-based asynchronous computer-mediated communications. Group Decis. Negot. **13**(1), 81–106 (2004)

Zou, S., Sun, H., Xu, G., Quan, R.: Ensemble strategy for insider threat detection from user activity logs. Comput. Mater. Continua **11**, 14 (2020)

The Impact of Policy on the Clustering of the Blockchain Industry
Evidence from GMM Estimates

Xiaolei Xu[1], Zhen Wu[1], Xiaotuo Qiao[2(✉)], Yuxi Zhang[3], and Haifeng Guo[4]

[1] National Internet Emergency Center, CNCERT/CC, Beijing 100044, China
[2] School of Finance, Zhongnan University of Economics and Law, Wuhan 430073, China
xiaotuoqiao@zuel.edu.cn
[3] Department of Finance, School of Management, Harbin Institute of Technology, Harbin 150001, China
[4] School of Finance, Southwestern University of Finance and Economics, Chengdu 610000, China

Abstract. Although local governments in China are encouraging the development of blockchain technology, the regional clustering of the blockchain industry still shows obvious differentiation. We use blockchain industry-related data during the period 2012–2019 to calculate the blockchain industrial clustering of each province in China. We find that the clustering state of the blockchain industry is quite different from the state of other industries and the situation of economic development in the same region. In less-developed regions, the blockchain industry is more prominent, which may benefit from local government management. We conduct an empirical analysis on the relationship between blockchain industrial clustering and regional government management using the generalized method of moments (GMM) of a dynamic panel. The results show that government management has a positive promoting effect on local blockchain industrial clustering as a whole, among which the promotion from economy, technology, infrastructure and policy is more significant.

Keywords: Blockchain industrial clustering · Government management · GMM method

1 Introduction

The transaction of virtual currency using blockchain technology can be traced back to 2011; however, it was not until 2016 that few provinces started to study blockchain technology. Since the Chinese government showed its attitude on supporting the development of the blockchain industry in 2016, large IT and internet enterprises have begun to arrange their blockchain empires one after another. Blockchain-related start-ups have entered a blowout mode, and the amount and frequency of investment and financing have increased sharply. The prototype of China's blockchain industry was initially formed in 2018. Based on the statistical data from the "Home of Blockchain" website, the Chinese blockchain industry can be divided into five categories: application products, underlying platforms, solutions, infrastructure and industry services. The blockchain industry continues to scale.

Most of the current literature on the blockchain industry focuses on the blockchain development of a certain region or certain kinds of enterprises. There is no research about the impact of government management on the blockchain industry from a macro perspective and regional level. Under this background, we tried to analyze the relationship between government management and blockchain industrial clustering. This is of great significance for government policy-making about scientific and technology development, industrial planning, resource allocation, etc.

2 Research Hypothesis and Model Setting

2.1 Research Hypothesis

There is a broad body of research on industrial clustering. Industrial clustering is a form of industrial organization manifested in the process of economic development, and it has become a prominent feature of various countries, regions or cities. The concept of industrial clustering was first proposed by Adam Smith [1]; he mentioned the idea of industrial clustering when he talked about the relationship between division of labor and market scope, industry development and market competition environment in his book "An Inquiry into the Nature and Causes of the Wealth of Nations". Marx [2] also pointed out in "Das Kapital" that a certain industry is al-ways located in a certain regional space, a certain labor space must have a certain industry corresponding to it, and the division of labor within an enterprise and the division of labor within the society are complementary and inseparable. Marshall [3] was the first to study the phenomenon of industrial spatial clustering. In the book "Principles of Economics", he adopted the concept of "clustering" to describe the proximity of regions and the concentration of enterprises and industries.

Another part of the literature focuses on the impact of government management on the formation of industrial clustering. The influencing factors are different for each development stage of industrial clustering, especially in the formation period. Michael Porter [4], the representative of new competitive economics, proposed the "diamond framework" of national competitive advantage theory, which emphasized that the government should give priority support to industries or enterprises on the four factors that determine national advantage: factor conditions; demand conditions; related or supporting industries; and firm strategy, structure and rivalry. In addition, the growth poles theory proposed by François Perroux [5] and the spatial economics theory represented by Paul Krugman [6] both advocate government intervention in industrial clustering; they believe that the government can promote the clustering of industries by investing in and cultivating leading industries, thereby promoting economic growth and development. Victor Gilsing and Marten J. Van. Sinderen [7] believes that government management has an obvious relationship with industrial clustering, and the role of government in industrial clustering policies should be redefined from the perspective of government function theory.

Although most scholars agree that government has a great influence on the formation of industrial clustering, most of the literature studies the impact of government management on industrial clustering from the level of government policy, and there is a lack of relevant research on the level of government functions. However, from empirical observation, we find that during the formation period of industrial clustering, especially for high-tech industry, the performance of government functions has a huge impact on the clustering process. The core of the blockchain industry is the development and application of blockchain technology, and the operation and development of most blockchain enterprises require the use of computer and internet technology. Computer and internet technology are part of the high-tech industry, and a permissive environment is the key to the formation of blockchain industrial clustering. Strengthening government behavior is the guarantee for creating a development environment for high-tech industry.

According to the released data of "Home of Blockchain", except for Beijing, the blockchain industry develops best in coastal areas, which mainly include Guangdong Province and the Yangtze River Delta region, with Jiangsu as the core. These regions not only have a solid economic foundation but also gather the top internet enterprises, providing a perfect environment for the development of the blockchain industry. In addition to the strong economic foundation and R&D capabilities, government management factors such as policy support, construction of supporting facilities, block-chain popularity and talent attraction are also important for the development of the blockchain industry. The excellent government management ability also makes four inland areas, Sichuan, Chongqing, Hubei and Shanxi, hotbeds of blockchain industry development. Based on the above background, we propose the first research hypotheses of this paper as follows:

H1: The formation of the blockchain industry is related to government management.

The regional governance effect of the government is not achieved overnight, especially in China, a country with a large population and complex government management mechanism. The execution of orders and policies must go through the process of conveying, comprehending and implementing to finally produce results. Each link also needs to experience different periods of time according to the actual situation. Therefore, the impact of government governance on the formation of blockchain industry agglomeration has a certain lag response. Based on this, the second hypothesis is proposed:

H2: The impact of current government management on blockchain industrial clustering will be reflected after at least one period.

2.2 Model Setting

We use panel data on China's blockchain industry from 2012 to 2019 as a sample to examine the impact of government management on blockchain industrial clustering. To measure blockchain industrial clustering, we choose the clustering degree of the blockchain industry in each province (Clustering) as the explained variable. The explanatory variables of government management can be divided into two categories: government function variables and government policy variables. Considering that government management is not the only influencing factor, we also include the maket variables as adjustment variables to measure the influence of the market factor and build a linear regression model:

$$clustering_{i,t} = \beta_1 gov_{i,t} + \beta_2 policy_{i,t} + \beta_3 market_{i,t} + \lambda_i + \varepsilon_{i,t} \qquad (1)$$

Considering the lag effect of government management, we optimize the above model by adding lagged variables. The speed of government policy impact is different from that of government function and market impact; the policy impact needs a longer process, so we introduce the 1–3 order lag terms of explanatory variables and control variables in the model. In addition, considering the inertia of the explanatory variable, a one-period lagged value will affect the current value of the explanatory variable to some extent, so we also introduce the first-order lag term of the explanatory variable to construct a dynamic panel model:

$$clustering_{i,t} = \alpha_1 clustering_{i,t-1} + \beta_{1,t-1} gov_{i,t-1} + \beta_{1,t-2} gov_{i,t-2}$$
$$+\beta_{1,t-3} gov_{i,t-3} + \beta_{2,t-1} policy_{i,t-1} + \beta_{2,t-2} policy_{i,t-2} + \beta_{2,t-3} policy_{i,t-3} \quad (2)$$
$$+\beta_{3,t-1} market_{i,t-1} + \beta_{3,t-2} market_{i,t-2} + \beta_{3,t-3} market_{i,t-3} + \lambda_i + \varepsilon_{i,t}$$

Due to the introduction of the first-order lag term of the explanatory variable, autocorrelation, heteroscedasticity and individual effects are unavoidable in this dynamic panel model. To solve their influence on the estimation and improve model accuracy, we adopt the generalized method of moments (GMM) for the regression analysis.

3 Casual Identification of Government Management and Blockchain Industry Clustering

3.1 Data and Variables

(1) Measure of Blockchain Industry Clustering

We use blockchain clustering as the explanatory variable. As an important part of industrial clustering research, the measurement method of industrial clustering has always been one of the topics that regional economists pay attention to. Since the 1930s, with the development of industrial clustering theory, the measurement of industrial clustering has been continuously improved. To objectively and reasonably measure China's blockchain industrial clustering, we adjust the location entropy measurement method to adapt to the characteristics of blockchain industry distribution.

Location entropy refers to the ratio of the proportion of relevant indications of a certain industry in a certain region to the relevant indicators of all industries in the same region and the proportion of the relevant indicators of this industry in all countries to the relevant indicators of all industries in the country.

$$LQ_{ij} = \frac{q_{ij}/q_j}{q_i/q} \quad (3)$$

In the context of this paper, LQ_{ij} refers to the location entropy of blockchain industry i in region j, q_i refers to the relevant indication (number of blockchain enterprises) of blockchain industry i in region j, q_j is the relevant indicator (number of all enterprises in the region) of all industries in region j, q_i is the relevant indicator (number of blockchain enterprises in the whole country) of blockchain industry in the country, and q is the relevant indicator (number of all enterprises in the whole country) of all industries in the country. The higher the LQij value is, the higher the regional block-chain industrial clustering level. Generally, LQij $\geq$ 1 means that blockchain industry i has higher clustering in region j, while LQij < 1 means no blockchain industrial clustering.

The blockchain industrial clustering calculated by the above formula is different from the figure presented by ArcGIS. The reason may be that the Chinese blockchain industry is in the initial formation stage, and the volume is too small compared to other industries. The results of the above formula are obtained by comparing all industry-related indicators, which will lead to bias with the actual distribution of block-chain industrial clustering. To solve this problem, we exclude all industry-related indicators in the formula and only keep the blockchain industry-related indicator. The modified formula is as follows:

$$LQ_{ij} = \frac{q_{ij}}{q_i} \tag{4}$$

The relevant indicator of blockchain industry is the number of enterprises in blockchain industry, and the data comes from "Home of Blockchain" website.

(2) Explanatory Variables

The explanatory variables can be divided into government function variables and government policy variables. The government function variables are investigated based on the economic foundation, scientific research inputs, infrastructure construction and talent development. We use local GDP (*gdp*) to measure economic foundation, local financial science and technology expenditure (*govrd*) and R&D expenditure (*rd*) to measure scientific and research inputs, number of broadband interfaces (*interface*) to measure infrastructure construction, local number of undergraduate students (student), local number of colleges and universities (*university*) and local number of IT practitioners (*employ*) to measure talent development. The government function is represented by the number of blockchain-related policies issued by each region (*policy*). To ensure the robustness of the estimation results, we divide the above indicators into three groups, and the difference in the three model groups lies in the measurement of scientific and research inputs and talent development. The specific grouping is shown in Table 1:

Table 1. Explanatory indicators and variables

Variables	Indicator	Model 1	Model 2	Model 3
Government function	Economic fundation	*gdp*	*gdp*	*gdp*
	Scientific & research inputs	*govrd*	*govrd*	*rd*
	Infrastructure construction	*Interface*	*Interface*	*Interface*
	Talent development	*Student*	*University*	*Employ*
Government policy	Number of policy issued	*Policy*	*Policy*	*Policy*

All of the above data come from the National Bureau of Statistics. To ensure the stability of data and avoid the impact of the dimensional difference of each variable, all data are analyzed using the ratio of local value to national value.

(3) Control Variables

In the new Structuralism, Justin Yifu Lin [8] proposed that the economic development of developing countries should develop industries and technologies based on comparative advantages. Give full play to comparative advantages. It can not only reduce industrial production costs and improve product competitiveness, but also help accumulate capital and provide a material basis for independent innovation and industrial upgrading [9].

Based on the above research views, we add the market variable as an adjustment variable in the model, choosing regional information market turnover (*revenue*) and one-period lagged industry clustering (explained variable) to measure the impact of the market on blockchain industrial clustering. The market turnover data also come from the National Bureau of Statistics. The specific grouping is shown in Table 2:

Table 2. Control indicators and variables

Variables	Indicator	Model 1	Model 2	Model 3
Market	Regional information market turnover	*Revenue*	*Revenue*	*Revenue*
	One-period lagged industry clustering	*Clustering*	*Clustering*	*Clustering*

To be consistent with the dimension of other explanatory variables, it is calculated using the ratio of local value and national value (Table 3).

Table 3. Descriptive statistics of variables

	Variable	Mean	Std. Dev.	Min	Max
Model 1	*Clustering*	.031	.067	0	.357
	gdp	.034	.026	.001	.11
	govrd	.019	.015	.003	.068
	Interface	.032	.023	.001	.098
	Policy	.032	.035	0	.152
	Student	.934	.295	.377	2.016
	Revenue	.032	.048	0	.176
Model 2	*Clustering*	.031	.067	0	.357
	gdp	.034	.026	.001	.11
	govrd	.019	.015	.003	.068
	Interface	.032	.023	.001	.098

(continued)

Table 3. (*continued*)

	Variable	Mean	Std. Dev.	Min	Max
	Policy	.032	.035	0	.152
	University	.032	.016	.002	.071
	Revenue	.032	.048	0	.176
Model 3	*Clustering*	.031	.067	0	.357
	gdp	.034	.026	.001	.11
	rd	.032	.036	0	.14
	Interface	.032	.023	.001	.098
	Policy	.032	.035	0	.152
	Employed	.032	.039	.001	.236
	Revenue	.032	.048	0	.176

3.2 Model Checking

To maintain the validity of the estimation results and rationality of the model setting, we use the Arellano–Bond test and Sargan test to check the autocorrelation and overidentification of our GMM model. As presented in Table 4, the residual of the model exhibits first-order serial correlation but has no second-order autocorrelation, indicat-ing that the model setting is appropriate.

Table 4. Test of autocorrelation and overidentification

	Order	z	Prob > z	Sargan-test
Model 1	1	−1.542	0.123	0.8756
	2	1.918	0.055	
Model 2	1	−1.426	0.154	0.9074
	2	2.037	0.042	
Model 3	1	−1.501	0.133	0.7266
	2	1.032	0.302	

The Sargan test values were all greater than 0.05, and the null hypothesis of "overidentifying restriction is effective" was accepted, indicating that the model has no overidentification problem, the instrumental variables are correlated with the disturbance term, and the model setting is reasonable.

3.3 Empirical Analysis

The estimation results of our dynamic panel GMM model are listed in Table 5:

Table 5. Estimation results

Model 1				Model 2				Model 3			
Clustering	Coef.	p-value	Sig.	Clustering	Coef.	p-value	Sig.	Clustering	Coef.	p-value	Sig.
L.clustering	.274	0	***	L.clustering	.251	0	***	L.clustering	.315	0	***
L.gdp	−1.91	0	***	L.gdp	−1.96	0	***	L.gdp	−.45	.532	
L2.gdp	2.716	0	***	L2.gdp	2.493	0	***	L2.gdp	1.117	.017	**
L3.gdp	−1.08	0	***	L3.gdp	−1.31	.001	***	L3.gdp	−1.23	.002	***
L.govrd	−3.31	0	***	L.govrd	−2.77	.001	***	L.rd	.036	.873	
L2.govrd	.745	.234		L2.govrd	1.134	.078	*	L2.rd	.066	.81	
L3.govrd	3.51	0	***	L3.govrd	3.004	0	***	L3.rd	1.425	.039	**
L.interface	.163	.325		L.interface	.243	.001	***	L.interface	.304	.005	***
L2.interface	.222	.232		L2.interface	.105	.434		L2.interface	.264	.008	***
L3.interface	.268	.003	***	L3.interface	.201	.028	**	L3.interface	.263	0	***
L.policy	.009	.774		L.policy	.003	.911		L.policy	.031	.321	
L2.policy	−.026	.115		L2.policy	−.023	.114		L2.policy	−.059	.186	
L3.policy	.123	.005	***	L3.policy	.129	.001	***	L3.policy	.178	.007	***
L.student	−.002	.194		L.university	.834	.406		L.employ	.418	0	***
L2.student	.003	.524		L2.university	.791	.096	*	L2.employ	−.329	0	***
L3.student	0	.969		L3.university	−1.17	.023	**	L3.employ	−.09	.095	*
L.reveue	.15	.002	***	L.reveue	.145	.018	**	L.reveue	.07	.406	
L2.reveue	.096	.448		L2.revenue	.087	.426		L2.reveue	−.238	.206	
L3.reveue	−.146	.277		L3.reveue	−.162	.108		L3.reveue	−.037	.738	
Constant	−.026	.146		Constant	−.028	.144		Constant	−.036	.172	

In general, blockchain industrial clustering is related to government management. The coefficients of government scientific and research inputs, infrastructure construction and three-period lagged government policy are positive and significant, indicating that government management on scientific and research, infrastructure and policy making three years ago have a promoting effect on local blockchain industrial clustering. The coefficients of the two lagged economic foundation variables also pass the significance test, which reflects that government economic management has a positive impact on local blockchain industrial clustering. In addition, government talent management also plays a role; the impact is mainly reflected in the number of colleges and universities and the number of information industry practitioners, and the influence of the undergraduate student scale is limited. This result shows that the talent demand of the blockchain industry has timeliness and is highly dependent on scientific and research talent. In addition, the coefficients of one-period lagged clustering and one-period lagged market variables are both significantly positive, which indicates that market adjustment positively promotes blockchain industrial clustering.

For the robustness check, we use three different sets of indicators for model analysis. As seen from Table 5, the estimation results of the three models are basically similar. In model 3, the coefficient of the one-period lagged market variable is quite different from the other two models. The impact of market factors on blockchain industrial clustering is not significant. It may be affected by the talent indicator, which is represented by the number of talent in the information industry. The talent indicators in model 1 and model 2 are the number of undergraduate students and the number of colleges and universities, respectively, and both variables are not related to the market.

Different variables have different influences on the number of lag periods. The results show that only the economic basis has a continuous impact on the agglomeration of the blockchain industry. This shows that the importance and influence of the economy on the development of the blockchain industry is far-reaching. However, the impact of policies can only be reflected after three periods, which is related to the complicated transmission mechanism of policies. Market variables have a relatively fast impact on blockchain industrial agglomeration. Last year's market situation can have an impact on this year's blockchain industrial agglomeration, but the persistence is not strong.

4 Conclusion

In this paper, we studied the relationship between government management and block-chain industrial clustering and chose the dynamic panel GMM method for empirical analysis. The results show that government management has a positive promoting impact on local blockchain industrial clustering, and this promotion requires at least three years to work. Among all the indicators, the impact of government management from economic, infrastructure, policy making and scientific and research talent is most prominent.

Currently, the development of China's blockchain industry is still in a start-up period. Although the blockchain industry in developed regions led by Guangdong is developing rapidly, there is still a catch-up in regions with relatively underdeveloped economies. The government needs to make relevant adjustments based on local conditions and develop distinctive blockchain industrial clustering according to local industrial characteristics.

Acknowledgments. This work was supported by The National Key Research and Development Program of China (2020YFB1006104), and the Financial support from the Innovation and Talent Base for Digital Technology and Finance (B21038).

References

1. Smith, A.: An Inquiry into the Nature and Causes of the Wealth of Nations, Britain (1776)
2. Marx, D.K.: Germany (1818)
3. Marshall. Principles of Economics 8th edn, Liberty Fund, Inc. Publishing, Britain (1890)
4. Poter, M.E.: Clusters and New Ecnomics Competition, no. 11. Harvard Business Review (1998)
5. Perroux, F.: A New Concept of Development. Routledge Library Editions Publishing, France (1983)
6. Krugman, P.: Increasing returns and economic geography. J. Politic. Econ. **99**(3), 483–499 (1991)
7. Roelandt, Th.J.A.A., Gilsing, V.A., van Sinderen, J.: New Policies for the New Economy (2003)

8. Lin, Y.: New structural economics: reconstructing the framework of development economics. Econ. Quart. **1**, 1–32 (2010)

9. Liu, G., Zhang, X., Deng, G.: Factors replacement, economic growth and unbalanced regional development. J. Quant. Tech. Econ. **7**, 35–56 (2017)

10. Lee, J.: The Role of a University in Cluster Formation: Evidence from a National Institute of Science and Technology in Korea, Regional Science and Urban Economics, vol. 86 (2021)

11. Liu, Z., Zeng, S., Jin, Z., Shi, J.J.: Transport infrastructure and industrial clustering: Evidence from manufacturing industries in China. Transp. Policy **121**, 100–112 (2022)

12. Jiangyong, L., Tao, Z.: Trends and determinants of China's industrial clustering. J. Urban Econ. **65**(2), 167–180 (2009)

13. Canh, N.P., Schinckus, C., Thanh, S.D.: Do economic openness and institutional quality influence patents? Evidence from GMM systems estimates. Int. Econ. **157**, 134–169 (2019)

14. Steinle, C., Schiele, H.: When do industries cluster? A proposal on how to assess an industry's propensity to concentrate at a single region or nation. Res. Policy **31**(6), 849–858 (2002)

15. You, S., Zhou, K.Z., Jia, L.: How does human capital foster product innovation? The contingent roles of industry cluster features. J. Bus. Res. **130**, 335–347 (2021)

16. Li, X.: Legal effect of smart contracts based on blockchain. In: Zeng, J.., Jing, W.., Song, X.., Lu, Z.. (eds.) ICPCSEE 2020. CCIS, vol. 1257, pp. 166–186. Springer, Singapore (2020). https://doi.org/10.1007/978-981-15-7981-3_12

17. Huang, J., et al.: Survey on blockchain incentive mechanism. In: Cheng, X., Jing, W., Song, X., Lu, Z. (eds.) ICPCSEE 2019. CCIS, vol. 1058, pp. 386–395. Springer, Singapore (2019). https://doi.org/10.1007/978-981-15-0118-0_30

18. Li, W., Guo, W.: The competence of volunteer computing for MapReduce big data applications. In: Zhou, Q., Gan, Y., Jing, W., Song, X., Wang, Y., Lu, Z. (eds.) ICPCSEE 2018. CCIS, vol. 901, pp. 8–23. Springer, Singapore (2018). https://doi.org/10.1007/978-981-13-2203-7_2

19. Wu, H., Li, Q., Li, X.: Research and simulation of mass random data association rules based on fuzzy cluster analysis. In: Zeng, J., Qin, P., Jing, W., Song, X., Lu, Z. (eds.) ICPCSEE 2021. CCIS, vol. 1451, pp. 80–89. Springer, Singapore (2021). https://doi.org/10.1007/978-981-16-5940-9_6

20. Song, Y., Wang, J., Yang, S., Zhu, X., Yin, K.: A blockchain-based scheme of data sharing for housing provident fund. In: Zeng, J., Qin, P., Jing, W., Song, X., Lu, Z. (eds.) ICPCSEE 2021. CCIS, vol. 1451, pp. 3–14. Springer, Singapore (2021). https://doi.org/10.1007/978-981-16-5940-9_1

21. Mijiyawa, A.G.: Drivers of structural transformation: the case of the manufacturing sector in Africa. World Develop. **99**, 141–159 (2017)

22. Zhang, H., Liu, Z., Zhang, Y.-J.: Assessing the economic and environmental effects of environmental regulation in China: the dynamic and spatial perspectives. J. Clean. Prod. **334** (2022)

23. Uddin, M.A., Ali, M.H., Masih, M.: Political stability and growth: an application of dynamic GMM and quantile regression. Econ. Model. **64**, 610–625 (2017)

24. Trinugroho, I., Law, S.H., Lee, W.C., Wiwoho, J., Sergi, B.S.: Effect of financial development on innovation: roles of market institutions. Econ. Model. **103** (2021)

25. Ullah, A., Pinglu, C., Ullah, S., Qaisar, Z.H., Qian, N.: The dynamic nexus of E-Government, and sustainable development: moderating role of multi-dimensional regional integration index in Belt and Road partner countries. Technol. Soc. **68** (2022)

Credit Risk Analysis of Chinese Companies by Applying the CAFÉ Approach

George X. Yuan[1,2,3,4](✉), Chengxing Yan[1,2](✉), Yunpeng Zhou[2], Haiyang Liu[2], Guoqi Qian[5], and Yukun Shi[6]

[1] Business School, Chengdu University, Chengdu 610106, China
george_yuan99@suda.edu.cn, 103757217@qq.com
[2] Shanghai Hammer Digital Technology Co., Ltd. (Hammer), Shanghai 200093, China
[3] Business School, Sun Yat-Sen University, Guangzhou 510275, China
[4] Business School, East China University of Science and Technology, Shanghai 200093, China
[5] School of Mathematics and Statistics, University of Melbourne, Melbourne, VIC 3010, Australia
[6] Adam Smith Business School, University of Glasgow, Glasgow G12 8QQ, UK

Abstract. It is known that the current Credit Rating in financial markets of China is facing at least three problems: 1) the rating is falsely high; 2) the differentiation of credit rating is insufficient; and 3) the poor performance of predicting early warning, thus we must consider how to create a reasonable new credit risk analysis approach to deal with issues for financial markets in China for those listed companies' performance.

This report shows that by using a new method called the "Hologram approach" as a tool, we are able to establish a so-called "CAFÉ Risk Analysis System" (in short, "CAFÉ Approach", or "CAFÉ") to resolve three issues for credit rating in China. In particular, the main goal in this paper is to give a comprehensive report for credit risk assessments for eight selected list companies by applying our "CAFÉ" from different industry sectors against actual market performance with the time period from the past one to three years through our one-by-one interpretation for event screening and true occurrence and related events. In this way, we show how "CAFÉ" is able to resolve current three major problems of "rating is falsely high, the differentiation of credit rating grades is insufficient, and the poor performance of predicting early warning" in the current credit market in China's financial industry in practice.

Keywords: Credit risk analysis · Digital economy · Big data financial technology · Hologram · Unstructured features · Credit rating · CAFÉ assessment system

1 The Background and Related Current Issues of Financial Markets in China

It is well known that credit rating and analysis is one of the most important things in today's financial market economy (see Altman et al. [2, 3], Hull [6, 7], etc.). However,

Y. Wang et al. (Eds.): ICPCSEE 2022, CCIS 1629, pp. 475–502, 2022.
https://doi.org/10.1007/978-981-19-5209-8_33

we also know that the current domestic credit rating market in China is now facing at least three main problems (see [1]): 1) the rating is falsely high; 2) the differentiation of credit rating grades is insufficient; and 3) the poor performance of predicting early warning.

Since the world's first Credit Rating agency by Moody in the early twentieth century, the Credit Rating (CR) industry has played an important intermediate role in promoting market development, revealing and preventing credit risks, reducing transaction costs, and assisting the government in financial supervision, and of course, a has faced many adjustments (Dun and Bradstreet [4], FICO [5], Hull [6], Anderson [13], Chi et al. [14], Thomas et al. [15], Witzling [16], Yuan and Wang [17], Yuan et al. [29]). At the same time, the development and growth of the CR industry and the formation of a system depend to a large extent on the development of the financial market, especially the bond and securities market (see Jing et al. [8], Du [9], Zhang [10], Ma et al. [11], Fitch Ratings [12] and related materials wherein).

Incorporating modern rating theory and approaches into the practice of international Credit Rating agencies, China's Credit Rating companies are gradually exploring Credit Rating methods and technologies that are suitable for China's national conditions and have initially formed rating methods that can cover basic rating theories, Credit Rating models, and Credit Rating systems based on the classification of industries, products, and subjects.

Nevertheless, the gap between China's CR companies and their international counterparts is also very prominent: for instance, CR is a necessary disclosure factor to promote the issuance of credit bonds in China and an important reference basis for bond issuance pricing, but during the rapid development of China's bond market, a large number of potential risks have accumulated, default events have occurred frequently, and the risks have shown normalization of default events, diversification of the nature of the subject, diversification of bond varieties, diffusion of industry distribution, and diversification of default area distribution.

Taking into account the fact that the available number of default (also called "bad") entitles observed for defaulted entities (companies or enterprises) in the market is very small, we must consider finding a new path to establish a reasonable credit rating method suitable for Chinese markets with international standards. On the other hand, in the current era of the digital economy (ecology), especially in today's rapid development of big data with financial technology (Fintech), under the premise of fully considering the information provided by both traditional structure and unstructured data, using a new approach in dealing with nonstructure data, which is called the "Hologram" approach (see Yuan and Wang [17]) as a fundamental tool, we are able to extract (nonstructured) risk feature factors based on unstructured data (instead of only traditional structure data) as breakthroughs to establish the so-called "CAFÉ Risk Assessment System" (in short, CAFÉ system) to conduct ratings for almost 10,000 companies in China by including all listed companies and bonds/debt issuers (see Yuan [19, 20] for more information). At the same time, combining the international standards that must be considered in the financial credit market, the basic investment level recognized in the financial industry is with "BBB" grade as the starting level. We are able to resolve the issue for the problem without enough default (also called "bad") samples by creating enough required "bad samples" under the category of nonstructure data types, which would help us to establish a so-called "CAFÉ Credit Rating system" (in short, "CAFÉ", or "CAFÉ System") for China's corporate entities and bonds (debts) that are in the line with international standards (Yuan [19, 20], Yuan et al. [29]).

In this report, we first point out the shortcomings of China's current ratings and then discuss the idea how the framework the "CAFÉ Risk Assessment System" can be used to conduct credit risk analysis for listed companies in China by applying the so-called "Hologram approach" (see Yuan and Wang [17]). The foundation of our CAFÉ system is a multidimensional risk assessment under the framework of big data analysis by using the so-called Hologram approach (as discussed by Yuan and Wang [17]) applied to "heterogeneous" data by combining the concept so-called "dynamic ontology" to achieve the extraction of entities' (corporate companies') risk genes by using AI algorithms (mainly the Gibbs sampling method) to resolve the issue of "not sufficient (defaulted) bad samples". In this way, we are able to achieve the comprehensive dynamic assessments for companies' credit risk from the four dimensions which consist of "Corporate structure hologram" (denoted by "C"), "Accounting behavior hologram" (denoted by "A"), "Financial behavior hologram" (denoted by "F") and "Ecosystem Hologram" (denoted by "E"), thus in short, "CAFÉ" system to form the "CAFÉ Risk Assessment System" for financial markets in China.

In this report, we focus on the application of the CAFÉ System (mainly the "Intelligence Stone Rating System", in short, IS) to conduct specific analysis for eight real entities from different industries, especially by combining the actual market performance for each case in the past one to three years back in history to against our risk assessment results with one-to-one interpretation of event screening and risk assessments derived by our Credit Risk system "IS", to show how the framework of "CAFÉ Risk Assessment System" has at least the ability to overcome the current three major problems in Chinese markets, which are as follows: "1) The rating is falsely high; 2) The differentiation of credit rating grades is not sufficient; and 3) The poor performance of predicting early warning".

This report consists of four parts. The Sect. 1 is an introduction to the background and issues we face in current Chinese markets. The Sect. 2 discusses the basic framework and the key ideas of our "CAFÉ Risk Assessment System" established under the framework of big data by applying the Hologram approach as a tool. The Sect. 3 is the case study for eight entities from the markets, in which we conduct eight case studies one by one against each company's actual performance with the risk assessment derived by our CAFÉ system. Finally, the Sect. 4 contains the conclusion and the comment.

2 Credit Risk Analysis by Applying the CAFÉ Approach Under a Big Data Framework

China currently maintains more than 50,000 bonds in the capital market and approximately 4800 listed companies, but approximately 92% of the bonds have a credit rating of AA or above. This clearly does not match the reality of the market, that is, the majority of "AAA" or "AA" companies have credit ratings that do not match their true performance!

Therefore, it is urgent to establish a set of international credit evaluation systems suitable for China's financial market. This paper attempts to construct enough bad samples required by the credit rating model, clearly defines the credit rating of "BBB" as the basic investment level, gives the rating division from AAA to C, and gives the corresponding "default probability" and "transfer matrix" of each level. In this way, we construct approximately 1200 bad samples since 2017, which was approximately 20% of listed companies in China's exchange stock markets which provides the basis for the "BBB" rating as the underlying investment level. Finally, we are able to establish a general framework called "CAFÉ Risk Assessment" and apply it to credit ratings, called the "Intelligence System" (IS) (see Yuan [19, 20], Yuan et al. [29]).

In summary, our "CAFE" approach no longer uses the "AA" credit rating as the starting point for the basic investment-grade standard that has been popular in China in the past almost 30 years; instead, our IS system solves the problem mentioned above, and this is also verified by the discussion given in Sect. 3. We will see the discussion given supported by the case study below.

2.1   The Framework of the CAFÉ Hologram Risk Analysis

To resolve three big issues of credit risk analysis for companies listed in China, the starting point of our implementation is to define the credit rating of entities (companies and their bonds/debts) by the following four dimensions (see Yuan etal. [29]), which effectively differentiates ratings and shows the company's actual credit standing in the market.

Here, we would like to share with readers that the two features of our CAFÉ system are: 1) it is able to convert static analysis into dynamic analysis; and 2) then it allows us to combine dynamic analysis from the perspective of corporate ecology to output more objective rating results.

To have related useful risk factors with characterization, we need to use the hologram approach by integrating heterogeneous and heterogeneous big data, which was developed by the implementation of the "data fusion" to extract the risk feature factors (Yuan & Wang [17], and the discussion by Yuan et al. [22–24] for more details).

In this report by applying the CAFE Risk Analysis, we use 11 categories to classify the fraud type of China's listed companies and nonlisted companies from 2016 to 2020. These 11 categories of fraud classification support us in completing the construction of "bad" samples and form dynamic enterprise risk assessment based on the company's audit information. The classified data come from the relevant penalties of the CSRC (China Securities Regulatory Commission) for the features extracted based on 11 categories of data, please see Yuan [19, 20] for more in details.

2.2   Extraction of Risk Features Based on the Gibbs Sampling Algorithm

The core problem of credit system evaluation construction of Chinese enterprises is: the available default samples are not enough to support reliable big data analysis. To solve such problems, we need to build a reasonable number of default samples based on the unstructured samples to support an evaluation of approximately 9000 companies (including approximately 4,800 listed companies, or approximately 5,500 issuers (for issuing bonds/debts))!

As mentioned above, we need to construct approximately 1,000 or 2,000 bad samples for approximately 4,800 listed companies to develop our credit risk analysis. However, the actual situation is that until the end of 2020 (from 2007), the default samples can be used to describe company failures in China are not more than 200 cases, so we have to consider approximately 2,700 information disclosure violations (mainly nonstructural data) issued by the regulator of the China Banking and Insurance Regulatory Commission (CBIRC) and the punishment data of the CSRC on listed companies and bond issuers as the raw (unstructured) data! These raw samples basically consist of 11 categories as mentioned above, and indeed, they basically appeared either embedded in the financial report in the form of document statements or in other forms, thus qualified as big data samples. Approximately 2,700 bad samples are the original sources of at least 1000 or 2000 bad samples that we need to model in practice using the big data approach. Once we have these 11 types of big data samples, we need to extract the risk characteristics embedded in these 11 types of bad samples to construct at least 1,000 or approximately 2,000 bad samples to support modeling for the CAFÉ system (see, Yuan [19, 20]).

Second, the Gibbs sampling algorithm is needed to complete feature extraction under big data (Yuan et al. [23, 24]), which helps us extract the characteristics of highly correlated financial fraud risks to establish a risk assessment system, distinguish good and bad samples, and test with "the Receiver Operating Characteristic Curve" (ROC). The ROC curve value is between 0.72 and 0.75! To put it simply, an ROC test result above 0.7 indicates that the model is highly usable and interpretable (see the literature of [26, 27] for more discussion details), which thus supports the CAFÉ system having the ability to evaluate companies and debts more effectively.

Then, considering the classification and analysis of the company's major shareholders, management, board of directors, and board of supervisors according to the proportion of shareholding, identity, and internal and external ratios, using a machine learning algorithm with the weight of evidence (in short, WOE) and information value (in short, IV) to explain the impact of the amount of information on the risk of fraudulent behavior for the entities, we then have the following general conclusion with the basic assessment criteria (see [22] and [24] for more details): The company's shareholding structure affects the company's financial fraud risk Important factors, and the following four characteristics can be used to warn the typical performance characteristics that may lead to fraud from the perspective of the corporate governance framework:

1) The shareholding ratio for major shareholders and corporate legal representatives is in the range between 5 and 50%;
2) The shareholding ratio of major shareholders is less than 60%;
3) The shareholding ratio of major shareholders who are management team members is less than 1%; and
4) The proportion of major shareholders who are the board of directors is less than 12%.

By using the above four kinds of risk features in describing the structure of companies' management for their daily management, we are able to establish the so-called CAFÉ assessment system by incorporating companies' internal and external audit data, plus daily management and business performance.

Indeed, by applying our CAFÉ system, the following is one case study on the financial fraud of a listed company in China called "Guangzhou Langqi Industrial Co., Ltd." (in short, denoted by "**Guangzhou Langqi**", or "**Langqi**") in the financial year of 2020.

Actually, by comparing the risk characteristics of financial fraud with the industry medians we established, we can clearly see the following two findings:

First Finding: In the three years of 2019, 2018, and 2017, its financial indicator data are far from the median of the industry (see Table 1 below):

Table 1. 2017–2019 Guangzhou Langqi financial index data and industry median comparison

No.	Items	Median (Chemical sector)	2019	2018	2017
1	Deduction of return on non net assets	6.64%	0.49%	0.57%	1.04%
2	Operating cash flow liability ratio	21.44%	−8.37%	−8.14%	−6.44%
3	Total asset turnover days	502.27	244.53	172.61	138.80
4	Prepayment turnover days	6.38	28.00	19.25	13.88
5	Growth rate of other receivables	16.33%	−11.23%	111.56%	−47.40%
6	Total growth rate of owner's equity (or shareholder's equity)	7.78%	−14.95%	21.69%	3.08%
7	Operating cost/total operating income	75.65%	95.15%	95.88%	97.18%
8	Interest expense (financial expense)/total operating income	0.95%	1.02%	0.61%	0.39%
9	Non operating net income/total operating income	1.04%	0.22%	0.00%	0.03%
10	Monetary capital/total assets	12.88%	11.61%	11.50%	9.33%
11	Other receivables (including interest and dividends)/total assets	0.42%	0.50%	0.65%	0.49%
12	Payroll payable/total assets	0.79%	0.34%	0.33%	0.38%
13	Paid in capital (or share capital)/total assets	13.35%	7.06%	8.13%	10.87%

(continued)

Table 1. (*continued*)

No.	Items	Median (Chemical sector)	2019	2018	2017
14	Undistributed profit/total assets	19.19%	3.45%	3.40%	4.11%
15	Subtotal of cash paid to and for employees/cash inflow from operating activities	9.27%	1.24%	1.37%	0.85%
16	Subtotal of cash received from other financing activities/cash inflows from financing activities	0.00%	5.51%	11.10%	0.00%

Second Finding: Based on the characteristic indicators that we have established to identify the company's possible fraudulent behaviors extracted from the corporate governance structure, we found that Guangzhou Langqi has a high risk of financial fraud in the following four characteristic indicators: 1) The shareholding ratio of major shareholders and corporate legal persons is between 5%–50%, but Langqi Company is 45%; 2) The cumulative shareholding ratio of major shareholders does not exceed 60%, but Langqi Company's 49%; 3) The Management proportion of major shareholders holding less than 1%, but of Langqi Company is 0; 4) The proportion of major shareholders on the board of directors does not exceed 12%, but the proportion of Langqi Company is 0.

Therefore, although the audit opinion did not reflect the risk of financial fraud, through the analysis of the company's statements and governance structure data, Langqi has a higher risk of fraud (actually, it happenend!).

2.3 The Construction of a Credit Transition Matrix for Companies in China's Financial Market

Credit rating adjustment is one of the most important rating actions for CRAs and mainly includes considerations such as "upscaling, downscaling, and maintenance" and related issues. In practice, the credit transition matrix in a certain period can reflect the changes in the credit quality of debt issuers (see references [4, 21] and related literature on the study for the construction of a general Credit Transition Matrix), and we summarize the construction method of the credit transition matrix for CAFÉ assessment here below.

A) The Credit Transition Matrix for CAFÉ Credit Risk Assessment

We process the construction of the rating transition matrix to maintain the necessary stability (Table 2) by following three parts:

1. Divide the default model into one-year and two-year periods and conduct ROC verification to determine the final model;

2. Give different rating results based on the 1-year and 2-year default models;

3. The 1-year and 2-year rating results are integrated to give the initial rating. The rules are as follows for the mapping method:

Table 2. CAFÉ's initial rating mapping table based on 1 year and 2 year period data

Items	Credit rating in 1 year						
Credit Rating in 2 Years	AAA	AA	A	BBB	BB	B	CCC-C
AAA	A	A	A	BBB	BB	BB	CCC-C
AA	A	A	A	BBB	BB	BB	CCC-C
A	A	A	A	BBB	BB	B	CCC-C
BBB	A	BBB	BBB	BBB	BB	B	CCC-C
BB	BBB	BBB	BBB	BBB	BB	B	CCC-C
B	BB	BB	BB	BB	BB	B	CCC-C
CCC-C	BB	B	B	B	CCC-C	CCC-C	CCC-C

B) The Brief Summary of Data used for CAFÉ's Credit Transition Matrix

Due to the limitation of data acquisition, the observation limit of CAFÉ on the sample of listed companies is limited to the period from 2014 to 2019. We obtained 3000 companies with complete annual financial information from 2014 and 4,500 companies from 2019. Taking December 31, 2020, as the observation day, the number of default samples returned for one year was 44; the number of default samples returned for the second year was 89 (here, we emphasize that we have actually observed "three years of return". The number of default samples in this report is 115"; however, in the construction and analysis of the credit transition matrix in this report, the reason is that other sample information needs to be improved when the default samples of three years backed back are used for feature extraction. Therefore, only the backed one-year and two-year default samples) (Jarrow et al. [21]).

Based on the use of these mapping rules and the rating results at 1 year and 2 years after default model integration, we will observe the following basic conclusion: based on BBB as the investment level, AAA-A to CCC-C maintain good stability to solve the problem of the transfer matrix (especially for emerging markets such as China) thereby solving the problem of instability of the AAA, AA and A-level migration matrix (especially for emerging markets such as China). Therefore, from the perspective of the credit transition matrix, if we divide AAA to A into an "AAA-A" category and adjust the mapping threshold, we have the following credit transition matrix results (see Table 3):

Table 3. Summary information of the CAFÉ 2015–2020 transition matrix

	AAA-A	BBB	BB	B	CCC-C
AAA-A	71.55%	24.76%	3.24%	0.39%	0.04%
BBB	15.11%	69.19%	13.77%	1.76%	0.15%
BB	3.68%	22.62%	55.66%	16.52%	1.51%
B	2.05%	8.94%	36.3%	42.93%	9.74%
CCC-C	1.52%	5.05%	16.41%	39.39%	37.37%

By Table 3 above, it seems that 1) the matrix that the AAA-A grades in China maintain maintains good stability after merging; and 2) the transition matrix also reflects a monotone decline in stability at AAA-C levels, which supports the relative stability of our CAFÉ system stability issues, which shows its different feature for companies listed in Chinese markets compared with those in mature markets in western countries and regions.

3   The Study of Listed Companies in China by Applying CAFÉ Risk Analysis

This is the key part as we focus on the study of eight listed companies in Chinese markets as applications of CAFÉ system with one-by-one risk assessments (with emphasis on the credit risk assessment for companies), through the analysis of eight companies' case-studies. The CAFÉ Risk Assessment System is used to interpret the actual market performance of eight companies in the past 1–3 years, we have that the CAFE risk assessment system can effectively solve the three problems in China's credit market: "1) the inflated credit ratings; 2) the insufficient credit rating discrimination, and 3) the weak performance for Credit rating's early warning functions in the practice".

First, we note that the companies/enterprises involved in these 8 cases have crossed 8 sectors: 1) Cross-border communication (002,640), from the sector of the commercial trade industry; 2) Guangzhou Langqi (000,523), from the sector of the chemical industry; 3) East Asahi Optoelectronics (000,413), from the sector of the electronics industry; 4) Kangdexin (002,450), from the sector of the chemical industry; 5) Kangmei Pharmaceutical (600,518), from the sector of the biomedical industry; 6) Tianxiang Environment (300,362), from the sector of the mechanical equipment industry; 7) Shenwu Environmental protection (300,156), from the sector of the public utility industry; and 8) Yongmei Group and its parent company "Yunenghua", from the sector mining industry and commercial trade industry.

3.1   The Risk Assessment for the Listed Company "Cross-Border Communication (002,640)"

The listed company belongs to the "commercial trade" sector (applying the "Sector Classification" designed by "**Shenwan**" in China, the same below), and its main risk events in the stock exchange market starting from April 28, 2021 are summarized as follows.

1) **The disclosure of major risk events for the company**

On April 28, 2021, Cross-Border Communication announced that "Xi Z resigned as a director of the company and a member of the Strategy Committee" [The announcement information of listed companies comes from the legal disclosure platform of the Shenzhen Stock Exchange-www.cninfo.com.cn (http://www.cninfo.com.cn), the following announcement information sources are the same.]:

On April 29, the company announced that "Zhang B resigned as a director of the company and a member of the audit committee".

On April 30, the company announced that "it is expected that regular reports will not be disclosed on time (2020 annual report)" and revised its performance downward. The estimated net profit for 2020 has been revised from 100–150 million to 3.0–38.0 billion.

On April 30, the company issued 35 announcements in the evening (8:19). Among them, the more important announcements disclosed the following information:

[1] The company had a loss of 3.374 billion RMB in 2020 and an asset impairment provision of 3.056 billion RMB;

[2] The audit result of Z Certified Public Accountants was "unable to express an opinion", and the result of the company's internal control verification was a "negative opinion";

[3] Director Lin Y and independent director Li Z issued a statement of objection to the annual report (the company's annual report cannot be guaranteed to be true), and Lin Y declared that "the company's internal control system has major defects, there is no full-time financial officer, and financial management is chaotic";

[4] The company's shareholders Yang J, Fan M and his wife and Xinyu Ruijing Enterprise Management Services Co., Ltd. lifted their voting rights, and the company changed to a state of no actual controller.

Company director Liang Y resigned from the company's director and strategic committee positions, and director Fang J resigned from the company's directorship.

2) **The risk assessment given by applying the CAFÉ System**

On May 1, 2019, the CAFÉ Hologram Credit Risk Assessment System (hereinafter referred to as "CAFÉ" or "system") lowered the level of cross-border communication from A to BBB.

On May 1, 2020, CAFÉ lowered the level of cross-border communication again, from BBB to CC, and issued an early warning of the risks of cross-border communication.

3) **The Interpretation of market performance with CAFÉ's risk assessment**

CAFÉ successively lowered the assessment level of cross-border communication from 2018 to 2020. In particular, it gave a CC level for cross-border communication in May 2020, revealing the risks of cross-border communication in advance. The assessment results in 2020 show that the scores of the various modules of cross-border communication are all underperforming. Among them, listed companies with financial performance scores of less than 95%, listed companies with financial report quality scores of less than 90%, and listed companies with governance structure scores of less than 70% have varying degrees of risk in the daily operation, financial management, financial management of cross-border communication, corporate governance, and other aspects (Table 4).

Table 4. Cross-border communication assessment level and performance of each module (2017–2020)

	2017	2018	2019	2020
Financial performance score	89.98%	90.14%	73.74%	4.80%
Financial report quality score	52.11%	80.79%	67.85%	8.38%
Governance structure score	33.55%	31.27%	30.67%	29.01%
CAFÉ comprehensive score	84.39%	88.71%	76.89%	5.99%
CAFÉ assessment rating	A	A	BBB	CC

The percentage in the table indicates the percentage of listed companies that the company's score exceeds. This percentage is a theoretical ratio adjusted by a normal distribution, not a true sample ratio.

The actual situation of cross-border communication also verifies the judgment of the CAFÉ Hologram Credit Risk Assessment System:

(1) The financial performance score is not high; the company's cumulative loss from 2019 to 2020 exceeds 6 billion RMB (more than the company's total owner's equity);

(2) The quality of the financial report is not high; the annual report is issued and cannot express opinions, and the directors issue a statement of objection to it;

(3) The governance structure score is not high; the company's internal control has major defects (no full-time financial manager, financial management chaotic).

4) **The company's stock price performance after the disclosure of its risk events**

After the risk event of the cross-border communication was disclosed, the company's stock fell by the "one-word" limit on April 30, 2021 (closing price of 3.51 RMB per share). On May 6, cross-border trading was suspended for one day. On May 7, the company's stock once again reached the "one-character" down limit (the closing price was 3.33 RMB/share).

After CAFÉ gave the CC rating, the company's stock price fell from "5.63 RMB per share" on April 30, 2020 to "3.33 RMB per share" on May 7, 2021, a cumulative drop of 40.85% (Fig. 1).

Fig. 1. Changes in cross-border communication levels and stock price trends (2016–2021)

3.2 The Risk Assessment for the Listed Company "Guangzhou Langqi (000,523)"

The listed company belongs to the chemical industry. Starting from September 25, 2020, the main risk events in the stock exchange market are as follows.

1) **The disclosure of major risk events for the company**

On September 25, 2020, Guangzhou Langqi announced that some of the company's debts were overdue, with a total of 395 million RMB of overdue debts, accounting for 20.74% of the company's most recent audited net assets.

On September 28, 2020, the company issued the "Reminder Announcement Concerning Some Inventory Goods May Involve Risks", and the whereabouts of 572 million RMB of inventory are unknown.

On October 23, 2020, the company announced that some new debts were overdue. The accumulated overdue debts totaled 524 million RMB, accounting for 27.52% of the company's most recent audited net assets.

On November 17, 2020, the company announced that some new debts were overdue. The accumulated overdue debts totaled 704 million RMB, accounting for 36.88% of the company's most recent audited net assets. In addition, the company's funds of 98 million RMB were frozen.

On December 26, 2020, the company disclosed that the amount of inventory in third-party warehouses and other accounts inconsistent in the trading business has accumulated to 898 million RMB, an increase of 316 million RMB from September 28.

On March 10, 2021, the company was filed for investigation by the Securities Regulatory Commission for alleged violations of information disclosure.

2) **The risk assessment given by applying the CAFÉ System**

On May 1, 2016, CAFÉ downgraded Guangzhou Langqi grade from BBB (investment grade) to B (speculative grade) and issued an early warning of the risks of cross-border communication.

May 1, 2018 [On May 1, 2017, CAFÉ gave a B grade to Guangzhou Langqi, which is consistent with the grade given on May 1, 2016. The following will not separately explain the situation in which the grade remains unchanged but only explain the level of change.], CAFÉ lowered the assessment level of Guangzhou Langqi from B to CC, and the company's risk further deteriorated.

On May 1, 2020, CAFÉ gave Guangzhou Langqi a CCC rating. The company's risk is still very high.

3) **The Interpretation of market performance with CAFÉ's risk assessment**

CAFÉ downgraded Guangzhou Langqi to B in 2016 and again in 2018. Guangzhou Langqi's assessment level has been maintained at CCC-C since then until Guangzhou Langqi disclosed the actual inventory inconsistency. The 2017–2020 assessment results show that Guangzhou Langqi's various module scores have long-term underperformance (see Table 5). Taking the assessment results in 2020 as an example, Guangzhou Langqi's financial performance score is lower than approximately 80% of listed companies. Listed companies whose financial report quality scores are lower than approximately 75% and those whose governance structure scores are lower than approximately 90%. This means that Guangzhou Langqi's daily operations, financial management, corporate governance, etc., all have varying degrees of risks, and the risks have been hidden for a long time because of the failure to disclose them in time.

The actual situation of Guangzhou Langqi also verified CAFÉ's judgment:

(1) The financial performance score is not high; the company's return on net assets has been below 5% for a long time (including a major loss in 2020), and a large amount of debt is overdue;

(2) The financial report quality score is not high-the whereabouts of 898 million inventory is unknown, and the authenticity of some warehousing contracts is in doubt;

(3) The governance structure score is not high—a number of senior executives left before the problem was exposed, and the vice chairman Chen Jianbin served as the general manager for a long time.

Table 5. Guangzhou Langqi assessment level and performance of each module (2017–2020)

	2017	2018	2019	2020
Financial performance score	41.71%	9.42%	7.82%	20.34%
Financial report quality score	17.12%	16.25%	12.79%	25.52%
Governance structure score	13.03%	11.83%	11.51%	10.68%
CAFÉ comprehensive score	24.15%	10.80%	8.75%	15.14%
CAFÉ assessment rating	B	CC	CC	CCC

4) **The company's stock price performance after the disclosure of risk events**

After Guangzhou Langqi's risk event was disclosed, the company's stock price fell to the "one-word" limit on September 28 and 29, 2020. The stock price fell from "5.70 RMB per share" on September 25 to "4.62 RMB on September 29 /share".

As of May 7, 2021, the closing price of the company's stock was "2.56 RMB per share", a cumulative decrease of 55.09% from September 25, 2020.

After the CAFÉ Hologram Credit Risk Assessment System gave a B grade, the stock price of Guangzhou Langqi dropped from "9.54 RMB per share" on April 29, 2016 to "2.56 RMB per share" on May 7, 2021, a cumulative decline of 73.17% (Fig. 2).

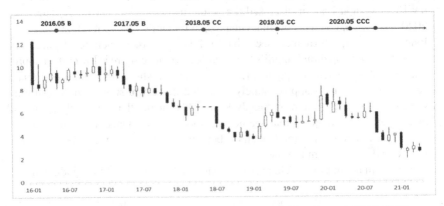

Fig. 2. Grade changes and stock price trend of Guangzhou Langqi (2016–2021)

3.3 The Risk Assessment for the Listed Company "Dongxu Optoelectronics (000,413)"

The listed company belongs to the electronics industry, and the main risk events in the stock exchange market from November 19, 2019, are as follows.

1) **The disclosure of major risk events for the company**

On November 19, 2019, Dongxu Optoelectronics Technology Co., Ltd. (hereinafter referred to as "Dongxu Optoelectronics" or the "Company") announced that "16 Dongxu Optoelectronics MTN001A" failed to complete the resale and interest payment as scheduled, and the bonds had substantially defaulted.

On December 2, 2019, the company announced that "16 Dongxu Optoelectronics MTN002" failed to pay the interest on schedule, and the bond has substantially defaulted.

On December 7, 2019, the company announced that "3.5 billion idle funds raised to supplement liquidity cannot be returned on schedule".

On May 19, 2020, the company failed to sign extension agreements with all "15 Dongxu Bond" investors, and 3.84% of the bond principal and interest defaulted.

On June 24, 2020, Zhongxing C Certified Public Accountants (Special General Partnership) (hereinafter referred to as "Zhongxing Cai") issued a qualified opinion on the company's 2019 annual report and disclosed that the company violated regulations to provide guarantees to related parties totaling 3.85 billion RMB;

On April 30, 2021, ZTE Cai issued a reservation on Dongxu Optoelectronics' 2020 annual report.

2) **The risk assessment given by applying the CAFÉ System**

On May 1, 2017, CAFÉ lowered the assessment level of Dongxu Optoelectronics, from investment grade BBB to speculative BB, and issued an early warning of Dongxu Optoelectronics' risks.

On May 1, 2018, CAFÉ downgraded Dongxu Optoelectronics' assessment level to B.

On May 1, 2019, the CAFÉ level of Dongxu Optoelectronics dropped to the CCC for the first time, and the company's risk further deteriorated.

On May 1, 2020, Dongxu Optoelectronics' assessment level was lowered to the lowest C level.

3) **The Interpretation of market performance with CAFÉ's risk assessment**

CAFÉ evaluated Dongxu Optoelectronics as BB in 2017 and then continuously lowered Dongxu Opto's assessment level, especially before the debt default occurred in May 2019. The system's assessment result of Dongxu Optoelectronics was already CCC. In addition, Dongxu Optoelectronics' financial report quality scores and governance structure scores for the four years from 2017 to 2020 have been underperforming, financial performance scores have continued to deteriorate, and the company's risk level has continued to increase. See Table 6. Before the debt default, in May 2019, Tunghsu Optoelectronics's financial performance score was lower than 70% of listed companies, financial report quality scores were lower than approximately 90% of listed companies, and governance structure scores were lower than approximately 80% of listed companies. This means that the company's risk level is already quite high.

Table 6. Dongxu Optoelectronics assessment level and performance of each module (2017–2020)

	2017	2018	2019	2020
Financial performance score	53.93%	40.14%	28.54%	1.11%
Financial report quality score	23.13%	13.93%	9.80%	1.38%
Governance structure score	22.44%	20.65%	20.18%	18.91%
CAFÉ comprehensive score	36.42%	23.70%	14.70%	4.20%
CAFÉ assessment rating	BB	B	CCC	C

The actual situation of Dongxu Optoelectronics also verified CAFÉ's judgment:

(1) The financial performance score is not high; the company continues to lose money after the debt default, and the accumulated loss by the end of 2020 exceeds 4.5 billion RMB.

(2) The financial report quality score is not high-the company's 3.5 billion fund-raising funds are occupied, and the annual report is issued with a qualified opinion.

(3) The governance structure score is not high—the company's management is controlled by the major shareholders, occupying the company's funds in disguise, infringing on the interests of the company and other shareholders.

4) **The company's stock price performance after the disclosure of risk events**

After Dongxu Optoelectronics was disclosed in the risk event, the company's stock was suspended on the same day (November 19, 2019). After the resumption of trading on December 10, 2019, the company's stock had a "one-word" limit for three consecutive days. As of May 7, 2021, the company's stock price was only "1.92 RMB per share", which is a cumulative drop of 60.00% from the "4.80 RMB per share" on November 18, 2019.

After the CAFÉ Hologram Credit Risk Assessment System gave a BB rating, the company's stock price fell from "10.32 RMB per share" on April 28, 2017 to "1.92 RMB per share" on May 7, 2021, a cumulative drop of 81.40% (Fig. 3).

Fig. 3. The Dongxu Optoelectronics grade changes and stock price trend (2016–2021)

3.4 Risk Assessment for the Listed Company "Kangdexin (002,450)"

The listed company belongs to the chemical industry, and the main risk events in the stock exchange market from January 15, 2019, are as follows.

1) **The disclosure of major risk events for the company**

On January 15, 2019, Kangdexin Composites Group Co., Ltd. (hereinafter referred to as "Kangdexin" or the "Company") announced that the company's ultrashort-term financing bill "18 Kangdexin SCP001" failed to repay the principal and interest in full on time Constitute a substantial breach of contract.

On January 21, 2019, the company announced that "18 Kangdexin SCP002" failed to repay the principal and interest in full, which constituted a substantial breach of contract.

On January 23, 2019, the company was filed and investigated by the Securities Regulatory Commission for suspected violations of information disclosure.

On February 11, 2019, the company's actual controller Zhong Y resigned from the company's chairman, director, and members of the professional committees under the board of directors.

On February 15, 2019, the company announced that "17 Kangdexin MTN001" could not pay the interest in full on time, which constituted a substantial breach of contract.

On March 18, 2019, the company announced that the company's offshore bonds failed to pay the interest in full on time, which constituted a substantial default.

On April 29, 2019, R Certified Public Accountants issued an "unable to express an opinion" on the company's 2018 annual report.

On May 13, 2019, Zhong Y, the actual controller of the company, was taken by the police for compulsory measures for suspected crime.

On July 5, 2019, the company received an administrative penalty notice from the China Securities Regulatory Commission, which disclosed that "the company had inflated profits of 11.921 billion RMB from January 2015 to December 2018".

2) **The risk assessment given by applying the CAFÉ System**

On May 1, 2017, CAFÉ downgraded Kangde's new assessment level from investment-grade A to speculative-grade BB and issued an early warning of the company's risks.

On May 1, 2018, CAFÉ downgraded Kangdexin's assessment level again, from BB to CC, and Kangdexin's risk further deteriorated.

On May 1, 2020, CAFÉ downgraded Kangdexin's assessment level to C.

3) **The Interpretation of market performance with CAFÉ's risk assessment**

The CAFÉ Hologram Credit Risk Assessment System gave Kangdexin a BB level in May 2017, warning of the risks of Kangdexin in advance. In May 2018, Kangdexin's risk profile further deteriorated, the grade dropped to CC, and its debt defaulted in January of the following year. The 2018 scoring results showed that the quality of Kangdexin's financial reports was low and that the company's financial quality had problems. This was the main reason CAFÉ warned Kangdexin. At the same time, while the company's financial quality score rebounded slightly in 2019, its financial performance performance declined significantly, which was consistent with the time when the CSRC initiated the investigation. Subsequently, the notice of the China Securities Regulatory Commission disclosed the behavior of "Kangdexin inflated profits of 11.921 billion RMB between January 2015 and December 2018", which further verified the rationality of CAFÉ's assessment results (Table 7).

Table 7. Kangde's new assessment level and performance of each module (2017–2020)

	2017	2018	2019	2020
Financial performance score	81.24%	58.64%	2.56%	0.13%
Financial report quality score	7.12%	0.07%	9.58%	56.85%
CAFÉ assessment rating	BB	CC	CC	C

4) **The company's stock price performance after the disclosure of risk events**

After Kangdexin's risk event was disclosed, the company's stock opened at the lower limit on January 15, 2019, and closed at the lower limit price of "6.46 RMB per share". As of May 7, 2021, the company's stock price was only "0.72 RMB per share", which is a cumulative decrease of 89.97% from the "7.18 RMB per share" on January 14, 2019.

After the CAFÉ Hologram Credit Risk Assessment System gave a BB rating, the company's stock price rose to the highest point (26.71 RMB/share) on November 22, 2017, and then began to fall. As of May 7, 2021, it was 96.36% lower than the "19.78 RMB per share" on April 28, 2017 (Fig. 4).

Fig. 4. Kangde's new grade changes and stock price trend (2016–2021)

3.5 The Risk Assessment for the Listed Company "Kangmei Pharmaceutical (600,518)"

The listed company belongs to the biomedical industry. The main risk events in the stock exchange market from December 28, 2018, are as follows.

1) **The disclosure of major risk events for the company**

On December 28, 2018, Kangmei Pharmaceutical Co., Ltd. (hereinafter referred to as "Kangmei Pharmaceutical" or the "Company") announced that "The company has been investigated by the Securities Regulatory Commission for suspected violations of information disclosure."

On April 30, 2019, Kangmei Pharmaceutical announced that "before 2018, the company's operating income, operating costs, expenses, and payment receipts and payments were inconsistent with the actual accounts, and the monetary funds, operating income, and operating costs were reduced by 29.944 billion. RMB 8.898 billion, RMB 7.662 billion, and increase inventory, other receivables, accounts receivable, and construction in progress amounts to RMB 19.546 billion, RMB 5.714 billion, RMB 641 million and RMB 632 million".

On August 16, 2019, Kangmei Pharmaceutical announced that "The company and related parties received the "Administrative Punishment and Market Ban Advance Notice" from the China Securities Regulatory Commission, and the company was punished by the China Securities Regulatory Commission for false records in the financial report."

On August 23, 2019, Kangmei Pharmaceutical announced that "China Chengxin Securities Appraisal Co., Ltd. lowered the company's main credit rating to BBB".

On January 31, 2020, the company failed to pay the principal and interest of the "15 Kangmei Bond" repurchase on schedule, which constituted a substantial default.

2) **The risk assessment given by applying the CAFÉ System**

On May 1, 2018, CAFÉ lowered the assessment level of Kangmei Pharmaceutical, the company's grade dropped from investment-grade BBB to speculative BB, and the system issued a risk warning to Kangmei Pharmaceutical.

On May 1, 2019, CAFÉ further downgraded Kangmei Pharmaceutical's assessment rating, and the company's rating was downgraded to the extremely risky CC.

On May 1, 2020, CAFÉ gave Kangmei Pharmaceutical a grade of C.

3) **The Interpretation of market performance with CAFÉ's risk assessment**

In May 2018, before the investigation by the China Securities Regulatory Commission, the CAFÉ Hologram Credit Risk Assessment System downgraded Kangmei Pharmaceutical's assessment level to speculative level (BB level) in advance. The company's 2018 financial performance score, financial report quality score, and governance structure score were similar. Compared with the previous year, there are different degrees of decline, indicating that the company's risk level needs attention (see Table 8).

The assessment results in May 2019 showed that the company's level and module performance declined significantly. Among them, the listed companies whose financial performance score is only higher than 2.54% after the decline and the listed companies whose financial quality score is only higher than 14.68% after the decline, the risk level of the company is already quite high. In May 2020, Kangmei Pharmaceutical's financial performance score and financial report quality score further declined, and their performance was lower than 99% of listed companies.

Table 8. Kangmei Pharmaceutical's assessment level and performance of each module (2017–2020)

	2017	2018	2019	2020
Financial performance score	58.81%	45.77%	2.54%	0.60%
Financial report quality score	71.82%	66.20%	14.68%	0.83%
Governance structure score	28.02%	25.96%	25.41%	23.93%
CAFÉ comprehensive score	75.24%	62.76%	6.14%	4.06%
CAFÉ assessment rating	BBB	BB	CC	C

4) **The company's stock price performance after the disclosure of risk events**

After the disclosure of the risk events of Kangmei Pharmaceutical, the company's stocks had a "one-character" limit on January 2 and 3, 2019, and closed at "7.44 RMB per share" on January 3. As of May 7, 2021, the company's stock price was only "2.07 RMB per share", which is a cumulative decrease of 77.48% from the "9.19 RMB per share" on December 28, 2018.

After the CAFÉ Hologram Credit Risk Assessment System gave a BB rating, the company's stock price rose to the highest point (27.99 RMB per share) on May 29, 2018, and then began to fall. As of May 7, 2021, it fell by 92.60% from the highest point (Fig. 5).

Fig. 5. Kangmei Pharmaceutical's grade changes and stock price trend (2016–2021)

3.6 The Risk Assessment for the Listed Company "Tianxiang Environment (300,362)"

The listed company belongs to the machinery and equipment industry. The main risk events in the stock exchange market from October 8, 2018, are as follows.

1) **The disclosure of major risk events for the company**

On October 8, 2018, Chengdu Tianxiang Environment Co., Ltd. (hereinafter referred to as "Tianxiang Environment" or the "Company") announced that "the company's financial situation caused some debts to be overdue, and the total amount of overdue debts of the company and its subsidiaries was approximately 3.81. 100 million RMB, accounting for 21.09% of the company's most recent audited net assets".

On January 4, 2019, the company announced that "the company and its subsidiaries will accumulate overdue debts of approximately 1.317 billion RMB, accounting for 72.82% of the company's most recent audited net assets".

On March 25, 2019, Tianxiang Environment failed to fully redeem the principal and interest of "16 Tianxiang 01" as scheduled, which constituted a substantial breach of contract.

On April 30, 2019, ShineWing Certified Public Accountants issued an "unable to express an opinion" on Tianxiang Environment's 2018 annual report.

On May 23, 2019, Tianxiang Environmental announced that "the company was punished by the China Securities Regulatory Commission for the nonoperating use of the company's funds by the actual controller from January 1, 2018 to July 17, 2018".

On September 30, 2019, Tianxiang Environment issued an announcement stating that "As of December 31, 2019, the company and its subsidiaries will accumulate a total overdue amount of approximately 3.472 billion RMB, accounting for 2633.23% of the company's most recent audited net assets."

2) **The risk assessment given by applying the CAFÉ System**

On May 1, 2016, CAFÉ downgraded Tianxiang's environmental assessment level to B and issued an early warning of the company's risks.

On May 1, 2017, CAFÉ lowered the Tianxiang environmental assessment level again, and the company level was lowered from B to CCC.

On May 1, 2018, CAFÉ downgraded Tianxiang's environmental assessment level to C, which further improved the company's risk profile.

On May 1, 2019, CAFÉ continued to maintain the assessment level of Tianxiang Environmental C.

3) **The Interpretation of market performance with CAFÉ's risk assessment**

The CAFÉ Hologram Credit Risk Assessment System provided early warning of the risks of the Tianxiang Environment two years in advance. The 2016 assessment results showed that the financial performance score of Tianxiang Environment was lower than 61.45% for listed companies, and the financial report quality score was lower than 80.11% for listed companies. It is already relatively high; see Table 9. From 2017 to 2019, the company's situation further deteriorated. As of May 1, 2019, the Tianxiang Environmental Financial Performance Score was lower than 99.46% for listed companies, and the financial report quality scores were lower than 99.81% for listed companies. The company's risks are no longer controllable.

The actual situation of the Tianxiang environment also verified the assessment results of the CAFÉ Hologram Credit Risk Assessment System:

(1) The financial performance score is not high-undertaking a large number of projects approximately 2017, coupled with overseas investment made the company's cash flow vulnerable;

(2) The financial report quality score is not high, and the governance structure score is not high. Deng Qinhua, the actual controller of the company, in January-July 2018, through the signing of false procurement contracts, borrowing, private bridge lending, etc., nonoperating occupation of Tianxiang's environment. The capital is 2.091 billion RMB.

Table 9. The Tianxiang environmental assessment level and performance of each module (2017–2020)

	2017	2018	2019	2020
Financial performance score	38.55%	39.11%	2.77%	0.54%
Financial report quality score	19.89%	10.47%	6.04%	0.19%
Governance structure score	31.17%	28.02%	25.96%	25.41%
CAFÉ comprehensive score	21.29%	16.47%	5.25%	4.11%
CAFÉ assessment rating	B	CCC	C	C

4) The company's stock price performance after the disclosure of risk events

After Tianxiang Environment's debt overdue on October 18, 2018, the company's stock fell to the "one-character" limit on the resumption day of December 10, and the process continued until December 13, and the company's stock price closed at "6.34 RMB per share." As of April 29, 2020 (the last trading day before delisting), the company's stock price is only "1.49 RMB/share", which is a cumulative decrease of 84.58% from the "9.66 RMB/share" on June 7, 2018.

After the CAFÉ Hologram Credit Risk Assessment System gave a BB grade, the company's stock price rose to the highest point (25.60 RMB/share) on October 28, 2016, and then began to fall. As of the trading day before the delisting, the company's stock fell by 94.18% (Fig. 6).

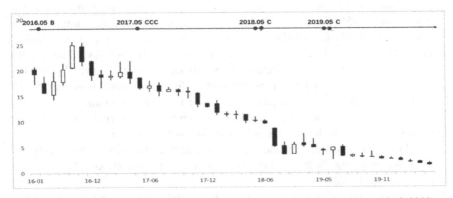

Fig. 6. Changes in Tianxiang's environmental rating and stock price trends (2016–2020)

3.7 Risk Assessment for the Listed Company "Shenwu Environmental (300,156)"

The listed company belongs to the public utility industry. The main risk events in the stock exchange market from March 14, 2018, are as follows.

1) **The disclosure of major risk events for the company**

On March 14, 2018, Shenwu Environmental Protection Technology Co., Ltd. (hereinafter referred to as "Shenwu Environmental Protection" or the "Company") failed to redeem the principal and interest of the "16 Environmental Bonds" sale back as scheduled, which constituted a substantial breach of contract.

On April 28, 2018, Shenwu Environmental Protection issued an announcement on the revision of the performance bulletin. The revised operating income of the company was 2.809 billion RMB (reduced by 1.307 billion RMB), and the revised total profit was 398 million RMB (reduced by 299 million RMB).

On April 28, 2018, D Certified Public Accountants issued a "qualified opinion" on the company's 2017 annual report.

On May 18, 2018, Shenwu Environmental Protection announced that "The company's 100 million RMB bank loan was overdue".

On January 29, 2019, Shenwu Environmental Announcement "The company is listed as a dishonest person to be enforced".

On April 30, 2019, D Certified Public Accountants issued an "unable to express an opinion" on the company's 2017 annual report.

On September 6, 2019, Shenwu Environmental Protection received the "Decision on Administrative Supervision Measures" from the China Securities Regulatory Commission. The decision disclosed that "The company's 2017 first quarterly, half-yearly, and third quarterly reports, respectively increased the number of monetary funds reported by no less than 1.575 billion. RMB, 835 million RMB, 1.247 billion RMB".

On August 25, 2020, the company's stock was delisted by the Shenzhen Stock Exchange.

2) **Risk assessment given by the CAFÉ system**

On May 1, 2015, CAFÉ lowered the environmental protection assessment level of Shenwu, and the company level was reduced from BBB to B, warning that the company may be at risk.

On May 1, 2016, CAFÉ upgraded Shenwu Environmental Protection's assessment level to BB, and the company is still a speculative company with higher risks.

On May 1, 2017, CAFÉ lowered the company's risk level to CCC, and the company's risk increased again.

On May 1, 2018, CAFÉ continued to lower the assessment level of Shenwu environmental protection and set the environmental protection level of Shenwu as CC.

On May 1, 2019, CAFÉ gave Shenwu Environmental Protection a grade C, which is also the highest risk grade in the CAFÉ system.

3) **The Interpretation of market performance with CAFÉ's risk assessment**

After Shenwu's environmental protection grade dropped to B on May 1, 2015, the overall company grade showed a downward trend. The 2017–2019 assessment results show that the performance of Shenwu Environmental's financial report quality score is not good, with less than 99.99% of listed companies in 2017, less than 97.19% of listed companies in 2018, and less than 97.48% of listed companies in 2019. Finance performance scores dropped significantly in 2018. As of May 2019, Shenwu

Environmental's financial performance scores were basically the lowest among listed companies.

The actual situation is also consistent with CAFÉ's assessment results. In 2018, the company revised its financial report, and the time of revision was consistent with the time when the financial performance score fell. The "Unable to Express Opinion" issued by the accounting firm and the "Decision on Administrative Supervision Measures" issued by the China Securities Regulatory Commission verified that there are problems with the quality of financial reports, and CAFÉ's score on the quality of financial reports of Shenwu Environmental is correct (Table 10).

Table 10. Shenwu's assessment grade, financial and quality scoring performance (2016–2019)

	2017	2018	2019	2020
Financial performance score	98.70%	11.46%	0.00%	98.70%
Financial report quality score	0.01%	2.81%	2.52%	0.01%
CAFÉ assessment rating	CCC	CC	C	CCC

4) **The company's stock price performance after the disclosure of risk events**

On the first trading day (March 23, 2018) after the default of the "16 Environmental Bonds", the company's stock price opened at "12.52 RMB per share", which is higher than the closing price of the previous trading day (February 1, 2018) (13.49 RMB/share) fell 7.19%. On the last trading day before delisting (August 24, 2020), the company's stock price was only "0.12 RMB per share", down 99.11% from February 1, 2018.

After CAFÉ gave a B grade on May 1, 2015, the company's stock price rose to the highest point (59.02 RMB per share) in November 2015 and then began to fall until it was delisted. As of the last trading day before delisting, the company's stock price fell by 99.80% from the highest point (Fig. 7).

Fig. 7. Shenwu environmental protection grade changes and stock price trend (2015–2020)

3.8 The Risk Assessment for the Issuer "Yongmei Group and Yunenghua Group" in the Capital Market

Regarding "Yongmei Group" and "Yunenghua Group", we know that "Yongmei Group" belongs to the mining industry; "Yunenghua Group" belongs to the commercial trade industry. Starting from November 14 near the end of 2020, its main risk events worth recording in China's capital market are as follows.

1) **The disclosure of major risk events for the company**

On November 13, 2020, Yongcheng Coal and Electricity Holding Group Co., Ltd. (hereinafter referred to as "Yongmei Group") announced that "20 Yongmei SCP003" failed to redeem the principal and interest on November 10, which constituted a substantial default.

On November 23, 2020, the Interbank Market Clearing House Co., Ltd. (hereinafter referred to as the "Clearing House") announced that it had not received the interest payment funds of Yongmei Group "20 Yongmei SCP004" and "20 Yongmei SCP007".

On December 15, 2020, Yongmei Group announced that the "20 Yongmei SCP005" bond was extended (delayed payment of principal or interest).

On December 21, 2020, Yongmei Group announced that the "19 Yongmei CP003" bond was extended.

On December 25, 2020, Yongmei Group announced that the "20 Yongmei SCP006" bond was extended.

On January 5, 2021, Henan Energy and Chemical Group Co., Ltd. (hereinafter referred to as "Yunenghua") announced that the company's "17 Yunenghua PPN001" bond was extended.

On January 26, 2021, Yongmei Group announced that the company's "18 Yongmei PPN001" bond was extended.

On February 3, 2021, Yongmei Group announced that the company's "18 Yongmei MTN001" bond was extended.

On March 10, 2021, Yongmei Group announced that the company's "20 Yongmei CP001" bond was extended.

On March 23, 2021, Yunenghua announced that the company's "20 Yunenghua CP001" bond was extended.

On March 29, 2021, Yongmei Group announced that the company's "18 Yongmei PPN002" bond was extended.

On April 9, 2021, Yongmei Group announced that the company's "20 Yongmei SCP008" bond was extended.

On April 20, 2021, Yunenghua announced that the company's "20 Yunenghua CP002" bond was extended.

2) **Risk assessment given by the CAFÉ system**

On May 1, 2017, CAFÉ gave the Yongmei Group a BB grade, indicating that the company's operating conditions are at risk. In the same period, Yongmei Group's parent company Yunenghua has a grade of B.

On May 1, 2018, CAFÉ gave Yongmei Group a BB rating, indicating that the company's operating conditions are at risk. In the same period, the rating of the parent company Yongmei Group "Yunenghua" rose to BB.

On May 1, 2019, both Yongmei Group and Yunenghua maintained their ratings at BB.

On May 1, 2020, Yongmei Group's grade continued to be maintained at BB, and Yunenghua's grade dropped to B again.

3) **The Interpretation of market performance with CAFÉ's risk assessment**

The results of the CAFÉ Hologram Credit Risk Assessment System show that Yongmei Group's assessment level has been maintained at BB for a long time from 2017 to 2019 (see Table 11), which is far lower than the AAA level given by third-party organizations. The main reason is that Yongmei Group's finances performance scores have been underperforming for a long time.

In addition, it is worth noting that the performance of Yongmei Group's parent company Yunenghua is even worse than that of Yongmei Group (see Table 12). In the case of Yongmei Group's default, Yunenghua should have a fairly high default risk.

The subsequent situation also confirmed the judgment of the CAFÉ Hologram Credit Risk Assessment System: on January 5, 2021, the "17 Yunenghua PPN001" issued by Yunenghua Group failed to repay the principal and interest on schedule, which constituted a substantial default.

Table 11. Yongcheng coal and electricity assessment and performance of each module (2017–2020)

	2017	2018	2019	2020
Financial performance score	3.36%	20.73%	23.47%	7.77%
Financial report quality score	28.85%	49.86%	34.34%	73.74%
CAFÉ assessment rating	BB	BB	BB	BB

Table 12. The assessment level of Yunenghua and the performance of each module (2017–2020)

	2017	2018	2019	2020
Financial performance score	1.64%	8.64%	17.66%	5.16%
Financial report quality score	77.31%	75.81%	57.33%	67.77%
CAFÉ assessment rating	B	BB	BB	B

4) **The company's breach of contract after the disclosure of risk events**

After "20 Yongmei SCP003" defaulted, as of May 7, 2021, 11 bonds under Yongmei Group had defaulted, with a default bond balance of 6.5 billion RMB; under Yuneng's pseudonym, 3 bonds had defaulted, with a default bond balance of 1.25 billion RMB.

4 Conclusion and Discussion

The credit rating is one of the most important parts of today's financial market, but the current domestic credit rating market in China is facing at least these three problems: 1) the rating is falsely high; 2) the degree of differentiation is insufficient; and 3) the performance of weak prewarning functions in practice and related other issues. Second, considering the simple fact that available default samples for companies observed from the market (i.e., the bad default number of samples) have not improved for a long time in Chinese markets since the 1980s, we must consider a new way to establish a credit rating system with international standards, and at the same time, it is suitable for the Credit Assessment System for China's financial market! Based on this idea, in the era of digital economics, especially in today's rapid development of big data for Fintech, as applications of the Hologram approach for companies as a tool, and under the help for the extraction of features for which embedded in the samples of unstructured data based on the both structure and unstructured sample for companies as breakthroughs we are able to establish the so-called "CAFÉ Risk Analysis System" to work for financial markets in China (see also a partially discussion related part for results given by Yuan et al. [29]).

In this report, our focus is to show the application of the CAFÉ Risk Assessment System for the selected eight listed companies or issuers from different sectors of industries, especially by combining the actual market performance of companies in the past from one to three years with one-to-one interpretation of event screening against risk assessment results based on the CAFÉ system, to show that the "CAFÉ Risk Assessment System" established based on the hologram approach can effectively resolve the current three major problems in China, as mentioned a few times above.

In summary, we hope to share that the CAFÉ Risk Assessment System established under the framework of big data can reveal the company's risks, in particular, comprehensive study for eight companies shows that the CAFÉ system based on big data thinking can provide a solution that is suitable for China's capital market in line with international standards to support the development of the credit rating system for the domestic financial industry.

Finally, we would like to point out that the definition of credit rating grades for the CAFÉ system from "AAA" to "CCC-C" is given in detail by Yuan et al. [29] and related literature there in.

References

1. Five ministries and commissions including the people's Bank of China: Notice on promoting the high-quality and healthy development of the credit rating industry in the bond market (Exposure Draft), 28 March 2021. http://www.pbc.gov.cn/tiaofasi/144941/144979/3941920/4215457/index.html
2. Altman, E.I.: Financial ratios, discriminant analysis and the prediction of corporate bankruptcy. J. Finance 23(4), 589–609 (1968)
3. Altman, EI., Sabato, G.: Modeling credit risk for SMEs: evidence from the US market. Abacus 43(3), 332–357 (2007)

4. Dun and Bradstreet: Dun and Bradstreet on credit rating [EB/OL] (2014). http://www.dandb.com/glossary/d-b-rating/
5. FICO. What is a credit score. https://www.myfico.com/credit-education/credit-scores/
6. Hull, J.: Options, Futures and Other Derivatives, 10th edn. Pearson, New York (2017)
7. Hull, J.: Risk Management and Financial Institutes, 5th edn. Pearson, New York (2018)
8. Jing, X.C., Li, D., Wang, J.: Current situation and development prospect of China's credit rating agencies. China Financ. Bimonthly **21**, 47–48 (2003)
9. Du, L.H.: Plan ahead: the internationalization challenge of Chinese rating agencies. Financ. Mark. Res. **62**(7), 119–127 (2017)
10. Zhang, H.: Review and prospect of the development of China's credit rating market. Financ. Dev. Res. **29**(10), 29–35 (2018)
11. Feng, GH., et al.: Principles and Pragmatism of Credit Rating. China Finance Press (2019)
12. Baidu Encyclopedia: The Introduction to the world's three major rating agencies (2020). https://baike.baidu.com/item/%E4%B8%96%E7%95%8C%E4%B8%89%E5%A4%A7%E8%AF%84%E7%BA%A7%E6%9C%BA%E6%9E%84/8368411?fr=aladdin
13. Anderson, R.: The Credit Scoring Toolkit: Theory and Practice for Retail Credit Risk Management and Decision Automation, 1st edn. Oxford University Press, Oxford (2007)
14. Chi, G.T., Yu, S.L., Yuan, G.X.: Facility rating model and empirical for small industrial enterprises based on LM test. J. Ind. Eng. Manag. **33**, 170–181 (2019)
15. Thomas, L., Crook, J., Edelman, D.: Credit Scoring and Its Applications, 2nd edn. SIAM, Philadelphia (2017)
16. Witzling, D.: Financial complexity: accounting for fraud. Science **352**(6283), 301–301 (2016)
17. Yuan, G., Wang, H.Q.: The general dynamic risk assessment for the enterprise by the hologram approach in financial technology. Int. J. Financ. Eng. **6**(1), 1–48 (2019)
18. Ministry of Finance of the People's Republic of China: Accounting Standards for Business Enterprises No. 36 - disclosure by related parties, promulgated on 18 June (2008)
19. Yuan, G.X.: Using big data approach to improve the quality of credit rating in China: the CAFÉ risk assessment system. Tsinghua Financ. Rev. **98**, 70–74 (2022)
20. Yuan, G.X.: Overview for the credit rating system for the financial industry in line with international standards and suitable for Financial Markets in China. Reported by Xinhua Finance and economics. https://bm.cnfic.com.cn/sharing/share/articleDetail/2603844/1. Accessed 25 Dec 2021
21. Jarrow, R.A., Lando, D., Turnbull, S.M.: A Markov model for the term structure of credit risk spreads. Rev. Financ. Stud. **10**(2), 481–523 (2004)
22. Yuan, G.X., Zhou, Y.P., Yan, C.X., et al.: New method for corporate financial fraud early warning and risk feature screening: based on artificial intelligence algorithm. In: Proceedings of the 15th China Annual Management Conference, pp. 709–724 (2020)
23. Yuan, G.X., Liu, H.Y., Zhou, Y.P., et al.: The extraction of risk features by applying stochastic search algorithm FOF under the framework of bigdata. Manag. Sci. **33**(6), 41–53 (2020)
24. Yuan, G.X., Zhou, Y.P., Li, D., et al.: The framework for risk characteristic factor base on AI algorithms in Fintech. J. Anhui Univ. Eng. **35**(04), 1–13 (2020)
25. Geman, S., Geman, D.: Stochastic relaxation, Gibbs distributions, and the Bayesian restoration of images. IEEE Trans. Pattern Anal. Math. Intell. **6**, 721–741 (1984)
26. Fawcett, T.: An introduction to ROC analysis. Pattern Recogn. Lett. **27**(8), 861–874 (2006)
27. Hanley, J.A., McNeil, B.J.: A method of comparing the areas under receiver operating characteristic curves derived from the same cases. Radiology **148**(3), 839–843 (1983)
28. Altman, E.I.: Corporate Financial Distress. A Complete Guide to Predicting, Avoiding, and Dealing with Bankruptcy. Wiley Interscience, Wiley, New York (1983)
29. Yuan, G.X., Zhou, Y.P., Liu, H.Y., Yan, C.X.: The framework of CAFE credit risk assessment for financial markets in China. Procedia Comput. Sci. **202**, 33–46 (2022)

Author Index

Printed in the United States
by Baker & Taylor Publisher Services